Eco-Restoration of the Polluted Environment

The book *Eco-Restoration of the Polluted Environment: A Biological Perspective* explores recent advances in biological strategies for the remediation of polluted environments, including soil, water, and air. It covers bioremediation of heavy metals, radioactive waste, and waste gases, which are believed to be bottleneck problems for researchers working in this field. The book provides deep insight into biotechnological advances in eco-restoration of the polluted environment, with separate chapters on genetic engineering technology for enhancement of the bioremediation potential of bioresources and the role of biosurfactants, enzymes, and exo-polysaccharides for bioremediation of polluted environments, along with basic aspects of eco-restoration by microorganisms. The book summarizes the significant developments of many years of research in bioremediation technology and discusses them critically by presenting selected examples, while also considering future research directions in the area.

Features:

- Deep insight into the modes of action of various bioremediation strategies, as well as the status and progress of bioremediation technology for sustainable developmental practices.
- A research overview of bioremediation strategies using engineered biological resources for remediation of contaminants. The book will also accelerate the application of suitable engineered microbes and plants for field applications.
- A survey of interdisciplinary findings and insights on the impact of pollution on the ecosystem and human health, climate, and other global changes, with individual solutions to the pollution issue.
- Comprehensive information for relevant stakeholders such as global leaders, agriculturists, investors, innovators, farmers, policymakers, extension workers, agro-industrialists, environmentalists, and the education and health sectors, as well as students and researchers in the field.

Eco-Restoration of the Polluted Environment
A Biological Perspective

Edited by Sandip V. Rathod

CRC Press
Taylor & Francis Group
Boca Raton London New York

CRC Press is an imprint of the
Taylor & Francis Group, an **informa** business

Designed cover image: Shutterstock

First edition published 2025
by CRC Press
2385 NW Executive Center Drive, Suite 320, Boca Raton FL 33431

and by CRC Press
4 Park Square, Milton Park, Abingdon, Oxon, OX14 4RN

CRC Press is an imprint of Taylor & Francis Group, LLC

© 2025 selection and editorial matter, Sandip V. Rathod; individual chapters, the contributors

Reasonable efforts have been made to publish reliable data and information, but the author and publisher cannot assume responsibility for the validity of all materials or the consequences of their use. The authors and publishers have attempted to trace the copyright holders of all material reproduced in this publication and apologize to copyright holders if permission to publish in this form has not been obtained. If any copyright material has not been acknowledged please write and let us know so we may rectify in any future reprint.

Except as permitted under U.S. Copyright Law, no part of this book may be reprinted, reproduced, transmitted, or utilized in any form by any electronic, mechanical, or other means, now known or hereafter invented, including photocopying, microfilming, and recording, or in any information storage or retrieval system, without written permission from the publishers.

For permission to photocopy or use material electronically from this work, access www.copyright.com or contact the Copyright Clearance Center, Inc. (CCC), 222 Rosewood Drive, Danvers, MA 01923, 978–750–8400. For works that are not available on CCC please contact mpkbookspermissions@tandf.co.uk

Trademark notice: Product or corporate names may be trademarks or registered trademarks and are used only for identification and explanation without intent to infringe.

ISBN: 978-1-032-72997-8 (hbk)
ISBN: 978-1-032-72998-5 (pbk)
ISBN: 978-1-003-42339-3 (ebk)

DOI: 10.1201/9781003423393

Typeset in Times LT Std
by Apex CoVantage, LLC

Contents

Foreword xiii
Preface.. xv
Biography.. xvii
List of Contributors xix

Chapter 1 Environmental Pollution: Threats and Challenges for Management............................1

Sandip V. Rathod, Piyushkumar Saras, and Shradhdha M. Gondaliya

1.1 Introduction 1
1.2 Major Types of Pollution 3
    1.2.1 Air Pollution..3
    1.2.2 Water Pollution 4
    1.2.3 Soil Pollution 7
1.3 Causes of Environmental Pollution 8
    1.3.1 Expansion of Urban Areas and Development of Real Estate..............8
    1.3.2 Mining and Exploration 9
    1.3.3 Agricultural Activities..9
    1.3.4 Burning of Fossil Fuels 14
    1.3.5 Particulate Matter...14
    1.3.6 Utilization of Plastics 14
1.4 Effects of Environmental Pollution 15
    1.4.1 Effects on the Environment..15
    1.4.2 Effects on Crop Production 16
    1.4.3 Effects on Human Health 18
    1.4.4 Effects on Animal Health 18
    1.4.5 Effects on Microorganisms 19
    1.4.6 Effects on Water Quality..20
    1.4.7 Effects on Soil Quality and Health 21
    1.4.8 Effects on Biodiversity 22
1.5 Remedies 22
1.6 Conclusion 24
References 24

Chapter 2 Phytoremediation of Heavy Metal-Contaminated Soil 35

*Innocent Ojeba Musa, Udeme Joshua Josiah Ijah, Olabisi Peter Abioye,
Mustapha Abdulsalam, Sanjoy Kumar Pal, Ikhumetse Agatha Abamhekhelu,
Asmau M. Maude, Oluwasola Olatunji Yusuf, and Akande Sikirula A.*

2.1 Introduction 35
2.2 Environmental Pollution and Sources of Contamination.........................36
    2.2.1 Pollution from Heavy Metals 37
    2.2.2 Heavy Metals and Their Effects...38
    2.2.3 Soil Contamination by Heavy Metals 40
    2.2.4 Heavy Metals in Water...42

v

2.3	Heavy Metal Remediation	43
	2.3.1	Rhizofiltration vs. Phytofiltration	43
	2.3.2	Phytostabilization	43
	2.3.3	Phytovolatilization	44
	2.3.4	Plant Decomposition	44
	2.3.5	Extraction of Plants	44
2.4	Hyperaccumulator Plant Selection	45
	2.4.1	Metal Accumulation Criteria in Plants	45
2.5	Advantages of Phytoremediation	47
2.6	Disadvantages of Phytoremediation	47
2.7	Conclusion	47
References		47

Chapter 3 Mycoremediation of Metallic Pollutants 53

Roshni J. Patel, Swati Mohapatra, and Arti Hansda

3.1	Introduction	53
3.2	Mycoremediation	53
3.3	Heavy Metal Toxicity	54
	3.3.1	Arsenic	54
	3.3.2	Lead	54
	3.3.3	Cadmium	54
	3.3.4	Chromium	55
	3.3.5	Nickel	55
	3.3.6	Zinc	55
	3.3.7	Copper	55
	3.3.8	Manganese	55
3.4	Mechanism of Fungi in Reducing Heavy Metal Toxicity	55
	3.4.1	Biosorption	57
	3.4.2	Bioaccumulation	57
	3.4.3	Mobilization	57
	3.4.4	Immobilization	57
	3.4.5	Biotransformation	58
3.5	Plant–Fungal Interaction as a Boon for Heavy Metal Stress Mitigation	58
3.6	Fungi Genes Involved in Heavy Metal Remediation	59
3.7	Environmental Factors Affecting Detoxification	60
3.8	Conclusion and Future Perspectives	60
References		61

Chapter 4 Biological Solutions for Metal-Contaminated Environments: Role of Engineered Microbes and Plants 66

Pooja Sharma, Ambreen Bano, and Surendra Pratap Singh

4.1	Introduction	66
4.2	HM Decontamination: Alarm for Human and Animal Health	67
4.3	Need for Engineered Organisms	69
4.4	Engineered Microbes	71
4.5	Engineered Plants	73
4.6	Risks Related to Genetically Modified Microbes	74
4.7	Managing Risks from GMOs	74

Contents

vii

4.8 Suggestions and Future Recommendations...75
4.9 Conclusions..76
References 76

Chapter 5 Phytoremediation of Organic Contaminants 84

*Idris Abdullahi Dabban, Oluwafemi Adebayo Oyewole, Bello Aisha Bisola,
Abdullahi Muhammad Asma'u, Bawa Muhammed Muhammed,
and Alfa Suleiman*

5.1 Introduction 84
5.2 Organic Contaminants...85
   5.2.1 Intentionally Persistent Organic Pollutants 85
   5.2.2 Unintentionally POPs...90
5.3 Plants Used for Phytoremediation 91
5.4 Current Phytoremediation Techniques 93
   5.4.1 Phytostabilization 93
   5.4.2 Phytoextraction...94
   5.4.3 Phytovolatalization 95
5.5 Merits of Phytoremediation 96
5.6 Limitations/Factors That Affect Phytoremediation 98
   5.6.1 Site Characteristics...98
   5.6.2 Soil Type...98
   5.6.3 Moisture Content..98
   5.6.4 Nature of the Contaminant..98
   5.6.5 Time Constraints...99
   5.6.6 Volatilization 99
   5.6.7 Phytostimulation...99
   5.6.8 Bioaugmentation..101
5.7 Sampling and Monitoring..101
5.8 Conclusion and Future Prospects 101
References 102

Chapter 6 Microbial and Phytoremediation of Crude Oil-Contaminated Soil........................110

Odangowei Inetiminebi Ogidi and Udeme Monday Akpan

6.1 Introduction 110
6.2 Soil Environment...111
   6.2.1 Soil Physical Properties..111
   6.2.2 Soil Chemical Properties..112
   6.2.3 Soil Biological Properties...113
6.3 Petroleum-Contaminated Soil 114
   6.3.1 Sources of Petroleum Pollutants..114
   6.3.2 Composition of Petroleum Pollutants...114
6.4 Toxic Effects of Petroleum on the Environment 114
   6.4.1 Toxic Effects of Petroleum on Soil..115
   6.4.2 Toxic Effects of Petroleum on Plants 115
   6.4.3 Toxic Effects of Petroleum on Human Health 115
6.5 Biological Remediation of Petroleum Hydrocarbon–Polluted Soil...............116
   6.5.1 Microbial Remediation...116
   6.5.2 Phytoremediation 117

viii   Contents

6.5.3   Rhizoremediation (Plant–Microbe Assisted Remediation:
         Recent Technology)... 118
6.5.4   Combined Microbial Method Remediation 119
6.6   Factors Influencing Petroleum Hydrocarbon Soil Remediation.................... 121
       6.6.1   Biological Factors... 121
       6.6.2   Physico-Chemical Factors.. 121
6.7   Conclusion 122
References 122

Chapter 7   The Biological Remediation of Water and Wastewaters Using Different
                 Treatment Techniques.. 129

*Mohamed Saad Hellal, Aleksandra Ziembińska-Buczyńska, Mohamed Azab
El-Liethy, and Joanna Surmacz-Górska*

7.1   Introduction 129
7.2   Biological Wastewater Treatment.. 131
       7.2.1   Importance of Wastewater Treatment 131
       7.2.2   Anaerobic Treatment.. 132
       7.2.3   Aerobic Treatment.. 132
       7.2.4   Natural Treatment Systems 132
7.3   Bioremediation Techniques in Wastewater Treatment 133
       7.3.1   Water and Wastewater Bioremediation Definition 133
       7.3.2   Bioaugmentation and Biostimulation 135
       7.3.3   Biofilters and Bioreactors 136
       7.3.4   Factors Affecting Biofilter Performance 142
7.4   Phytoremediation Techniques in Wastewater Treatment 144
       7.4.1   Phytoextraction... 144
       7.4.2   Phytodegradation.. 144
       7.4.3   Phytofiltration... 144
       7.4.4   Phytostabilization 145
7.5   Bacterial Community Characteristics and Roles in Various Systems of
       Wastewater Bioremediation and Phytoremediation 145
7.6   Opportunities and Challenges of Bioremediation and
       Phytoremediation.. 148
       7.6.1   Opportunities of Bioremediation and Phytoremediation 148
       7.6.2   Challenges of Bioremediation and Phytoremediation.................... 149
7.7   Conclusion 149
References 150

Chapter 8   Effective Microorganisms: A Microbial Technology to Improve Polluted
                 Water Quality 161

Mohit Yadav

8.1   Introduction 161
       8.1.1   Lactic Acid Bacteria... 163
       8.1.2   Photosynthetic Bacteria.. 163
       8.1.3   Yeasts.. 163
       8.1.4   Actinomycetes 163
       8.1.5   Fermenting Fungi 164
8.2   Activation of the EM 164
8.3   Approach to Using EM in Wastewater Improvement.................................. 165

Contents

ix

		8.4	Parameters Used to Access EM Technology	165
		8.4.1	pH	166
		8.4.2	BOD	166
		8.4.3	EC	166
		8.4.4	TA	166
		8.4.5	TS	167
		8.4.6	TSS	167
		8.4.7	TDS	167
		8.4.8	COD	167

8.4 Parameters Used to Access EM Technology 165
    8.4.1 pH 166
    8.4.2 BOD 166
    8.4.3 EC 166
    8.4.4 TA 166
    8.4.5 TS 167
    8.4.6 TSS 167
    8.4.7 TDS 167
    8.4.8 COD 167

8.5 Typical Testing of Effective Microorganisms 167
8.6 Applications 169
    8.6.1 Agricultural Applications 170
    8.6.2 Environmental Management 170
8.7 India's Approach to Polluted Water Treatment 173
8.8 New Developments in EM Technology 173
8.9 Conclusions 174
References 174

Chapter 9 Microbe-Assisted Remediation of Pesticide Residues from Soil and Water............ 179

Odangowei Inetiminebi Ogidi and Nkechinyere Richard-Nwachukwu

9.1 Introduction 179
9.2 Classification of Pesticides 180
    9.2.1 Classification by Origin 180
    9.2.2 Classification by Target 182
9.3 Significance of Pesticide Residues in Soil and Water 183
9.4 Environmental Fate of Pesticides in Soil and Water 183
    9.4.1 Sources of Pesticides Found in Soil and Water 183
    9.4.2 Fate of Pesticides in Soil and Water 184
9.5 Impacts of Pesticides on Ecosystem 184
    9.5.1 Important Mechanisms Involved in Pesticides
          Accumulation in Soil Ecosystem 185
9.6 Biodegradation of Pesticides 186
9.7 Mechanisms of Microbial Degradation of Pesticides 186
    9.7.1 Bacterial Degradation 187
    9.7.2 Fungal Degradation 187
    9.7.3 Enzymatic Degradation 187
    9.7.4 Mineralization 188
    9.7.5 Co-Metabolism 188
    9.7.6 Algal–Bacterial–Fungal (Multitaxon) Consortia 188
9.8 Factors Affecting Microbial Degradation of Pesticides 189
    9.8.1 Soil pH and Salinity 189
    9.8.2 Soil Moisture 189
    9.8.3 Pesticide Structure 190
    9.8.4 Pesticide Concentration and Solubility 190
    9.8.5 Temperature 190
    9.8.6 Soil Organic Matter 190
9.9 Potential of Microorganisms in Pesticide Residue Remediation 190
9.10 Application of Microbial Remediation 190
    9.10.1 Biostimulation 191
    9.10.2 Bioaugmentation 191

x   Contents

        9.10.3 Application of Natural Attenuation, Biostimulation, and
                  Bioaugmentation... 192
        9.10.4 Bioventing.. 192
        9.10.5 Biosparging 192
        9.10.6 Composting 192
        9.10.7 Landfarming... 193
        9.10.8 Biopiles.. 193
        9.10.9 Slurry Bioreactors 193
  9.11   Conclusion 193
  References 193

Chapter 10 Rejuvenation of Ponds through Phytoremediation: A Sustainable
Approach for Water Quality Enhancement 198

*Ritambhara K. Upadhyay, Naval Kishore, Mukta Sharma, Kenate Worku,
Chandra Shekhar Dwivedi, and Gaurav Tripathi*

  10.1   Introduction 198
        10.1.1 Background and Significance... 198
        10.1.2 Objectives of the Chapter 199
  10.2   Phytoremediation: An Overview 200
        10.2.1 Principles of Phytoremediation of Ponds 200
        10.2.2 Phytoremediation Mechanisms 201
        10.2.3 Advantages and Limitations of Phytoremediation 201
  10.3   Types of Pond Contaminants.. 203
        10.3.1 Nutrient Enrichment (Eutrophication) 203
        10.3.2 Heavy Metals... 204
        10.3.3 Organic Pollutants 206
  10.4   Phytoremediation Techniques for Pond Rejuvenation...................... 207
        10.4.1 Phytoextraction... 207
        10.4.2 Phytostabilization 207
        10.4.3 Rhizofiltration 207
        10.4.4 Phytodegradation.. 207
        10.4.5 Phytovolatilization.. 208
        10.4.6 Phytostimulation/Rhizodegradation 208
  10.5   Selection of Suitable Plant Species... 208
        10.5.1 Characteristics of Phytoremediation Plants 208
        10.5.2 Native and Non-Native Plant Species.................................. 209
        10.5.3 Plant Tolerance and Accumulation...................................... 210
  10.6   Management Strategies for Phytoremediation in Ponds.................... 211
        10.6.1 Planting Design and Arrangement 211
        10.6.2 Nutrient Management.. 212
        10.6.3 Water Level and Flow Management...................................... 213
        10.6.4 Harvesting and Disposal of Plant Biomass 214
  10.7   Case Studies and Success Stories... 216
        10.7.1 Phytoremediation in Nutrient-Enriched Ponds 216
        10.7.2 Phytoremediation of Heavy Metal–Contaminated Ponds............... 217
        10.7.3 Phytoremediation of Organic Pollutants in Ponds 218
  10.8   Challenges and Limitations... 219
        10.8.1 Site-Specific Considerations... 219
        10.8.2 Plant Stress and Mortality.. 220
        10.8.3 Long-Term Effectiveness.. 221

Contents   xi

10.9   Future Directions and Research Needs 222
     10.9.1   Integrating Phytoremediation with Other Restoration Techniques 222
     10.9.2   Genetic Engineering and Plant Breeding for Enhanced Phytoremediation 223
     10.9.3   Ecological and Socioeconomic Impacts of Phytoremediation 224
10.10   Conclusion 225
References 225

Chapter 11   Simple Techniques for Isolation and Characterisation of Bacteria with Potential for Degradation of DDT from Contaminated Soil 227

Murtala Ya'u

11.1   Introduction 227
11.2   Choice of Soil Horizon for Isolation of DDT-Degrading Bacteria 229
     11.2.1   Spatio-Vertical Distribution of DDT in the Soil Horizons 229
11.3   Soil Sample Collection and Processing 229
11.4   Growth of Bacterial Isolates 230
     11.4.1   Culture Media Choice and Conditions for the Growth of Bacteria from a Soil Sample 230
     11.4.2   Preparation of SEM for Bacterial Growth 230
     11.4.3   Preparation of LB Medium for Bacterial Growth 230
     11.4.4   Isolation of Bacteria from Soil Using LB Medium 230
11.5   Isolation of DDT-Degrading Bacteria 231
     11.5.1   Preparation of DDT–Minimal Salt Enrichment Medium (DDT-MSM) 231
     11.5.2   Isolation of DDT-Degrading Isolates Using MSM 231
11.6   Genetic Identification of the Bacterial Isolate 232
     11.6.1   Simple Procedure for the Extraction of Bacterial DNA 232
     11.6.2   Protocol for 16S Ribosomal RNA Gene Amplification 232
     11.6.3   Protocol for Agarose Gel Purification of Amplified 16S rRNA Gene 233
     11.6.4   Phylogenetic Analysis of the Isolate 234
11.7   Growth Characterisation of an Isolate in DDT Enrichment Medium Using One-Factor-at-a-Time Method 234
11.8   Conclusion 235
References 235

Chapter 12   Bacterial Reduction of Molybdenum as a Tool for Its Bioremediation 238

Mohd Yunus Abd Shukor

12.1   Introduction 238
     12.1.1   Molybdenum (Mo) 239
     12.1.2   Animals' Molybdenum Entry Pathways 239
     12.1.3   Molybdenum Acute and Chronic Toxicities 240
     12.1.4   Bioremediation 241
     12.1.5   Molybdenum Pollution 242
     12.1.6   Molybdenum Bioremediation 244
12.2   RSM vs OFAT Approach 256
12.3   Mathematical Models of the Kinetics of Molybdenum Reduction 256

xii   Contents

12.4   Characterization of Partially Purified Molybdenum-Reducing Enzymes from Various Bacteria 261

12.5   Molybdenum Reduction Location Characterization with Electron Transport Chain Inhibitors 263

12.6   Bioremoval of Molybdenum from Aqueous Environments 264

12.7   Conclusion 267

References 267

Chapter 13   Health Hazards and Bacterial Bioremediation of Endocrine-Disrupting Chemicals—A Concise Discussion on Phthalic Acid Esters and the Organophosphorus Pesticide Malathion 278

Shalini Chandel, Rishi Mahajan, and Subhankar Chatterjee

13.1   Introduction 278

13.2   Environmental Occurrence and Health Effects of Endocrine-Disrupting Chemicals 278

      13.2.1   Phthalic Acid Esters and the Organophosphate Pesticide Malathion: Two Potent Endocrine-Disrupting Chemicals 279

13.3   Bacterial Bioremediation 286

      13.3.1   Remediation of Phthalic Acid Esters by Bacteria 286

      13.3.2   Bacterial Bioremediation of the Organophosphate Pesticide Malathion 286

13.4   Evaluating the Enzymatic Biodegradation of EDC (Organophosphate Pesticide Malathion) 287

      13.4.1   Organo Phosphorus Hydrolases in Malathion Bioremediation 287

13.5   Conclusion and Future Prospects 290

13.6   Acknowledgments 290

References 291

Chapter 14   Bacterial Ammonia Oxidation: A Way towards Environment Remediation 294

Vijaylakshmi, Arti Chamoli, Anne Bhambri, Neetu Pandey, and Santosh Kumar Karn

14.1   Introduction 294

      14.1.1   Ammonia in Soil and Its Effects 295

      14.1.2   Ammonia in Aquatic Water and Its Effects 295

14.2   Methods Used in Ammonia Removal from Wastewater 296

      14.2.1   Ammonia Stripping and Distillation 296

      14.2.2   Breakpoint Chlorination for Removal of Ammonia 296

      14.2.3   Ion Exchange Method 296

14.3   Biological Oxidation of Ammonia 296

      14.3.1   The Contribution of Anammox Bacteria to N Cycling 299

14.4   Conclusion 300

14.5   Acknowledgments 300

References 300

Foreword

The ecosystem is the base of life, composed of three basic building blocks of life, soil, water, and air, that sustain all living beings. Pollution of the environment through various anthropogenic activities is a serious crisis faced by both developed and developing economies. Compromised environmental health leads to an imbalanced ecosystem, which in turn makes life on Earth unsustainable. Irreversible damage to the ecosystem is indicated by global warming, melting of ice, unrestricted floods, land erosion, eutrophication, polluted water and air, irregular rainfall, crop failures, lack of clean drinking water, frequent epidemics, and compromised human immunity.

Global climate change has made a number of threats to arable land and food security worse, including extreme drought conditions, extremely high and low temperatures, extended periods of flooding and submergence, lavish use of agrochemicals (fertilizers and pesticides), compaction and salinization of the soil, deteriorated soil and rhizospheric microbial health, and a declining water table. The health of the ecosystem is a key factor in achieving optimal crop output. Therefore, in order to guarantee soil quality, crop development, and production in a sustainable way, it becomes essential to look for ways to improve the health of the soil and water. The current environmental catastrophe is the result of harm done to the global ecology, which requires rapid attention and correction. Various solutions for environmental cleanup have been developed, but they are both costly and energy intensive. Exploring nature's potential of widely distributed biological resources for eco-restoration of damaged ecosystems provides an *in situ*, low-cost, eco-friendly, broadly accepted, and easily applicable large-scale remediation technique.

The book *Eco-Restoration of the Polluted Environment: A Biological Perspective*, edited by Sandip V. Rathod, is an attempt to bring out a comprehensive review of knowledge- and experience-based efforts of scientists from various national and international institutes of repute, working hard to devise technologies for bioremediation of emerging environmental pollutants, in a single source. The book will be a great resource for students, teachers, and scientists working in the area of environmental restoration through eco-friendly strategies and would be useful to realize the vital role of this promising science in combating the threat posed to humanity by environmental pollution.

(K. B. Kathiria)
Vice-Chancellor
Anand Agricultural University
Anand, Gujarat, India

Preface

Environmental pollution has become a major global problem as a result of the rapid rise of industrialization, urbanization, enhanced agricultural output, and electricity generation. These have also resulted in the indiscriminate exploitation of natural resources in order to suit human interests and demands, ultimately contributing to the disruption of the ecological stability on which the quality of our environment depends. Advanced technologies for sustaining global population produce toxic pollutants that are beyond the self-cleaning potential of nature. One of the most serious issues confronting the developing world today is the contamination of water, soil, and air with hazardous and destructive chemicals. Environmental pollution will pose a serious threat to all lives on earth, and to cope with this issue, scientific communities and legislative authorities are working intensively to find sustainable solutions for pollution without affecting the current economical growth. A large number of strategies have been kept at the forefront for restoration of polluted environments, among which exploring biological resources, including plants and microbes, was found to be promising technology for eco-restoration. Bioremediation strategies focus on the natural potential of plants and microorganisms to treat the polluted environment at the site of contamination. The evolution of plants and microbes at the contaminated site provides them the benefit of diverse metabolic pathways due to short- and long-term relationships with pollutants, enabling them to efficiently transform or degrade the pollutants.

This book examines the many versions and combinations of diverse procedures that will enhance knowledge about biological remediation strategies, taking into account the impact of pollution on various habitats. Contributors to this book include experts in soil science, agronomy, microbiology, edaphology, agriculture, biotechnology, and environment science. The contributors to the book belong to a variety of institutions, universities, and government laboratories, with backgrounds in basic, applied, and industrial science. This book brings together a diverse variety of topics, making it a valuable resource for undergraduate and post-graduate students studying soil science, agricultural science, environmental sciences, biotechnology, microbiology, biochemistry, and environment engineering. I hope that the information, including its basic and practical parts, will be useful to students; scientists; and engineers in academia, industry, and government.

Biography

Mr. Sandip V. Rathod (Assistant Research Scientist, Soil Science Division) has eight years of teaching and research experience. His research interests include the mitigation of greenhouse gas emissions from agroecosystems, the use of effluent water for agricultural purposes, remediation of heavy metal–contaminated soil and nano fertilizers, and its impact on soil properties. He has handled research projects on the reuse of wastewater and nanofertilizers. He is actively involved in the dissemination of agricultural technologies to farmers through the organization of various training programs and through TV and radio talks. He acted as co-coordinator for a training program on soil and water testing for agriculture. His publication profile includes 15 research papers, 1 book chapter, 2 books, 2 teaching manuals, and 15 popular articles.

Contributors

Akande Sikirula A
Department of Mathematics
Federal University of Technology Minna
Niger State, Nigeria

Mustapha Abdulsalam
Department of Microbiology
Federal University of Technology Minna
Niger State, Nigeria

Olabisi Peter Abioye
Department of Microbiology
Federal University of Technology Minna
Niger State, Nigeria

Udeme Monday Akpan
Department of Science Laboratory Technology,
  School of Applied Sciences
Federal Polytechnic Ekowe
Bayelsa State, Nigeria

Abdullahi Muhammad Asma'u
PMB 55, Bida Niger State
Nigeria

Ambreen Bano
IIRC-3, Plant-Microbe Interaction, and
  Molecular Immunology Laboratory,
  Department of Biosciences, Faculty of
  Sciences, Integral University
Lucknow, UP, India

Anne Bhambri
Department of Biotechnology, Shri Guru
  Ram Rai University, Patel Nagar
Dehradun, Uttarakhand, India

Bello Aisha Bisola
PMB 55, Bida Niger State
Nigeria

Arti Chamoli
Department of Biochemistry and
  Biotechnology, Sardar Bhagwan Singh
  University Balawala
Dehradun, India

Shalini Chandel
C/o Dr. Rishi Mahajan, Assistant Professor,
  Department of Microbiology, College of
  Basic Sciences, CSK Himachal Pradesh
  Agricultural University
Palampur, India

Dr. Subhankar Chatterjee
Associate Professor, Department of Ecology &
  Environmental Sciences, School of
  Life Sciences, Pondicherry University,
  R.V. Nagar
Kalapet, Puducherry, India

Idris Abdullahi Dabban
PMB 55, Bida Niger State
Nigeria
Ikhumetse Agatha Abamhekhelu, Department
  of Microbiology, Federal University of
  Technology Minna
Niger State, Nigeria

Chandra Shekhar Dwivedi
Department of Geoinformatics, Central
  University of Jharkhand, Ranchi,
  Jharkhand, India

Mohamed Azab El-Liethy
Water Pollution Research Department,
  National Research Centre
Dokki, Giza, Egypt

Dr. Shradhdha M. Gondaliya
Assistant Professor, MVM Science & Home
  Science college
Rajkot, Gujarat, India

Arti Hansda
Department of Life Sciences, School of
  Science, GSFC UniversityVadodara
Gujarat, India

Mohamed Saad Hellal
Department of Microbiology, Federal
  University of Technology, PMB 65 Minna
Niger State, Nigeria

Udeme Joshua Josiah Ijah
Department of Microbiology, Federal
University of Technology Minna
Niger State, Nigeria

Santosh Kumar Karn
Department of Biochemistry and Biotechnology,
Sardar Bhagwan Singh University
Balawala, Dehradun, India

Naval Kishore
Department of Geology, Panjab University
Chandigarh, India

Dr. Rishi Mahajan
Assistant Professor, Department of
Microbiology, College of Basic Sciences,
CSK Himachal Pradesh Agricultural
University
Palampur, India

Asmau M. Maude
Department of Microbiology, Federal
University of Technology Minna
Niger State, Nigeria

Swati Mohapatra
Department of Life Sciences, School of
Science, GSFC UniversityVadodara
Gujarat, India

Bawa Muhammed Muhammed
PMB 55, Bida Niger State
Nigeria

Innocent Ojeba Musa
Department of Microbiology, Skyline
University Nigeria
Kano State, Nigeria

Odangowei Inetiminebi Ogidi
Department of Biochemistry, Faculty of
Basic Medical Sciences, Bayelsa Medical
University
Bayelsa State, Nigeria

Oluwafemi Adebayo Oyewole
PMB 65 Minna, Niger State
Nigeria

Sanjoy Kumar Pal
Department of Microbiology, Skyline
University Nigeria
Kano State, Nigeria

Neetu Pandey
Department of Chemistry and
Applied Science, Sardar Bhagwan Singh
University Balawala
Dehradun, India

Roshni J. Patel
Department of Life Sciences, School of
Science, GSFC UniversityVadodara
Gujarat, India

Sandip V. Rathod
Assistant Research Scientist, Tribal
Research cum Training Centre,
AAU, Devgadhbaria
Gujarat, India

Nkechinyere Richard-Nwachukwu
Department of Biology, Ignatius
Ajuru University of Education,
Port Harcourt
Rivers State, Nigeria

Piyushkumar Saras
Department of Agronomy, Sardar Krushinagar
Dantiwada University, Banaskantha
Gujarat, India

Mukta Sharma
Department of Civil Engineering, IKGPTU,
Jalandhar
Punjab, India

Pooja Sharma
NUS Environmental Research Institute,
National University of Singapore, #02–01,
T-Lab Building, 5A Engineering Drive
1, Singapore; Energy and Environmental
Sustainability for Megacities (E2S2) Phase
II, Campus for Research Excellence and
Technological Enterprise (CREATE),
1 CREATE Way
Singapore, Singapore

Contributors

Mohd Yunus Abd Shukor
Department of Biochemistry, Faculty of
Biotechnology and Biomolecular Sciences,
Universiti Putra Malaysia, Serdang
Selangor, Malaysia

Surendra Pratap Singh
Plant Molecular Biology Laboratory,
Department of Botany, Dayanand
Anglo-Vedic (PG) College, Chhatrapati
Shahu Ji Maharaj University
Kanpur, India

Alfa Suleiman
Department of Biological Sciences, The
Federal Polytechnic, PMB 55
Bida Niger State, Nigeria

Joanna Surmacz-Górska
Department of Environment Biotechnology,
Silesian University of Technology,
Akademicka St. 2, 44–100
Gliwice, Poland

Gaurav Tripathi
Centre for Climate Change and Water
Research, Suresh Gyan Vihar
University
Jaipur, India

Ritambhara K. Upadhyay
Department of Geology, Panjab University
Chandigarh, India

Vijaylakshmi
Department of Biochemistry and
Biotechnology, Sardar Bhagwan
Singh University Balawala
Dehradun, India

Kenate Worku
Department of Geography and Environmental
Sciences, Jimma University
Jimma, Ethiopia

Dr. Mohit Yadav
Department of Molecular Biology and
Biotechnology, Tezpur University, Napaam,
Tezpur, Sonitpur
Assam, India

Murtala Ya'u
Department of Biochemistry, Bayero University
Kano, Nigeria

Oluwasola Olatunji Yusuf
Department of Industrial Education
Technology, Federal University of
Technology Minna
Niger State, Nigeria

Aleksandra Ziembińska-Buczyńska
Environmental Biotechnology Department,
Faculty of Power and Environmental
Engineering, Silesian University of
Technology
Gliwice, Poland

1 Environmental Pollution
Threats and Challenges for Management

Sandip V. Rathod, Piyushkumar Saras, and Shradhdha M. Gondaliya

1.1 INTRODUCTION

An unfavorable change in the physical, chemical, and biological properties of the air, water, and soil that has an impact on human life, the lives of other beneficial living plants and animals, industrial development, living conditions, and cultural assets can be referred to as environmental pollution. A pollutant is something that negatively affects people's health, comfort, property, or environment. The majority of pollutants, in general, enter the environment through waste, sewage, or unintentional discharge or as byproducts or residues from the manufacture of anything valuable. As a result, the land, water, air, and biosphere are becoming polluted. Environmental pollution is generally higher in poor countries than in developed ones, because of poverty, inadequate legislation, and a lack of awareness of pollution forms. It could be that humans come into contact with pollution on every single day without noticing it or that in our hectic daily lives, we have become immune to it (Muralikrishna and Manickam, 2017). Pollution-related illness was estimated to account for 9 million premature deaths in 2015, which is more than three times the number of deaths from tuberculosis, AIDS, and malaria combined (Landrigan *et al.*, 2017).

Deforestation, bush burning, waste from domestic and agricultural sectors in bodies of water, use of pesticides in aquatic animal harvesting, and inappropriate disposal of electronic waste, for example, all are involved in air, land, and water pollution. Furthermore, harmful materials such as air pollutants, heavy and carcinogenic metals, and particulate matter are introduced into the biosphere; sewage and sludge, industrial waste substances, agricultural pollutants, and electronic waste are introduced into water bodies; and activities such as removal of forest areas, poor management of waste from domestic landfills, mining, and intensive farming cause soil pollution. Furthermore, when human populations rise, so do human activities, with a corresponding increase in environmental effects. The effects affect not only people but also other aquatic animal and ocean creatures, including microbes, which, due to their quantity and diversity, maintain the biogeochemical functions required for ecosystem survival.

The contaminants that are discharged into the environment as a byproduct of cultivating and raising food crops and livestock are known as agricultural pollutants. These pollutants are biotic and abiotic byproducts of farming methods (fertilizers, pesticides, and livestock dung) that cause degradation or pollution of the environment and ecosystem, causing harm to humans and their economic interests.

Emerging contaminants can develop in the environment from both natural products and those created by biochemical processes from manufactured substances. When manure, fertilizers, biosolids, or other solid waste items are applied to the soil, emerging agricultural contaminants are either released into the environment or indirectly make a channel to spread into the soil. They may be carried to water bodies by leaching, runoff, and drainage processes after they penetrate the soil. Land used for agricultural production is lost as a result of subsequent degradation such as erosion, landslides, soil mineral leakage, and crop and soil contamination by pathogens from humans or animals, salt, or pesticides (Lindgren *et al.*, 2011). Flooding or drought-related declines in crops and cattle are examples of direct repercussions, as is damage to infrastructure that may have an impact on food production. After extreme weather events, a number of infectious illnesses that harm livestock, including anthrax and blackleg, may appear (Bezirtzoglou *et al.*, 2011).

It may be necessary to employ phosphate fertilizers to boost or maintain food production, but doing so results in rising soil cadmium levels, which primarily pose a threat to food safety. Additionally, tainted water is used for agriculture in nations with a shortage of freshwater. Thus, there is a significant likelihood that agricultural land in the USA, Mexico, Japan, and Canada will become more chemically contaminated. A type B nation will probably experience an even bigger expansion of its metal and chemical sectors, along with an increase in the hazards that go along with it. In developing nations like China, there are places with high amounts of chemical pollution. Again, the use of cadmium-contaminated fertilizers poses the biggest hazard to food safety. Due to their past and continuing release into the environment, organic pollutants and metals can be found in a type C nation (such as those in Asia and Europe) in various amounts, but neither the ongoing leakage of toxic chemicals nor serious accidents are anticipated to have a significant impact on agricultural output. Nuclear incidents in type C countries should be evaluated similarly to those in type B nations.

Previous studies have mainly focused on one direction or stage in the pathways connecting environment, food, and health, for example, focusing on the impact of environmental change on crop production or the effects of various diets on health. The effects of climate change on staple crops have received the majority of the attention in research on the effects of environmental change on food production (Porter *et al.*, 2014; Challinor *et al.*, 2014), while the effects on other foods and of other environmental stressors have received less attention. A few studies (Myers *et al.*, 2017; Springmann *et al.*, 2016a) that focus primarily on significant staple crops and/or meat have combined environmental alteration, health, agriculture, nutrition, and markets.

According to recent research, eating more fruits and vegetables still lowers the risk of cancer, cardiovascular disease, and all-cause mortality even after exceeding the WHO limit of 400 grams of fruit and vegetables combined per day (Aune *et al.*, 2017). According to research, socioeconomic status predicted the amount of fruits and vegetables consumed per person.

According to Miller *et al.* (2016a), people in low-income countries consume less per person than high-income ones, and within countries, consumption has been found to be lower in underprivileged areas compared to wealthy ones (Pessoa *et al.*, 2015; Dubowitz *et al.*, 2008). Many fruit and vegetable crops, however, have been shown to be relatively susceptible to environmental pollution (Backlund *et al.*, 2008), increasing the risk of a diminished fruit and vegetable supply in the future, with attendant public health implications.

The world population will continue to face environmental changes that will offer greater challenges to our food systems, health, and well-being in the coming decades. Climate change, increased carbon dioxide fertilization, ground-level ozone, deforestation, soil degradation, changes in water availability, and intensive use of land are all examples of changes that can have a direct and significant impact on agricultural production. Furthermore, variations in the abundance and spread of diseases, pests, and pollinators, which are all linked to environmental change, may have an indirect impact on agriculture.

Since 1950, the earth's temperature has been rising above the average. Climate change is another term for this. Climate change refers to a rise in the global average temperature and changes in the climate that are observed statistically, such as mean surface temperature, and can last from one year to millions of years. Global warming is defined as an average increase in global temperature and is one of the aspects of climate change. Plants' DNA, RNA, proteins, and membranes have been reported to be damaged by UV-B radiation, which hinders photosynthesis (Caldwell *et al.*, 2007). The majority of vegetables, such as beans, tomatoes, spinach, radishes, carrots, cucumbers, and gourds, as well as many fruits, like strawberries and sea-buckthorn, showed a more significant

Environmental Pollution

reduction in production due to UV-B exposure than woody plants, according to a meta-analysis of the effect of increases in UV-B on yields (Li *et al.*, 2010).

Numerous environmental, behavioral, and economic variables will influence how much environmental change will affect food systems and health. First, the level and trends of various environmental stressors as well as the mitigation measures adopted by both individual nations and the global community as a whole will determine how much of an environmental shift there will be. Second, the mechanisms for adaptation that are created and used will determine the effects of environmental change. Third, marketplaces are essential for moving food between areas of production and consumption. Food systems in regions that are heavily reliant on local markets may be more susceptible to environmental change than agricultural systems that are globally integrated and may be better equipped to adapt to changes in the environmental situation for food production. Fourth, the consumption of particular foods is substantially more susceptible to price fluctuations than the consumption of other foods. Finally, due to variations in pre-existing dietary patterns and price responsiveness, the impact of varying food availability on diet and health is expected to vary between nations and demographic groups.

Understanding the causes and threat of environmental contamination is crucial, yet taking action has a high price. Different physical, chemical, and biological methods have been applied to reduce soil, water, and air pollution, but most chemical and physical methods are costly and contribute to new environmental issues. Microbial bioremediation is one of the methods that have drawn attention from all around the world, presumably because it is an effective and environmentally beneficial way to restore the environment. Biological methods remain sustainable to reduce and control pollution.

1.2 MAJOR TYPES OF POLLUTION

Pollution in the environment is a hot topic these days. Air, water, and soil are all polluted.

1.2.1 Air Pollution

Air pollution refers to a chemical compound, metals, or any other substance that leads to reduced air quality and thereafter the atmosphere. Carbon monoxide (CO), nitrogen oxides (particularly NO and NO_2), sulfur oxides (especially SO_2), and volatile organic compounds are common gaseous pollutants. The health of people all around the world is seriously threatened by air pollution.

Researchers and decision-makers are interested in how air pollution negatively affects agriculture. Realizing that air pollution is a concern on a global scale, and that sustainable food and agricultural growth is a global objective, it is vital to investigate the influence of air pollution on agriculture using global data. Meanwhile, agricultural performance must be measured more thoroughly. Initially, numerous studies used biochemical and field tests to evaluate the processes and toxicology that explain how air pollution affects particular types of livestock and agricultural crops. Nitrogen dioxide, ammonia, sulfur dioxide, particulate matter, and ozone are some of the pollutants that have been discovered to have an effect (Das *et al.*, 2021; Sillmann *et al.*, 2021).

Producing animal manure aids in the large release of methane into the environment, creating difficult circumstances, while intensive farming helps to deal with the world's food shortages. Additionally, a fungi and algae bloom amplifies the emission of methane into the environment, which has a negative impact on greenhouse gases. There are variations in the global warming potential (GWP) and atmospheric lifespan of greenhouse gases. The warming potential of other gases is based on and represented in relation to the CO_2 GWP, which has been given 1 GWP (Paustian *et al.*, 2006). For instance, the Intergovernmental Panel for Climate Change (IPCC) estimates that over 100- and 20-year periods, respectively, 1 ton of CH_4 has a warming effect of around 25 and 72 tons of CO_2. According to studies, CH_4 is more potent than CO_2; hence reducing CH_4 emissions will have a greater immediate and significant impact on mitigating climate change than reducing CO_2 emissions (Moore and MacCracken, 2009). N_2O is another significant greenhouse gas, remaining in the atmosphere for 114 years (Solomon *et al.*, 2007) and being 298 times as potent as CO_2 over 100 years (Forster *et al.*, 2007). As greenhouse gas emissions such as those of CH_4, CO_2, N_2O, and other gases, have an adverse effect on the environment and cause global warming, which in turn causes changes in the climate and environmental degradation, there has been an increase in concern (Patra and Saxena, 2010). Nitrous oxide (N_2O) is the main greenhouse gas (GHG) released by agriculture, making up 38% of all global emissions. It is produced through the processes of nitrification and denitrification from human activities (such as the use of nitrogen fertilizer, the growth of crops and forage that fix nitrogen, the retention of crop residues, and the application of livestock manure), either through direct additions or through indirect additions. Enteric fermentation, a term used to describe the natural digestive processes in ruminants, is the primary source of methane production in the livestock sector, making it the second-greatest contributor of total agricultural emissions globally, after rice agriculture, which accounts for 11% of emissions. The level of emissions of greenhouse gases has been rising due to both natural and human sources, which is causing global warming and promoting climate change.

A warming globe causes climate change, which has a larger number of negative effects on the weather. It is anticipated that increased floods, storms, droughts, heat waves, and other extreme weather events may result from climate change. Therefore, as the climate changes, severe weather patterns pose serious risks to human civilization. Sea surface temperature, land temperature, snow cover on hills, land temperature, and humidity are all signs of global warming. First, air pollution impairs the biological mechanisms that support animal growth and development. Agriculture is impacted by air pollution in a number of ways, directly as well as indirectly. According to the literature, there are three main ways by which air pollution affects agricultural productivity.

1. As an aggressive oxidant, ozone, for instance, reduces photosynthesis, modifies carbon fixation, directs synthesis toward chemical defenses, decreases water intake, and hastens senescence (Ainsworth, 2017). According to Wang *et al.*'s (2021) estimate, ozone caused a 10% yield loss in winter wheat in the Henan province of China between 2010 and 2012. According to Mills *et al.* (2018), the mean ozone level in 2010 and 2012 decreased annual global yields of maize, rice, wheat, and soybeans by 6.1%, 4.4%, 7.1%, and 12.4%, respectively.
2. Air pollution leads to deterioration of the quality of water and soil. For instance, metals and acidic precipitation from air pollution will end up in the soil and water. Therefore, soil and water's chemical compositions are altered (Aragón and Rud, 2016). Acidification of the soil and water, greater loss of plant nutrients, slower degradation of organic matter, metal pollution, and other impacts are possible outcomes (Vázquez-Arias *et al.*, 2023; Luo *et al.*, 2019). Through ozone layer loss and global warming, air pollution also has a negative impact on how people live on Earth.
3. Air pollution impairs the health of agricultural workers, which reduces their productivity. One of the biggest dangers to the health of the world's population is air pollution. Medical research has shown that air pollution increases the risk of a wide range of disorders, including heart disease, lung disease, cancer, insomnia, and depressive disorders, among many others. (Dominski *et al.*, 2021; Almetwally *et al.*, 2020). It makes sense to assume that air pollution lowers farmers' labor productivity and performance (Shah *et al.*, 2022).

1.2.2 WATER POLLUTION

Water pollution is a global issue that has gotten worse in both rich and emerging nations, threatening both the physical and environmental health of billions of people as well as economic progress.

According to FAO *et al.* (2012), human settlements, industries, and agriculture are the main causes of water pollution. Millions of tons of toxic sludge, solvents, heavy metals, and other wastes are dumped into water bodies each year by industry, and 80% of municipal wastewater released into water bodies worldwide is untreated (WWAP, 2017). Significant quantities of untreated wastewater from municipalities and industries are a major concern in low-income nations and growing economies. Pesticides, pollutants coming from livestock management, VOCs, food processing waste, chemical waste, medical waste, and heavy metals from various inputs are examples of human-made sources of contamination. Chemicals like pesticides residue; hydrocarbon compounds; POPs; or heavy metals like cadmium, lead, and arsenic can cause cancer, hormone imbalances, reproductive problems, and severe liver and kidney damage, among other harmful health impacts.

Additionally, according to Ewuzie *et al.* (2020), the chemical structure of the water system is significantly impacted by the geological structural formations of various areas. As a result, this could be the reason for the elevated concentrations of the specific compound or structure that is causing water pollution.

However, agricultural pollution is also becoming a problem, made worse by increased sediment outflow and groundwater salinization. Agriculture is the primary cause of contamination in streams and rivers, the secondary source in wetlands, and the third primary source in lakes in the United States (USEPA, 2016). Thirty-eight percent of the water bodies in the European Union are seriously impacted by agricultural pollution (WWAP, 2015). Water pollution is largely caused by agriculture, which uses 70% of the world's water resources. Large amounts of organic matter, agrochemicals, drug remnants, saline drainage, and sediments are released into water bodies by farms. As a result, there are clear threats to human health, aquatic ecosystems, and economic activity (UNEP, 2016). By 2050, 9.8 billion people are expected to inhabit the planet (UNDESA, 2017). Production of more (and more varied) food is necessary due to population expansion, shifting dietary preferences, and changes in consumption habits. This in turn is bringing about new negative environmental impacts, including effects on water quality, and encouraging agricultural development and intensification.

Global crop output has increased mostly because of the extensive usage of additives such as pesticides and artificial fertilizers. The growth of agricultural land has exacerbated the trend, with irrigated playing a crucial role in enhancing production and livelihoods in rural areas while also transmitting agricultural pollution to aquatic bodies. In response to the growing demand for food, agricultural systems have grown and intensified. In absolute terms, clearing of land and expanding agriculture have led to increasing pollution loads in water, but it is likely that some unsustainable trends in agricultural intensification have had the greatest influence.

Irrigation is a critical component in agricultural intensification. Large projects for irrigation have been critical in enhancing global food security, particularly in underdeveloped countries. Nonetheless, the facilities of irrigation have frequently been linked to water system degradation caused by salt, pesticide, and fertilizer drainage and leaching. In the recent era, irrigation area has more than quadrupled (from 139 million hectares [Mha] in 1961 to 320 Mha in 2012; FAO, 2014), while livestock population has increased by over threefold (from 7.3 billion units in 1970 to 24.2 billion units in 2011; FAO, 2016a). Chemical fertilizers like urea and DAP have been used to add to natural sources of nutrients and cycling to raise crops and animals since the 19th century, but their use has risen dramatically in recent decades. The world now consumes ten times as much mineral fertilizer as it consumed in the 1960s (FAO, 2016a). According to Rockstrom *et al.* (2009), nutrient mobilization may already have exceeded limits that will cause drastic changes in the environment in continental- to planetary-scale systems. Nutrients, pesticides, metals, organic carbon, sediments, salts, microbes, and medication residues are the main agricultural contributors to water pollution (and the main objectives for water pollution treatment). The significance of various types of agricultural pollution varies depending on the specific circumstances, and harmful effects like eutrophication result from a confluence of stressors. When fertilizers are used in crop production at a rate higher than they are fixed by soil particles or exported from the soil profile, fertilizer-related water pollution develops. Extra phosphates and nitrogen can seep into the groundwater or enter streams through surface runoff. Since phosphate is less soluble than nitrate and ammonia, it often binds to soil particles and seeps into water sources through soil erosion.

Lakes, reservoirs, ponds, and coastal waterways may become eutrophic due to high nitrogen loads, which can result in algal blooms that crowd out other aquatic life. Despite data shortages, 415 coastal regions throughout the world have been identified as being affected by eutrophication in some way, 169 of which are hypoxic (WRI, 2008). The most frequent chemical contamination in the world's groundwater aquifers is nitrate from agriculture (WWAP, 2013). Agriculture-related water pollution has a direct negative impact on human health, as seen by the well-known blue-baby syndrome, in which excessive levels of nitrates in water cause methemoglobinemia—a potentially fatal condition—in infants.

Manure is typically collected to be utilized as organic fertilizer, but if too much of it is used, it can cause diffuse water contamination. Significant water contaminants include organic matter from animal waste, uneaten animal feed, the animal-processing industry, and improperly managed crop residues. Manure is frequently not stored in secure locations, and after heavy rainstorms, it may enter watercourses via runoff from the ground. As organic matter breaks down, it uses up the dissolved oxygen in the water, significantly causing hypoxia in aquatic environments. Biological oxygen demand (BOD) is the greatest for wastes related to livestock. In contrast to the normal BOD of home sewage, which ranges from 200 to 500 milligrams per liter, pig slurry, for instance, has a BOD of between 30,000 and 80,000 milligrams per liter (FAO, 2006). Aquaculture can have a significant role in the localization of organic burdens in water. Bangladesh's shrimp farming produces 600 tons of garbage per day (SACEP, 2014). The likelihood of eutrophication and blooms of algal organism in lakes, reservoirs, and coastal areas is further increased by the release of organic materials.

Pesticide management in developing nations is extremely difficult due to factors such as the rapid increase in pesticide use, reliance on broad-spectrum pesticides, lax institutional frameworks, lax enforcement of laws, and little knowledge and awareness among farmers regarding the use of dangerous chemicals. Some broad-spectrum and persistent pesticides (like many organophosphates and DDT) were widely banned as a result of the accumulation of pesticides in water and the food chain, which had been shown to have harmful effects on humans. However, some of these pesticides continue to be applied in poorer countries, where they have acute and likely long-term health effects. Countries have increasingly embraced a pest management strategy based on the use of synthetic pesticides as land usage intensifies. The global market for pesticides is now a multi-billion dollar sector, valued at more than USD 35 billion annually (FAO, 2016a). Numerous countries use insecticides, herbicides, and fungicides heavily in agriculture (Schreinemachers and Tipraqsa, 2012). The largest pesticide use intensities worldwide are found in Costa Rica, Colombia, Japan, and Mexico (Schreinemachers and Tipraqsa, 2012). According to Zhang *et al.* (2011), the share of herbicides in the world's pesticide usage climbed quickly, while the percentage of fungicides and insecticides decreased.

Nevertheless, agriculture uses millions of tons of active pesticide components (FAO, 2016a). Globally, acute pesticide poisoning significantly increases morbidity and mortality in humans, particularly in underdeveloped nations where subsistence farmers frequently use extremely dangerous pesticide formulations. They can contaminate water supplies with carcinogens and other poisonous materials that can harm humans if improperly chosen and managed. By destroying plants and insects and having detrimental effects up the food chain, pesticides can also have an adverse impact on biodiversity.

Agricultural pollution also has an impact on aquatic ecosystems; for instance, eutrophication brought on by the buildup of nutrients in lakes and coastal waterways affects fisheries and biodiversity. Aquaculture relies heavily on carnivorous species, which demand enormous quantities of fishmeal and other pelleted food. Numerous non-fed aquaculture practices, like mussel farming, can clean and filter water, but others, like intense caged crab culture, may disturb natural nutrient cycles and worsen water quality. Since the 1980s, aquaculture has increased more than 20-fold, primarily in Asia and inland fed aquaculture (FAO, 2016b). According to FAO (2016b), 167 million tons of aquatic animals were produced globally in 2014, of which 146 million tons were reportedly directly

Environmental Pollution 7

consumed by people. Nearly 90% of the world's aquacultural production is in Asia, with China leading the way with 45.5 million tons produced annually (FAO, 2016b). From the last few decades, the demand for fish and shellfish has increased tremendously, more than that for any other agricultural product. The greater production intensity and concentrations of one species are being caused by market forces and differentiation. Because of these changes, people are using more medications (such as antibiotics, fungicides, and anti-fouling agents), which leads to contamination further down the food chain.

1.2.3 Soil Pollution

Soil pollution is described as the accumulation of persistent toxic substances, chemicals, salts, radioactive elements, or disease-causing agents in soils, which has a negative impact on plant development and animal health (Okrent D., 1999). Soil is the foundation of agriculture. All crops for human consumption and animal feed rely on it. To some extent, we are depleting this crucial natural resource due to increasing erosion. Furthermore, massive amounts of human-made garbage, sludge, and other products from new waste management plants, as well as polluted water, are creating or contributing to soil pollution. Pollutants may likewise enter the food chain through plant absorption. There are numerous methods for soil to become contaminated, including seepage from a garbage dump; releasing industrial trash into the ground; allowing contaminated water to seep into the ground; underground storage tank rupture; excessive pesticide, herbicide, or fertilizer use; and seepage of solid waste

Heavy metals, organic and inorganic solvents, fossil fuels from petrochemical plants, oil refineries, hydrocarbons, and power plants are some examples of soil pollutants. The main causes of soil pollution are inadequate landfills, open-air disposal, and waste burning. Soil pollution is frequently a byproduct of petroleum discovery, refinement, and distribution via vehicle transport. Petroleum hydrocarbons, pesticides, heavy metals, and solvents are the most frequent chemical pollutants of soil. It is difficult work with many related issues to evaluate the ecological hazards of contaminated soil, application of chemically formulated pesticides, sewage and sludge amendment, and other anthropogenic activities that expose the environment to poisonous compounds. Terrestrial assessment of ecological risk is not only a young scientific topic that has advanced quickly just since the middle of the 1980s, but it is also made difficult by the fact that, unlike most aquatic habitats, soil is frequently on small, marginal, and medium enterprises.

1.2.3.1 Organic Wastes

Different kinds of organic waste offer pollution risks. When left in piles or disposed of inappropriately, household waste, municipal sewage, and industrial waste adversely impact the health of people, plants, and animals (Crane and Giddings, 2004). Borates, phosphates, and detergents are abundant in organic waste. They will have an impact on plants' vegetative growth if left untreated. Organic compound likes coal and phenols are the principal organic pollutants.

1.2.3.2 Sewage and Sludge

According to Tarazona *et al.* (2005) and Evans *et al.* (2006), uncontrolled waste wastes from domestic water use, urban drainage, irrigation water runoff, animal husbandry liquid waste, and industrial untreated as well as treated effluent have a number of pollutants which are responsible for soil pollution. When crops are irrigated with sewage water, heavy metals and other hazardous substances accumulate, changing the soils' physical and chemical characteristics. Among the many changes that occur in the soil as a result of sewage water are physical changes such as porosity, leaching, and decline in bulk density and chemical changes like soil reaction; salinity; base exchange capacity; and the content and form of nutrients such as nitrogen, phosphorus, and potash. Sewage sludges contaminate the land by accumulating heavy metals such as lead and cadmium, which can cause plant phytotoxicity.

1.2.3.3 Inorganic Compounds

Industrial waste containing inorganic wastes poses major disposal challenges. They include metals with high toxicity potential. Fluorides, arsenic, and sulfur dioxide (SO_2) are other significant emissions from industrial operations (Richardson *et al.*, 2006). The superphosphate, aluminum, steel and ceramic, phosphoric acid, and aluminum industries all contribute fluorides to the atmosphere. Acidic soils may result from sulfur dioxide emissions from industry and thermal plants. These metals damage leaves and kill vegetation.

The elements that can build up in the soil include copper, mercury, cadmium, lead, nickel, and arsenic if they enter through sewage irrigation and industrial waste. Additionally, some fungicides that include heavy metals worsen soil pollution. Lead, which is hazardous to plants and gets absorbed by soil particles, is present in the smoke from automobiles. By application of organic manure, amending soils with lime, and maintaining soil alkalinity, the detrimental effect of particular elements can be reduced (Van Zorge, J. A., 1996).

1.2.3.4 Heavy Metals

Heavy metals are metals with an elemental density greater than 5. They mainly find particular absorption sites on the soil particles, at which they are strongly held on either inorganic (clay particles) or organic colloids (humus). These heavy metals are abundant in the environment, soils, animals, and plants, as well as in plant tissues. In trace amounts, they are required by plants and animals. Heavy metal pollution is mostly caused by urbanization and industrialization, livestock waste including solid and liquid, human excreta, fuel combustion, mining byproducts, agrochemicals, and so on.

1.2.3.5 Organic Pesticides

Today, many different species of pests are controlled by the application of pesticides. It is found that pesticide application may have detrimental effects on soil macro- and microorganisms, which could impact plant growth. These issues could be caused by pesticides that do not break down quickly. Higher quantities of pesticide residue accumulation are hazardous. Pesticides can enter foods and pose a health risk due to their persistence in soil and migration into water streams. Pesticides, especially aromatic chemical compounds, have a lengthy persistence time because they do not break down quickly.

1.3 CAUSES OF ENVIRONMENTAL POLLUTION

1.3.1 EXPANSION OF URBAN AREAS AND DEVELOPMENT OF REAL ESTATE

Since the industrial revolution, we have rapidly identified and delivered a variety of elements, toxic compounds, and dangerous products into the environment. Urbanization; industrialization; economic development; and natural resources like soil, water, the biosphere, and the environment are all linked by a variety of beneficial and bad impacts. Globally, urbanization and fast socioeconomic development have an impact on people's willingness to migrate.

Even though rapid urbanization promotes soil pollution, deteriorates soil health, disturbs water bodies, deteriorates ecosystems, and lowers air quality, developing countries do not take it as a serious issue. Many non-biodegradable materials, including plastic bags, polythene bags, plastic water bottles, plastic residue, glass bottles, glass items, stone/cement pieces, vegetable waste, livestock wastes, papers, furniture waste, carcasses, plant material, and textile industry waste, is considered soil pollution (Nawrot *et al.*, 2006). According to estimates, Indian cities generate 50,000–80,000 metric tons of solid trash per day. If it is not collected and degraded, it can create a variety of issues, including: (1) major drainage issues, such as burst or leaky drainage lines, which can lead to health risks. (2) Solid wastes have substantially harmed how water normally moves, leading to flooding issues, damage to building foundations, and risks to the public's health. (3) A considerable amount

Environmental Pollution

of methane and other gases are produced by the decomposition of organic material, which pollutes the soil and water. Small and multispecialist hospitals produce a larger quantity of solid waste, which is responsible for health issues and a number of diseases. Beside this, dangerous drugs and medicine also promote health problems. (4) Foul odor is produced due to disposing of garbage.

1.3.1.1 Pollution of Underground Soil

Cities' underground soil is likely to contain pollutants: (1) chemical waste releases from industries or (2) materials made of sanitary waste that has partially or completely degraded. Many hazardous substances, such as chromium, cadmium, lead, selenium, and arsenic, are likely to accumulate in subsurface soil. Similarly, sanitary waste–polluted subsurface soils produce a plethora of hazardous compounds. These can disrupt typical subterranean soil activity and ecological equilibrium.

Solid waste, in general, encompasses garbage and waste materials from commercial sources, home refuse, agricultural practices, and industrial byproducts. It is increasingly made up of paper, cardboard, plastics, glass, expired raw construction products, packaging material, and toxic or other harmful substances. The bulk of urban solid waste is degradable or biodegradable in landfills since food and paper waste make up a sizable portion of it. Similar to how mining explorer material is left on location, the majority of agricultural and animal waste is recycled. We must pay close attention to the hazardous portions of solid waste, such as heavy metals, metals from smelting industries, oils, and organic solvents. Long-term deposition of these can contaminate nearby soils by changing their chemical and biological properties (Patterson *et al.*, 2007).

1.3.2 MINING AND EXPLORATION

From the mining industries, various pollutants are released like dust from surface mining and greenhouse gas emission by coal industries, and other heavy metals are released in the atmosphere, which is able to negatively influence the soil, water, and air quality. The concentration of a particular pollutant depends on which level of mining is carried out. Heavy metals like lead are prominent in polluting the environment. Precious metal mining, such as gold mining, is necessary to do, but at the same time, it causes heavy metal pollution in the environment as a byproduct. The depreciation of soil characteristics, water ecosystem, and air quality has accelerated due to large-scale exploration.

1.3.3 AGRICULTURAL ACTIVITIES

Any nation, developed or developing, depends on its agriculture industry for economic success. But, as we know for the production of every product, there are some byproducts. For produce, a number of pesticides and agrochemicals are utilized, which leaves their residue in the soil. These residues are not easily degradable and remain in the soil for a long time. In this way, it can pollute the soil; by leaching, it can contaminate water and reduce air quality. This pollution directly affects human health. It is caused by the burning of waste by products such as cotton husk and rice straw from agricultural practices such as clearing land for sowing the next crop, providing more fertilizer than plants require, and using chemically strong herbicides which have high persistence. So, these agrochemical residues contaminate natural resources.

Various industries, such as agriculture and animal husbandry, can pollute the soil. Some farming techniques contaminate the soil. They include the use of fertilizers; some agricultural methods; and long-lasting insecticides, fungicides, herbicides, and nematicides. Vital nutrients, such as nitrogen, phosphorus, potassium, sulfur, magnesium, calcium, and others, must be received from the soil. Fertilizers are commonly used by farmers to supply soil essential nutrient deficiencies. Heavy metals present in fertilizers contaminate the soil with contaminants derived from the basic materials used in their production. Nitrogen, phosphorous, and potassium are common components of mixed fertilizers. Arsenic, lead, and cadmium, for example, which are present in very small quantities in rock phosphate mineral, are transported to super phosphate fertilizer.

Because the metals are not biodegradable, they accumulate in the soil to dangerously high amounts due to overuse of phosphate fertilizers, which poisons the soil and reduces crop yields. Additionally, nitrates produced by agricultural operations are well-known chemical contaminants in aquifers of groundwater. Due to lack of awareness, farmers apply excess quantities of nitrogenous fertilizer, which leads to leaching of nitrate in the ground water or surface runoff. This may promote the eutrophication process and thereby build up unnecessary growth of plants and organisms which contaminate the water ecosystem. Also, leached nitrates go into ground water, which will get into the food system through drinking water or as irrigation.

Food-producing plants must fight with weeds for nutrition while being attacked by insects, fungi, bacteria, viruses, rodents, and other animals. Farmers employ insecticides to eradicate undesirable populations that are present in or on their crops. At the conclusion of World War II, dichlorodiphenyltrichloroethane (DDT) and gammaxene were first widely used as insecticides. DDT was quickly overcome by insect resistance, and because it took a long time to degrade, it remained in the environment. It affected calcium metabolism in birds, generating thin and brittle eggshells, and biomagnified up the food chain because it was soluble in fat rather than water. Large raptors like the brown pelican, ospreys, falcons, and eagles became threatened as a result. Most Western countries have now outlawed DDT. Ironically, a lot of them, including the USA, continue to make DDT for export to other developing countries whose needs outweigh the issues it causes (Toccalino and Norman, 2006).

When weathered soil particles are displaced and moved away by wind or water, soil erosion occurs. This erosion is a result of deforestation; agricultural expansion; temperature extremes; precipitation, particularly acid rain; and human activities. Through construction, mining, lumber harvesting, overcrowding, and overgrazing, humans hasten this process. Floods and soil erosion are the results. The soil is kept clean and healthy by the great binding properties of grasslands and forests. The fragile rainforest environments of South America, tropical Asia, and Africa are under threat from population increase and development, particularly in the areas of agriculture, forestry, and building. Many scientists think that these trees contain a variety of medicinal compounds, including cancer and AIDS treatments. The most productive areas of flora and wildlife in the world, which also make up large tracts of an extremely valuable CO_2 sink, are slowly being destroyed by deforestation (Leon Paumen, M., 2008).

1.3.3.1 Emerging Contaminants

Emerging contaminants (ECs) have numerous meanings. Since their presence and significance are now understood, they may be substances that have been present in the environment for a long time instead of the typical novel chemicals. They could be chemicals or microbes that are not typically monitored in the environment yet have the potential to harm the environment or negatively affect people. Although there are numerous definitions of what "emerging contaminants" are, it is crucial to clarify that the term refers to chemical substances that are either unknown or that have not undergone considerable study. Another definition of emerging contaminants is chemical substances or compounds that are distinguished by a perceived threat to human health or the environment without meeting established health standards. A new human exposure, a new detection method, or a new detection technology could all be used to identify them (Murnyak *et al.*, 2011).

There are numerous and regionally specific definitions of emerging pollutants. They could be substances that, through various channels, have been demonstrated to pose a risk to the environment or to human health without sufficient information to examine the magnitude of that risk.

Once more, identified chemicals with undiscovered negative effects on human health and the environment can be classified as emerging contaminants. Emerging pollutants can infiltrate the ecosystem and have negative biological and ecological effects, even if their detection can be more difficult (Snow *et al.*, 2007). They are described by Boxall (2012) as contaminants from a chemical class that hasn't been thoroughly studied, where scientists, regulators, NGOs, or other stakeholders are worried that the contaminant class may have an impact on human health or the

Environmental Pollution

environment or that current environmental assessment paradigms aren't appropriate for the contaminant class.

It has long been understood that contaminants from agricultural operations are caused by both natural and artificial human activity. It is alarming to think about their growing impact on the natural system and the ensuing greenhouse effects. Although agricultural toxins can generally be divided into a few primary categories, animal manure and the methane gas (CH_4) it releases into the atmosphere are a serious environmental hazard. Through the reuse of effluent water for irrigation purposes and the application of sewage to fields as fertilizers, agriculture not only contributes to the introduction of such pollutants into aquatic ecosystems, but it is also a source of developing pollutants. According to Thebo *et al.* (2017), wastewater is used indirectly on an estimated 35.9 Mha of agricultural land. It is important to pay attention to the possible dangers to human health provided by contact with developing contaminants through contaminated food products.

The three basic classifications that can be used for different new contaminant release routes are as follows.

1.3.3.1.1 Use of Livestock

This includes all veterinarian composts, artificial fertilizers, hormones used in livestock, manure and flatulent gas (methane) generated directly from animals, and their compost that releases methane gas into the atmosphere. In the past 20 years, new agricultural contaminants have arisen, including antibiotics, vaccinations, growth boosters, and hormones. These pollutants are increased by the careless treatment of organic manures in aquaculture and animal husbandry, and they may also spread through runoff and leaching processes (OECD, 2012b). Heavy metal residues in agricultural inputs like insecticides and animal feed are another rising danger. More than 700 emerging pollutants, along with their metabolites, are common in aquatic environments across Europe, according to NORMAN (2016).

1.3.3.1.2 Human Use Activities

These include substances that are emitted both directly and indirectly, such as novel chemical compounds and medications created by humans. Normally, these substances move through wastewater treatment, producing sludge and biosolids, or through irrigation, land, producing wastewater effluent. This group also includes pesticides that have higher absorption rates, higher solubilities, or higher toxicities and may be dispersed as nanoparticles.

Higher rates of fertilizer application to field crops resulted in either adsorption by clay lattices or leaching through soil into ground water, which contaminated the water with nutrients. Extra phosphates and nitrogen can seep into the groundwater or enter streams through surface runoff. Since phosphate is less soluble than nitrate and ammonia, it often binds to soil exchange sites and seeps into water sources through soil erosion. High nitrogen loads can eutrophize lakes, reservoirs, ponds, and coastal areas when combined with other stresses, resulting in algal blooms that end many aquatic lives. Despite data shortages, 415 coastal regions throughout the world have been acknowledged as being affected by eutrophication in some way, 169 of which are hypoxic (WRI, 2008). The overabundance of nutrients may potentially increase harmful health effects, such blue baby syndrome, as a result of the high amounts of nitrate in drinking water.

1.3.3.1.3 Plant Protection Activities

Pesticides, such as rodenticides, fungicides, bactericides, weed killers, zootoxins that kill small animals like rodents, phycotoxins that prevent algae growth, and various personal and household products, all have the ability to form new substances that are subsequently released either directly or indirectly into the environment as emerging contaminants, particularly through air and wastewater. These mostly consist of nanoparticles created to function as "smart" chemicals or pesticides with the capacity for selective toxicities. Unfortunately, these nanomaterials practically go undetected as they penetrate the natural system.

Numerous nations use insecticides, herbicides, and fungicides heavily in agriculture (Schreinemachers and Tipraqsa, 2012). They can contaminate water supplies with carcinogens and other substances that can harm humans if improperly chosen and managed. By destroying plants and insects and having detrimental effects on the food channel, pesticides can also have an adverse impact on biodiversity.

While older broad-spectrum pesticides are still widely used and produced in the USA and Europe, there is a trend toward the use of newly created pesticides that are more selective and only need a small amount to cover a large area. Nevertheless, agriculture uses millions of tons of active pesticide components (FAO, 2016a). Globally, pesticide poisoning significantly increases the number of deaths in humans through various diseases, particularly in underdeveloped nations where subsistence farmers frequently use extremely dangerous pesticide formulations.

1.3.3.2 Emerging Contaminants from Agricultural Activities

Different channels allow emerging contaminants to enter the agricultural environment. When manure, biosolids, or other solid waste is supplied to the soil, it can enter the soil indirectly or directly (in the case of veterinary medications used for animals in pasture). The persistence of the EC and its interactions with soil and air determine the level of transfer.

1.3.3.2.1 Agricultural Soils

The main portion of the agricultural sector's emissions of greenhouse gases (GHGs) comes from the management and handling of manure and enteric fermentation of agricultural soil N_2O emissions. As a result, the sources of GHG emissions from agriculture are as follows: nitrous oxide from fertilizers (37%), methane from livestock (32%), residue burning/forest cleaning (13%), methane from rice cultivation (11%), and methane and nitrous oxide from management of manure (7%) (US Environmental Protection Agency). Nitrous oxide (N_2O), which contributes 37% of the agricultural sector's GHG emissions, is the main source of these emissions. It is made naturally from soil through the processes of nitrification and denitrification. Anthropogenic agricultural practices may directly or indirectly enrich soils with nitrogen. Due to increased nutrition and the strong demand for agricultural products, these emissions are anticipated to rise in the future (Delgado et al., 1999). Direct addition of fertilizers, both synthetic and organic, including nitrogen, may play a significant role in the rise in N_2O emissions, with developing countries using 36 million tons more fertilizer than industrialized countries (Bumb and Baanante, 1996). The construction of equipment, the utilization of pesticides, fertilizers, on-farm fuel consumption, and the transportation of agricultural goods are additional anthropogenic sources of GHG emissions from agriculture (Rosegrant et al., 2008).

1.3.3.2.2 Livestock

Globally, there is a significant rise in both the demand and production of cattle products, although the following areas take the lead: the central and eastern United States of America, northern Argentina, southern Brazil, Uruguay, Europe, China, and India. Thirty percent of the world's land surface is covered by livestock production, and 70% of all agricultural land is used for this purpose. The cattle sector is one of the top three causes of the most significant environmental problems, including the decline in water quality, on all scales, from the local to the global (FAO, 2006).

Major structural changes in the livestock sector are associated with the growth of modern and extensive livestock production methods, which often involve large numbers of animals concentrated in relatively small regions. Intensive livestock systems increasingly rely on domestically and internationally traded feed concentrates. The vast majority of waste products such as manure, liquid, and wastewater that is used to wash cattle is released back into the environment. These developments are placing growing pressure on the environment, and specifically on the quality of the water. Excreta from livestock contain significant amounts of nutrients, compounds that deplete oxygen, infections, heavy metals, medication remnants, hormones, and antibiotics. When livestock accumulates, the

Environmental Pollution

associated waste generation frequently exceeds the capacity of the ecosystem to serve as a buffer, which is polluting the surface waters and underground.

Livestock-related emissions account for 9% of the CO_2 equivalent produced by all human-related activities and are responsible for 64% of ammonia emissions, 37% of methane, and 65% of nitrous oxide, mostly from manure. According to Steinfeld *et al.* (2006), the livestock industry accounts for close to 80% of total emissions. Due to all of these pollutants, livestock is a top target for mitigation. The main source of methane production in this category is enteric fermentation in sheep and cattle, which accounts for 34% of worldwide agricultural emissions followed by rice farming, which accounts for 11% of global agricultural emissions. Horses, swine, and poultry are some other domesticated species that release methane (methanogenesis) as a byproduct of enteric fermentation. About 80,000 Gg of enteric methane from ruminants is thought to be produced globally (Carlos *et al.*, 2020).

1.3.3.2.3 Aquaculture

Greater production intensity and greater concentrations of one species are being caused by market forces and differentiation. Because of these changes, people are using more medications (such as antibiotics and fungicides), which leads to contamination further down the food channel. The largest amount of aquaculture development has taken place in developing nations, which produce 91% of the world's aquaculture; low-income developing countries have the greatest concentration of aquaculture. Nearly 90% of the world's aquacultural production is in Asia, with China leading the way with 45.5 million tons produced annually (FAO, 2016b).

1.3.3.2.4 Salts

Irrigation can release trapped salts in the soil, which can be carried by water from drainage systems to receiving bodies of water and cause salinization. Overwatering can also raise the water tables in saline aquifers and facilitate the seepage of saltwater from the ground into watercourses. The penetration of saline into aquifers, which typically occurs as a result of overuse of groundwater for agriculture, is another important factor contributing to the salinization of coastal areas (Mateo-Sagasta and Burke, 2010). Australia, Argentina, India, China, the Sudan, the United States of America, and numerous nations in Central Asia have all reported serious water-salinity issues (FAO, 2011). According to IGRAC (2009), 1.1 billion people in 2009 resided in areas with shallow or intermediate depths of saline groundwater.

Lorenz (2014) says that when salinity increases, there is typically a loss in the biodiversity of bacteria, algae, plants, and animals. Herbert *et al.* (2015) claim that excessively salinized waters have a broad effect on ecosystems by altering the cycles of significant elements such as carbon, iron, nitrogen, phosphorus, silicon, and sulfur.

1.3.3.2.5 Sediments

Many types of contaminants are found in rivers and lakes, which are rich with inorganic and organic material. The capacity of dams can be decreased by sediments, which can also block fish gills and cover and ruin fish spawning beds. Sedimentation can clog streams, harm watercourses, and require filtration to use for irrigation purposes and urban water supplies. Additionally, it may have an impact on delta dynamics and navigational potential. Sedimentary clay particles have surfaces that can adsorb a wide range of substances, including inorganic and organic pollutants. Therefore, one important method for bringing such pollutants to water bodies is through sediment.

An enormous amount of soil is lost and carried to aquatic bodies annually as a result of improper tillage, unsustainable land use, and inadequate management of soil in agriculture. These elements promote the release of sediment into streams, lakes, and reservoirs, causing erosion. Approximately 193 kilogram of soil organic carbon per ha per year is the global rate of cropland erosion. Approximately 1.7 Mg per ha per year and about 40.4 kg of soil organic carbon per hectare per year accounted for pastureland. According to Doetterl *et al.* (2012), Asia accounts for 43% of agricultural sediment flux.

1.3.3.2.6 Management of Manure and Organic Matter

Manure handling, treatment, and storage are responsible for 7% of agricultural emissions. Methane is produced during the anaerobic decomposition of manure (methanogenesis), whereas nitrous oxide is produced during the aerobic processing of manure (nitrification), which is subsequently converted to nitrogen dioxide for use by plants (denitrification) (Urzelai *et al.*, 2000). Methane emissions from enteric fermentation are anticipated to rise by 32% because of the high demand for beef and dairy products anticipated worldwide, particularly from emerging countries (US Environmental Protection Agency).

1.3.3.2.7 Pathogens

Despite the fact that humans are constantly exposed to a wide variety of microbes in their surroundings, only a small percentage of these bacteria are able to interact with the host in a way that leads to disease. Numerous multicellular parasites and zoonotic bacteria that can be dangerous to the health of humans are present in livestock excrement. Waterborne or foodborne pathogenic bacteria are both possible. The pathogen typically has to develop inside or on the host in order to cause sickness. The incubation period is the period of time between contamination and the onset of medical signs and symptoms. When excrement is left on land, some germs can linger there for days or even weeks before contaminating water sources through runoff (WHO, 2012; FAO, 2006).

1.3.4 BURNING OF FOSSIL FUELS

Our energy needs are met by burning gas, coal, and oil, which is what is generating the present global warming problem. Burning fossil fuels release a number of air pollutants that harm the ecosystem and the surrounding environment. Fossil fuels have the potential to release harmful air pollutants years before they are burned. Several major and secondary pollutants are released during the burning of fossil fuels, such as airborne particles, hydrocarbons, chemicals, nitrogen oxides (NOx), CO_2, CO, SO_2, and organic compounds. The greenhouse gases, like nitrous oxide and carbon dioxide, are all present in the emissions from burning fossil fuels.

1.3.5 PARTICULATE MATTER

Particulate matter concentration and type of particulate material are considered very seriously when we talk about the biosphere and atmosphere. From natural sources, a number of tons of this matter is released into the environment each year. Among them, some examples are volcanic eruptions, rock material, deforestation, dust storms, greenhouse gas emission, and soil degradation. On other hand, anthropogenic activities also contribute to disturbing the atmosphere through contamination of particulate matter. The waste from steel industries, petroleum refinery byproducts, mining of coal and its exploration, agrochemical residues in the soil and water bodies, industries related to chemical production, emissions from power plant systems, and combustion of fuels and petroleum products all are responsible for contamination of the environment through producing particulate matter.

1.3.6 UTILIZATION OF PLASTICS

Acrylic, polyethylene, polyvinyl chloride, polyester, and polycarbonates are all examples of types of plastic materials. Because plastic bags are inexpensive and long-lasting, they are frequently used in developing nations for carrying, buying, and storing food. The United States' municipal solid waste generation between 1960 and 2013 increased by 188%, whereas plastic generation increased

Environmental Pollution

by 8238% (Tsiamis *et al.*, 2018). However, the production of metallic and glass garbage decreased, while the production of plastic waste increased. While secondary microplastics (MPs) are created as larger plastic waste breaks down, MPs are primarily found in consumer goods, including paints, cosmetics, and fibers in cleaned synthetic garments (Auta *et al.*, 2017). People are becoming aware of the environmental damage caused by plastic pollution. However, regulating commercial use and reducing the use of plastic is extremely difficult. There are no alternative materials or products on the market that can rival the plastic carry bag.

1.4 EFFECTS OF ENVIRONMENTAL POLLUTION

The majority of developing countries, which are the most severely impacted by pollution, still lack adequate documentation on its effects due to a lack of understanding of the potential harm that pollution can have for the environment and public health as well as unstable database management systems. In low-income nations, where people prioritize food and shelter over their health and the environment, pollution and its effects are intensifying. Public choices for conserving the environment are known to be highly associated with socioeconomic determinants of health, including education and income. Some African regions completely attribute certain health problems, such as birth defects, cancer, stunted growth, pregnancy loss, and early death, to bad luck and "acts of the gods," which draws attention to pollution and its effects.

1.4.1 EFFECTS ON THE ENVIRONMENT

The biosphere, water, soil, and atmosphere form the environment, which serves as a repository for all harmful substances. Environmental pollution is so called because the environment always faces the greatest damage as a consequence of an increase in pollution. Increases in GHGs have the potential to drastically alter our civilization, either positively or badly, but the full extent of these effects is unknown. A warming globe causes climate change; it has a number of negative impacts on the weather. Therefore, as the climate alters, the worst weather poses serious risks to human civilization. The temperature of the ocean, height of the sea surface, sea ice, temperature of the biosphere, heat level of the surface water of aquabodies, snow area on hills, and tropospheric temperature are all indicators of global warming. It is anticipated that increased droughts, floods, storms, and heatwaves may result from climate change. According to IPCC estimates, temperatures could increase by 2 to 6°C in years to come (Singh and Singh, 2012).

Despite recent evidence suggesting that the stratospheric ozone layer is healing because of reduced chlorofluorocarbon emissions, the stratospheric ozone layer decreasing in the past decades might be due to emissions of nitrous oxide and chlorofluorocarbons, which protect the earth from solar ultraviolet (UV) radiation (Solomon *et al.*, 2016). The Arctic ozone, on the other hand, exhibits substantial year-to-year fluctuation, while ozone depletion continues to occur each year in Antarctica (Andrady *et al.*, 2015).

The loss of land for agricultural production is a result of subsequent damage such as erosion, soil mineral leakage, landslides, and soil and crop contamination by pathogens from humans or animals, salt, or chemicals (Lindgren *et al.*, 2011; Miraglia *et al.*, 2009). Flooding or drought-related declines in crops and cattle are examples of direct repercussions, as is damage to infrastructure that may have an impact on food production. After extreme weather events, a number of infectious illnesses that harm livestock, including anthrax and blackleg, may appear (Bezirtzoglou *et al.*, 2011; Skovgaard, 2007). Furthermore, a rise in insects that could serve as infection vectors may be seen following floods, particularly when they occur in combination with high temperatures. For instance, Rift Valley fever infects animals through eggs, which rely on still water (Nardone *et al.*, 2010; Githeko *et al.*, 2000). Extreme weather conditions may also additionally affect the dynamics of how pathogens spread and the existence of insects and pests, which can harm agricultural productivity (Jaggard *et al.*, 2010; Miraglia *et al.*, 2009).

The repercussions for land include waste littering, tree damage, wildlife species, soil sterility reducing crop yield, degradation of roofing sheets, influence on historical monuments and structures, and the staining of automobiles. Continuous mining, for example, devastates soil–plant systems and decreases soil productivity (Feng *et al.*, 2019), whereas anthropogenic activities cause landscape damage such as soil upper layer loss, habitat disturbances, loss of animal productivity, and resource loss, such as wetlands ecosystems (Vallero and Vallero, 2019). Food crises cause hunger and, in extreme cases, death of living organisms. The pH of the soil decreases due to changing soil chemical properties, and crucial cationic nutrients like magnesium, potassium, and calcium are lost. All of these generate scarcity of food for both humans and other living things, which can lead to starvation and even death.

Pollution typically modifies the biological, chemical, microbiological, and mechanical characteristics of bodies of water. The spreading of oil on the water's surface also limits sunshine and oxygen. Other instances include the presence of heavy metal contaminants in products, heated water bodies from UV radiation, improper irrigation water management that encourages salt sedimentation on the land's surface, and runoff from drainage that enriches the water ecosystem with nutrients like nitrogen and phosphorus. All these events affect the biological oxygen demand for aquaculture, deteriorate the quality of water, and cause unnecessary growth of vegetation. Several pollutants have been identified as being delivered by air and subsequently deposited on soil and in water bodies. Water bodies become odiferous and unpleasant as a result of the introduction of sulfur- and nitrogen-containing compounds, which leads to loss of the aesthetically pleasing qualities of water, and they are abandoned.

A main or significant source of anthropogenic CH_4 and CO_2 greenhouse gas emissions has been identified as livestock production, which is an agricultural food-based industry (Audsley and Wilkinson, 2014). Because of the significant amount of greenhouse gases created during ruminal fermentation of feeds, they significantly contribute to global warming, pollution, and environmental deterioration. Therefore, the livestock industry accounts for approximately 18% of total CH_4 and 9% of total CO_2 emissions (FAO, 2013), with methane accounting for 50–60% of emitted gases during livestock production (Mirzaei-Ag *et al.*, 2012). For instance, greenhouse gases emitted into the environment are responsible for climate change at every step of the making of eggs, meat, and milk in agriculture, which disturbs the temperature, weather, and ecosystem health. It will be necessary to alter agricultural methods and cattle consumption in order to mitigate these issues.

1.4.2 Effects on Crop Production

It is possible to isolate a number of unique elements that have somewhat diverse effects on agricultural output within the framework of human climate change. According to experimental data, a boost in atmospheric CO_2 concentration boosts the vegetation growth and photosynthetic activity in a variety of crop (C3) species (Ainsworth and Long, 2005). A research experiment indicated that high temperatures have a detrimental effect on the physiological processes, and its combined effect of high temperatures and increased CO_2 may result in reduced photosynthesis and biomass production (Ruiz-Vera *et al.*, 2013). Furthermore, heat stress during pollination will make some commodity crops more vulnerable (Semenov and Shewry, 2011), particularly in places where crops are planted near the temperature that is the critical limit for photosynthesis (Ruiz-Vera *et al.*, 2013). Higher temperatures, on the other side, may extend cultivation seasons and result in greater yields in northern latitudes and colder locations (Eckersten *et al.*, 2011). Ainsworth *et al.* (2012) found that as temperatures rise, tropospheric ozone levels rise, which puts plants under oxidative stress and inhibits photosynthesis and plant growth.

According to Porter *et al.* (2014) and Smith *et al.* (2014), there are numerous direct and indirect ways that climate change will affect agricultural production. Crop yields will be directly impacted by changes in temperature and water availability, as well as by greater weather variability and more frequent episodic weather events (Lobell and Gourdji, 2012). Faster crop growth, shorter cropping seasons, and lower yields are all effects of rising temperatures. The rate of photosynthesis and respiration are also impacted by temperature. The optimum temperature for photosynthesis is higher for C4 crops than for C3 crops (cereals and the majority of vegetables and fruits), including maize, sorghum, and sugarcane.

The production of important crops will be significantly impacted by increased climatic variability (Lobell *et al.*, 2008). Increased inter-annual weather variability might increase the probability of crop failures by making crop management strategies that aim to maximize yield and quality while minimizing environmental consequences more challenging. Furthermore, if biotic stressors increase due to, among other things, pests and the invasion of alien weed species, crop yield may suffer (Garrett *et al.*, 2011; Anderson *et al.*, 2004). Therefore, the influence of climate change on crop production can be either direct or indirect. The degree to which certain nations and areas are successful in adapting to climate change will depend on their capacity to create and use new technology (Varshney *et al.*, 2011). In addition to the direct consequences, rising temperatures may have an indirect impact on fruit and vegetable production due to lower labor productivity of farmers (Kjellstrom *et al.*, 2016). Since many fruit and vegetable crops demand significant labor inputs, particularly during planting and harvest, the industry may be disproportionately affected by heat stress caused by climate change.

The number and susceptibility of the host (crop plant), the abundance and virulence of the pathogenic organism, and favorable climatic conditions are crucial components of plant disease epidemics that define the occurrence and severity of a specific plant disease (Agrios, 2004). According to Ayliffe *et al.* (2008) and Stuthman *et al.* (2007), agricultural practices that increase host density, such as increasing field aggregation, field size, and crop species uniformity, tend to increase the severity of plant disease epidemics because they both make hosts more vulnerable and make it easier for the plant pathogen to spread. Additionally, little genetic variety is linked to few features that provide resistance to a particular pathogen, and genetic uniformity of cultivars makes the host more vulnerable (Tadesse *et al.*, 2010). Therefore, if the disease evolves to beat the genetic resistance, the outcome could be widespread crop loss (Forbes and Jarvis, 1994). International trade in seed and planting stock also has a considerable effect on the abundance of plant diseases. In actuality, diseases have spread into different regions of the world where they were previously nonexistent due to global trade and interaction (Zadoks, 2008). Pathogens may also spread because of people traveling to and from low- and middle-income nations while bringing their own food and avoiding border checks. Therefore, the huge fields devoted to homogeneous crop cultivars, higher planting densities, and increased use of fertilizers typical of specialist agriculture in the industrial world may increase the possibility of the spread of plant disease (Stuthman *et al.*, 2007). Nevertheless, it is typically challenging to forecast the infestation of plant diseases, and the severity of their effects depends on both environmental factors and interactions between plants and pathogens (Wellings, 2007).

Animal species are also impacted by climate change, and a decline in plant pollinator numbers, for instance, might have a variety of effects on agricultural output (Pacifici *et al.*, 2015). Pests, diseases, fungus, and weeds are also predicted to cause more crop losses and damage due to climate change (Flood, 2010). According to estimates made by Bebber *et al.* (2013), between 1960 and 2012, hundreds of pests and pathogens traveled closer to the poles on average by 2.7 km yr^{-1}.

In some regions, losses in agricultural yield resulting from surface ozone exposure and heavy metal contamination may be significant (Chepurnykh and Osmanov, 1988). The importance of ozone in this area could increase in the future (Avnery *et al.*, 2011). According to experimental data, some plants' ability to photosynthesize can be negatively impacted by cadmium at concentrations of less than 1 micro molar, which is observed in some soils (Prasad, 1995). Additionally, Kalantari (2006) hypothesized that decreased rice yields in Iran are related to concurrent increases in cadmium burden. Agricultural production and soil cadmium loads are directly related to one another because phosphate rock fertilizer is the main source of cadmium. According to Pan *et al.* (2010), sewage sludge used as fertilizer also contains cadmium.

1.4.3 Effects on Human Health

Could the fact that the results of humans' actions have come back to find them constitute "karma"? The bulk of human ailments have been connected to environmental pollution because of its correlation with human health. More information about the relationship between pollution and a number of serious health issues is being uncovered by recent studies. The quantity of studies examining the negative impacts of air pollution on health is alarmingly rising. The World Health Organization's report made it abundantly evident that indoor air pollution from fires used for cooking and heating was responsible for 3.8 million fatalities (WHO, 2018). This percentage varied, as one might expect, from 10% in developing countries to 0.2% in high-income nations.

Many people worldwide breathe air that contains elevated levels of pollutants that are beyond WHO guideline limits, according to the World Health Organization, raising the risk of various illnesses, including strokes, heart disease, lung cancer, respiratory illnesses, cognitive decline, and many more. There are several social and economic effects of atmospheric pollution that cannot be disregarded.

Additionally, according to the global burden of disease, one aspect of ambient (or outdoor) air pollution, PM2.5, was the fifth most important cause of death globally in 2015, accounting for 4.2 million deaths and more than 103 million disability-adjusted life years lost (Schraufnagel *et al.*, 2018). According to research by Song *et al.* (2019), third-trimester maternal exposure to PM2.5, PM10, CO, and SO_2 is linked to shorter infant telomere length. This suggests that these pollutants not only put us in danger but also provide serious risks to unborn children.

1.4.4 Effects on Animal Health

Heat stress can worsen metabolic disorders and increase mortality in farm animals. Additionally, it can lower fertility, feed intake, and immune response, all of which tend to lower output (Nardone *et al.*, 2010; Thornton *et al.*, 2009). Pig and chicken intensive indoor production is particularly susceptible to temperature increases since there may be an increase in mortality if additional cooling is not provided. The modern, high-yielding dairy cow is sensitive to heat stress because of its high metabolic rate (Black *et al.*, 2008; Sartori *et al.*, 2002). Long droughts may also directly result in feed and water shortages, which would further reduce output.

Vector-borne diseases are also impacted by rising temperatures. For instance, they boost the quantity and intensity of female mosquitoes' blood meals, the vector's reproductive rate, and the virus's rate of replication inside the vector (Pinto *et al.*, 2008). Ticks and biting midges, vectors of Lyme disease and blue tongue, respectively, have previously been observed moving northward in the northern hemisphere (Forman *et al.*, 2008; Van den Bossche and Coetzer, 2008). Furthermore, it is dangerous to store food and feed because of the increased prevalence of hazardous mycotoxins brought on by humidity and a warm temperature. Additionally, due to lignification, changes in the content of grass species triggered by climate change possibly will affect the productivity of grazing animals (Thornton *et al.*, 2009).

Transboundary animal illnesses are highly infectious and spread quickly within and among populations of animals. As a result, they endanger the farmers' way of life, the cattle industry's financial stability, and ultimately global food security. For reasons related to public health, livestock productivity may be hampered by zoonotic transboundary animal illnesses. The worldwide outbreak of the highly virulent avian influenza (H5N1) that started in East Asia in 2003 serves as a strong illustration of this (Kaufman, 2008; Sims *et al.*, 2005).

According to Harrus and Baneth (2005), changes in ecosystems can also make it easier for domestic and wild animals to contract the same diseases. The spread of the Nipah virus from fruit bats to farm pigs and ultimately to humans in Malaysia in 1999 is a prime illustration of this (Chua, 2003). Through worldwide travel and trade in animals, animal products, and consumables, an illness can quickly spread to nations with weak livestock populations once it has been established, endangering livestock output (Sherman, 2010; Thornton, 2010). In 1986, a swine flu outbreak in Great Britain made the value of trade clear. According to Williams and Matthews (1988), the sickness was believed to have been brought on by feeding animals with unprocessed food waste that contained imported pig flesh.

The devastating foot-and-mouth disease outbreak in Great Britain in 2001, which resulted in losses estimated to be over 3.1 billion GBP (Thompson *et al.*, 2002), is a good example of the threat posed by transboundary animal illnesses to livestock productivity and food security. The severity of a contagious disease depends on the pathogen's virulence, farm and livestock density, biosecurity practices, the production system, the volume of trade in animals and animal products, the availability of veterinary services, and the population densities of people and wildlife and how close they are to livestock (Rossiter and Al Hammadi, 2009). The relative importance of these characteristics varies and could be affected by governance (Graham *et al.*, 2008) and economic development (Forman *et al.*, 2009).

In East Asia, there is a noticeable rise in the population of farmed animals, particularly poultry and pigs raised in enclosed production systems (Thornton, 2010). According to Steinfeld *et al.* (2006), large-scale intensive animal production plants are typically created in highly inhabited areas. Infection outbreaks could have catastrophic effects in these large-scale systems, which emphasizes the value of strong biosecurity (Sherman, 2010). There will unavoidably be human–animal interactions due to the rise in small-scale, backyard animal production that urbanization in type A and type B countries entails.

Moreover, genetic diversity and biodiversity of the ecosystem are affected by contaminants present in the atmosphere. It is demonstrated that the quantity of ribosomal duplicates of DNA regularly increases in response to environmental changes. This happens because these sequences are essential for preserving the integrity of the genome (Araujo da Silva *et al.*, 2019). Studies show that fish living in heavily polluted environments have incredibly complex ribosomal sequences in their genomes.

The issues that plastics have created in the environment have been discussed recently. Animals are harmed by plastics either directly or indirectly. It also harms ecosystems and limits biodiversity. Ultimately, it could affect the lives of mostly fish, birds, lobsters, sea turtles, and other kinds of marine animals (Barboza *et al.*, 2019). Ingestion stress issues can cause lesions, lacerations, and internal damage. Additionally, plastics have the ability to entangle and choke aquatic life; hinder photosynthesis in the principal food providers, such as algae; and have an impact on the growth and reproduction of crustaceans (Barnes, 2019).

Oil spills during refining, drilling, and transfer on the ground through transmission lines, including underwater, can have sub-lethal health effects on wildlife and aquatic organisms. When these animals inhale or eat substances from petroleum that contain hazardous compounds, their respiratory, digestive, and circulatory systems suffer. Beside these, seabirds are severely impacted by oil spills, yet they are often not recorded. According to studies, oil fouling is killing birds. Many oil spill–related deaths go unreported, despite the fact that certain oil-fouled birds are recognized and recorded when they die (Walker *et al.*, 2019).

1.4.5 EFFECTS ON MICROORGANISMS

Microbial pollution is the term used to describe pathogen contamination, which includes those caused by bacteria, viruses, and parasites. Infections may be species specific or zoonotic, meaning that they harm both humans and animals, and they can get into agricultural systems in a variety of ways. They may be spread by contaminated water or organic fertilizer (Tirado *et al.*, 2010). Following an epidemic of a disease that produces significant amounts of pathogen-contaminated animal feces, pathogens of animal origin can build up in the environment. The nutrient cycle and transfer of energy in the aquatic food webs are critically dependent on microscopic populations in flowing water habitats, such as zooplankton (Xiong *et al.*, 2019). Consequently, biotic reactions of microscopic organisms to their ambient condition could be used to accurately detect changes in the environment in aquatic habitats. However, pollution has had a considerable impact on the zooplankton community's geographic spread, which has decreased their effectiveness. Such garbage could contaminate water supplies or the land it is dumped on or used as fertilizer after being collected, stored, or buried. As a result, microbial pollution of an environment used for agriculture may make it impossible to engage in agricultural activity and provide potential hazards to both humans and animals.

1.4.6 EFFECTS ON WATER QUALITY

A serious danger to the quality of irrigation water is salinization. From crop to crop, salt tolerance levels might differ significantly. Salinization primarily reduces crop yields, but it has a mixed effect on crop quality (Hoffman *et al.*, 1989). Salinity has a negative impact on a variety of vegetable crops and can significantly lower their market value. However, salinity can enhance the sugar content of some crops, like carrots and asparagus, while also increasing the amount of soluble solids in others, like tomatoes and melons. However, salinity-related yield reductions typically outweigh any positive effects (Hoffman, 2010).

Climate change may make salinity issues more severe, which will have an effect on health via nutrition and drinking water (Scheelbeek *et al.*, 2017; Khan *et al.*, 2014). The rising incidence of tropical cyclones and flooding is able to considerably impact sodium and other salts contained in soils and ground and surface water in a number of low-lying coastal locations. When farmers shift away from saltwater irrigation supplies and acquire water from deeper groundwater layers in coastal regions that are sensitive to climate change, like Bangladesh, an additional issue arises since significant arsenic concentrations have been recorded in these groundwater sources. After harvest, arsenic may still be present on the crop's surface, posing a major health risk to consumers (Su *et al.*, 2014; Das *et al.*, 2004). The quality of irrigation and drinking water may be impacted further inland by considerable increases in salt concentrations brought on by shifting precipitation patterns and drought (Jeppesen *et al.*, 2015).

Irrigation water contamination has a substantial impact on agricultural yield and quality. In low-income nations with dry and semi-arid climates, more than 10% of the world's population consumes food that has received irrigation from untreated wastewater or lakes or reservoirs water which is contaminated by feces (WHO, 2006). The main causes of the rising use of contaminated water for irrigation include the shortage of freshwater, growing populations, and awareness of the potential of wastewater as fertilizer. Food-borne disease outbreaks have been connected to the use of pathogen-contaminated municipal wastewater for irrigation and post-harvest procedures (Antwi-Agyei *et al.*, 2015). This is especially problematic for fruits and vegetables, which are frequently consumed raw.

The presence of excessive nutrients in irrigation water, particularly nitrogen, is a serious danger to water quality. This is frequently the result of over-fertilizing agricultural land, where excess fertilizer ends up in irrigation water sources and could harm marine ecosystems. High nitrogen concentrations cause excessive vegetative growth and a delay in maturity in crops that are susceptible to it, including apricot, citrus, and avocado. This reduces the amount of produce that can be harvested from leafy vegetables and may have a detrimental impact on fruit quality indicators such sugar content (Ayers and Westcot, 1985). Crops may become taller as a result, making them more susceptible to lodging during severe weather events like tropical storms.

High quantities of some harmful ions, such sodium, boron, and chloride, can cause damage to crops and lower yields when they are taken up by plants and accumulate in irrigation water (Banon *et al.*, 2011). Toxin concentrations in water are influenced by both industrial and agricultural factors, including the release of chemical wastes into irrigation watersheds and the disposal of agrochemicals on farms. The majority of irrigation water sources have element quantities below toxicity criteria; nonetheless, the majority of vegetable crops have a rather limited tolerance to boron, and even very low boron concentrations can harm crops (Hoffman, 2010). The severity of the harm varies

Environmental Pollution

depending on the crop, and permanent perennial-type crops are thought to be the most vulnerable (WHO, 2006).

In general, water contaminated by human and animal pathogens is unfit for consumption since it may result in illnesses and subsequent loss of output. For a similar reason, this kind of water should not be used to irrigate crops meant for human or animal consumption. Internationally, the significance of preserving freshwater's good microbiological quality is generally accepted (Fewtrell *et al.*, 2005). Similar to this, applying pathogen-contaminated fertilizers to crops meant for direct human or animal consumption might be dangerous. Pathogens can be found in both human and animal feces (Barrett *et al.*, 2001). Exposures may occur in type C nations when untreated sewage water leaks into the water supply system because of severe weather or mishaps such as burst sewage pipes (Cabral, 2010). According to Bartram and Cairncross (2010), in both type B and type A countries, inadequate or absent wastewater treatment, poor sanitation, and outdoor defecation can all contribute to the discharge of human diseases.

1.4.7 Effects on Soil Quality and Health

A scarce natural resource is agricultural land. According to estimates, during the past 40 years, soil erosion and pollution have caused the loss of over a third of the world's arable land (Cameron *et al.*, 2015). Urbanization, sea level rise, the need for space for biofuels and other non-food crops, as well as the production of renewable energy (such as solar panels on agricultural land), are other factors contributing to the loss of agricultural land. Forests, on the other hand, have been turned into agricultural land, mainly because of rising meat consumption and a need for space for the production of animal feed. As a result, throughout the past few decades, the proportion of worldwide land that is used for agriculture has remained largely unchanged. Deforestation, on the other side, has a detrimental indirect influence on food security since it accelerates a number of environmental processes, such as global warming and biodiversity failure.

Acid rains or, in some cases, the usage of synthetic nitrogen fertilizers, contribute to soil acidification. Acid showers often come about due to an atmospheric reaction between water molecules and sulfur dioxide or nitrogen oxide, which are mostly produced by human activities such as energy production and industrial processes (Klimont et al., 2013). Except in alkaline soils, where moderate acidification can be advantageous, soil acidification can change the availability of nutrients and generally has a detrimental impact on plant growth. Crop losses brought on by acidification can be lessened with the use of lime and balanced fertilizers (Mason *et al.*, 1994). Phytotoxicity is the hazardous impact that substances like trace metals, allelochemicals, pesticides, phytotoxins, or salt have on plants. Both crop productivity and people's health are harmfully affected by toxic metal contamination of soil, such as that caused by cadmium and high levels of aluminum (Khan *et al.*, 2015). Plants experience oxidative stress from metals, which hinders the accumulation of biomass.

Persistent organic pollutants include low-use pesticides like DDT, industrial toxins, and some industrial compounds like polychlorinated biphenyls. According to research (Wang *et al.*, 2010; Holoubek *et al.*, 2009), certain pollutants, such as PCBs, can be hazardous to plants, however only in quantities that are several orders of magnitude higher than those discovered in soil that was irrigated with chemical-containing water. From the standpoint of food safety, these chemicals are a serious concern since they could contaminate foods of animal origin, notably seafood, and then make their way to people (Guo *et al.*, 2009; Zhao *et al.*, 2006).

The constant discharge of chemicals and radioactive materials from various sources, as well as more dramatic occurrences like industrial accidents or the purposeful transport or release of toxic waste, can all contribute to chemical and radioactive pollution of the environment. Chemical pollutants can enter agricultural soils through a variety of channels, including air, rain, irrigation, and direct application as pesticides (Montanarella, 2007). However, even in nations with sophisticated monitoring systems, it is difficult to find the extent of soil contamination.

According to Gibbs and Salmon (2015), the term "soil degradation" often refers to a number of processes, including desertification, salinization, erosion, compaction, and the invasion of exotic species. In order to keep soils productive over the long term, soil organic matter is crucial. One of the main causes of loss in soil organic matter levels is the increased use of industrial farming techniques, such as monocropping, minimum use of organic fertilizers, and removal of crop wastes from fields. According to experts' estimates, land degradation affects roughly 15% of the world's land surface severely, 46% of it moderately, and 38% of it gently (Bridges and Oldeman, 1999). These estimates have come under fire for being arbitrary and overstating the degree of land degradation, particularly in dry and semi-arid areas (Nkonya et al., 2011). The time series of the normalized difference vegetation index (NDVI) (1981–2006), as shown by more recent measures of plant cover (Bai et al., 2008), demonstrated ongoing land degradation in humid regions. Australia is the only country where dryland areas stood out. Sub-Saharan Africa later received confirmation of this global trend in land degradation (Vlek et al., 2010). With 13% of the world's current land degradation, Africa south of the equator experienced the most deterioration. Climate change is likely to be responsible for some of what was once thought to be anthropogenic land degradation near the Sahara (De Jong et al., 2011). From a methodological perspective, the use of NDVI as a stand-in for land degradation has also drawn criticism. New worldwide estimates of land degradation based on expert opinion are currently being developed, taking into account soil variables, ecosystem services, and land-use classifications (FAO, 2011).

1.4.8 Effects on Biodiversity

Because of how complicated ecosystem activities are, it is currently impossible to model the level of biodiversity needed to support agricultural production. Agro-ecosystems are therefore considered more resilient to environmental changes when a high level of biodiversity is maintained (Koohafkan et al., 2012; Lin, 2011). Diversification of agro-ecosystems, high genetic variety of crops, control of soil organic matter, integration of livestock and crop production, and water conservation are all farming techniques that lessen sensitivity to environmental change. Crop variety boosts resilience to increasing climate variability and extreme events while reducing pest, disease, and weed outbreaks. In low-income settings, farms with a high level of biodiversity have been found to be more resilient to climate disasters, such as hurricanes and droughts (Altieri et al., 2015). Smallholder farmers in tropical regions are particularly vulnerable to climate variability, including erratic rainfall, and as a coping mechanism, they rely on agricultural biodiversity, such as planting a high diversity of crops each year, including many varieties of the same crop, using drought-tolerant crop varieties, changing the locations of crops, and planting trees to provide shade and to maintain humidity (Meldrum et al., 2013).

In some instances, the availability of food can be directly impacted by biodiversity loss in regions whose diets largely consist of wild foods, such as wild fruits and vegetables. Numerous ecosystem services, such as pollination, natural pest control, and functions offered by soil macro- and microorganisms, are strongly reliant on by field-grown crops and cattle. Pollinator populations have decreased during the past ten years because of a stressor that includes parasites, insecticides, and habitat degradation (Goulson et al., 2015). Since so many fruit and vegetable species depend on pollinators, a total loss of pollinators has been anticipated to result in a 23% reduction in the global fruit supply, a 16% reduction in vegetable production, and a 22% reduction in nut and seed production (Smith et al., 2015).

1.5 REMEDIES

It has been recommended to employ several remediation strategies to stop pollution in order to facilitate the speedy and efficient restoration of the environment that has already been impacted. Chemical treatments degrade contaminants and further modify their physicochemical properties, hence reducing the ecological danger associated with them. Beside this, physical approaches to soil reclamation have no effect on the physicochemical properties of the impurities accumulated in the surroundings that need to be removed. More crucially, biological approaches that depend on the biological activity of higher plants and microorganisms have the power to break down accumulated contaminants and ultimately result in their mineralization, immobilization, or elimination.

Regulations for the environment should be effective. Essential environmental regulating systems and policies, such as pollutant emission limits, pollution fees, and emission trading schemes, must be developed and improved by governments (Zhou *et al.*, 2018; Tai *et al.*, 2014). Renewable energy technology development and use must be firmly supported. Governments have a duty to the public to assist the renewable energy industry with resources, money, and technology. Agenda 2030 provides a framework for both the creation of a more environmentally friendly future for humanity and the responsible utilization of the resources of nature on which we depend (Barboza *et al.*, 2019). Several studies have indicated specific areas for investigation and creativity, including understanding the consequences on human and animal health, developing alternative resources for cleaning beaches and oceans, and reducing the use of plastics (Barnes, 2019). In summary, workshops, meetings, training sessions, and media use can all help to educate the public regarding the way to manage and strengthen the relationship between human community as well as the surroundings in a sustainable and integrated way.

Green agriculture has to be promoted and practiced. Green agriculture needs robust and sustainable agroecosystems that can handle long-term difficulties, together with well-organized management of natural resources, ecosystem services, and biodiversity (Tan *et al.*, 2022). Precision farming and other technical advancements in agriculture can be advantageous. For example, drought-resistant crops (Hu and Xiong, 2013) or crop varieties with greater concentrations and bioavailability of micronutrients (Bhullar and Gruissem, 2013) can be produced using novel plant breeding approaches. Utilizing geographic information systems, remote sensing, and GPS, precision farming technologies enable farmers to target the application of fertilizers and pesticides where they are most required.

Limiting and optimizing the type, amount, and timing of crop treatments are two management strategies for lowering the risk of water contamination caused by organic and inorganic fertilizers and pesticides. It has been demonstrated that setting up protection zones along surface watercourses, inside farms, and in buffer zones around farms can effectively stop pollutants from migrating to water bodies. Pesticide waste and empty containers need to be stored and disposed of according to safety regulations. Additionally, effective irrigation plans will decrease water return flows, which will significantly lessen the migration of pesticides and fertilizers into water bodies (Mateo-Sagasta and Burke, 2010). Measures to reduce soil erosion include contour plowing and banning the cultivation of soils with steep slopes (USEPA, 2003). The ability of conservation agriculture to reduce erosion has also been demonstrated.

Farmers have the option of changing farm management procedures, such as crop varieties, planting dates, irrigation techniques, and residue management, or implementing significant systemic changes, such as switching to different crop species and changing farming systems, or even moving agriculture to new areas (Challinor *et al.*, 2014). Farmers and societies have a variety of options for adjusting to and minimizing environmental changes (FAO, 2010; FAO, 2012). These procedures can involve small changes up to significant system-level alterations, and they can take place at different levels. Agriculture and food production sectors can put adaptation strategies into place that will guarantee increasing production of high-quality food while putting less strain on the environment. Food production could be revolutionized by cellular agriculture, or adaptation of cell culturing methods for agricultural production. Acellular and cellular goods are both produced by cellular agriculture. Live cells are used to manufacture cellular goods like cultured beef or leather, while yeast or bacteria are cultivated to produce the protein that ends up in the product, such as milk protein or egg albumin (Post, 2012).

Research needs to focus more on the paths of emerging agricultural pollutants such as animal hormones, antibiotics, and other pharmaceuticals and the threats they represent to human community and the biosphere. For instance, more knowledge is required regarding how animal medications contribute to the growing issue of pathogen antibiotic resistance. The following strategies can be applied to reduce emerging contaminants: (1) GHG emissions from rumen fermentation and stored manure will be reduced by increasing the intake of digestible fodder (Hristov *et al.*, 2013). (2) Encouragement should be given to reducing the application of fertilizers and pesticides and introducing natural pest-control techniques. (3) Due to the ease with which they can enter a fragile ecosystem, pollution releases at the point of production should be minimized. (4) Methane is converted into things like sugars before it is emitted into the environment by methanotrophs, sometimes referred to as methane-eating bacteria, which exist in the surroundings where methane is created. (5) Biochemical techniques for reducing greenhouse gas emissions. (6) Restoration of degraded pastureland using conventional methods or procedures. (7) The creation of sound environmental regulations to prevent the unintentional discharge of developing toxins, such as water quality requirements, enhanced pollution and emission testing, and analyses of the environmental impact of farms and irrigation systems.

It is obvious that preventing or limiting the export of pollutants from where they are applied is the most efficient strategy to reduce stresses on aquatic ecosystems and rural ecosystems more broadly. Once pollutants are in an environment, the costs of mitigation skyrocket. Remediation of contaminated waters, such as lakes and aquifers, is a lengthy, expensive, and occasionally impractical project.

1.6 CONCLUSION

An overview of pollution and its sources, impacts, and prevention strategies has been provided in this chapter. Developed and developing nations share the burden of pollution, but because of weak laws, low knowledge, and extreme poverty, the latter are more impacted than the former. The most vulnerable people in middle-class and lower-class nations suffer disproportionately from pollution. Moreover, air pollution seems to be the type of contamination that has received the most attention and study. This may be owing to increasing premature deaths and illnesses associated with air pollution. To be able to repair an already destroyed ecosystem, pollution awareness must be raised, and all hands must be on board to stop activities which contribute to environmental pollution.

Many anthropogenic activities, including experiments that contaminate the land, are the source of agricultural contamination. An urgent need exists for a tiered strategy in the evaluation of the breakdowns of ecosystems and biodiversity with polluted soils. Emerging agricultural contaminants contribute to climate alteration through global warming in both positive and negative ways. By addressing the root causes, such as the occasions that emitted important greenhouse gases, the impact can be lessened. Biological remediation techniques that utilize microorganisms have been deemed among the safest for both the environment and people among all other remediation techniques.

REFERENCES

Adelekan, B. & K. Abegunde. 2011. "Heavy metals contamination of soil and groundwater at automobile mechanic villages in Ibadan, Nigeria". *International Journal of Physical Sciences*, 6, no. 2: 1045–1058.

Agrios, G. N. 2004. *Plant Pathology*. 5th ed. Amsterdam: Elsevier, Academic Press.

Ainsworth, Elizabeth A. 2017. "Understanding and Improving Global Crop Response to Ozone Pollution." *Plant Journal: For Cell & Molecular Biology* 90, no. 5: 886–97. https://doi.org/10.1111/tpj.13298.

Ainsworth, Elizabeth A., and Stephen P. Long. 2005. "What Have We Learned from 15 Years of Free-Air CO2 Enrichment (FACE)? A Meta-analytic Review of the Responses of Photosynthesis, Canopy Properties and Plant Production to RisingCO_2." *New Phytologist* 165, no. 2: 351–71. https://doi.org/10.1111/j.1469-8137.2004.01224.x.

Ainsworth, Elizabeth A., Craig R. Yendrek, Stephen Sitch, William J. Collins, and Lisa D. Emberson. 2012. "The Effects of Tropospheric Ozone on Net Primary Productivity and Implications for Climate Change." *Annual Review of Plant Biology* 63: 637–61. https://doi.org/10.1146/annurev-arplant-042110-103829.

Almetwally, Alsaid Ahmed, May Bin-Jumah, and Ahmed A. Allam. 2020. "Ambient Air Pollution and Its Influence on Human Health and Welfare: An Overview." *Environmental Science & Pollution Research International* 27, no. 20: 24815–30. https://doi.org/10.1007/s11356-020-09042-2.

Altieri, M. A., C. I. Nicholls, A. Henao, and M. A. Lana. 2015. "Agro-Ecology and the Design of Climate Change-Resilient Farming Systems." *Agronomy for Sustainable Development* 35: 869–90.

Anderson, Pamela K., Andrew A. Cunningham, Nikkita G. Patel, Francisco J. Morales, Paul R. Epstein, and Peter Daszak. 2004. "Emerging Infectious Diseases of Plants: Pathogen Pollution, Climate Change and Agrotechnology Drivers." *Trends in Ecology & Evolution* 19, no. 10: 535–44. https://doi.org/10.1016/j.tree.2004.07.021.

Andrady, A. L., P. J. Aucamp, A. T. Austin *et al.* 2014. "Environmental Effects of Ozone Depletion and Its Interactions with Climate Change." *Assessment 2015*. https://www.ncbi.nlm.nih.gov/pmc/articles/PMC6400464/

Antwi-Agyei, P., A. J. Dougill, and L. C. Stringer 2015. "Barriers to Climate Change Adaptation: Evidence from Northeast Ghana in the Context of a Systematic Literature Review." *Climate Development* 7: 297–309.

Aragón, F. M., and J. P. Rud. 2016. "Polluting Industries and Agricultural Productivity: Evidence from Mining in Ghana." *Economic Journal* 126, no. 597: 1980–2011. https://doi.org/10.1111/ecoj.12244.

Araújo da Silva, Francijara, Eliana Feldberg, Natália Dayane Moura Carvalho, Sandra Marcela Hernández Rangel, Carlos Henrique Schneider, Gislene Almeida Carvalho-Zilse, Victor Fonsêca da Silva, and Maria Claudia Gross. 2019. "Effects of Environmental Pollution on the rDNAomics of Amazonian Fish." *Environmental Pollution* 252, no. A: 180–7. https://doi.org/10.1016/j.envpol.2019.05.112.

Audsley, E., and M. Wilkinson. 2014. "What Is the Potential for Reducing National Greenhouse Gas Emissions from Crop and Livestock Production Systems?" *Journal of Cleaner Production* 73: 263–8. https://doi.org/10.1016/j.jclepro.2014.01.066.

Aune, Dagfinn, Edward Giovannucci, Paolo Boffetta, Lars T. Fadnes, NaNa Keum, Teresa Norat, Darren C. Greenwood, Elio Riboli, Lars J. Vatten, and Serena Tonstad. 2017. "Fruit and Vegetable Intake and the Risk of Cardiovascular Disease, Total Cancer and All-Cause Mortality: A Systematic Review and Dose–Response Meta-analysis of Prospective Studies." *International Journal of Epidemiology* 46, no. 3: 1029–56. https://doi.org/10.1093/ije/dyw319.

Auta, H. S., C. U. Emenike, and S. H. Fauziah. 2017. "Distribution and Importance of Microplastics in the Marine Environment: A Review of the Sources, Fate, Effects, and Potential Solutions." *Environment International* 102: 165–76. https://doi.org/10.1016/j.envint.2017.02.013.

Avnery, S., D. L. Mauzerall, J. F. Liu, and L. W. Horowitz. 2011. "Global Crop Yield Reductions Due to Surface Ozone Exposure: 1. Year 2000 Crop Production Losses and Economic Damage." *Atmospheric Environment* 45, no. 13: 2284–96. https://doi.org/10.1016/j.atmosenv.2010.11.045.

Ayers, R. S., and D. W. Westcot. 1985. *Water Quality for Agriculture:* 174. FAO Irrigation and Drainage Paper 29 Rev 1. Rome, Italy: Food and Agriculture Organization of the United Nations.

Ayliffe, Michael, Ravi Singh, and Evans Lagudah. 2008. "Durable Resistance to Wheat Stem Rust Needed." *Current Opinion in Plant Biology* 11, no. 2: 187–92. https://doi.org/10.1016/j.pbi.2008.02.001.

Backlund, P., D. Schimel, A. Janetos *et al.* 2008. *Introduction: The Effects of Climate Change on Agriculture, Land Resources, Water Resources, and Biodiversity in the United States*: 362. A Report by the US Climate Change Science Program. Washington, DC: Subcommittee on Global Change Research.

Bai, Z. G., D. L. Dent, L. Olsson, and M. E. Schaepman. 2008. "Proxy Global Assessment of Land Degradation." *Soil Use and Management* 24: 223–34.

Banon, S., J. Miralles, J. Ochoa, J. A. Franco, and M. J. Sanchez-Blanco. 2011. "Effects of Diluted and Undiluted Wastewater on the Growth. Physiological Aspects and Visual Quality of Potted Lantana and Polygala Plants." *Scientific Horticulture* 129: 869–76.

Barboza, L. G. A., A. Cozar, B. C. G. Gimenez, T. L. Barros, P. J. Kershaw, and L. Guilhermino. 2019. "Macroplastics Pollution in the Marine Environment." In *World Seas: An Environmental Evaluation*, edited by C. Shepicprard: 305328. Cambridge, MA: Academic Press. http://doi.org/10.1016/b978-0-12-805052-1.00019-x.

Barnes, Stuart J. 2019. "Understanding Plastics Pollution: The Role of Economic Development and Technological Research." *Environmental Pollution* 249: 812–21. https://doi.org/10.1016/j.envpol.2019.03.108.

Barrett, C. B., T. Reardon, and P. Webb. 2001. "Nonfarm Income Diversification and Household Livelihood Strategies in Rural Africa: Concepts, Dynamics and Policy Implications." *Food Policy* 26, no. 4: 315–31.

Bartram, J., and S. Cairncross 2010. "Hygiene, Sanitation, and Water: Forgotten Foundations of Health." *PLoS Med* 7, no. 11: e1000367. https://doi.org/10.1371/journal.pmed.1000367.

Bebber, D. P., M. A. T. Ramotowski, and S. J. Gurr. 2013. "Crop Pests and Pathogens Move Polewards in a Warming World." *Nature Climate Change* 3, no. 11: 985–8. https://doi.org/10.1038/nclimate1990.

Bezirtzoglou, Christos, Konstantinos Dekas, and Ekatherina Charvalos. 2011. "Climate Changes, Environment and Infection: Facts, Scenarios and Growing Awareness from the Public Health Community Within Europe." *Anaerobe* 17, no. 6: 337–40. https://doi.org/10.1016/j.anaerobe.2011.05.016.

Bhullar, Navreet K., and Wilhelm Gruissem. 2013. "Nutritional Enhancement of Rice for Human Health: The Contribution of Biotechnology." *Biotechnology Advances* 31, no. 1: 50–7. https://doi.org/10.1016/j.biotechadv.2012.02.001.

Black, P. F., J. G. Murray, and M. J. Nunn. 2008. "Managing Animal Disease Risk in Australia: The Impact of Climate Change." *Revue Scientifique & Technique* 27, no. 2: 563–80. https://doi.org/10.20506/rst.27.2.1815.

Boxall, A. B. A. 2012. "New and Emerging Water Pollutants Arising from Agriculture." In *Water Quality and Agriculture: Meeting the Policy Challenge.* OECD Report. http://www.oecd.org/agriculture/water.

Bridges, E. M., and L. R. Oldeman. 1999. "Global Assessment of Human-Induced Soil Degradation." *Arid Soil Research and Rehabilitation* 13, no. 4: 319–25. https://doi.org/10.1080/089030699263212.

Bumb, B., and C. Baanante. 1996. *World Trends in Fertilizer Use and Projections to 2020.* Washington, DC: International Food Policy Research Institute.

Cabral, J. P. 2010. "Water Microbiology. Bacterial Pathogens and Water." *International Journal of Environmental Research and Public Health* 7: 3657–703.

Caldwell, M. M., J. F. Bornman, C. L. Ballaré, S. D. Flint, and G. Kulandaivelu. 2007. "Terrestrial Ecosystems, Increased Solar Ultraviolet Radiation, and Interactions with Other Climate Change Factors." *Photochemical & Photobiological Sciences* 6, no. 3: 252–66. https://doi.org/10.1039/b700019g.

Cameron, D., C. Osborne, P. Horton, and M. Sinclair 2015. "A Sustainable Model for Intensive Agriculture." Technical Report. The University of Sheffield, Sheffield.

Carlos, Ku-Vera Juan, Rafael Jiménez-Ocampo, Sara Stephanie Valencia-Salazar, María Denisse Montoya-Flores, Isabel Cristina Molina-Botero, Jacobo Arango, Carlos Alfredo Gómez-Bravo, Carlos Fernando Aguilar-Pérez, and Francisco Javier Solorio-Sánchez. 2020. "Role of Secondary Plant Metabolites on Enteric Methane Mitigation in Ruminants." *Frontiers in Veterinary Science* 7, no. 584. https://doi.org/10.3389/fvets.2020.0058.

Challinor, A. J., J. Watson, D. B. Lobell, S. M. Howden, D. R. Smith, and N. Chhetri. 2014. "A Meta-analysis of Crop Yield Under Climate Change and Adaptation." *Nature Climate Change* 4, no. 4: 287–91. https://doi.org/10.1038/nclimate2153.

Chepurnykh, N. V., and O. I. Osmanov. 1988. "Two Approaches to Calculating Economic Losses in Crop Production Due to Contamination of the Atmosphere by Industry." *Vestnik Sel'skokhozyaistvennoi Nauki Moscow. USSR* 4: 73–81.

Chua, Kaw Bing. 2003. "Nipah Virus Outbreak in Malaysia." *Journal of Clinical Virology* 26, no. 3: 265–75. https://doi.org/10.1016/s1386-6532(02)00268-8.

Crane, M., and J. M. Giddings. 2004. "'Ecologically Acceptable Concentrations' When Assessing the Environmental Risks of Pesticides Under European Directive 91 414/EEC." *Human & Ecological Risk Assessment* 10, no. 4: 733–47. https://doi.org/10.1080/10807030490484237.

Das, H. K., A. K. Mitra, P. K. Sen Gupta, A. Hossain, F. Islam, and G. H. Rabbani. 2004. "Arsenic Concentrations in Rice, Vegetables, and Fish in Bangladesh: A Preliminary Study." *Environment International* 30, no. 3: 383–7. https://doi.org/10.1016/j.envint.2003.09.005.

Das, S., D. Pal, and Abhijit Sarkar. 2021. "Chapter 4." *Particulate Matter Pollution and Global Agricultural Productivity* V: 79–107. https://doi.org/10.1007/978-3-030-63249-6_4.

De Jong, R., S. De Bruin, M. Schaepman, and D. Dent. 2011. "Quantitative Mapping of Global Land Degradation Using Earth Observations." *International Journal of Remote Sensing* 32, no. 21: 6823–53.

Delgado, C. M., H. Rosegrant, S. Steinfeld, and C. C. Ehui. 1999. "Livestock to 2020: The Next Food Revolution." In *Journal of Food, Agriculture & Environment.* Discussion Paper No. 28. Washington, DC: International Food Policy Research Institute.

Doetterl, S., K. Van Oost, and J. Six. 2012. "Towards Constraining the Magnitude of Global Agricultural Sediment and Soil Organic Carbon Fluxes." *Earth Surface Processes & Landforms* 37, no. 6: 642–55. http://doi.org/10.1002/esp.3198.

Dominski, Fábio Hech, Joaquim Henrique Lorenzetti Branco, Giorgio Buonanno, Luca Stabile, Manuel Gameiro da Silva, and Alexandro Andrade. 2021. "Effects of Air Pollution on Health: A Mapping Review of Systematic Reviews and Meta-analyses." *Environmental Research* 201: 111487. https://doi.org/10.1016/j.envres.2021.111487.

Dubowitz, Tamara, Melonie Heron, Chloe E. Bird, Nicole Lurie, Brian K. Finch, Ricardo Basurto-Dávila, Lauren Hale, and José J. Escarce. 2008. "Neighborhood Socioeconomic Status and Fruit and Vegetable Intake Among Whites, Blacks, and Mexican Americans in the United States." *American Journal of Clinical Nutrition* 87, no. 6: 1883–91. https://doi.org/10.1093/ajcn/87.6.1883.

Eckersten, H., A. Herrmann, A. Kornher, M. Halling, E. Sindhøj, and E. Lewan. 2012. "Predicting Silage Maize Yield and Quality in Sweden as Influenced by Climate Change and Variability." *ActaAgriculturaeScandinavica, Section B—Soil & Plant Science* 62, no. 2: 151–65. https://doi.org/10.1080/09064710.2011.585176.

Ediin, G., E. Golantu and M. Brown. 2000. "Essentials for health and wellness". Bartlett publishers, Toronto, Canada pp: 368.

Evans, Jens, Graham Wood, and Anne Miller. 2006. "The Risk Assessment-Policy Gap: An Example from the UK Contaminated Land Regime." *Environment International* 32, no. 8: 1066–71. https://doi.org/10.1016/j.envint.2006.06.002.

Ewuzie, U., I. C. Nnorom, and S. O. Eze. 2020. "Lithium in Drinking Water Sources in Rural and Urban Communities in Southeastern Nigeria." *Chemosphere* 245. https://doi.org/10.1016/j.chemosphere.2019.125593.

FAO (Food and Agriculture Organization). 2011. *The State of the World's Land and Water Resources for Food and Agriculture (SOLAW)—Managing Systems at Risk.* Accessed January 30, 2013. http://www.fao.org/docrep/015/i1688e/i1688e00.pdf.

FAO. 2012. *FAOSTAT.* Accessed May 10, 2012. http://www.faostat.fao.org.

FAO. 2013. *Tackling Climate Change Through Livestock: A Global Assessment of Emissions and Mitigation Opportunities.* Rome: Food and Agricultural Organization of the United Nations (FAO).

FAO. 2014. *Building a Common Vision for Sustainable Food and Agriculture: Principles and Approaches.* Rome: Food and Agriculture Organization of the United Nations (FAO).

FAO, WFP, and IFAD. 2012. *The State of Food Insecurity in the World 2012. Economic Growth Is Necessary but Not Sufficient to Accelerate Reduction of Hunger and Malnutrition.* Rome: Food & Agriculture Organization of the United Nations (FAO). World Food Programme and International Fund for Agricultural Development.

FAO. 2006. *Livestock's Long Shadow.* Rome: Food and Agriculture Organization of the United Nations (FAO).

FAO. 2016a. *FAOSTAT: Database.* Rome: Food and Agriculture Organization of the United Nations (FAO). Accessed July 2016. http://faostat3.fao.org/browse/R/RP/E.

FAO. 2016b. *The State of World Fisheries and Aquaculture: Contributing to Food Security and Nutrition for All.* Rome: Food and Agriculture Organization of the United Nations (FAO).

FAO. 2010. *Climate Smart Agriculture: Policies, Practices and Financing for Food Security, Adaptation and Mitigation.* Rome, Italy: Food and Agriculture Organization of the United Nations (FAO).

FAO. 2018. "The future of food and agriculture. Alternative pathways to 2050". FAO, FAO, Rome. http://www.fao.org/3/I8429EN/i8429en.pdf

Feng, Y., J. Wang, Z. Bai, and L. Reading. 2019. "Effects of Surface Coal Mining and Land Reclamation on Soil Properties: A Review." *Earth-Science Reviews* 191: 12–25. https://doi.org/10.1016/j.earscirev.2019.02.015.

Fewtrell, L., D. Kay, W. Enanoria, L. Haller, R. B. Kaufmann, and J. M. Colford 2005. "Water, Sanitation and Hygiene Interventions to Reduce Diarrhoea in Developing Countries: A Systematic Review and Metaanalysis." *Lancet Infectious Diseases* 5, no. 1: 42–52.

Flood, J. 2010. "The Importance of Plant Health to Food Security." *Food Security* 2, no. 3: 215–31. https://doi.org/10.1007/s12571-010-0072-5.

Forbes, G. A., and M. C. Jarvis. 1994. "Host Resistance for Management of Potato Late Blight." In *Advances in Potato Pest Biology and Management*, edited by G. W. Zehnder, M. L. Powelson, R. K. Jansson, and K. V. Raman: 439–57. St. Paul, MN: American Phytopathological Society.

Forman, S., N. Hungerford, M. Yamakawa, T. Yanase, H. J. Tsai, Y. S. Joo, D. K. Yang, and J. J. Nha. 2008. "Climate Change Impacts and Risks for Animal Health in Asia." *Revue Scientifique & Technique* 27, no. 2: 581–97. https://doi.org/10.20506/rst.27.2.1814.

Forman, S., F. Le Gall, D. Belton, B. Evans, J. L. François, G. Murray, D. Sheesley, A. Vandersmissen, and S. Yoshimura. 2009. "Moving Towards the GlobalControl of Foot and Mouth Disease: An Opportunity for Donors." *Revue Scientifique & Technique* 28, no. 3: 883–96. https://doi.org/10.20506/rst.28.3.1935.

Forster, P., V. Ramaswamy, P. Artaxo *et al.* 2007. "Changes in Atmospheric Constituents and in Radiative Forcing." In *Climate Change*, edited by S. Solomon, D. Qin, M. Manning *et al.*: 212, Table 2.14. The Physical Science Basis. Contribution of Working Group I to the Fourth Assessment Report of the Intergovernmental Panel on Climate Change. Cambridge, UK and New York, NY: Cambridge University Press.

Garrett, K. A., G. A. Forbes, S. Savary, P. Skelsey, A. H. Sparks, C. Valdivia, A. H. C. van Bruggen, L. Willocquet, A. Djurle, E. Duveiller, H. Eckersten, S. Pande, C. Vera Cruz, and J. Yuen. 2011. "Complexity in Climate-Change Impacts: An Analytical Framework for Effects Mediated by Plant Disease." *Plant Pathology* 60, no. 1: 15–30. https://doi.org/10.1111/j.1365-3059.2010.02409.x.

Gibbs, H. K., and J. M. Salmon 2015. "Mapping the World's Degraded Lands." *Applied Geography* 57: 12–21.

Githeko, A. K., S. W. Lindsay, U. E. Confalonieri, and J. A. Patz. 2000. "Climate Change and Vector-Borne Diseases: A Regional Analysis." *Bulletin of the World Health Organization* 78, no. 9: 1136–47.

Goulson, D., B. Nicholls, C. Botías, and E. L. Rotheray. 2015. "Bee Declines Driven by Combined Stress from Parasites, Pesticides, and Lack of Flowers." *Science* 347: 1435.

Graham, Jay P., Jessica H. Leibler, Lance B. Price, Joachim M. Otte, Dirk U. Pfeiffer, T. Tiensin, and Ellen K. Silbergeld. 2008. "The Animal-Human Interface and Infectious Disease in Industrial Food Animal Production: Rethinking Biosecurity and Biocontainment." *Public Health Reports* 123, no. 3: 282–99. https://doi.org/10.1177/003335490812300309.

Guo, Y. H., Y. P. Yu, D. Wang, C. A. Wu, G. D. Yang, J. G. Huang, and C. C. Zheng. 2009. "GhZFP1, a Novel CCCH-Type Zinc Finger Protein from Cotton, Enhances Salt Stress Tolerance and Fungal Disease Resistance in Transgenic Tobacco by Interacting with GZIRD21A and GZIPR5." *New Phytology* 183, no. 1: 62–75. https://doi.org/10.1111/j.1469-8137.2009.02838.x.

Harrus, S., and G. Baneth. 2005. "Drivers for the Emergence and Reemergence of Vector-Borne Protozoal and Bacterial Diseases." *International Journal for Parasitology* 35, no. 11–12: 1309–18. https://doi.org/10.1016/j.ijpara.2005.06.005.

Herbert, E. R., P. Boon, A. J. Burgin, S. C. Neubauer, R. B. Franklin, M. Ardón, K. N. Hopfensperger, L. P. M. Lamers, and P. Gell. 2015. "A Global Perspective on Wetland Salinization: Ecological Consequences of a Growing Threat to Freshwater Wetlands." *Ecosphere* 6, no. 10: 1–43. https://doi.org/10.1890/ES14-00534.1.

Hoffman, G. J., P. B. Catlin, R. M. Mead, R. S. Johnson, L. E. Francois, and D. Goldhamer. 1989. "Yield and Foliar Injury Responses of Mature Plum Trees to Salinity." *Irrigation Science* 10, no. 3: 215–29. https://doi.org/10.1007/BF00257954.

Hoffman, G. J. 2010. *Salt Tolerance of Crops in the Southern Sacramento-San Joaquin Delta*. Final Report. Sacramento, CA: For California Environmental Protection Agency State Water Resources Control Board, Division of Water Rights.

Holoubek, I., D. Ladislav, S. Milan, H. Jakub, P. Čupr, J. Jiří, J. Zbíral, and J. Klánová. 2009. "Soil Burdens of Persistent Organic Pollutants – Their Levels, Fate and Risk. Part I. Variation of Concentration Ranges According to Different Soil Uses and Locations." *Environmental Pollution* 157, no. 12: 3207–17.

Hristov, A. N., J. Oh, C. Lee, R. Meinen, F. Montes, T. Ott, J. Firkins, A. Rotz, C. Dell, A. Adesogan, W. Yang, J. Tricarico, E. Kebreab, G. Waghorn, J. Dijkstra, and S. Oosting. 2013. "Mitigation of Greenhouse Gas Emissions in Livestock Production—A Review of Technical Options for Non-CO$_2$ Emissions." In *Makkar: FAO Animal Production and Health*, edited by Pierre J. Gerber, Benjamin Henderson, and P. S. Harinder: Paper no. 177. Rome, Italy: Food and Agriculture Organization.

Hu, C., and W. Xiong. 2013. *Are Commodity Futures Prices Barometers of the Global Economy?* NBER Working Paper Series. 19706. National Bureau of Economic Research, Inc., Cambridge, MA 02138, USA.

IGRAC. 2009. *Global Overview of Saline Groundwater Occurrence and Genesis*. Report No. GP 2009-1. Utrecht, The Netherlands: International Groundwater Resources Assessment Centre.

Jaggard, Keith W., Aiming Qi, and Eric S. Ober. 2010. "Possible Changes to Arable Crop Yields by 2050." *Philosophical Transactions of the Royal Society of London. Series B, Biological Sciences* 365, no. 1554: 2835–51. https://doi.org/10.1098/rstb.2010.0153.

Jeppesen, E., S. Brucet, L. Naselli-Flores, E. Papastergiadou, K. Stefanidis, T. Nõges, P. Nõges, J. L. Attayde, T. Zohary, J. Coppens, T. Bucak, R. F. Menezes, F. R. S. Freitas, M. Kernan, M. Søndergaard, and M. Beklioğlu. 2015. "Ecological Impacts of Global Warming and Water Abstraction on Lakes and Reservoirs Due to Changes in Water Level and Related Changes in Salinity." *Hydrobiologia* 750, no. 1: 201–27. https://doi.org/10.1007/s10750-014-2169-x.

Kalantari, M. R. 2006. "Soil Pollution by Heavy Metals and Its Remediation Mazandaran (Iran)." *Journal of Applied Sciences* 6: 2110–6.

Kaufman, Joan A. 2008. "China's Heath Care System and Avian Influenza Preparedness." *Journal of Infectious Diseases* 197(Suppl. 1): S7–13. https://doi.org/10.1086/524990.

Khan, Aneire Ehmar, Pauline Franka Denise Scheelbeek, Asma Begum Shilpi, Queenie Chan, Sontosh Kumar Mojumder, Atiq Rahman, Andy Haines, and Paolo Vineis. 2014. "Salinity in Drinking Water and the Risk of (Pre)Eclampsia and Gestational Hypertension in Coastal Bangladesh: A Casecontrol Study." *PLOS ONE* 9, no. 9: E108715. https://doi.org/10.1371/journal.pone.0108715.

Khan, A., S. Khan, M. A. Khan, Z. Qamar, and M. Waqas. 2015. "The Uptake and Bioaccumulation of Heavy Metals by Food Plants, Their Effects on Plants Nutrients, and Associated Health Risk: A Review." *Environment Science & Pollution Research International* 22, no. 18: 13772–99. https://doi.org/10.1007/s11356-015-4881-0.

Kjellstrom, Tord, David Briggs, Chris Freyberg, Bruno Lemke, Matthias Otto, and Olivia Hyatt. 2016. "Heat, Human Performance, and Occupational Health: A Key Issue for the Assessment of Global Climate Change Impacts." *Annual Review of Public Health* 37: 97–112. https://doi.org/10.1146/annurev-publhealth-032315-021740.

Klimont, Z., S. J. Smith, and J. Cofala. 2013. "The Last Decade of Global Anthropogenic Sulfur Dioxide: 2000–2011 Emissions." *Environment Research Letters* 8: 014003. https://doi.org/10.1088/1748-9326/8/1/014003.

Koohafkan, P., M. A. Altieri, and E. H. Gimenez. 2012. "Green Agriculture: Foundations for Biodiverse, Resilient and Productive Agricultural Systems." *International Journal of Agriculture & Sustainability* 10, no. 1: 61–75. https://doi.org/10.1080/14735903.2011.610206.

Landrigan, P. J., R. Fuller, N. J. R. Acosta, O. Adeyi, R. Arnold, N. Basu *et al.* 2017. "The Lancet Commission on Pollution and Health." *Lancet* 391, no. 10119, 462512: 32345. https://doi.org/10.1016/S0140-6736(17).

Leon Paumen, M. 2008. *Invertebrate Life Cycle Responses to PAC Exposure*. PhD thesis. University of Amsterdam, Amsterdam.

Li, F. R., S. L. Peng, B. M. Chen, and Y. Hou. 2010. "A Meta-analysis of the Responses of Woody and Herbaceous Plants to Elevated Ultraviolet-B Radiation." *Acta Oecologica* 36, no. 1: 1–9. https://doi.org/10.1016/j.actao.2009.09.002.

Lin, B. B. 2011. "Resilience in Agriculture Through Crop Diversification: Adaptive Management for Environmental Change." *Bioscience* 61, no. 3: 183–93. https://doi.org/10.1525/bio.2011.61.3.4.

Lindgren, E., A. Albihn, and Y. Andersson. 2011. "Climate Change, Water Related Health Impacts, and Adaptation: Highlights from the Swedish Government's Commission on Climate and Vulnerability." In *Climate Change Adaptation in Developed Nations—from Theory to Practice*. Vol. 42, edited by J. D. Ford and L. Berrang-Ford. Dordrecht: Springer Science C Business Media B.V. https://doi.org/10.1007/978-94-007-0567-8_12.

Lobell, David B., and Sharon M. Gourdji. 2012. "The Influence of Climate Change on Global Crop Productivity." *Plant Physiology* 160, no. 4: 1686–97. https://doi.org/10.1104/pp.112.208298.

Lobell, David B., Marshall B. Burke, Claudia Tebaldi, Michael D. Mastrandrea, Walter P. Falcon, and Rosamond L. Naylor. 2008. "Prioritizing Climate Change Adaptation Needs for Food Security in 2030." *Science* 319, no. 5863: 607–10. https://doi.org/10.1126/science.1152339.

Lorenz, J. J. 2014. "A Review of the Effects of Altered Hydrology and Salinity on Vertebrate Fauna and Their Habitats in Northeastern Florida Bay." *Wetlands* 34, no. S1: 189–200. https://doi.org/10.1007/s13157-013-0377-1.

Luo, Xiaosan, Haijian Bing, Zhuanxi Luo, Yujun Wang, and Ling Jin. 2019. "Impacts of Atmospheric Particulate Matter Pollution on Environmental Biogeochemistry of Trace Metals in Soil-Plant System: A Review." *Environmental Pollution* 255, no. 1: 113138. https://doi.org/10.1016/j.envpol.2019.113138.

Mason, R. P., W. F. Fitzgerald, and F. M. M. Morel. 1994. "The Biogeochemical Cycling of Elemental Mercury: Anthropogenic Influences." *Geochimica et Cosmochimica Acta* 58: 3191–8.

Mateo-Sagasta, J., and J. Burke. 2010. *Agriculture and Water Quality Interactions: A Global Overview*. SOLAW Background Thematic Report-TR08. Rome: Food and Agriculture Organization of the United Nations (FAO).

Meldrum, J., S. Nettles-Anderson, G. Heath, and J. Macknick. 2013. "Life Cycle Water Use for Electricity Generation: A Review and Harmonization of Literature Estimates." *Environmental Research Letters* 8, no. 1: 015031.

Miller, Victoria, Salim Yusuf, Clara K. Chow, Mahshid Dehghan, Daniel J. Corsi, Karen Lock, Barry Popkin, Sumathy Rangarajan, Rasha Khatib, Scott A. Lear, Prem Mony, Manmeet Kaur, Viswanathan Mohan, Krishnapillai Vijayakumar, Rajeev Gupta, Annamarie Kruger, Lungiswa Tsolekile, Noushin Mohammadifard, Omar Rahman, Annika Rosengren, Alvaro Avezum, Andrés Orlandini, Noorhassim Ismail, Patricio Lopez-Jaramillo, Afzalhussein Yusufali, Kubilay Karsidag, Romaina Iqbal, Jephat Chifamba, Solange Martinez Oakley, Farnaza Ariffin, Katarzyna Zatonska, Paul Poirier, L. Wei, B. Jian, Chen Hui, Liu Xu, Bai Xiulin, Koon Teo, and Andrew Mente. 2016a. "Availability, Affordability, and Consumption of Fruits and Vegetables in 18 Countries Across Income Levels: Findings from the Prospective Urban Rural Epidemiology (PURE) Study." *Lancet. Global Health* 4, no. 10: E695–703. https://doi.org/10.1016/S2214-109X(16)30186-3.

Mills, Gina, Katrina Sharps, David Simpson, Håkan Pleijel, Michael Frei, Kent Burkey, Lisa Emberson, Johan Uddling, Malin Broberg, Zhaozhong Feng, Kazuhiko Kobayashi, and Madhoolika Agrawal. 2018. "Closing the Global Ozone Yield Gap: Quantification and Cobenefits for Multistress Tolerance." *Global Change Biology* 24, no. 10: 4869–93. https://doi.org/10.1111/gcb.14381.

Miraglia, M., H. J. Marvin, G. A. Kleter, P. Battilani, C. Brera, E. Coni, F. Cubadda, L. Croci, B. De Santis, S. Dekkers, L. Filippi, R. W. Hutjes, M. Y. Noordam, M. Pisante, G. Piva, A. Prandini, L. Toti, G. J. van den Born, and A. Vespermann. 2009. "Climate Change and Food Safety: An Emerging Issue with Special Focus on Europe." *Food& Chemical Toxicology* 47, no. 5: 1009–21. https://doi.org/10.1016/j.fct.2009.02.005.

Mirzaei-Ag, A., S. Alireza Sy, H. Fathi, S. Rasouli, M. Sadaghian, and M. Tarahomi. 2012. "Garlic in Ruminants Feeding." *Asian Journal of Biological Sciences* 5, no. 7: 328–40. https://doi.org/10.3923/ajbs.2012.328.340.

Montanarella, L. 2007. "Trends in land degradation in Europe". In Climate and Land Degradation; *Environmental Science and Engineering*, Sivakumar, M.V.K. & N. Ndiang'ui Eds.; Springer: Berlin/Heidelberg, Germany, pp. 83–104.

Moore, F. C., and M. C. MacCracken. 2009. "Lifetime-Leveraging: An Approach to Achieving International Agreement and Effective Climate Protection Using Mitigation of Short-Lived Greenhouse Gases." *International Journal of Climate Change Strategies & Management* 1, no. 1: 42–62. https://doi.org/10.1108/17568690910934390.

Muralikrishna, I. V., and V. Manickam. 2017. "Analytical Methods for Monitoring Environmental Pollution." In *Environmental Management*: 495570. Oxford: Butterworth-Heinemann, Elsevier. https://doi.org/10.1016/b978-0-12-811989-1.00018-x.

Murnyak, George, John Vandenberg, Paul J. Yaroschak, Larry Williams, Krishnan Prabhakaran, and John Hinz. 2011. "Emerging Contaminants: Presentations at the 2009 Toxicology and Risk Assessment Conference." *Toxicology & Applied Pharmacology* 254, no. 2: 167–9. https://doi.org/10.1016/j.taap.2010.10.021.

Myers, Samuel S., Matthew R. Smith, Sarah Guth, Christopher D. Golden, Bapu Vaitla, Nathaniel D. Mueller, Alan D. Dangour, and Peter Huybers. 2017. "Climate Change and Global Food Systems: Potential Impacts on Food Security and Undernutrition." *Annual Review of Public Health* 38: 259–77. https://doi.org/10.1146/annurev-publhealth-031816-044356.

Nardone, A., B. Ronchi, N. Lacetera, M. S. Ranieri, and U. Bernabucci. 2010. "Effects of Climate Changes on Animal Production and Sustainability of Livestock Systems." *Livestock Science* 130, no. 1–3: 57–69. https://doi.org/10.1016/j.livsci.2010.02.011.

Nawrot, Tim, Michelle Plusquin, Janneke Hogervorst, Harry A. Roels, Hilde Celis, Lutgarde Thijs, Jaco Vangronsveld, Etienne Van Hecke, and Jan A. Staessen. 2006. "Environmental Exposure to Cadmium and Risk of Cancer: A Prospective Population-Based Study." *Lancet. Oncology* 7, no. 2: 119–26. https://doi.org/10.1016/S1470-2045(06)70545-9.

Nkonya, E., N. Gerber, P. Baumgartner, J. von Braun, A. De Pinto, V. Graw, E. Kato, J. Kloos, and T. Walter. 2011. *The Economics of Land Degradation – Towards an Integrated Global Assessment*. Switzerland: Peter Lang.

NORMAN. 2016. *List of Emerging Substances: Network of Reference Laboratories, Research Centres and Related Organisations for Monitoring of Emerging Environmental Substances*. NORMAN. http://www.norman-network.net/?q=node/19.

OECD. 2012b. *New and Emerging Water Pollutants Arising from Agriculture, Prepared by Alistair B. A. Boxall*. Paris: Organization for Economic Co-operation and Development (Organization for Economic Co-operation and Development) Publishing.

Okrent, D. 1999. "On Intergenerational Equity and Its Clash with Intragenerational Equity and on the Need for Policies to Guide the Regulation of Disposal of Wastes and Other Activities Posing Very Long Time Risks." *Risk Analysis* 19, no. 5: 877–901. https://doi.org/10.1023/a:1007014510236.

Pacifici, M., W. B. Foden, P. Visconti, J. E. M. Watson, S. H. M. Butchart, K. M. Kovacs, B. R. Scheffers, D. G. Hole, T. G. Martin, H. R. Akçakaya, R. T. Corlett, B. Huntley, D. Bickford, J. A. Carr, A. A. Hoffmann, G. F. Midgley, P. Pearce-Kelly, R. G. Pearson, S. E. Williams, S. G. Willis, B. Young, and C. Rondinini. 2015. "Assessing Species Vulnerability to Climate Change." *Nature Climate Change* 5, no. 3: 215–24. https://doi.org/10.1038/nclimate2448.

Pan, Jilang L., Jane A. Plant, Nikolaos Voulvoulis, Christopher J. Oates, and Christian Ihlenfeld. 2010. "Cadmium Levels in Europe: Implications for Human Health." *Environmental Geochemistry & Health* 32, no. 1: 1–12. https://doi.org/10.1007/s10653-009-9273-2.

Patra, Amlan K., and Jyotisna Saxena. 2010. "A New Perspective on the Use of Plant Secondary Metabolites to Inhibit Methanogenesis in the Rumen." *Phytochemistry* 71, no. 11–12: 1198–222. https://doi.org/10.1016/j.phytochem.2010.05.010.

Patterson, Bradley M., Elizabeth Cohen, Henning Prommer, David G. Thomas, Stuart Rhodes, and Allan J. McKinley. 2007. "Origin of Mixed Brominated Ethane Groundwater Plume: Contaminant Degradation Pathways and Reactions." *Environmental Science & Technology* 41, no. 4: 1352–8. https://doi.org/10.1021/es0615674.

Paustian, K., J. M. Antle, J. Sheehan, and E. A. Paul. 2006. *Agriculture's Role in Greenhouse Gas Mitigation*: 3. Arlington, VA: Pew Center on Global Climate Change.

Pessoa, Milene Cristine, Larissa Loures Mendes, Crizian Saar Gomes, Paula Andréa Martins, and Gustavo Velasquez-Melendez. 2015. "Food Environment and Fruit and Vegetable Intake in a Urban Population: A Multilevel Analysis." *BMC Public Health* 15: 1012. https://doi.org/10.1186/s12889-015-2277-1.

Pinto, J., C. Bonacic, C. Hamilton-West, J. Romero, and J. Lubroth. 2008. "Climate Change and Animal Diseases in South America." *Revue Scientifique & Technique* 27, no. 2: 599–613. https://doi.org/10.20506/rst.27.2.1813.

Porter, J. R., L. Xie, A. J. Challinor *et al.* 2014. "Food Security and Food Production Systems." In *Climate Change: Impacts, Adaptation, and Vulnerability. Part, A. 2014. Global and Sectoral Aspects. Contribution of Working Group II to the Fifth Assessment Report of the Intergovernmental Panel on Climate Change*, edited by C. B. Field, V. R. Barros, D. J. Dokken *et al.*: 485–533. Cambridge, UK and New York, NY: Cambridge University Press.

Porter, J. R., L. Xie, A. J. Challinor *et al.* 2014. "Food Security and Food Production Systems." In *Climate Change: Impacts, Adaptation, and Vulnerability. Part, A. 2014. Global and Sectoral Aspects. Contribution of Working Group II to the Fifth Assessment Report of the Intergovernmental Panel on Climate Change*, edited by C. B. Field, V. R. Barros, D. J. Dokken *et al.*: 485–533. Cambridge, UK and New York, NY: Cambridge University Press.

Post, Mark J. 2012. "Cultured Meat from Stem Cells: Challenges and Prospects." *Meat Science* 92, no. 3: 297–301. https://doi.org/10.1016/j.meatsci.2012.04.008.

Prasad, M. N. V. 1995. "Cadmium Toxicity and Tolerance in Vascular Plants." *Environmental & Experimental Botany* 35, no. 4: 525–45. https://doi.org/10.1016/0098-8472(95)00024-0.

Richardson, G. M., D. A. Bright, and M. Dodd. 2006. "Do Current Standards of Practice in Canada Measure What Is Relevant to Human Exposure at Contaminated Sites? II: Oral Bioaccessibility of Contaminants in Soil." *Human & Ecological Risk Assessment* 12, no. 3: 606–16. https://doi.org/10.1080/10807030600561824.

Rockström, J., W. Steffen, K. Noone, Å. Persson, F. S. I. Chapin, III, E. Lambin, T. M. Lenton, M. Scheffer, C. Folke, H. J. Schellnhuber, B. Nykvist, C. A. de Wit, T. Hughes, S. van der Leeuw, H. Rodhe, S. Sörlin, P. K. Snyder, R. Costanza, U. Svedin, M. Falkenmark, L. Karlberg, R. W. Corell, V. J. Fabry, J. Hansen, B. Walker, D. Liverman, K. Richardson, P. Crutzen, and J. Foley. 2009. "Planetary Boundaries: Exploring the Safe Operating Space for Humanity." *Ecology & Society* 14, no. 2: 32. https://doi.org/10.5751/ES-03180-140232.

Rosegrant, M. W., M. Ewing, G. Yohe, I. Burton, S. Huq, and R. Valmonte Santos. 2008. *Climate Change and Agriculture Threats and Opportunities*: 1–36. http://www.gtz.de/climate.

Rossiter, Paul B., and Najib Al Hammadi. 2009. "Living with Transboundary Animal Diseases (TADs)." *Tropical Animal Health & Production* 41, no. 7: 999–1004. https://doi.org/10.1007/s11250-008-9266-7.

Ruiz-Vera, Ursula M., Matthew Siebers, Sharon B. Gray, David W. Drag, David M. Rosenthal, Bruce A. Kimball, Donald R. Ort, and Carl J. Bernacchi. 2013. "Global Warming Can Negate the Expected CO2 Stimulation in Photosynthesis and Productivity for Soybean Grown in the Midwestern United States." *Plant Physiology* 162, no. 1: 410–23. https://doi.org/10.1104/pp.112.211938.

SACEP. 2014. *Nutrient Loading and Eutrophication of Coastal Waters of the South Asian Seas—A Scoping Study*. Washington, DC: South Asian Co-Operative Environmental Programme (SACEP).

Sartori, R., R. Sartor-Bergfelt, S. A. Mertens, J. N. Guenther, J. J. Parrish, and M. C. Wiltbank. 2002. "Fertilization and Early Embryonic Development in Heifers and Lactating Cows in Summer and Lactating and Dry Cows in Winter." *Journal of Dairy Science* 85, no. 11: 2803–12. https://doi.org/10.3168/jds.S0022-0302(02)74367-1.

Scheelbeek, Pauline F. D., Muhammad A. H. Chowdhury, Andy Haines, Dewan S. Alam, Mohammad A. Hoque, Adrian P. Butler, Aneire E. Khan, Sontosh K. Mojumder, Marta A. G. Blangiardo, Paul Elliott, and Paolo Vineis. 2017. "Drinking Water Salinity and Raised Blood Pressure: Evidence from a Cohort Study in Coastal Bangladesh." *Environmental Health Perspectives* 125, no. 5: 057007. https://doi.org/10.1289/EHP659.

Schraufnagel, D. E., J. Balmes, C. T. Cowl, S. De Matteis, S.-H. Jung, K. Mortimer *et al.* 2018. "Air Pollution and Non-communicable Diseases: A Review by the Forum of International Respiratory Societies' Environmental Committee, Part 1: The Damaging Effects of Air Pollution." *Chest* 155, no. 2: 409416. https://doi.org/10.1016/j.chest.2018.10.042.

Schreinemachers, P., and P. Tipraqsa. 2012. "Agricultural Pesticides and Land Use Intensification in High, Middle and Low Income Countries." *Food Policy* 37, no. 6: 616–26. https://doi.org/10.1016/j.foodpol.2012.06.003.

Semenov, M. A., and P. R. Shewry. 2011. "Modelling Predicts That Heat Stress, Not Drought, Will Increase Vulnerability of Wheat in Europe." *Scientific Reports* 1, no. 1: 66. https://doi.org/10.1038/srep00066.

Shah, M. I., U. Muhammad, O. O. Hephzibah, and A. Shujaat. 2022. "Nexus between Environmental Vulnerability and Agricultural Productivity in BRICS: What Are the Roles of Renewable Energy, Environmental Policy Stringency, and Technology?" *Environmental Science & Pollution Research. Online* 30, no. 6, 15756–74.

Sherman, David M. 2010. "A Global Veterinary Medical Perspective on the Concept of One Health: Focus on Livestock." *ILAR Journal* 51, no. 3: 281–7. https://doi.org/10.1093/ilar.51.3.281.

Sillmann, J., K. Aunan, L. Emberson, P. Büker, B. Van Oort, C. O'Neill, N. Otero, D. Pandey, and A. Brisebois. 2021. "Combined Impacts of Climate and Air Pollution on Human Health and Agricultural Productivity." *Environmental Research Letters* 16, no. 9: 093004. https://doi.org/10.1088/1748-9326/ac1df8.

Sims, L. D., J. Domenech, C. Benigno, S. Kahn, A. Kamata, J. Lubroth, V. Martin, and P. Roeder. 2005. "Origin and Evolution of Highly Pathogenic H5N1 Avian Influenza in Asia." *Veterinary Record* 157, no. 6: 159–64. https://doi.org/10.1136/vr.157.6.159.

Singh, B. R., and O. Singh. 2012. "Study of Impacts of Global Warming on Climate Change: Rise in Sea Level and Disaster Frequency." In *Global Warming-Impacts and Future Perspective*, edited by Bharat Raj Singh. InTech (online). https://doi.org/10.5772/50464.

Skovgaard, Niels. 2007. "New Trends in Emerging Pathogens." *International Journal of Food Microbiology* 120, no. 3: 217–24. https://doi.org/10.1016/j.ijfoodmicro.2007.07.046.

Smith, P., M. Bustamante, H. Ahammad *et al.* 2014. "Agriculture, Forestry and Other Land Use (AFOLU)." In *Climate Change. Mitigation of Climate Change. Contribution of Working Group III to the Fifth Assessment Report of the Intergovernmental Panel on Climate Change*, edited by O. Edenhofer, R. Pichs-Madruga, and Y. Sokona. Cambridge, UK and New York, NY: Cambridge University Press.

Smith, S. M., T. E. Nichols, D. Vidaurre, A. M. Winkler, T. E. J. Behrens, M. F. Glasser, K. Ugurbil, D. M. Barch, D. C. Van Essen, and K. L. Mille 2015. "A Positive-Negative Mode of Population Covariation Links Brain Connectivity, Demographics and Behavior." *Nature Neuroscience* 18, no. 11: 1565–7.

Snow, D. D., S. L. Bartelt-Hunt, S. E. Saunders, and D. A. Cassada. 2007. "Detection, Occurrence, and Fate of Emerging Contaminants in Agricultural Environments Water." *Environmental Research* 79, no. 10, Literature Reviews [CD-ROM content]: 1061–84(24 pages). Published By: Wiley. https://doi.org/10.2307/29763261; http://www.jstor.com/stable/29763261.

Solomon, Susan, Diane J. Ivy, Doug Kinnison, Michael J. Mills, Ryan R. Neely, and Anja Schmidt. 2016. "Emergence of Healing in the Antarctic Ozone Layer." *Science* 353, no. 6296: 269–74. https://doi.org/10.1126/science.aae0061.

Solomon, S., D. Qin, M. Manning *et al.* 2007. "The Physical Science Basis: Technical Summary." In *Climate Change*, edited by S. Solomon, D. Qin, M. Manning *et al.*: 33, Table TS.2. Contribution of Working Group I to the Fourth Assessment Report of the Intergovernmental Panel on Climate Change. New York, NY: Cambridge University Press. http://www.ipcc.ch/pdf/assessmentreport/ar4/wg1/ar4-wg1-ts.pdf.

Song, Lulu, Bin Zhang, Bingqing Liu, Mingyang Wu, Lina Zhang, Lulin Wang, Shunqing Xu, Zhongqiang Cao, and Youjie Wang. 2019. "Effects of Maternal Exposure to Ambient Air Pollution on Newborn Telomere Length." *Environment International* 128: 254–60. https://doi.org/10.1016/j.envint.2019.04.064.

Springmann, Marco, Daniel Mason-D'Croz, Sherman Robinson, Tara Garnett, H. Charles J. Godfray, Douglas Gollin, Mike Rayner, Paola Ballon, and Peter Scarborough. 2016a. "Global and Regional Health Effects of Future Food Production Under Climate Change: A Modelling Study." *Lancet* 387, no. 10031: 1937–46. https://doi.org/10.1016/S0140-6736(15)01156-3.

Steinfeld, H., T. Wassenaar, and S. Jutzi. 2006. "Livestock Production Systems in Developing Countries: Status, Drivers, Trends." *Revue Scientifique & Technique* 25, no. 2: 505–16. https://doi.org/10.20506/rst.25.2.1677.

Stuthman, D. D., K. J. Leonard, and J. Miller-Garvin. 2007. "Breeding Crops for Durable Resistance to Disease." *Advances in Agronomy* 95: 319–67. https://doi.org/10.1016/S0065-2113(07)95004-X.

Su, Shaw-Wei, Chun-Chih Tsui, Hung-Yu Lai, and Zueng-Sang Chen. 2014. "Food Safety and Bioavailability Evaluations of Four Vegetables Grown in the Highly Arsenic-Contaminated Soils on the Guandu Plain of Northern Taiwan." *International Journal of Environmental Research & Public Health* 11, no. 4: 4091–107. https://doi.org/10.3390/ijerph110404091.

Tadesse, W., Y. Manes, R. P. Singh, T. Payne, and H. J. Braun. 2010. "Adaptation and Performance of CIMMYT Spring Wheat Genotypes Targeted to High Rainfall Areas of the World." *Crop Science* 50, no. 6: 2240–8. https://doi.org/10.2135/cropsci2010.02.0102.

Tai, A. P., M. V. Martin, and C. L. Heald. 2014. "Threat to Future Global Food Security from Climate Change and Ozone Air Pollution." *National Climate Change* 4: 817–21.

Tan, Y., W. Xu, S. Li, and K. Chen. 2022. "Augmented and Virtual Reality (AR/VR) for Education and Training in the AEC Industry: A Systematic Review of Research and Applications." *Buildings* 12, no. 10: 1529.

Tarazona, J. V., M. D. Fernandez, and M. M. Vega. 2005. "Regulation of Contaminated Soils in Spain—A New Legal Instrument (4 Pp)." *Journal of Soils & Sediments* 5, no. 2: 121–4. https://doi.org/10.1065/jss2005.05.135.

Thebo, A. L., P. Drechsel, E. F. Lambin, and K. L. Nelson. 2017. "A Global, Spatially Explicit Assessment of Irrigated Croplands Influenced by Urban Wastewater Flows." *Environmental Research Letters* 12, no. 7: 074008. https://doi.org/10.1088/1748-9326/aa75d1.

Thompson, D., P. Muriel, D. Russell, P. Osborne, A. Bromley, M. Rowland, S. Creigh-Tyte, and C. Brown. 2002. "Economic Costs of the Foot and Mouth Disease Outbreak in the United Kingdom in 2001." *Revue Scientifique & Technique* 21, no. 3: 675–87. https://doi.org/10.20506/rst.21.3.1353.

Thornton, Philip K. 2010. "Livestock Production: Recent Trends, Future Prospects." *Philosophical Transactions of the Royal Society of London. Series B, Biological Sciences* 365, no. 1554: 2853–67. https://doi.org/10.1098/rstb.2010.0134.

Thornton, P. K., J. van de Steeg, A. Notenbaert, and M. Herrero. 2009. "The Impacts of Climate Change on Livestock and Livestock Systems in Developing Countries: A Review of What We Know and What We Need to Know." *Agricultural Systems* 101, no. 3: 113–27. https://doi.org/10.1016/j.agsy.2009.05.002.

Tirado, M. C., R. Clarke, L. A. Jaykus, A. McQuatters-Gollop, and J. M. Frank. 2010. "Climate Change and Food Safety: A Review." *Food Research International* 43, no. 7: 1745–65. https://doi.org/10.1016/j.foodres.2010.07.003.

Toccalino, Patricia L., and Julia E. Norman. 2006. "Health-Based Screening Levels to Evaluate US Geological Survey Groundwater Quality Data." *Risk Analysis* 26, no. 5: 1339–48. https://doi.org/10.1111/j.1539-6924.2006.00805.x.

Tsiamis, Demetra A., Melissa Torres, and Marco J. Castaldi. 2018. "Role of Plastics in Decoupling Municipal Solid Waste and Economic Growth in the US" *Waste Management* 77: 147–55. https://doi.org/10.1016/j.wasman.2018.05.003.

US Environmental Protection Agency. *Frequently Asked Questions About Global Warming and Climate Change: Back to Basics*: 3. http://www.epa.gov/climatechange/downloads/Climate_Basics.pdf.

UNDESA. 2017. *World Population Prospects. 2017 Revision, Key Findings and Advance Tables.* Working Paper No. ESA/P/WP/248. New York: United Nations Department of Economic and Social Affairs Population Division.

UNEP. 2016. *A Snapshot of the World's Water Quality: Towards a Global Assessment.* Nairobi: United Nations Environment Programme (UN Environmental Program).

Urzelai, A., M. Vega, and E. Angulo. 2000. "Deriving Ecological Risk-Based Soil Quality Values in the Basque Country." *Science of the Total Environment* 247, no. 2–3: 279–84. https://doi.org/10.1016/s0048-9697(99)00497-0.

US EPA. 2016. *Water Quality Assessment and TMDL Information.* Washington, DC: United States Environmental Protection Agency (US EPA). https://ofmpub.epa.gov/waters10/attains_index.home.

US EPA. 2003. *National Management Measures to Control Nonpoint Source Pollution from Agriculture.* Washington, DC: United States Environmental Protection Agency (US EPA).

Vallero, D. J., and D. A. Vallero. 2019. "Land Pollution." In *Waste*, edited by M. T. Letcher and D. A. Vallero: 631648. Cambridge, MA: Academic Press. http://doi.org/10.1016/b978-0-12-815060-3.00032-3.

Van den Bossche, P., and J. A. Coetzer. 2008. "Climate Change and Animal Health in Africa." *Revue Scientifique & Technique* 27, no. 2: 551–62. https://doi.org/10.20506/rst.27.2.1816.

Van Zorge, J. A. 1996. "Exposure to Mixtures of Chemical Substances: Is There a Need for Regulations?" *Food & Chemical Toxicology* 34, no. 11–12: 1033–6. https://doi.org/10.1016/s0278-6915(97)00071-9.

Varshney, Rajeev K., Kailash C. Bansal, Pramod K. Aggarwal, Swapan K. Datta, and Peter Q. Craufurd. 2011. "Agricultural Biotechnology for Crop Improvement in a Variable Climate: Hope or Hype?" *Trends in Plant Science* 16, no. 7: 363–71. https://doi.org/10.1016/j.tplants.2011.03.004.

Vázquez-Arias, A., F. J. Martín-Peinado, and A. Parviainen. 2023. "Effect of Parent Material and Atmospheric Deposition on the Potential Pollution of Urban Soils close to Mining Areas." *Journal of Geochemical Exploration* 244: 107131. https://doi.org/10.1016/j.gexplo.2022.107131.

Vlek, P. L. G., Q. B. Le, and L. Tamene. 2010. "Assessment of Land Degradation, Its Possible Causes and Threat to Food Security in Sub-Saharan Africa." In *Food Security and Soil Quality*, edited by R. Lal and B. A. Stewart, 57–86. Boca Raton, FL: CRC/Taylor and Francis.

Walker, T. R., O. Adebambo, M. C. Del AguilaFeijoo, E. Elhaimer, T. Hossain, S. J. Edwards *et al.* 2019. "Environmental Effects of Marine Transportation." In *World Seas: An Environmental Evaluation*, edited by C. Sheppard: 505530. Cambridge, MA: Academic Press. http://doi.org/10.1016/b978-0-12-805052-1.00030-9.

Wang, T., L. Zhang, S. Zhou, T. Zhang, S. Zhai, Z. Yang, D. Wang, and H. Song. 2021. "Effects of Ground-Level Ozone Pollution on Yield and Economic Losses of Winter Wheat in Henan, China." *Atmospheric Environment* 262: 118654. https://doi.org/10.1016/j.atmosenv.2021.118654.

Wang, W. T., S. M. Simonich, M. Xue, J. Y. Zhao, N. Zhang and R. Wang. 2010. "Concentrations, sources and spatial distribution of polycyclic aromatic hydrocarbons in soils from Beijing, Tianjin and surrounding areas, North China". *Environmental Pollution* 158: 1245–51.

Wellings, C. R. 2007. "Puccinia Striiformis in Australia: A Review of the Incursion, Evolution, and Adaptation of Stripe Rust in the Period 1979–2006." *Australian Journal of Agricultural Research* 58, no. 6: 567–75. https://doi.org/10.1071/AR07130.

WHO (World Health Organization). 2018. *Global Health Observatory (GHO) Data, Mortality from Household Air Pollution.* Geneva, Switzerland: World Health Organization.

WHO. 2012. *Animal Waste, Water Quality and Human Health.* Geneva, Switzerland: World Health Organization.

WHO. 2006. "Guidelines for the Safe Use of Wastewater, Excreta and Greywater." In *Wastewater Use in Agriculture.* Vol. II. Geneva, Switzerland: World Health Organization.

Williams, D. R., and D. Matthews. 1988. "Outbreaks of Classical Swine Fever in Great Britain in 1986." *Veterinary Record* 122, no. 20: 479–83. https://doi.org/10.1136/vr.122.20.479.

WRI. 2008. *Eutrophication and Hypoxia in Coastal Areas: A Global Assessment of the Sate of Knowledge.* WRI Policy Note. Washington, DC: World Resources Institute (Washington Research Institute).

WRI. 2008. *Eutrophication and Hypoxia in Coastal Areas: A Global Assessment of the Sate of Knowledge.* WRI Policy Note. Washington, DC: World Resources Institute (Washington Research Institute).

WRI. 2013. *The United Nations World Water Development Report 2013.* Paris: United Nations.

World Water Assessment Programme, United Nations Educational, Scientific and Cultural Organization (WWAP). 2015. *The United Nations World Water Development Report 2015: Water for a Sustainable World. United Nations World Water Assessment Programme (WWAP).* Paris: United Nations Educational, Scientific and Cultural Organization.

WWAP. 2017. *The United Nations World Water Development Report 2017: Wastewater, the Untapped Resource. United Nations World Water Assessment Programme (WWAP).* Paris: United Nations Educational, Scientific and Cultural Organization.

Xiong, Wei, Ping Ni, Yiyong Chen, Yangchun Gao, Shiguo Li, and Aibin Zhan. 2019. "Biological Consequences of Environmental Pollution in Running Water Ecosystems: A Case Study in Zooplankton." *Environmental Pollution* 252, no. B: 1483–90. https://doi.org/10.1016/j.envpol.2019.06.055.

Zadoks, J. C. 2008. "The Potato Murrain on the European Continent and the Revolutions of 1848." *Potato Research* 51, no. 1: 5–45. https://doi.org/10.1007/s11540-008-9091-4.

Zhang, W., F. Jiang, and J. Ou. 2011. "Global Pesticide Consumption and Pollution: With China as a Focus." *Proceedings of the International Academy of Ecology and Environmental Sciences* 1, no. 2: 125–44.

Zhao, Y., S. Wang, K. Aunan, H. M. Seip, and J. Hao 2006. "Air Pollution and Lung Cancer Risks in China: A Meta-Analysis." *Science of Total Environment* 366: 500–13.

Zhou, J., Y. Yang, X. Qiu, X. Yang, H. Pan, B. Ban *et al.* 2018. "Serial Multiple Mediation of Organizational Commitment and Job Burnout in the Relationship Between Psychological Capital and Anxiety in Chinese Female Nurses: A Cross-Sectional Questionnaire Survey." *International Journal of Nursing Studies* 83: 75–82. https://doi.org/10.1016/j.ijnurstu.2018.03.016.

2 Phytoremediation of Heavy Metal-Contaminated Soil

Innocent Ojeba Musa, Udeme Joshua Josiah Ijah,
Olabisi Peter Abioye, Mustapha Abdulsalam,
Sanjoy Kumar Pal, Ikhumetse Agatha Abamhekhelu,
Asmau M. Maude, Oluwasola Olatunji Yusuf,
and Akande Sikirula A.

2.1 INTRODUCTION

The proliferation of heavy metal pollution in the biosphere has arisen as a major environmental problem since the beginning of the industrial revolution. Heavy metal pollution poses significant dangers to human health as well as the integrity of soil and water resources, according to Yoon *et al.* (2016). According to Singh and Singh (2017), yearly heavy metal releases in previous decades totaled 22,000 tons of Cd, 939,000 tons of Cu, 135,000 tons of Zn, and 738,000 tons of Pb. This pollution is caused by a mix of natural and human processes, such as fuel generation, mining, smelting, industrial emissions, military actions, and pesticide usage in agriculture (Jadia and Fulekar, 2009). Furthermore, the use of phosphate fertilizers has resulted in higher levels of Cd, Cu, Zn, and As in agricultural soil (Zarcinas *et al.*, 2014). This tendency is compounded by rising agricultural product demand, which demands the use of agrochemicals such as wetsuits, herbicides, and fertilizers. However, overuse of these agrochemicals adds to environmental problems by allowing hazardous compounds to accumulate in the soil and be absorbed by plants (Sahibin *et al.*, 2015; Musa *et al.*, 2021a).

Microorganisms, unlike living organisms, lack the ability to counteract the presence of these heavy metals. Heavy metal toxicity's long-term environmental persistence is a major problem, with a half-life of more than 20 years (Ruiz *et al.*, 2018). According to the United States Environmental Action Group (USEAG), this environmental problem endangers the health of roughly ten million individuals in several countries. Heavy metal pollution has become a worldwide issue, owing to the fact that heavy metals are composed of 53 unique elements and are prevalent contaminants in industrial settings, with concentrations nearing 5 g cm^{-3}. Throughout history, human civilizations have faced a variety of environmental threats. Nonetheless, toxic metal poisoning of the biosphere has been on the increase since the beginning of the industrial revolution, presenting significant hazards to human health and the ecology. Contaminants migrate as dust or leachate into previously uncontaminated regions as a consequence of both managed and uncontrolled waste disposal, accidental spills, mining, smelting of metalliferous ores, and the use of sewage sludge in agriculture. Heavy metals are organic and inorganic compounds that include petroleum products, hazardous waste, combustible chemicals, putrescible materials, and explosives. Logan (2017) and Alloway (2016) show that the ecosystem is more sensitive to metal contamination than organic pollution.

Because organic contaminants may be biodegraded by soil bacteria, heavy metal pollutants must be physically removed or immobilized. These heavy metals may act as substitutes for key elements in pigments or enzymes, impairing normal function and leading to heavy metal toxicity (Henry, 2000). Heavy metal poisoning has been linked to livestock losses, which may have a negative economic impact on a country, since these poisons tend to accumulate in plants and animals,

DOI: 10.1201/9781003423393-2

accumulating throughout the food chain and negatively impacting certain human organs (Bondada and Ma, 2003). The existence of heavy metal toxicity is still a persistent problem in the present industrial period. Soil pollution or contamination is defined as the level of chemical modification in soil that is harmful to the surrounding environment and its people (FAO, 2015). Organic molecules, heavy metals, salts, and oils are examples of chemicals that either do not break down naturally or disintegrate at such a slow pace that they accumulate in the biosphere, inflicting permanent damage to the environment (Haller, 2017).

While natural occurrences such as air deposition and water seeping in dry places may contribute to soil pollution, human activity is the predominant cause. Human-caused soil pollution, such as inappropriate waste management, leaching, industrial spills, and previous fertilizer usage, is one of the primary contributors to environmental challenges in developing countries (FAO, 2018).

There are several soil remediation processes available that may either completely eradicate contaminant concentrations in the soil or lower them to ecologically acceptable levels. Polluted soils were previously cleaned using in-situ procedures such as pollution containment with caps and liners or ex-situ methods such as excavation. Nonetheless, these methods are often costly, resource intensive, and harmful to natural ecosystems (Jonsson and Haller, 2014; Musa *et al.*, 2021a).

Plants were pushed for wastewater treatment roughly 300 years ago; therefore, the idea of using plants to repair damaged settings is not a new one. *Thlaspi caerulescens* and *Viola calaminaria* were the first plant species to be discovered for their potential to collect large metal levels in their leaves in the late 1800s (Baumann, 2015). Subsequent study has shown that Astragalus plants' dry shoot biomass may retain up to 0.6% selenium (Byers, 2017). The occurrence of plants that hyperaccumulate metals other than Cd, Ni, Se, and Zn, on the other hand, is still under research, with reports claiming the finding of Co, Cu, and Mn hyperaccumulators (Salt *et al.*, 2017). Chaney (1989) originally proposed and investigated the use of plants to remove metals from polluted soil, with the first field investigation on Zn and Cd phytoextraction undertaken in 1991 (Baker *et al.*, 1991). While we have made tremendous progress in understanding the biology of metal phytoextraction over the past decade, our knowledge of the plant processes that promote metal absorption remains limited. Furthermore, important practical problems remain unsolved, specifically how agronomic practices influence a plant's ability to absorb metals. The commercialization of phytoextraction will almost certainly need a thorough knowledge of plant systems as well as the use of suitable agronomic strategies. The investigation of these systems is especially exciting owing to the prevalence of plant species that are natively suited to collect very large amounts of metal.

Phytoremediation is a term that refers to a group of natural processes that include phytovolatilization, phytoextraction, phytostabilization, and phytodegradation. The process by which a plant absorbs contaminants via its roots and then translocates and deposits them in its leaves, blooms, and stems is referred to as phytoextraction. Following that, the plant is harvested to eliminate any contaminants that have been absorbed. The gathered biomass may be disposed of in a waste management facility by burning, compaction, or composting, possibly yielding bioenergy. The resultant ash, known as "bio-ore," might be a substantial source of revenue. This method is often used for toxins that are non-biodegradable, have little environmental effect, and are reasonably cheap to handle (in non-time-sensitive settings).

Plant roots stabilize and confine the movement of pollutants inside the soil during phytostabilization, also known as in-situ inactivation. This method efficiently inhibits the formation of leachate and consequent soil erosion while retaining contaminants in the soil.

2.2 ENVIRONMENTAL POLLUTION AND SOURCES OF CONTAMINATION

The presence of dangerous quantities of toxins in the air, water, and soil causes environmental pollution. This pollution arises when waste is purposefully or unintentionally discharged into the environment as a result of human activity. It includes both organic and inorganic toxins that have

2.2.1 Pollution from Heavy Metals

The buildup of considerable amounts of heavy metals may be harmful to both plant and animal life. While heavy metals such as iron (Fe), zinc (Zn), copper (Cu), cobalt (Co), chromium (Cr), manganese (Mn), and nickel (Ni) are required for specific metabolic processes in organisms, excessive doses of these metals are hazardous. Metals such as lead (Pb), mercury (Hg), cadmium (Cd), and arsenic (As) are especially hazardous since no known biological activities need their presence. Human exposure to these metals may occur via a variety of sources, including contaminated food, water, or air, and can result in health problems such as emphysema, osteoporosis, renal failure, and brain damage. Even modest amounts of cadmium exposure, for example, may elicit flu-like symptoms, but chronic contact can cause considerable lung damage (Jadia and Fulekar, 2009).

The intensity and duration of heavy metal poisoning symptoms are determined by both the degree of exposure and the time of exposure. Mining, for example, is a typical source of heavy metal discharge into the environment. Heavy metal pollution is created by human activities such as mining, smelting, agricultural fertilizer runoff, and industrial waste. These sources may emit heavy metals into the soil and water, contaminating neighboring places over time. Because of the process of biomagnification, inhabitants of contaminated ecosystems, including flora and animals, transfer heavy metals up the food chain (Liphadzi and Kirkham, 2015; Jadia and Fulekar, 2009).

To address the health hazards connected with these metals, most nations create soil standards that determine the maximum permitted quantities of heavy metals in their soils. If the concentration of heavy metals in soil exceeds a certain limit, the soil is deemed contaminated and must be remedied to a safe level. Determining the acceptable limit for heavy metals in soil may be difficult since it is dependent on a number of variables. The projected future use of the contaminated land is one of the factors to consider, with lower limits necessary if people are expected to have frequent access. Furthermore, the threshold concentration is determined by the land's function, whether it is for human consumption, animal feed, or biofuel generation in agriculture. Other factors that influence threshold concentration determination include soil pH and profile. Because heavy metals do not decay naturally, many nations and environmental or governmental authorities use significantly disparate criteria to assess acceptable amounts of heavy metal contamination in soil. As a result, phytoremediation procedures such as phytodegradation and phytovolatilization are rendered useless. Plants must absorb heavy metals via phytoextraction or immobilize them through phytostabilization in the event of phytodegradation. The chemical stability of heavy metals stays intact when cleanup activities are initiated. As a result, heavy metal pollution must still be regulated, especially in the case of phytoextraction, while phytostabilization just avoids leaching and confines heavy metals to the soil (Baker and Crompton, 2000).

It is critical to identify and use hyperaccumulator plant species to ensure the effectiveness of heavy metal soil phytoremediation activities. These plants, known as hyperaccumulators, have an unusual capacity to thrive and absorb significant amounts of pollutants from contaminated soils, even in situations where these compounds would normally inhibit development. They may absorb heavy metals at up to 500 times higher quantities than other plants (Qiu *et al.*, 2014). According to Baker and Brooks (1989), a plant's shoots must be capable of absorbing 100 mg/kg^{-1} of Cd to be designated as a cadmium (Cd) hyperaccumulator.

According to Bose and Hemantaranjan (2005), heavy metal pollution is the presence of dangerously concentrated heavy metal pollutants. Heavy metals have an elemental density more than 5 g/cm^3 (Bose and Hemantaranjan, 2005; Mishra *et al.*, 2019). Because of the negative consequences heavy metal pollution has on living creatures, researchers throughout the globe are paying close attention to it (Mishra *et al.*, 2019).

Heavy metal poisoning has resulted in enormous mortality in the field of human biology (Shrivastav *et al.*, 2014). While an excess of any heavy metal is harmful to living organisms, certain heavy metals, such as iron (Fe), copper (Cu), cobalt (Co), manganese (Mn), and zinc (Zn) for plants, and chromium (Cr), nickel (Ni), and tin (Sn) for animals, are required in small but critical amounts for normal development and reproduction (Bose and Hemantaranjan, 2005). However, in order for organisms to operate efficiently, very low and particular quantities of these metals must be maintained in living tissues (Mishra *et al.*, 2019). According to Voet *et al.* (2018), heavy metals are non-biodegradable and stay in the environment for long periods of time once introduced. They pose major dangers to people and other forms of life in the environment due to their toxicity, persistence, and non-biodegradability. Particulate matter contains the vast bulk of heavy metals in the atmosphere. As a result, the first step in the life cycle of atmospheric heavy metals is their deposition onto land or aquatic surfaces by dry, wet, or occult methods (Shrivastav *et al.*, 2014).

Heavy metals have limited environmental mobility, according to Adeleken and Abegunde, suggesting that a single incident of pollution might result in persistent exposure to heavy metals for microbes, plants, fauna, and other soil-dwelling animals. The problem of heavy metal pollution in the environment is not likely to go away fast but rather to remain for many generations owing to the legacy of massive industrial operations. In this context, it is critical to create historical and current databases of heavy metal concentrations in the atmosphere (Shrivastav *et al.*, 2014).

2.2.2 HEAVY METALS AND THEIR EFFECTS

2.2.2.1 Lead

Lead is a heavy metal with detrimental effects on people of all age groups. Particularly concerning are its consequences on the mental and physical development of children (Simeonov, 2010). Children exposed to lead may exhibit a lower intelligence quotient, increased fatigue, and irritability. Pregnant women exposed to lead face a higher risk of infertility, miscarriage, and still birth (Ediin *et al.*, 2000). Prolonged exposure to lead can result in adverse effects on physical development, including anemia, renal damage, headaches, speech difficulties, hearing loss, fatigue, and irritability (Simeonov *et al.*, 2013). Lead poisoning carries various biological consequences, such as enzyme deactivation, competition with calcium for bone integration, and interference with nerve impulses and brain growth (Ediin *et al.*, 2000).

The primary sources of lead (Pb) in the environment are numerous. These include tap water from soldered pipes, dust from leaded paints found in older homes, and historical use of leaded gasoline (Ediin *et al.*, 2000). Additionally, indoor pollutants and indoor smoking contribute to the presence of lead in the environment (Simeonov *et al.*, 2013).

The World Health Organization has established a maximum allowable pollutant level for lead in water, which is set at 0.01 mg/l. Research studies on heavy metal contamination in drinking water have reported varying levels of lead. For instance, Mebrahtu and Zerabruk (2011) used atomic absorption spectroscopy to analyze drinking water samples from urban areas of the Tigray region in northern Ethiopia, and their findings ranged from 1.347 mg/l at Indasilase to below the detection limit in other areas. Importantly, over 70.15% of these samples had lead concentrations within the WHO's maximum permissible threshold for lead in drinking water. Similarly, Kaplan *et al.* (2011) found lead levels in drinking water below the maximum allowable limit. It's important to note that lead levels can vary in different locations and are influenced by a range of factors.

Lead concentrations in soil also vary depending on the location. For example, a study by Micó *et al.* (2016) in a Mediterranean semiarid segura river valley in Spain found soil lead levels ranging from 8.9 to 34.5 mg/kg. Similarly, studies in Nigeria by Adei-Atiemo *et al.* (2015) measured lead levels in road soils ranging from 10.06 to 29.71 mg/kg and from 33.640 to 117.45 mg/kg, respectively. In urban areas like the Kirisia commercial area in Maralal town, metal plating and tire rubber were considered likely sources of lead (Cd) (Jaradat *et al.*, 2015). These variations highlight the influence of local factors and human activities on soil lead levels.

Phytoremediation of Heavy Metal–Contaminated Soil

2.2.2.2 Cadmium (Cd)

Cadmium is a highly mobile heavy metal in biological systems, primarily released into the atmosphere during combustion processes (Wieczorek *et al.*, 2014). Chronic exposure to cadmium has been linked to an increased risk of certain cancers, renal impairment, and disruptions in calcium metabolism. Cadmium can cause oxidative stress in plant cells and deactivate some enzymes (Wieczorek *et al.*, 2014). Plants initially accumulate cadmium in their roots before distributing it to other parts (Wieczorek *et al.*, 2014).

Environmental sources of cadmium include mining, smelting of metal ores, burning of fossil fuels, and the use of phosphate fertilizers. Additionally, cadmium is used in the production of nickel-cadmium rechargeable batteries, which can contribute to its presence in the environment. Cultivating tobacco, which accumulates cadmium in its tissues, can also raise environmental cadmium levels.

The World Health Organization has set a maximum contamination level for cadmium in water at 0.003 mg/l. Studies have reported varying cadmium levels in water, with some studies finding levels below the maximum allowable limit, such as those by Wogu and Okaka (2011) and Raji *et al.* (2015). On the other hand, research by Wogu and Okaka (2011) in the Delta area of Nigeria found cadmium levels in the water ranging from 0.0 to 0.04 mg/l, exceeding the allowed limit. In soil, Delbari and Kulkarni (2013) reported cadmium levels ranging from 0.000 to 0.004 mg/kg in the summer and 0.001 to 0.004 mg/kg in the winter.

2.2.2.3 Chromium

The concentration of chromium in the environment is gradually rising due to industrial expansion, particularly in the tanning, chemicals, and metals sectors. Two primary forms of chromium exist: chromium VI, which is carcinogenic, and chromium III, an essential element in human and animal diets. Chromium III plays a role in fat and carbohydrate metabolism, and its deficiency can lead to metabolic disruptions. Chronic exposure to chromium VI can result in health issues, including lung cancer, dermatitis, skin ulcers, bronchial asthma, and more (Sarkar *et al.*, 2015).

Chromium is found in various products, including cement, leather, plastics, dyes, textiles, paints, printing ink, cutting fluids, photography materials, detergents, and wood preservatives. It can also be emitted into the environment through water-eroded rocks, power plants, liquid fuels, hard and brown coal, and industrial and municipal waste. The non-biodegradability of chromium contributes to its persistence in the environment.

The maximum allowable limit for chromium in drinking water is 0.4 mg/l, per the WHO. Studies have reported varying levels of chromium in drinking water, with some exceeding the WHO limit, such as Mebrahtu and Zerabruk (2011) in Ethiopia. In soils, Adeleken and Abegunde (2011) reported chromium levels ranging from 2.0 to 29.75 mg/l, which fell within the limit set by the United Kingdom.

2.2.2.4 Manganese

Manganese is a necessary trace element for humans and animals, crucial for normal physiological function. However, exposure to higher doses over an extended period can lead to harmful health effects. Chronic exposure to elevated manganese levels can result in manganism, characterized by symptoms like tremors, weakness, and psychological disorders. The nervous system is the primary target organ for manganese toxicity.

Manganese can naturally occur in surface and groundwater and can also enter these sources through soil erosion. Human activities, such as power stations, iron and steel production, wood burning, and coal combustion, contribute to manganese emissions. Soil leaching, landfills, and underground injection from industrial facilities are the main sources of manganese emissions into surface and ground waters.

The WHO sets a maximum allowable level of 0.4 mg/l for manganese in drinking water. Research studies have reported manganese levels in drinking water, and findings vary by location.

For example, Mebrahtu and Zerabruk (2011) found manganese levels below the WHO's maximum acceptable range in northern Ethiopia. In Sokoto, Nigeria, Raji *et al.* (2016) reported varying manganese levels in drinking water at different stations, all within the WHO's allowable limit.

Soil manganese levels also differ by region. A study by Micó *et al.* (2016) in Spain reported varying soil manganese levels, while a study by Delbari and Kulkarni (2013) in Tehran found different concentrations in agricultural soils.

2.2.2.5 Zinc

Zinc is an essential trace element present in most food and drinkable water (Swaminathan *et al.*, 2019). Although zinc poisoning is rare, symptoms can occur at concentrations of up to 40 mg/l, including discomfort, muscle stiffness, and irritability (Al-Weher, 2008). Zinc plays a vital role in connective tissue formation. It's worth noting that the WHO has not specified a recommended quantity of zinc in drinking water.

Zinc in drinking water can be influenced by plumbing materials, and levels may occasionally exceed 0.1 mg/l due to this source (Swaminathan *et al.*, 2019). Anthropogenic sources of zinc in soil and water include the use of commercial products containing zinc, as well as industrial discharges and smelter slag.

Studies have reported varying zinc levels in both water and soil. For instance, research by Raji and Gopchandran (2017) in Sokoto, Nigeria, found zinc levels within the WHO's allowable limit. In other studies, zinc levels were reported at varying concentrations, such as in China by Jia *et al.* (2015) in soil and by Kar *et al.* (2017) in surface water. These findings illustrate the variability of zinc concentrations influenced by location and local factors.

2.2.3 Soil Contamination by Heavy Metals

Heavy metals find their way into the environment through two primary avenues: naturally occurring sources and human-induced activities. Interestingly, even natural occurrences of heavy metal pollution often have their roots in human interventions, creating a paradox (Wei *et al.*, 2018).

Heavy metals are elements characterized by an atomic number exceeding 20 and distinguished by metallic properties such as high density, electrical conductivity, and the capacity to exist as stable cations (Raskin *et al.*, 2014).

Heavy metal contamination is a global threat to ecosystems. It permeates various environmental components, including sediments, water bodies, soils, and biological organisms. Notably, substantial environmental contamination by heavy metals can be traced back to the late 19th and early 20th centuries, primarily stemming from mining and industrial activities. The United States Environmental Protection Agency (2017) underlines the detrimental effects of heavy metals such as cadmium (Cd), copper (Cu), chromium (Cr), zinc (Zn), nickel (Ni), and lead (Pb), particularly in areas with high human activity.

While heavy metals naturally occur in the Earth's crust, human actions have significantly disrupted their ecological balance and geological processes (Bacio *et al.*, 2013). Given their resistance to natural decomposition, these elements undergo transformations into new forms. Copper (Cu), iron (Fe), zinc (Zn), molybdenum (Mo), and manganese (Mn) are regarded as essential micronutrients for biological systems' ability to sustain life. Nevertheless, at higher concentrations, these metals transform into highly toxic compounds, endangering human and animal health by contaminating food, water, and air. Notably, cadmium (Cd), copper (Cu), nickel (Ni), and mercury (Hg) exhibit higher phytotoxicity compared to zinc and lead (Raskin *et al.*, 2014).

Soil pollution has emerged as a compelling concern, garnering substantial public attention in recent years. A significant portion of land has been rendered hazardous and unsuitable for human and animal habitation due to extensive pollution. Importantly, heavy metals are rarely found in pure form within soils, and their concentrations rise due to natural processes and human activities, significantly impacting ecosystems (Turan and Esringu, 2017).

Various sources contribute to the introduction of heavy metals into soil, including volcanic activity, parent materials, and industrial processes. The concentration of these metals in soil can vary widely, ranging from detectable traces to levels as high as 100,000 mg kg^{-1}, contingent upon the specific metal and its location. When the quantity of an element in the soil surpasses permissible concentrations, it is deemed heavy metal pollution in soils, carrying detrimental consequences for the local ecosystem. It's important to recognize that every ecosystem inherently contains baseline levels of heavy metals. Nevertheless, anthropogenic emissions can elevate these concentrations beyond natural background levels, culminating in pollution. Surface soils may become contaminated over time as they accumulate heavy metals through mechanisms such as automobile emissions and anomalous enrichment (Fong *et al.*, 2018).

Research indicates that soils beneath or adjacent to landfills and agricultural fields irrigated with polluted water often host high concentrations of heavy metals. Furthermore, soil pollution, both short-term and long-term, exerts adverse effects on soil microbial and enzymatic activities. The toxicity and mobility of heavy metals within soils are influenced by various factors, including the unique chemical form of the metals, bonding states, metal properties, ambient conditions, soil features, and organic matter content.

Children, characterized by active digestive systems and heightened hemoglobin sensitivity, are particularly susceptible to heavy metal exposure. Such exposure significantly increases the ingestion of soil particles laden with metals through hand-to-mouth activities. However, adults are not exempt from risk, as inhalation serves as a more direct pathway for harmful metals to enter the body (Fong *et al.*, 2018).

2.2.3.1 Sources of Contamination and Remediation Strategies

Broadly, heavy metals, defined by atomic numbers exceeding 100 and metallic properties such as ductility, conductivity, cationic stability, and ligand selectivity, contribute to heavy metal pollution (Cd, Cr, Cu, Hg, Pb, and Zn) in soil (Kabata-Pendias and Pendias, 2001). These pollutants infiltrate soil through various industrial processes, such as mining, smelting, electroplating, metalliferous ore handling, gas emissions, energy production, fuel generation, pesticide and fertilizer use, and municipal waste disposal.

In response to soil contamination, remediation strategies are essential to restore land health. A primary objective in these scenarios is to establish a vegetative cover, preventing soil erosion and limiting the spread of contaminants, particularly in cases where soil coverage and plant development have been compromised, leading to metal toxicity, runoff, and wind-driven dust pollution (Fong *et al.*, 2018).

2.2.3.2 Risk Assessment

Soil remediation plays a crucial role in mitigating the adverse effects of toxic metals on both human health and the environment. Cadmium, for instance, has been associated with human health issues, as highlighted in studies by Nogawa *et al.* (2016) and Kobayashi (2018). Additionally, the toxic effects of certain metals, like cadmium, have led to harm in animals and wildlife, as evidenced by research conducted by Kopsell and Randle (2015), Ohlendorf *et al.* (2016), and Berti and Jacobs (1996). Moreover, soil pollution by zinc (Zn), nickel (Ni), and copper (Cu) from mining wastes and smelters has been shown to have phytotoxic effects on sensitive plant species, as demonstrated by Chaney (2014). Lead (Pb) exposure remains a significant threat to human health, with various routes of ingestion, such as through food, beverages, soil, or dust. Excessive lead exposure can result in convulsions, cognitive impairment, and abnormal behavior. Furthermore, limited environmental mobility can exacerbate lead poisoning, even in cases of heavy rainfall.

2.2.3.3 Metal Bioavailability in Soil

Metals interact with soil through several mechanisms, including (1) existing in the soil solution as free metal ions and soluble metal complexes; (2) adsorption onto inorganic soil components at ion exchange sites; (3) bonding to organic materials in the soil; (4) precipitation as oxides, hydroxides, and carbonates; and (5) incorporation into the structure of silicate minerals. Sequential soil extractions, as employed by Tessier *et al.* (2015), are utilized to identify and quantify components associated with specific soil constituents. The bioavailability of pollutants is crucial for phytoextraction, as it enables ready absorption by plant roots. The solubility of metals in the soil solution determines metal bioavailability. Some metals, like zinc and cadmium, are predominantly present in easily exchangeable forms, while others, such as lead (Pb), exist as soil precipitates, rendering them significantly less bioavailable.

The concept of phytoremediation is underpinned by the chemistry of metal interactions with the soil matrix. Sorption to soil particles often results in a reduction in the activity of metals within the system. The greater the extent of metal sorption and immobilization, the higher the cation exchange capacity (CEC) of the soil. In acidic soils, metal desorption from binding sites into the solution increases due to competition with H+ ions for these sites. The pH of the soil impacts both metal bioavailability and the process of metal root absorption, although this effect varies depending on the specific metal. For instance, while manganese (Mn) and cadmium (Cd) uptake appear to be more sensitive to soil acidity, zinc (Zn) absorption in the roots of *T. caerulescens* is not as affected, as observed by Brown and Green (2016).

2.2.3.4 Heavy Metal Toxicity in Plants

The toxicity of heavy metals affects both macro- and microorganisms by directly interfering with physiological and metabolic processes. This interference can impede growth, disrupt cellular organelles, and hinder photosynthesis. Notably, certain metals, particularly lead (Pb), tend to accumulate more in plant roots than in their above-ground parts due to barriers limiting the movement of metals from roots to aerial portions. Conversely, plants can readily absorb other metals such as cadmium (Cd), as indicated by studies by Garbisu and Alkorta (2003). Soil solutions typically provide essential elements required for plant growth and development, including copper (Cu), iron (Fe), zinc (Zn), calcium (Ca), potassium (K), magnesium (Mg), and sodium (Na). However, plants can also uptake non-essential elements like cadmium (Cd), arsenic (As), chromium (Cr), aluminum (Al), and lead (Pb) during this process.

2.2.4 Heavy Metals in Water

In recent decades, there has been growing concern regarding the contamination of freshwater ecosystems by a wide range of pollutants, as highlighted by Al-Weher (2008). According to the findings of Wogu and Okaka (2011), the influx of heavy metals into aquatic systems is influenced by a combination of human-driven activities, naturally occurring deposits, and inherent natural processes. The primary contributors to metal pollution in rivers and streams are human-related sources, stemming from activities such as municipal and industrial discharges, as well as non-point source runoff, as noted by Sonayi *et al.* (2009).

The presence of heavy metals in aquatic environments is of particular concern due to their toxic properties and their propensity to accumulate. When released into rivers and other water bodies, heavy metals can disrupt the diversity of aquatic species and have adverse effects on ecosystems, as emphasized by Al-Weher (2008). Moreover, heavy metals dissolving in water pose a risk to public health, especially when utilized for irrigation and drinking purposes. The use of water containing heavy metals in irrigation increases the likelihood of these metals entering the food chain and ultimately being consumed by individuals, as observed by Wogu and Okaka (2011).

Problems associated with long-term untreated wastewater application include the buildup of heavy metals in soils to hazardous levels. Consequently, soils irrigated with wastewater become repositories for heavy metals near the soil surface, as noted by Sonayi *et al.* (2009). As a result of frequent wastewater application, heavy metals may leach into groundwater or soil solutions

Phytoremediation of Heavy Metal–Contaminated Soil

accessible for absorption when the soil's capacity to retain them is compromised, as highlighted by Sonayi *et al.* (2009).

2.3 HEAVY METAL REMEDIATION

Some of the current standard approaches for treating heavy metal–contaminated soil and water include ex situ excavation, landfilling the most polluted soils (Zhou and Song, 2004), detoxification (Ghosh and Singh, 2015), and physico-chemical remediation. These approaches are labor intensive, costly, and time consuming and increase contaminant mobilization. They also jeopardize the soil's biotic and structural integrity. According to Bacio *et al.* (2013), these restoration strategies are consequently both financially and technically inadequate for large polluted areas. Bioremediation uses microorganisms to reduce toxins to less hazardous levels. When it comes to considerable metal and organic pollution, however, the use of living organisms to heal damaged areas was less effective. Plants may digest substances created by their environment. Phytoremediation is the use of plants to clean up polluted places (Garbisu and Alkorta, 2003; Mangkoedihardjo and Surahmaida, 2018). Plants are used to clean up polluted regions in a unique and promising technology known as phytoremediation. It is an inexpensive, long-term, environmentally and aesthetically acceptable method for immobilizing, breaking down, transferring, eliminating, or detoxifying contaminants such as metals, pesticides, hydrocarbons, and chlorinated solvents (Jadia and Fulekar, 2009; Zhang *et al.*, 2015). It has received universal acceptability as a technology for cleaning polluted soil and water during the last 20 years (USEPA, 2017). Phytoremediation has been viewed as a natural process since it was identified and shown more than three centuries ago. Through the use of certain plant and wild species, this approach effectively accumulates escalating quantities of harmful heavy metals (Ghosh and Singh, 2015; Brunet *et al.*, 2018). These plants are known as accumulators. They collect heavy metals at greater concentrations (100 times above ground) than non-hyperaccumulators growing under the same circumstances (Barceló and Poschenrieder, 2013). Phytoremediation may disinfect areas with minimal metal, nutrient, organic matter, or pollution contamination. Certain species, notably *Cassia auriculata* (Fabaceae family), *Dodonaea viscose* (Sapindaceae family), and *Jatrophacurcas* (Euphorbiaceae family), have the capacity to repair soils polluted with different trace and major elements, according to Nagaraju and Karimulla (2017). Among the applications of phytoremediation are phytofiltration, also known as rhizofiltration, phytostabilization, phytovolatilization, phytodegradation, and phytoextraction (Long *et al.*, 2014; Jada and Fulekar, 2019).

2.3.1 RHIZOFILTRATION VS. PHYTOFILTRATION

Phytofiltration or rhizofiltration is the use of plant roots to filter contaminants out of waste water, surface water, or extracted ground water (Pivetz, 2001). In a hydroponic experiment, Abhilash *et al.* (2019) examined the phytofiltration capability of *Limnocharis flava* (L.) Buchenau growing in contaminated water with low Cd contents. *L. flava* seedlings were sprayed with varied Cd doses (0.5, 1, 2, and 4 mg L1) at 45 days. The plant's roots contained the most Cd, followed by the peduncle and leaves. This indicated that *L. flava* was a good candidate for low-Cd phytofiltration of water.

2.3.2 PHYTOSTABILIZATION

Phytostabilization is a simple, cost-effective, and environmentally friendly method of using plants to stabilize pollutants and reduce their bioavailability. Actually, plant roots are used in this strategy to limit the quantity of soil pollutants that may travel around and become bioavailable (Jadia and Fulekar, 2009). Plants may help to minimize the quantity of pollution in the atmosphere by preventing contaminants from entering groundwater or flying into the sky. This method is useful when there isn't an urgent need to clean up polluted regions (for example, when a responsible firm is only there briefly or when the location isn't high on the remediation priority list). This method modifies the chemical and biological properties of degraded soils by boosting organic matter, cation exchange capacity (CEC), nutrient level, and biological activity (Alvarenga *et al.*, 2018). Plants play an important role in phytostabilization by reducing water percolation within the soil matrix, which can lead to the production of hazardous leachate; acting as a barrier to prevent direct contact with polluted soil; and preventing soil erosion, which can lead to toxic metals spreading to other sites (Raskin and Ensley, 2000). Phytostabilization is an effective remediation approach for Cd, Cu, As, Zn, and Cr. Alvarenga *et al.* (2019) evaluated the phytostabilization of a soil polluted with extremely acidic metals using three organic residues: sewage sludge, compost from municipal solid waste, and compost from garden waste. The plant material used in this experiment was perennial ryegrass (*Lolium perenne* L.). Organic residues were administered at rates of 25, 50, and 100 mg ha^{-1} (dry weight basis). These compounds immobilized Cu, Pb, and Zn by reducing their mobility percentage. It was established that ryegrass has the potential to be employed in phytostabilization for mine-polluted soil, municipal solid waste compost, and, to a lesser degree, sewage sludge at a rate of 50 mg ha^{-1}. It is also efficient in immobilizing metals in situ, improving soil chemical composition, and dramatically increasing plant biomass.

2.3.3 PHYTOVOLATILIZATION

Phytovolatilization is the process of extracting volatile pollutants from disturbed soils, such as mercury and selenium, and releasing them into the atmosphere via the leaves of green plants (Karami and Shamsuddin, 2015). In high-selenium settings, Bañuelos (2015) discovered that certain plants may convert selenium (Se) to dimethylselenide and dimethyldiselenide.

2.3.4 PLANT DECOMPOSITION

Phytodegradation is the process by which bacteria and plants consume, degrade, and remove organic pollutants. Bacteria and plant roots are used in this procedure to clean up soil that has been polluted by organic contaminants, also known as phytotransformation, by generating enzymes, some plants may detoxify soil, sludge, silt, groundwater, and surface water. This method includes inorganic and organic pollutants, such as pesticides, herbicides, and chlorinated solvents (Pivetz, 2001).

2.3.5 EXTRACTION OF PLANTS

Phytoextraction is a phytoremediation technology that employs plants to extract heavy metals such as Cd from soil, water, and sediments (Yanai *et al.*, 2016; Van Nevel *et al.*, 2017). It is a great technique for removing contaminants from soil without compromising the soil's qualities. Furthermore, the ash created by drying, ashing, and composting these harvestable plant parts may be easily utilized to recover metals that have collected in the plant components. Other terminology for phytoextraction include phytomining and biomining (Pivetz, 2001). This is a more complex kind of phytoremediation in which heavy metals are recovered and bioharvested via the cultivation of high-biomass crops in polluted soil. It might be used in the mining sector to grow metals for industrial purposes (Sheoran *et al.*, 2019).

Plants must be able to move and absorb heavy metals from the soil into their above-ground shoots and harvestable parts of their subterranean roots in order to extract them. A few field studies and commercial ventures have been conducted in the last ten years to investigate the viability of phytoextraction, according to Robinson *et al.* (2016). Furthermore, contaminated regions must be cleaned up to environmental legislation standards at a lower cost than traditional approaches in order for phytoextraction to be regarded as effective (Kos and Le tan, 2013). Phytoextraction, according to Nascimento and Xing (2006), may someday be regarded as a viable commercial strategy.

Numerous authors have documented in detail the phytoextraction capacity of numerous species in heavy metal–contaminated ecosystems (Grispen *et al.*, 2016; Daghan *et al.*, 2018; Zadeh *et al.*,

2018). Mangkoedihardjo and Surahmaida (2018) investigated *J. curcas* L.'s ability to detoxify Pb and Cd-contaminated soil. The garden soil was deliberately poisoned with $Pb(NO3)^2$ and $Cd(NO3)^2$. For one month, plants were treated with varying concentrations of these components in a specific compound. According to the findings, the primary concentration of these two elements (50 mg kg^{-1}) did not seem to have any negative impacts on the plants. The researchers discovered *J. curcas* L.'s phytoremediation capacity for Cd and Pb in polluted regions. Ang (2013) investigated the removal of Cd, Pb, As, and Hg from slime tailings at Malaysia's Forest Research Institute. *Hopea odorata*, *Acacia mangium*, *Swietenia macrophylla*, and *Intsia palembanica* were planted in slime tailing to test their potential to bioaccumulate Cd, Hg, Pb, and As. According to the findings of the experiment, *A. mangium* was the greatest choice for As removal, while *H. odorata* and *I. palembanica* had the most potential for fast eliminating Cd. Jiang *et al.* (2014) assessed the growth performance and Cu phytoextraction capacity of *Elsholtzia splendens*. $CuSO_{4.5}H_2O$ was supplied at different levels in a greenhouse study: 100, 200, 400, 600, 800, 1000, and 1200 mg kg^{-1}. When exposed to Cu concentrations as high as 80 mg kg^{-1}, the plant species demonstrated extraordinary resistance to Cu toxicity and continued to grow normally. Li *et al.* (2019) investigated the phytoextraction capability of *Averrhoa carambola*, a high-biomass tree species. In low-Cd soil, this species flourished over 170 days, generating greater shoot biomass (18.6 t ha^{-1}) and accumulating 213 g Cd per hectare. Cambola was discovered to be an effective plant for low levels of Cd pollution in soil by the researchers. Wu *et al.* (2009) studied the phytoextraction capability of a hybrid poplar (*Populus deltoids Populus nigra*) in two Cd-polluted alluvial and purple soils. It was discovered that when the quantity of this element grew in the two soils, so did the concentration of Cd in plant tissues. Cd buildup in plant roots was greater in purple soil than in shoots and leaves but not in alluvial soil.

Murakami *et al.* (2017) investigated the phytoextraction capabilities of rice (*Oryza sativa* L., cv. Nipponbare and Milyang 23), maize (*Zea mays* L., cv. Gold Dent), and soybeans (*Glycine max* L. Merr., cv. Enrei and Suzuyutaka). The species were planted for 60 days on two fluvisols and one andosol that were both significantly Cd polluted (0.83 to 4.29 mg Cd kg^{-1}). According to the data, the shoots of Milyang 23 rice accumulated 1 to 15% of the total Cd in the soil. These plants were shown to be capable of restoring low-concentration paddy soil.

2.4 HYPERACCUMULATOR PLANT SELECTION

Finding plant species capable of absorbing harmful amounts of metals while generating a considerable amount of biomass is critical for the effective application of phytoremediation (Odoemelam and Ukpe, 2018). Ideal plants for phytoextraction have fast growth rates, resistance to high salinity; high pH; and the ability to accumulate dangerous quantities of metals in their aerial parts, or shoots. These plants must also be easily grown and entirely harvestable, have a big dry biomass, and absorb and translocate metals to their aerial parts effectively. Native plant species indigenous to the region should be utilized. These species are less competitive in the region and will lower metal concentrations to levels appropriate for healthy plant development (Rajakaruna *et al.*, 2016).

2.4.1 Metal Accumulation Criteria in Plants

All plant species have the ability to absorb metals; however, some have a larger capacity than others (100 times more than the average plant in the same setting without displaying any detrimental impact). Woody or herbaceous plants that collect and tolerate large amounts of heavy metals in their tissues are known as hyperaccumulators (Barceló and Poschenrieder, 2013; Zhou and Song, 2004). Because hyperaccumulators can absorb heavy metals from polluted soil and collect them in their shoots, researchers have recently concentrated on utilizing them to clean up contaminated areas. The primary requirements for hyperaccumulators (TF) are (1) accumulating capacity, (2) tolerability, (3) removal efficiency (RE) based on plant biomass, (4) bioconcentration factor (BCF), and (5) transfer factor.

2.4.1.1 Capability to Accumulate

Accumulating capability is the natural ability of plants to accumulate metals within their elevated parts (the threshold concentration), 1000 mg kg^{-1} for Cu, Cr, Pb, and Co; 10 mg kg^{-1} for Hg; and 10,000 mg kg^{-1} dry weight of shoots for Ni and Zn.

2.4.1.2 Capability for Tolerance

Plants with a high resistance to heavy metals show no signs of chlorosis, necrosis, whitish-brown pigmentation, or a substantial drop in above-ground biomass (Sun *et al.*, 2019). Instead, they may thrive in heavy metal–contaminated environments.

2.4.1.3 Removal Effectiveness

The effectiveness of removal based on plant biomass is defined as the ratio of dry plant biomass and total metal concentrations to total metal load in the growing medium (Soleimani *et al.*, 2010). The Index BCF is defined as the heavy metal concentration in plant roots divided by the concentration in soil (Yoon *et al.*, 2016). The BCF for hyperaccumulators is larger than one and, under some situations, may surpass 100.

TF is the capacity of plants to transmit heavy metals from their roots to their above-ground portions (shoots). As a consequence, Liu *et al.* (2015) define it as the ratio of heavy metal concentrations in plant leaves to concentrations in roots. This particular requirement must be greater than 1 in order for hyperaccumulators to establish that their heavy metal levels are higher above ground than below ground (roots). Given that the primary purpose of phytoextraction is to collect plant aerial parts, this criterion is more important (Karami and Shamsuddin, 2010). Accumulators have a TF greater than 1, and excluders have a TF less than 1. The following formulae compute BCF, TF, and RE:

$$BCF = \left[\frac{\text{Metal Concentration in Root} \left(\frac{mg}{kg} \right)}{\text{Metal Concentration in Soil} \left(\frac{mg}{kg} \right)} \right]$$

$$TF = \left[\frac{\text{Metal Concentration in Shoot} \left(\frac{mg}{kg} \right)}{\text{Metal Concentration in Root} \left(\frac{mg}{kg} \right)} \right]$$

$$RE(\%) = \left[\frac{\left(\text{Metal Concentration in Shoot} \left(\frac{mg}{kg} \right) \times \text{Shoot Biomass} (kg) \right) + \left(\text{Metal Concentration in Root} \left(\frac{mg}{kg} \right) \times \text{Root Biomass} (kg) \right)}{\text{Total Added Metal per Pot} (mg)} \right]$$

As highlighted, hyperaccumulators exhibit bioconcentration factors (BCFs) and translocation factors (TFs) exceeding 1. These exceptional plants, capable of accumulating significant concentrations of metals, are a rare but fascinating botanical phenomenon. Worldwide, nearly 400 species belonging to 45 different plant families have been identified as hyperaccumulators, as reported by Sun *et al.* Sarma (2015) presented the most recent tally of metal hyperaccumulators, indicating that almost 500 plant species from 101 families, including Caryophyllace, Euphorbiaceae, Violaceae, Brassicaceae, Poaceae, Lamiaceae, Flacourtiaceace, Cunouniaceae, Asateraceae, and Cyperaceae, fall into this category.

It's essential to note, as pointed out by Zhou and Song, that there are relatively few plant groups demonstrating hyperaccumulation of elements like cadmium (Cd) and arsenic (As). This scarcity

Phytoremediation of Heavy Metal–Contaminated Soil

can be attributed to the fact that hyperaccumulators often have low shoot biomass, extended maturation periods, and prolonged growth seasons. Nevertheless, Baker *et al.* (2014) identified a substantial number of plant species that can be categorized as hyperaccumulators due to their ability to tolerate hazardous levels of various metals, including Cd, Cu, As, Co, Mn, Zn, Ni, Pb, and Se.

2.5 ADVANTAGES OF PHYTOREMEDIATION

Phytoremediation offers several notable advantages as an environmentally friendly and cost-effective remediation method. It eliminates the need for expensive machinery and expert personnel, making it an efficient and economical approach for cleaning up contaminated soils, as documented by the Environmental Protection Agency and Ghosh and Singh (2015). As a "green technology," phytoremediation is versatile, capable of addressing a wide range of organic and inorganic contaminants while also enhancing the aesthetic appeal of the environment by utilizing trees and creating green spaces. This dual benefit extends to social and psychological well-being, as discussed by Ghosh and Singh (2015) and Lewis (2006). When other remediation methods prove prohibitively expensive or less effective for large areas, this environmentally friendly approach becomes particularly advantageous, as noted by Prasad and Freitas (2013). Additionally, the potential for waste recycling makes it a low-impact environmental solution for land and water decontamination, as suggested by Schnoor (2012). Moreover, it helps reduce leaching and soil erosion, minimizing the transfer of contaminants into the air and water, as highlighted by Ghosh and Singh (2015).

2.6 DISADVANTAGES OF PHYTOREMEDIATION

While phytoremediation offers numerous benefits, it also presents some notable disadvantages. One of the most significant drawbacks is the protracted recovery period, often spanning several years, as emphasized by Vidali and Rajakaruna *et al.* Maintaining vegetation in highly contaminated areas is a challenging endeavor, as pointed out by Vidali, and there is a risk of animals consuming contaminated plants, introducing pollutants into the food chain and potentially endangering human health, as documented by Pivetz (2001). Phytoremediation may not be highly effective when the pollutant is only partially bioavailable to plants in the soil, as highlighted by Rajakaruna *et al.* Additionally, it is limited to areas within the root zone of the plants and is most effective in regions with low to high pollution levels, as discussed by Ghosh and Singh (2015).

2.7 CONCLUSION

For the treatment of heavy metal–contaminated soil, phytoremediation is a viable and eco-friendly option, to sum up. With the help of this novel method, polluted soil may be restored by taking use of some plants' innate capacities to absorb, collect, and detoxify heavy metals. Phytoremediation provides a long-term and affordable way to lessen the negative impacts of heavy metal contamination in soils, even if it might not be a universally applicable answer. For it to be successful, meticulous plant selection, site-specific factors, and continuous research to improve its effectiveness and applicability are necessary. Phytoremediation is still an important instrument in our toolbox for restoring and revitalizing polluted areas as we tackle environmental issues.

REFERENCES

Abhilash, Mavinakere R., Gangadhar Akshatha, and Shivanna Srikantaswamy. 2019. "Photocatalytic Dye Degradation and Biological Activities of the Fe 2 O 3/Cu 2 O Nanocomposite." *RSC Advances* 9, no. 15: 8557–68. https://doi.org/10.1039/c8ra09929d.

Adei-Atiemo, Eunice, Onike Rodrigues, and Ebenezer Badoe. 2015. "Classification and Risk Factors for Cerebral Palsy in the Korle Bu Teaching Hospital, Accra: A Case-Control Study." *Pediatrics* 135(Suppl. 1), no. Supplement: S7. https://doi.org/10.1542/peds.2014-3330K.

Adelekan, B. A., and K. D. Abegunde. 2011. "Heavy Metals Contamination of Soil and Groundwater at Automobile Mechanic Villages in Ibadan, Nigeria." *International Journal of the Physical Sciences* 6: 1045–58.

Alvarenga, Débora M., Matheus S. Mattos, Mateus E. Lopes, Sarah C. Marchesi, Alan M. Araújo, Brenda N. Nakagaki, Mônica Morais Santos, Bruna Araújo David, Viviane Aparecida De Souza, Érika Carvalho, Rafaela Vaz Sousa Pereira, Pedro Elias Marques, Kassiana Mafra, Hortência Maciel de Castro Oliveira, Camila Dutra Moreira de Miranda, Ariane Barros Diniz, Thiago Henrique Caldeira de Oliveira, Mauro Martins Teixeira, Rafael Machado Rezende, Maísa Mota Antunes, and Gustavo B. Menezes. 2018. "Paradoxical Role of Matrix Metalloproteinases in Liver Injury and Regeneration After Sterile Acute Hepatic Failure." *Cells* 7, no. 12: 247. https://doi.org/10.3390/cells7120247.

Alvarenga, Jeferson C., Robson R. Branco, André L. A. Guedes, Carlos A. P. Soares, and W. S. da Silveira. 2019. "The Project Manager Core Competencies to Project Success." *International Journal of Managing Projects in Business* 13, no. 2: 277–92. https://doi.org/10.1108/IJMPB-12-2018-0274.

Al-Weher, S. M. 2008. "Levels of Heavy Metal Cd, Cu and Zn in Three Fish Species Collected from the Northern Jordan Valley, Jordan. *Jordan Journal of Biological Sciences* 1: 41–46.

Ang, I. 2013. *Watching Dallas: Soap Opera and the Melodramatic Imagination*. London: Routledge.

Bacio, Guadalupe A., Vickie M. Mays, and Anna S. Lau. 2013 Mar. "Drinking Initiation and Problematic Drinking Among Latino Adolescents: Explanations of the Immigrant Paradox." *Psychology of Addictive Behaviors* 27, no. 1: 14–22. https://doi.org/10.1037/a0029996. Epub October 1 2012. PubMed: 23025707, PubMed Central: PMC3627496.

Baker, A. J. M., and R. R. Brooks. 1989. "Terrestrial Higher Plants Which Hyperaccumulate Metallic Elements—A Review of Their Distribution, Ecology and Phytochemistry." *Biorecovery* 1: 81–126.

Baker, A. J. M., R. D. A. Reeves, and S. M. Hajar. 1991. "Heavy Metal Accumulation and Tolerance in British Populations of the Metallophyte *Thlaspicaerulescens (Brassicaceae)*." *New Phytologist* 127: 61–8.

Baker, Dwayne A., and John L. Crompton. 2000. "Quality, Satisfaction and Behavioral Intentions." *Annals of Tourism Research* 27, no. 3: 785–804. http://doi.org/10.1016/S0160-7383(99)00108-5.

Baker, S., N. Lesaux, M. Jayanthi, J. Dimino, C. P. Proctor, J. Morris *et al.* 2014. *Teaching Academic Content and Literacy to English Learners in Elementary and Middle School*. IES Practice Guide. NCEE 2014–4012. What Works Clearinghouse. Washington, DC 20202, USA.

Bañuelos, G. S. 2015. "Phytoextraction of Selenium from Soils Irrigated with Selenium-Laden Effluent." *Plant & Soil* 224, no. 2: 251–8.

Barceló, J., and C. Poschenrieder. 2013. "Phytoremediation: Principles and Perspectives." *Contribution Science* 2, no. 3: 333–44.

Baumann, M. K. 2015. "Phytoremediation of Zinc- Contaminated Soil and Zinc-Biofortification for Human Nutrition." In *Phytoremediation & Biofortification*: 33–57. Dordrecht: Springer Netherlands.

Berti, W. R., and L. W. Jacobs. 1996. "Chemistry and Phytotoxicity of Soil Trace Elements from Repeated Sewage Sludge Applications." *Journal of Environmental Quality* 25, no. 5: 1025–32. https://doi.org/10.2134/jeq1996.00472425002500050014x.

Bondada, B. R., and L. Q. Ma. 2003. "Tolerance of Heavy Metals in Vascular Plants: Arsenic Hyperaccumulation by Chinese Brake Fern (*Pteris Vittata L.*)." In *Pteridology in the New Millennium*, edited by S. Chandra and M. Srivastava. Dordrecht: Springer. https://doi.org/10.1007/978-94-017-2811-9_28

Bose, B., and A. Hemantaranjan. 2005. *Developments in Physiology, Biochemistry and Molecular Biology of Plantsi*: 105. New Delhi: New India Publishing Agency.

Brown, Abbie, and Tim Green. 2016. "Virtual Reality: Low-Cost Tools and Resources for the Classroom." *TechTrends* 60, no. 5: 517–19. https://doi.org/10.1007/s11528-016-0102-z.

Brunet, Alain, Daniel Saumier, Aihua Liu, David L. Streiner, Jacques Tremblay, and Roger K. Pitman. 2018. "Reduction of PTSD Symptoms with Pre-Reactivation Propranolol Therapy: A Randomized Controlled Trial." *American Journal of Psychiatry* 175, no. 5: 427–33. https://doi.org/10.1176/appi.ajp.2017.17050481.

Byers, L. 2017. "Characteristics of Lead-Resistant and ACC Deaminase-Producing Endophytic Bacteria and Their Potential in Promoting Lead Accumulation of Rape." *Journal of Hazardous Materials* 186: 720–5.

Chaney, R. L. 1989. "Speciation, Mobility and Bioavailability of Soil Lead." *Environmental Geochemistry & Health* 11: 105–29.

Chaney, Thomas. 2014. "The Network Structure of International Trade." *American Economic Review* 104, no. 11: 3600–34. https://doi.org/10.1257/aer.104.11.3600.

Daghan, H., A. Schaeffer, R. Fischer, and U. Commandeur. 2018. "Phytoextraction of Cadmium from Contaminated Soil Using Transgenic Tobacco Plants." *International Journal Environmental Applied Science* 3, no. 5: 336–45.

Delbari, A. S., and D. K. Kulkarni. 2013. *Determination of Heavy Metal Pollution in Vegetables Grown Along the Roadside in Tehran-Iran*. Annals of Biological Research, 4, 224–233.

FAO. 2015. *Food and Agricultural Organization of the United Nations*. Rome: FAO.

Fong, Patrick S. W., Chenghao Men, Jinlian Luo, and Ruiqian Jia. 2018. "Knowledge Hiding and Team Creativity: The Contingent Role of Task Interdependence." *Management Decision* 56, no. 2: 329–43. https://doi.org/10.1108/MD-11-2016-0778.

Garbisu, C., and I. Alkorta. 2003. *European Journal of Mineral Processing & Environmental Protection* 3, no. 1: 58–66.

Ghosh, M., and S. P. Singh. 2015. "A Review on Phytoremediation of Heavy Metals and Utilization of It's by Products." *Applied Ecology & Environmental Research* 3: 1–18.

Grispen, V. M. J., H. J. M. Nelissen, and J. A. C. Verkleij. 2016. "Phytoextraction with *Brassica napus* L.: A Tool for Sustainable Management of Heavy Metal Contaminated Soils." *Environmental Pollutant* 144, no. 1: 77–83.

Haller, T. 2017. "Enhancement of Plant Growth and Decontamination of Nickel-Spiked Soil Using PGPR." *Journal of Basic Microbiology* 49: 195–204.

Henry, L. 2000. "Using Municipal Biosolids in Combination with Other Residuals to Restore Metal-Contaminated Mining Areas." *Plant & Soil* 249: 203–15.

Ijeoma, N. B. 2014. "The Effect of Creative Accounting on the Nigerian Banking Industry." *International Journal of Managerial Studies & Research* 2, no. 10: 13–21.

Jadia, C. D., and M. H. Fulekar. 2009. "Phytoremediation of Heavy Metals: Recent Techniques." *African Journal of Biotechnology* 8, no. 6: 921–8.

Jaradat, H. M., F. Awawdeh, S. Al-Shara, M. Alquran, and S. Momani. 2015. "Controllable Dynamical Behaviors and the Analysis of Fractal Burgers Hierarchy with the Full Effects of Inhomogeneities of Media." *Romanian Journal of Physics* 60, no. 3–4: 324–43.

Jia, Yali, Steven T. Bailey, Thomas S. Hwang, Scott M. McClintic, Simon S. Gao, Mark E. Pennesi, Christina J. Flaxel, Andreas K. Lauer, David J. Wilson, Joachim Hornegger, James G. Fujimoto, and David Huang. 2015. "Quantitative Optical Coherence Tomography Angiography of Vascular Abnormalities in the Living Human Eye." *Proceedings of the National Academy of Sciences of the United States of America* 112, no. 18: E2395–402. https://doi.org/10.1073/pnas.1500185112.

Jiang, G., J. Keller, and P. L. Bond. 2014. "Determining the Long-Term Effects of H2S Concentration, Relative Humidity and Air Temperature on Concrete Sewer Corrosion." *Water Research* 65: 157–69. https://doi.org/10.1016/j.watres.2014.07.026.

Jonsson, H., and T. Haller. 2014. "Cadmium-Induced Rhizospheric pH Dynamics Modulated Nutrient Acquisition and Physiological Attributes of Maize (*Zea mays L.*)." *Environmental Science & Pollution Research* 22: 9193–203.

Kabata-Pendias, A., and H. Pendias. 2001. *Trace Elements in Soils and Plants*. 3rd ed. Boca Raton, FL: CRC Press.

Kaplan, O., N. C. Yildirim, N. Yildirim, and M. Cimen. 2011. "Toxic Elements in Animal Products and Environmental Health." *Asian Journal of Animal & Veterinary Advances* 6, no. 3: 228–32. https://doi.org/10.3923/ajava.2011.228.232.

Kar, A., C. Häne, and J. Malik. 2017. "Learning a Multi-view Stereo Machine." *Advances in Neural Information Processing Systems* 30.

Karami, A., and Z. H. Shamsuddin. 2010. Phytoremediation of Heavy Metals with Several Efficiency Enhancer Methods. *African Journal of Biotechnology* 9(25): 3689–98.

Karami, A., and Z. H. Shamsuddin. 2015. "Phytoremediation of Heavy Metals with Several Efficiency Enhancer Methods." *African Journal of Biotechnology* 9, no. 25: 3689–98.

Kobayashi, A. 2018. "Bioactive Potentiality of POD Products Derived from Natural Simple Phenolics." In *Biochemistry and Physiology: Fourth International Symposium Proceedings*, edited by C. Obinger, U. Burner, R. Ebermann, C. Penel, and H. Greppin, Plant Peroxidases: 292–7. Vienna: University of Agriculture and Cham, Switzerland: University of Geneva.

Kopsell, T. S., and P. A. Randle. 2015. "Potential of Rhizobacteria for Improving Lead Phytoextraction in Ricinus Communis." *Remediation Journal* 24, no. 1: 99–106.

Kos, B., and D. Lestan. 2013. "Phytoextraction of Lead, Zinc and Cadmium from Soil by Selected Plants." *Plant, Soil & Environment* 49: 548–53.

Lewis, A. C. 2006. *Assessment and Comparison of Two Phytoremediation Systems Treating Slow-Moving Groundwater Plumes of TCE*: 158. Masters thesis. Ohio University, Athens.

Li, X., J. Feng, Y. Meng, Q. Han, F. Wu, and J. Li. 2019. "A Unified MRC Framework for Named Entity Recognition." *arXiv Preprint ArXiv: 1910.11476*.

Liphadzi, M., and L. Kirkham. 2015. "Strategies for Enhancing the Phytoremediation of Cadmium-Contaminated Agricultural Soils by (*Solanum nigrum L.*)." *Environmental Pollution* 159, no. 3: 762–8.

Liu, L., S. Oza, D. Hogan, J. Perin, I. Rudan, J. E. Lawn *et al*. 2015. Global, Regional, and National Causes of Child Mortality in 2000–13, with Projections to Inform Post-2015 Priorities: An Updated Systematic Analysis. *The Lancet* 385(9966): 430–40.

Logan, F. 2017. "Effects of Bacteria on Enhanced Metal Uptake of the Cd/Zn-Hyperaccumulating Plant, Sedum Alfredii." *Journal of Experimental Botany* 58: 4173–82.

Long, X. X., X. E. Yang, and W. Z. Ni. 2014. "Current Status and Perspective on Phytoremediation of Heavy Metal Polluted Soils." *Journal of Applied Ecology* 13: 757–62.

Mangkoedihardjo, S., and D. Surahmaida. 2018. "Jatropha Curcas L. for Phytoremediation of Lead and Cadmium Polluted Soil." *World Applied Sciences Journal* 4, no. 4: 519–22.

Mebrahtu, G., and S. Zerabruk. 2011. "Concentration of Heavy Metals in Drinking Water from Urban Areas of the Tigray Region, Northern Ethiopia." *Momona Ethiopian Journal of Science* 3, no. 1: 105–21. http://doi.org/10.4314/mejs.v3i1.63689.

Micó, Victor, Roberto Martín, Miguel A. Lasunción, Jose M. Ordovás, and Lidia Daimiel. 2016. "Unsuccessful Detection of Plant MicroRNAs in Beer, Extra Virgin Olive Oil and Human Plasma After an Acute Ingestion of Extra Virgin Olive Oil." *Plant Foods for Human Nutrition* 71, no. 1: 102–8. https://doi.org/10.1007/s11130-016-0534-9.

Mishra, S., R. N. Bharagava, A. Yadav, S. Zainith, and P. Chowdhary. 2019. "Heavy Metal Contamination: An Alarming Threat to Environment and Human Health." In *Environmental Biotechnology: For Sustainable Future*, edited by R. Sobti, N. Arora, and R. Kothari, 103–25. Singapore: Springer. https://doi.org/10.1007/978-981-10-7284-0_5.

Murakami, M., Y. Nakatani, G. I. Atsumi, K. Inoue, and I. Kudo. 2017. "Regulatory Functions of Phospholipase A 2." *Critical Review™ in Immunology* 37: 2–6.

Musa, O. I., U. J. J. Ijah, O. P. Abioye, and M. O. Adebola.2021a. "Microbial Determination of Hydrocarbon Polluted Soil in Some Parts of Niger State, Nigeria." *Biosciences & Bioengineering* 6, no. 3: 20–7.

Nagaraju, A., and S. Karimulla. 2017. "Accumulation of Elements in Plants and Soils in and Around Nellore Mica Belt, Andhra Pradesh, India: A Biogeochemical Study." *Environmental Geology* 41, no. 7: 852–60.

Nascimento, C. W. A., and B. Xing. 2006. "Phytoextraction: A Review on Enhanced Metal Availability and Plant Accumulation." *Scientia Agricola* 63, no. 3: 299–311. https://doi.org/10.1590/S0103-90162006000300014.

Nogawa, K., D. D. B. Miled, M. H. Ghorbal, and M. Zarrouk. 2016. "Itai-Itai Disease and Follow-Up Studies. Cadmium in the Environment." *Health Effects* 2: 1–37.

Odoemelam, S. A., and R. A. Ukpe. 2018. "Heavy Meal Decontamination of Polluted Soils Using *Bryophyllum pinnatum*." *African Journal of Biotechnology* 7, no. 23: 4301–3.

Ohlendorf, R., T. K. Prasad, and C. R. Stewart. 2016. "Changes in Isozyme Profiles of Catalase, Peroxidase, and Glutathione Reductase During Acclimation to Chilling in Mesocotyls of Maize Seedlings." *Plant Physiology* 109, no. 4: 1247–57.

Pivetz, P. 2001. *Phytoremediation of Contaminated Soil and Ground Water at Hazardous Waste Sites*: 36. EPA/540/S-01/500. Washington, DC: United States Environmental Protection Agency (EPA).

Prasad, M. N. V., and H. M. De Oliveira Freitas. 2013. "Metal Hyperaccumulation in Plants: Biodiversity Prospecting for Phytoremediation Technology." *Electronic Journal of Biotechnology* 6, no. 3: 285–321. https://doi.org/10.2225/vol6-issue3-fulltext-6.

Qiu, R. L., X. W. Zeng, L. Q. Ma, and Y. T. Tang. 2014. "Effects of Zn on Plant Tolerance and Nonprotein Thiol Accumulation in Zn Hyperaccumulator Arabispaniculata Franch." *Environmental & Experimental Botany* 70: 227–32.

Rajakaruna, N., K. M. Tompkins, and P. G. Pavicevic. 2016. "Phytoremediation: An Affordable Green Technology for the Clean-Up of Metal- Contaminated Sites in Sri Lanka." *Ceylon Journal of Science (Biological Sciences)* 35: 25–39.

Raji, Babak, Martin J. Tenpierik, and Van A. Den Dobbelsteen. 2015. "The Impact of Greening Systems on Building Energy Performance: A Literature Review." *Renewable & Sustainable Energy Reviews* 45: 610–23. https://doi.org/10.1016/j.rser.2015.02.011.

Raji, Babak, Martin J. Tenpierik, and Van A. Den Dobbelsteen. 2016. "An Assessment of Energy-Saving Solutions for the Envelope Design of High-Rise Buildings in Temperate Climates: A Case Study in the Netherlands." *Energy & Buildings* 124: 210–21. https://doi.org/10.1016/j.enbuild.2015.10.049.

Raji, R., and K. G. Gopchandran. 2017. "ZnO Nanostructures with Tunable Visible Luminescence: Effects of Kinetics of Chemical Reduction and Annealing." *Journal of Science: Advanced Materials & Devices* 2, no. 1: 51–8. https://doi.org/10.1016/j.jsamd.2017.02.002.

Raskin, I., V. Dushenkov, P. B. A. N. Kumar, and H. Motto. 2014. "The Use of Plants to Remove Heavy Metals from Aqueous Streams." *Environmental Science & Technology* 29: 1239–45.

Raskin, I. and B. D. Ensley. 2000. "Phytoremediation of Toxic Metals: Using Plants to Clean up the Environment". John Wiley & Sons, Inc., New York, USA. pp: 53–70.

Robinson, B., R. Schulin, B. Nowack, S. Roulier, M. Menon, B. Clothier, S. Green, and T. Mills. 2016. "Phytoremediation for the Management of Metal Flux in Contaminated Sites. Forest." *Snow Landscape Resources* 80, no. 2: 221–34.

Ruiz, J. M., B. Blasco, J. J. Ríos, L. M. Cervilla, M. A. Rosales, M. M. Rubio-Wilhelmi, E. Sánchez-Rodríguez, R. Castellano, and L. Romero. 2018. "Distribution and Efficiency of the Phytoextraction of Cadmium by Different Organic Chelates." *Terra Latino Americana* 27, no. 4: 296–301.

Sahibin, A. R., A. R. Zulfahmi, K. M. Lai, P. Errol, and M. L. Talib. 2015. "Heavy Metals Content of Soil Under Vegetables Cultivation in Cameron Highland." *Proceedings of the Regional Symposium on Environment and Natural Resources, Kuala Lumpur, Malaysia* 1: 660–7.

Salt, Carina, Penelope J. Morris, Alexander J. German, Derek Wilson, Elizabeth M. Lund, Tim J. Cole, and Richard F. Butterwick. 2017. "Growth Standard Charts for Monitoring Bodyweight in Dogs of Different Sizes." *PLoS One* 12, no. 9: e0182064. https://doi.org/10.1371/journal.pone.0182064.

Sarkar, S. K., S. K. Hazra, H. S. Sen, P. G. Karmakar, and M. K. Tripathi. 2015. *Sunnhemp in India*. Barrackpore, West Bengal: ICAR-Central Research Institute for Jute and Allied Fibres (Indian Council of Agricultural Research).

Sarma, M. 2015. "Measuring Financial Inclusion." *Economics Bulletin* 35: 604–11.

Schnoor, J. L. 2012. *Phytoremediation of Soil and Groundwater*: 252–630. Technology Evaluation Report TE-021. Ground-Water Remediation Technologies Analysis Center, Pittsburgh, PA 15238, Germany.

Sheoran, Sonia, Sarika Jaiswal, Deepender Kumar, Nishu Raghav, Ruchika Sharma, Sushma Pawar, Surinder Paul, M. A. Iquebal, Akanksha Jaiswar, Pradeep Sharma, Rajender Singh, C. P. Singh, Arun Gupta, Neeraj Kumar, U. B. Angadi, Anil Rai, G. P. Singh, Dinesh Kumar, and Ratan Tiwari. 2019. "Uncovering Genomic Regions Associated with 36 Agro-Morphological Traits in Indian Spring Wheat Using GWAS." *Frontiers in Plant Science* 10: 527. https://doi.org/10.3389/fpls.2019.00527.

Shrivastava, Saurabh R., Prateek S. Shrivastava, and Jegadeesh Ramasamy. 2014. "Exploring the Dimensions of Doctor–Patient Relationship in Clinical Practice in Hospital Settings." *International Journal of Health Policy & Management* 2, no. 4: 159–60. http://doi.org/10.15171/ijhpm.2014.40.

Simeonov, Plamen L. 2010. "Integral Biomathics: A Post-Newtonian View into the Logos of Bios." *Progress in Biophysics & Molecular Biology* 102, no. 2–3: 85–121. ISSN: 0079-6107. http://doi.org/10.1016/j.pbiomolbio.2010.01.005.

Simeonov, Plamen L., J. Gomez-Ramirez, and P. Siregar. 2013. "On Some Recent Insights in Integral Biomathics." *Journal Progress in Biophysics and Molecular Biology* 113, no. 1: 216–28. Special Theme Issue on Integral Biomathics: Can Biology Create a Profoundly New Mathematics and Computation? Elsevier. ISSN: 0079-6107. https://doi.org/10.1016/j.pbiomolbio.2013.06.001. also in: arXiv.org. http://arxiv.org/abs/1306.2843.

Singh, S., and B. B. Singh. 2017. "Nutritional Evaluation of Grasses and Top Foliages Through *In Vitro* System of Sheep and Goat for Silvipasture System." *Range Management and Agroforestry* 38, no. 2: 241–8.

Soleimani, M., M. Afyuni, M. A. Hajabbasi, F. Nourbakhsh, M. R. Sabzalian, J. H., and Christensen. 2010. Phytoremediation of an Aged Petroleum Contaminated Soil Using Endophyte Infected and Non-Infected Grasses." *Chemosphere* 81, no. 9: 1084–90.

Sonayi, Y., N. Ismail, and S. Talebi. 2009. "Determination of Heavy Metals in Zayandeh Road River, Isfahan, Iran." *World Applied Sciences Journal* 6, no. 9: 1209–14.

Sun, J., X. Dai, Q. Wang, M. C. Van Loosdrecht, and B. J. Ni. 2019. "Microplastics in Wastewater Treatment Plants: Detection, Occurrence and Removal." *Water Research* 152: 21–37.

Swaminathan, S., R. Hemalatha, A. Pandey, N. J. Kassebaum, A. Laxmaiah, T. Longvah *et al.* 2019. "The Burden of Child and Maternal Malnutrition and Trends in Its Indicators in the States Of India: The Global Burden of Disease Study 1990–2017." *The Lancet Child & Adolescent Health* 3, no. 12: 855–70.

Tessier, Mickael D., Dorian Dupont, Kim De Nolf, Jonathan De Roo, and Zeger Hens. 2015. "Economic and Size-Tunable Synthesis of InP/ZnE (E= S, Se) Colloidal Quantum Dots." *Chemistry of Materials* 27, no. 13: 4893–8. https://doi.org/10.1021/acs.chemmater.5b02138.

Turan, M., and A. Esringu. 2017. "Phytoremediation Based on Canola (*Brassica napus L.*) and Indian Mustard (*Brassica juncea L.*) Planted on Spiked Soil by Aliquot Amount of Cd, Cu, Pb, and Zn." *Plant Soil Environmental* 53, no. 1: 7–15.

US EPA (US Environmental Protection Agency). 2017. "National Primary Drinking Water Regulations; Arsenic and Clarifications to Compliance and New Source Contaminants Monitoring; Final Rule." *Federal Register* 66, no. 14: 6975–7066.

Van Nevel, L., J. Mertens, K. Oorts, and K. Verheyen. 2017. "Phytoextraction of Metals from Soils: How far from Practice?" *Environmental Pollutant* 150, no. 1: 34–40.

Voet, Sofie, Conor Mc Guire, Nora Hagemeyer, Arne Martens, Anna Schroeder, Peter Wieghofer, Carmen Daems, Ori Staszewski, Lieselotte Vande Walle, Marta Joana Costa Jordao, Mozes Sze, Hanna-Kaisa Vikkula, Delphine Demeestere, Griet Van Imschoot, Charlotte L. Scott, Esther Hoste, Amanda Gonçalves, Martin Guilliams, Saskia Lippens, Claude Libert, Roos E. Vandenbroucke, Ki-Wook Kim, Steffen Jung, Zsuzsanna Callaerts-Vegh, Patrick Callaerts, Joris de Wit, Mohamed Lamkanfi, Marco Prinz, and Geert van Loo. 2018. "A20 Critically Controls Microglia Activation and Inhibits Inflammasome-Dependent Neuroinflammation." *Nature Communications* 9, no. 1: 2036. https://doi.org/10.1038/s41467-018-04376-5.

Wei, W., L. Wang, P. Bao, Y. Shao, H. Yue, D. Yang *et al.* 2018. Metal-Free C (sp2)–H/N–H Cross-Dehydrogenative Coupling of Quinoxalinones with Aliphatic Amines Under Visible-Light Photoredox Catalysis. *Organic Letters* 20, no. 22: 7125–30.

Wieczorek, N., M. A. Kucuker, and K. Kuchta. 2014. Fermentative Hydrogen and Methane Production From Microalgal Biomass (*Chlorella vulgaris*) in a Two-Stage Combined Process. *Applied Energy* 132: 108–17.

Wogu, M. D., and C. E. Okaka. 2011. "Pollution Studies on Nigerian Rivers: Heavy Metals in Surface Water of Warri River, Delta State." *Journal of Biodiversity and Environmental Sciences* 1, no. 3.

Wu, S. C., X. L. Peng, K. C. Cheung, S. L. Liu, and M. H. Wong. 2009. "Adsorption Kinetics of Pb and Cd by Two Plant Growth Promoting Rhizobacteria." *Bioresource Technology* 100, no. 20: 4559–63. https://doi.org/10.1016/j.biortech.2009.04.037.

Yanai, J., F. J. Zhao, S. P. McGrath, and T. Kosaki. 2016. "Effect of Soil Characteristics on Cd Uptake by the Hyperaccumulator *Thlaspicaerulescens*." *Environmental Pollutant* 139, no. 1: 167–75.

Yoon, E., A. Babar, M. Choudhary, M. Kutner, and N. Pyrsopoulos. 2016. "Acetaminophen-Induced Hepatotoxicity: A Comprehensive Update." *Journal of Clinical and Translational Hepatology* 4, no. 2: 131.

Zadeh, B. M., G. R. Savaghebi-Firozabadi, H. A. Alikhani, and H. M. Hosseini. 2018. "Effect of Sunflower and Amaranthus Culture and Application of Inoculants on Phytoremediation of the Soils Contaminated with Cadmium." *American-Eurasian Journal of Agricultural & Environmental Sciences* 4, no. 1: 93–103.

Zarcinas, B. A., C. F. Ishak, M. J. McLaughlin, and G. Cozens. 2014. "Heavy Metals in Soils and Crops in Southeast Asian Environment." *Geochemical Health* 26, no. 3: 343–57.

Zhang, X., X. Yuan, H. Shi, L. Wu, H. Qian, and W. Xu. 2015. Exosomes in Cancer: Small Particle, Big Player. *Journal of Hematology & Oncology* 8: 1–13.

Zhou, Q. X., and Y. F. Song. 2004. *Principles and Methods of Contaminated Soil Remediation*. Vol. 568. Beijing: Science Press.

3 Mycoremediation of Metallic Pollutants

Roshni J. Patel, Swati Mohapatra, and Arti Hansda

3.1 INTRODUCTION

The environment contains variety of contaminants, including atoms, molecules, compounds, metals, and non-metals. The contamination of an ecosystem due to overabundance of any of these components causes imbalance and pollution throughout the ecosystem. Due to anthropogenic activities, the fundamental ecosystem components have deteriorated, such as air, water, and soil (Hansda *et al.*, 2022). Both organic and inorganic contaminants are responsible for polluting the environment. Anthropogenic activities introduce a variety of toxins into the environment, including agrochemicals, hydrocarbons, plastic debris, and e-waste. Elements with >5 g cm^{-3} are known as heavy metals (e.g. copper (Cu), gold (Au), cadmium (Cd), nickel (Ni), thallium (Tl), mercury (Hg), cobalt (Co), zinc (Zn), gallium (Ga), lead (Pb), chromium (Cr), arsenic (As), silver (Ag)) (Koller and Saleh, 2018). Some heavy metals are essential for the biochemical and physiological activities of living organisms; however, their presence over threshold limits results in a negative impact on biotic components of the surrounding environment (Hamba and Tamiru, 2016). Organic contaminants can be degraded by microbial actions; however, heavy metal/metalloid remediation involves various physicochemical practices. Prolonged exposure to heavy metals can cause various environmental, nutritional, and health concerns in animals and human beings (e.g. liver and kidney malfunctioning, cancer, and neural disorders) (Mitra *et al.*, 2022). Heavy metals enter the environment in three ways: environmental accumulation, industrial metal disposal in sludge, and metal mining.

Conventional methods like ex-situ excavation, soil detoxification, and remediation of soil by physicochemical techniques (e.g. electroplating, evaporation, ion exchange, etc.) are costly, time consuming, and non-eco-friendly because of elevated energy requirements, economic inefficiency, and environmental non-friendliness, which results in degradation of the biotic features of the environment (Singh and Prasad, 2015). The shortcomings of conventional methods promoted the development of a revolutionary biological technology called bioremediation. This process takes part in metal clean-up via extracellular adsorption, intracellular accumulation, complexation, and redox reactions (Hansda *et al.*, 2016). The application of biological materials for remediation of metal pollutants makes this technology an eco-friendly, feasible, and sustainable approach. In the current scenario, biological agents, particularly microorganisms like fungi, actinomycetes, and bacteria, have been used to remove toxic wastes through their enzymatic activity. Microbe-based technologies, such as biostimulation, bioaugmentation, bioaccumulation, biosorption, phytoremediation, and rhizoremediation, have been utilized for the complete mineralization of pollutants (Prasad and Prasad, 2012). Employing such microbial technologies for the economical and eco-friendly removal of heavy metals can be the best solution for pollution abatement.

3.2 MYCOREMEDIATION

Paul Stamets invented the word "mycoremediation" to describe the technique of employing fungi to reduce pollution in the environment (Stamets, 1999; Hamba and Tamiru, 2016). Mycoremediation holds promise over traditional methods due to the pervasive and preeminent nature of fungi, the

DOI: 10.1201/9781003423393-3

production of large biomass, and various metabolites (extracellular enzymes and organic acids). They also play an important role in the cleanup of polluted environments by enhancing plants' phytoremediation efficiency through nutrient uptake and phytoextraction (Hou *et al.*, 2017).

By virtue of their diversity and ability to live in harsh environments, fungi are abundant in soil. Through the existence of mycelium, fungi may propagate through materials very quickly. Hypha penetration offers a mechanical foundation for chemical breakdown, which is influenced by the released enzymes. The success of mycoremediation relies on selecting the appropriate fungal species that target specific contaminants. Several fungal species, including *Trichoderma* spp., *Aspergillus* spp., *Rhizopus* spp., and *Penicillium* spp., have demonstrated potential for heavy metal remediation (Priyadarshini *et al.*, 2021). However, the effectiveness of mycoremediation depends on favorable environmental parameters, for example, pH, temperature, initial metal concentration, agitation rate, flow rate, biomass, and contact time (Kumar and Dwivedi, 2021). The current chapter delves into the role of fungi in removal of heavy metal from polluted sites and their involvement in improving plant tolerance to heavy metals.

3.3 HEAVY METAL TOXICITY

Each heavy metal has its own toxicity route. In addition to the restricted biodegradation of organic contaminants in the environment, the excessive availability of inorganic contaminants such as heavy metals also negatively affects the quality of the soil and water (Oves *et al.*, 2016). Phytotoxicity and nutrient deficiency are issues related to crops grown in polluted soil (Shahid *et al.*, 2014). Heavy metals enter the food chain by using wastewater for irrigation purposes, which simultaneously results in a serious threat to human health (Mahmood and Malik, 2014).

3.3.1 ARSENIC

Arsenic (As) is a prevalent metalloid found naturally and anthropogenically (by mining, usage of pesticides, burning of fossil fuels, and irrigation with arsenic-contaminated water, etc.) in nature. Plants affected with As toxicity result in the reduction of biomass, productivity, and metabolism (Srivastava and Sharma, 2013). Humans are exposed to As mostly via contaminated drinking water, crops, and fodder.

3.3.2 LEAD

Lead (Pb) has been recognized as the second most hazardous metal after As due to its prevalent appearance, adverse effects, and exposure to humans (Jennrich, 2013). Pb contamination of the environment is due to a variety of human activities such as mining and smelting of lead ores, automotive exhaust, lead-containing paints, petrol and explosives, and effluents from storage battery businesses (Sharma and Dubey, 2005). Reduced germination index, stunted growth, damaged photosynthetic system, and decreased nutrient uptake are the outcomes of lead toxicity in plants (Emamverdian *et al.*, 2015). In humans, Pb causes various cardiovascular, neurotoxic, and carcinogenic detrimental effects (WHO, Fact Sheet, 2017).

3.3.3 CADMIUM

The toxicity, lengthy half-life of elimination, and high solubility in water make cadmium (Cd) a non-essential heavy metal contaminant. It is mostly released into the environment due to the usage of sewage sludge and phosphatic fertilizers in industrial, mining, and agricultural activities (Satarug *et al.*, 2003). Cd toxicity in plants results in chlorotic leaves, which eventually lead to a reduction in the photosynthetic rate (Perfus-Barbeoch *et al.*, 2002). The toxicity of Cd in humans results in detrimental effects on cardiovascular, neurological, and reproductive systems (Bernard, 2008).

3.3.4 Chromium

Chromium (Cr), a non-essential, toxic metal for biotic communities, exists in the states of Cr (III) and Cr (VI) (Shanker *et al.*, 2005). Cr toxicity results in the damaging of various physiological and metabolic activities of plants (Singh and Sharma, 2017). Shahid *et al.* (2017) reported that Cr causes toxicity in plants by interfering with plant growth, nutrition uptake, and photosynthesis; increasing reactive oxygen species generation; inducing lipid peroxidation; and changing the activities of antioxidant enzymes. It also alters the permeability of cell membranes, disrupts ion channel balance, and damages DNA and protein.

3.3.5 Nickel

Nickel (Ni) is an essential nutrient for plants which acts as a constituent of various metalloenzymes (hydrogenase, urease, methyl coenzyme M reductase, etc.) (Das *et al.*, 2018). It is required for redox reactions, which are important for several vital plant processes. Despite being crucial for plant metabolism, higher concentrations are toxic to plants. When the permissible limit is exceeded, it changes the natural metabolism of plants, resulting in cellular damage and, at the extreme, plant death (Yusuf *et al.*, 2011). Chlorosis, necrosis, retarded shoot root development, and a reduction in leaf area are among the signs of nickel poisoning in plants (Singh *et al.*, 2010). Ni toxicity in humans results in dermatitis, carcinogenicity, and epigenetic effects.

3.3.6 Zinc

Zinc (Zn) is an essential micronutrient for plants' growth, though, at higher concentrations, its toxicity results in chlorosis and tilted leaves (Kaur and Garg, 2021). Zn poisoning has an impact on root growth as well, which leads to short lateral roots that can even shrink and turn yellow; it also damages the photosynthetic apparatus (Vassilev *et al.*, 2011). Zn toxicity in humans may result in dermal, cardiovascular, gastrointestinal, and neurotoxic disorders (Nriagu, 2007).

3.3.7 Copper

Copper (Cu) is an important micronutrient necessary for the metabolic activities of flora and fauna. However, concentrations above the permissible limit result in negative effects in leaves such as chlorosis and wilting. Root growth is also affected by copper toxicity by producing short and blunted radicles; some cases have been reported where no root growth was observed in excess of Cu (Yruela, 2005). Cu toxicity results in gastrointestinal, cardiovascular, hematological, and hepatic disorders (Ashish *et al.*, 2013).

3.3.8 Manganese

Manganese (Mn) is a vital micronutrient for the growth of plants and animals. The presence of Mn above the permissible limit negatively affects the transportation process in plants (Pearson *et al.*, 1995; Herren and Feller, 1997). The toxic level of Mn damages the plants' leaves by browning and crinkling. The plant toxicity of Mn includes chlorosis, necrosis, and inactivation of enzymes (El-Jaoual and Cox, 1998). Mn toxicity in humans may result in neurological damage (Keen *et al.*, 2000).

3.4 MECHANISM OF FUNGI IN REDUCING HEAVY METAL TOXICITY

Depending on the complexity of the cell, various mechanisms govern how metal is assimilated by microbes. Though the mechanism is unclear, there are two recognized forms of heavy metal biosorption. The initial mechanism (passive mode), which includes the metal's absorption regardless of metabolic activity, involves the interaction between metal ions and functional groups present on the surface of cell walls and extracellular material by physical adsorption, precipitation, ion exchange, and complexation (Veglio and Beolchini, 1997). The second method involves moving metals throughout the cell membrane and is effective, active, and energy dependent. Fungi's superior ability to bind metals is made more favorable due to the existence of a significant amount of components found in cell walls (Abbas *et al.*, 2014). The availability of various ligands (amine, carboxyl, hydroxyl, sulfhydryl, phosphate groups, etc.) on the surface of fungi is responsible for the adsorption of metal ions via complexing with negatively charged reaction sites (Kumar *et al.*, 2000). Due to variations in the composition of their cell walls, bacteria, algae, fungi, and yeast all exhibit different types of metal chelation. Metal ion sequestration is a property of the chitin and chitosan-based fungal cell wall (Das *et al.*, 2008). Fungi can remediate heavy metal contamination by using various mechanisms such as biotransformation, biosorption, bioaccumulation, and biomineralization (Figure 3.1). Table 3.1 illustrates the various reports available on fungal mechanisms for heavy metal removal.

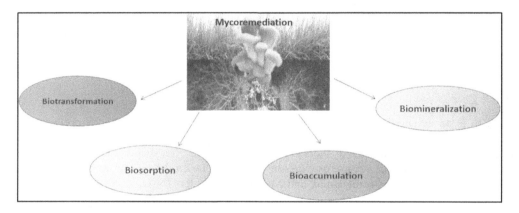

FIGURE 3.1 Mechanism of heavy metal remediation in fungi.

TABLE 3.1
Results Showing Involvement of Fungi for Remediation of Heavy Metals

Fungi	Heavy Metal	Mechanism Involved	Reference
Mucor sp.	Pb, Cd, Ni, Zn, Cu	Bioaccumulation	Zhang *et al.* (2017)
Pleurotus sp.	Co, Cu, Ni	Biosorption	Mohamadhasani and Rahimi (2022)
Pleurotus	Pb	Bioaccumulation	Dulay *et al.* (2015)
Pleurotus florida, Trichoderma viride	Pb	Biosorption	Prasad *et al.* (2013)
Aspergillus niger (M1DGR), *Aspergillus fumigatus* (M3Ai), *Penicillium rubens* (M2Aii)	Cd, Cr	Biosorption, bioaccumulation	Khan *et al.* (2019)
Pleurotus ostreatus	Pb, Cd	Bioaccumulation	Kapahi and Sachdeva (2017)
Hymenoscyphus ericae	Zn, Cd and Fe	Biosorption	Binsadiq (2015)
Rhizopus microspores	Cd, Cu, As, and Fe	ND	Oladipo *et al.* (2018)
Mucor indicus	Pb	Biosorption	Samadi *et al.* (2017)

ND = Not defined

Mycoremediation of Metallic Pollutants

3.4.1 Biosorption

The capacity of microbes for biosorption determines how microbes can influence the availability of metals. Ionic exchange, chelation, adsorption on cell surfaces, and concentration gradient-induced entrapment in inter- and intra-fibrillar capillaries are all instances of biosorption, which is described as the removal of materials (compounds, metal ions, etc.) by biotic processes (Wierzba, 2017).

3.4.2 Bioaccumulation

Microbes have evolved a number of defense mechanisms against the harmful effects of metals and metalloids, including accumulation, resistance, and—more interestingly—lowering the bioavailability or toxicity of heavy metals via biomethylation and transformation. Bioaccumulation is a different active technique that uses an energy-dependent metal inflow mechanism to remove metal from living cells (Iram *et al.*, 2015). It is a more gradual technique than biosorption and is reliant on good environmental status and the toxicity of the metal. Certain *Trichoderma* species have reportedly been shown to assemble several heavy metals in vitro, including Cu, Zn, Cd, and As (Errasquín and Vázquez, 2003; Zeng *et al.*, 2010). Based on enzymatic mechanisms, biomethylation transforms organic and inorganic metal compounds into their volatile cognates.

3.4.3 Mobilization

Metals' mobility and bioavailability are largely controlled by microbes through a number of processes in the biogeochemistry of metals. Metal mobilization and immobilization processes are influenced by the kind of microorganisms involved, their habitat, and physicochemical circumstances. The processes by which metals are mobilized include heterotrophic and autotrophic leaching, complexation/chelation by siderophores and metabolites, and their methylation-induced volatilization.

The ability of microbes to mobilize or dissolve insoluble metal compounds depends on several factors, including pH, the stability constants of the various complexes, and the relative ionic and metal concentrations in solution. By releasing protons through plasma membranes, maintaining charge balance, releasing organic acids, and accumulating carbon dioxide through respiration to build metal complexes, microbes acidify their surroundings. The most typical method for bacteria and fungi to acquire iron is by the synthesis of low-molecular-weight iron-chelating siderophores. Though siderophores are well known for improving iron uptake, they can also bind hazardous and trace amounts of heavy metals, though their affinity is only marginally high (Rajkumar *et al.*, 2010). Cd, Zn, and Pb's bioavailability were reduced as a result of siderophore-induced metal solubilization due to adsorbed or precipitated metal in the biomass (Asad *et al.*, 2019).

3.4.4 Immobilization

According to Gadd and Sayer (2000), fungi can immobilize metals via precipitation, intracellular sequestration, or transformation in both metabolically dependent and independent ways. The fungal cell wall is the initial site of interaction between metal and fungi. The polysaccharides present in the fungal cell wall include a number of functional groups (carboxylate, hydroxyl, sulfate, phosphate, and amino groups) that bind to various metal ions depending on their bioavailability. Rahman and Sathasivam (2015) reported that the binding of metal ions to ligands is influenced by a number of variables, including their electrostatic interactions, electronegativity, accessibility, ionic radius, and metal speciation. Microbes also secrete extracellular polymeric substances (EPSs), a complex mixture of proteins, mucopolysaccharides, and polysaccharides that can bind with potentially harmful metals, as a form of self-defense. To sequester positively charged heavy metal ions, the ionizable functional groups and noncarbohydrate substituents present on EPS provide the polymer with an overall negative charge (Gupta and Diwan, 2017). *Herminiimonas arsenicoxydans* caused the production of biofilms in response to exposure to arsenic and its removal via EPS (Marchal *et al.*, 2010). Fungi produce oxalic acids, which immobilize metal ions by forming compounds with them. Calcium oxalate is a common oxalate form in the environment associated with the precipitation of solubilized calcium (Gadd, 2004). In addition to calcium oxalates, fungi can produce a variety of metal oxalates such as Co, Cd, Mn, Sr, and Zn (Franceschi and Nakata, 2005). Heavy metal ions with the same charge and ionic radius can cross the cell membrane via the cation transport system (Javanbakht *et al.*, 2014).

3.4.5 BIOTRANSFORMATION

It has been demonstrated that a variety of fungi have the unique capacity to biotransform metals, which aids in the removal of heavy metals. They are capable of influencing the mobility and toxicity of metals, metalloids, and organometallic compounds by various chemical reactions, including oxidation-reduction, methylation, catalysis, and dealkylation. S-adenosylmethionine, methylcobalamin, or N-methyltetrahydrofolate are the three possible routes by which methylation of metals and metalloids takes place (Mason, 2012). The process of methylation involves adding methyl groups to metals to create molecules with a range of solubility, volatility, and toxicity. Carbon, oxygen, nitrogen, and sulfur function as methyl group acceptors in secondary metabolic activities. The transformation of arsenate into methylarsonic acid or dimethylarsinic acid is one of the potential detoxification processes. In the presence of arsenate reductase, arsenate is first reduced to arsenite and then transformed from the water-soluble arsenate to the volatile trimethylarsine. Following a sequence of methylation and reoxidation, the organo-arsenical intermediates are then reduced using S-adenosylmethionine as the typical methyl donor (Mukhopadhyay *et al.*, 2002). The same mechanism is used for the methylation of selenium, while the cobalamin pathway is used for the methylation of other metal ions, such as tin (Sn) and mercury (Hg), when they are in their most oxidized state (Mason, 2012). In contrast, some results have also been found suggesting that the cobalamin pathway is used to methylate arsenic and other metalloids (Wuerfel *et al.*, 2012). Several fungal strains, such as *Rhizopus* spp., *Neocosmospora* spp., and *Trichoderma* spp., have been known to effectively remove As from the growth medium (Srivastava *et al.*, 2011). *Penicillium* undergoes biotransformation to produce the organic forms of arsenic (MMA and DMA).

3.5 PLANT–FUNGAL INTERACTION AS A BOON FOR HEAVY METAL STRESS MITIGATION

A confined area of soil known as the rhizosphere surrounds plant roots and acts as a hub for microbial activity. These microorganisms are important in controlling the plants' development and health. Through a variety of methods, including biological nitrogen fixation, mineral solubilization, ACC deaminase activity, and the synthesis of siderophores and phytohormones, microorganisms aid in the encouragement of plant growth (Fayuan *et al.*, 2022). In addition to stimulating plant growth, microbes also protect plants from a number of abiotic and biotic stressors through a variety of approaches due to their broad metabolic and genetic capacities (Meena *et al.*, 2017). Numerous biochemical, physiological, and molecular investigations have shown that plants are heavily influenced by stressors through their relationship with microorganisms (Farrar *et al.*, 2016). The relationship between microorganisms and plants under stressful circumstances culminates in a variety of local and systemic reactions that aid plants in overcoming these challenges by enhancing their metabolic capacities (Nadeem *et al.*, 2014).

Plant–fungi associations differ greatly from other associations, they mainly include mycorrhizal associations (Genre *et al.*, 2020). Frank introduced the term "mycorrhiza" in 1885, emphasizing the importance of mycorrhizal interaction for the survival of both species and the absorption of nutrients from the soil (Harley and Smith, 1983). One of the most well-known types of mycorrhizal

Mycoremediation of Metallic Pollutants

associations is the vesicular arbuscular mycorrhizal (VAM) association in which fungi derive benefits from this association by obtaining nutrients from the plant while also enhancing the uptake of phosphorus and trace metals by the plant (Sylvia, 1994). VAM fungi are commonly found in the rhizosphere, but they may also occur in the stem and thallus of plants (Read *et al.*, 2000). The arbuscular mycorrhiza improves the nutritional state of the host plant, whereas fungal hyphae aid in the absorption and transport of heavy metals to plants.

Heavy-metal resistant rhizobacteria or mycorrhiza have drawn increasing attention in phytoremediation strategies because of their capacity to influence the mobility and bioavailability of heavy metals to the plant via release of chelating agents, mineral solubilization, and redox changes (Jiang *et al.*, 2016). The heavy metal–plant–mycorrhiza relationship plays an integral role in the cleanup of ecosystems by facilitating the removal of heavy metals from the soil. On the one hand, mycorrhizal plants exhibit an increased uptake of heavy metals, while on the contrary, VAM fungi help in the immobilization of heavy metals in the soil. This interaction helps in maintaining metal–ion equilibrium both internally within the plants and in their surrounding environment, especially during stressful conditions characterized by elevated concentrations of toxic heavy metals (Clemens, 2001; Hall, 2002). Plants employ several mechanisms in order to overcome these stress conditions and maintain the equilibrium of metal ions.

Arbuscular mycorrhizal fungi (AMF) and various fungal strains (e.g. *Trichoderma, Penicillium, Rhizopus, etc.*) are also known to influence a plant's mechanism for absorbing and decreasing the negative effects of heavy metals accumulated from the environment (Herath *et al.*, 2021). *Trichoderma harzianum* was found to enhance the growth of crack willow (*Salix fragilis*) in metal-contaminated soil, while *Trichoderma atroviride* affected the absorption and transportation of Zn, Ni, and Cd in *Brassica juncea* (Adams *et al.*, 2007).

According to Lynch and Moffat (2005), *Trichoderma harzianum* strains can neutralize the toxicity of potassium cyanide and encourage the root development of the arsenic hyperaccumulating fern *Pteris vittata*. The combination treatment of AM fungi and *Trichoderma harzianum* boosted the resistance and uptake of *Eucalyptus globulus* to elevated concentrations of aluminum and arsenic in soil (Arriagada *et al.*, 2009). AMFs' stress response is influenced by various factors, including plant and fungus species, metal availability, soil quality, and plant development circumstances. Additionally, they lessen the effects of metal toxicity on plants, or they accumulate the metals into their biomass.

3.6 FUNGI GENES INVOLVED IN HEAVY METAL REMEDIATION

Different operons are responsible for the bioremediation of heavy metal contamination. These genes contain a variety of pathways that aid in heavy metal detoxification. A number of ericoid mycorrhizal genes, including DNA damage repair proteins, metal transporters, and antioxidant enzymes, have been identified in *Saccharomyces cerevisiae* using functional complementation, which may be responsible for metal tolerance (Daghino *et al.*, 2016). According to Gonzalez Guerrero *et al.* (2010), exposure to cadmium and copper increased the expression of the ABC transporter gene (GintABC1) of *Gigaspora intraradices*. Hildebrandt *et al.* (2007) found that the presence of the Zn transporter, the heat-shock protein metallothionein, the enzyme glutathione S-transferase, and the expression of several genes producing proteins are responsible for the heavy metal stress tolerance in Glomus intraradices Sy167. Eukaryotic organisms are known to thrive and develop normally due to the availability of polyamines. There is also a ton of evidence that polyamines have a protective effect against stress. According to Tripathi *et al.* (2017), different *Trichoderma* species may help lower grain arsenic levels by altering gene expression, architecture, and nutrient and mineral uptake. Many microorganisms evolve metal-resistant genes as a form of adaptation when exposed to hazardous metals. These metal-resistant genes, which are organized in operon clusters, have an optimum availability for heavy metal bioremediation (Das *et al.*, 2016).

3.7 ENVIRONMENTAL FACTORS AFFECTING DETOXIFICATION

By modifying the local soil and water conditions, it may be feasible to stimulate biological activity and, as a result, the degradation/removal of toxic compounds via mycoremediation. The living and nonliving states of biomass; the types of biomaterials; the properties of metals; and other environmental circumstances like contact time, temperature, and pH all have an impact on biosorption (Abbas *et al.*, 2014). Luo *et al.* (2010) reported the highest removal of Cr (VI) and Ni (II) by *Aspergillus* sp. and *Micrococcus* sp. at pH 5.0 and pH 7.0, respectively. However, no sorption was observed at pH 2.0. Cd was shown to be more readily absorbed in mild acid conditions than strong acid conditions. A Pb-tolerant fungal strain, *Rhizopus arrhizus*, was discovered in addition to Cr and Cd, and it accumulated more lead in the hyphae, suggesting that the functional groups present on the cell surface of fungi may function as ligands for metal sequestration, responsible for recovery of metals from aqueous culture media (Pal *et al.*, 2010).

The use of *Penicillium canescens* has also demonstrated improved sorption effectiveness with an increase in pH for binding of heavy metals such as Pb and Cu (Say *et al.*, 2003). According to Damodaran *et al.* (2011), increased aeration and the availability of glucose as a carbon source encourage the biomass growth of *Saccharomyces cerevisiae*. This increases the biosorption of Cd and Pb. Thippeswamy *et al.* (2012) discovered the enhancement in metal extraction of metals in *Saccharomyces* spp. with a decrease in their biomass and proposed that low biomass results in high surface to volume, holding the maximum amount of heavy metals in both soluble and specific form by increasing metal interaction with active binding sites of the cell surface.

3.8 CONCLUSION AND FUTURE PERSPECTIVES

Both industrial development and new process implementation, which often require the usage of heavy metal pollutants, have an adverse effect on the ecosystem As a result, there is an immediate need for environmentally friendly, financially viable, and superior methods in terms of operational concerns, profitability, and efficacy. Mycoremediation is an environmentally friendly and sustainable technology with enormous potential for the removal of heavy metals and organic pollutants present in the environment. The symbiotic relationship between plants and fungi in mycoremediation processes can effectively eliminate harmful metals, transport them, and collect them in the shoots of plants.

Despite its tremendous potential, mycoremediation has not been widely marketed or used on a significant basis. Furthermore, several challenges must be addressed in order to fully realize the potential of mycoremediation. These challenges include optimizing fungal–plant interactions, selecting appropriate fungal and plant species for different contaminants, and addressing practical considerations for large-scale implementation. This can be made possible by collaboration between scientists, industry, and policymakers, along with continued research and investment in this field, which are crucial for the widespread adoption and utilization of mycoremediation as a viable and effective solution for environmental cleanup. However, more focus is needed to unravel the mechanisms involved in the transport and accumulation of heavy metals and metalloids in plants, employing transcriptomic and proteomic approaches. The knowledge gained from transcriptomic and proteomic studies can be utilized to enhance the efficiency and effectiveness of mycoremediation strategies, leading to a more successful implementation of this approach in the remediation of heavy metal and metalloid-contaminated sites. This integrative approach provides valuable insights into the metabolic changes and adaptations that occur under heavy metal and metalloid stress.

Overall, while significant strides have been made in understanding the molecular and metabolic aspects of mycoremediation, there is still work to be done to unlock its full potential for widespread commercial application in environmental cleanup. Future research should concentrate on creating methods that use genetic and metabolic engineering to improve the tolerance, absorption, and hyperaccumulation of heavy metals and metalloids.

REFERENCES

Abbas, S. H., I. M. Ismail, T. M. Mostafa, and A. H. Sulaymon. 2014. "Biosorption of Heavy Metals: A Review." *Journal of Chemical Science & Technology* 3, no. 4: 74–102.

Adams, P., F. A. De-Leij, and J. M. Lynch. 2007. "*Trichoderma harzianum* Rifai 1295–22 Mediates Growth Promotion of Crack Willow (*Salix fragilis*) Saplings in Both Clean and Metal-Contaminated Soil." *Microbial Ecology* 54, no. 2: 306–13. https://doi.org/10.1007/s00248-006-9203-0.

Arriagada, C., E. Aranda, I. Sampedro, I. Garcia-Romera, and J. A. Ocampo. 2009. "Contribution of the Saprobic Fungi *Trametes versicolor* and *Trichoderma harzianum* and the Arbuscular Mycorrhizal Fungi *Glomus deserticola* and *G. Claroideum* to Arsenic Tolerance of *Eucalyptus globulus*." *Bioresource Technology* 100, no. 24: 6250–7. https://doi.org/10.1016/j.biortech.2009.07.010.

Asad, Saeed Ahmad, Muhammad Farooq, Aftab Afzal, and Helen West. 2019. "Integrated Phytobial Heavy Metal Remediation Strategies for a Sustainable Clean Environment-A Review." *Chemosphere* 217: 925–41. https://doi.org/10.1016/j.chemosphere.2018.11.021.

Ashish, B., K. Neeti, and K. Himanshu. 2013. "Copper Toxicity: A Comprehensive Study." *Research Journal of Recent Science* 2277: 2502.

Bernard, A. 2008. "Cadmium and Its Adverse Effects on Human Health." *Indian Journal of Medical Research* 128, no. 4: 557–64.

Binsadiq, A. R. H. 2015. "Fungal Absorption and Tolerance of Heavy Metals." *Industrial Wastewater* 7: 11.

Clemens, S. 2001. "Molecular Mechanisms of Plant Metal Tolerance and Homeostasis." *Planta* 212, no. 4: 475–86. https://doi.org/10.1007/s004250000458.

Daghino, Stefania, Elena Martino, and Silvia Perotto. 2016. "Model Systems to Unravel the Molecular Mechanisms of Heavy Metal Tolerance in the Ericoid Mycorrhizal Symbiosis." *Mycorrhiza* 26, no. 4: 263–74. https://doi.org/10.1007/s00572-015-0675-y.

Damodaran, D., G. Suresh, and R. Mohan. 2011. *Bioremediation of Soil by Removing Heavy Metals Using Saccharomyces cerevisiae.* Vol. 2: 22–7. 2nd International Conference on Environmental Science and Technology (IPCBEE), Singapore.

Das, K. K., R. C. Reddy, I. B. Bagoji, S. Das, S. Bagali, L. Mullur, J. P. Khodnapur, and M. S. Biradar. 2018. "Primary Concept of Nickel Toxicity–an Overview." *Journal of Basic & Clinical Physiology & Pharmacology* 30, no. 2: 141–52. https://doi.org/10.1515/jbcpp-2017-0171.

Das, N., R. Vimala, and P. Karthika. 2008. "Biosorption of Heavy Metals–an Overview." *Indian Journal of Biotechnology* 7: 159–69.

Das, Surajit, Hirak R. Dash, and Jaya Chakraborty. 2016. "Genetic Basis and Importance of Metal Resistant Genes in Bacteria for Bioremediation of Contaminated Environments with Toxic Metal Pollutants." *Applied Microbiology & Biotechnology* 100, no. 7: 2967–84. https://doi.org/10.1007/s00253-016-7364-4.

Dulay, R. M. R., M. E. G. D. Castro, N. B. Coloma, A. P. Bernardo, A. G. D. Cruz, R. C. Tiniola, S. P. Kalaw, and R. G. Reyes. 2015. "Effects and Myco-Accumulation of Lead (Pb) in Five *Pleurotus* Mushrooms." *International Journal of Biology, Pharmacy & Allied Sciences* 4, no. 3: 1664–77.

Emamverdian, A., Y. Ding, F. Mokhberdoran and Y. Xie. 2015. "Heavy metal stress and some mechanisms of plant defense response". *Scientific World Journal.* 2015: 756120.

El-Jaoual, T., and D. A. Cox. 1998. "Manganese Toxicity in Plants." *Journal of Plant Nutrition* 21, no. 2: 353–86. https://doi.org/10.1080/01904169809365409.

Farrar, D., M. Simmonds, M. Bryant, T. A. Sheldon, D. Tuffnell, S. Golder, F. Dunne, and D. A. Lawlor. 2016. "Hyperglycaemia and Risk of Adverse Perinatal Outcomes: Systematic Review and Meta-Analysis." *BMJ* 13, no. 354: i4694. https://doi.org/10.1136/bmj.i4694. PMID: 27624087; PMCID: PMC5021824.

Fayuan, W., P. Cheng, S. Zhang, S. Zhang, and S. Yuhuan. 2022. "Contribution of Arbuscular Mycorrhizal Fungi and Soil Amendments to Remediation of Heavy Metal-Contaminated Soil Using Sweet Sorghum." *Pedosphere* 32, no. 6: 844–55. https://doi.org/10.1016/j.pedsph.2022.06.011.

Franceschi, Vincent R., and Paul A. Nakata. 2005. "Calcium Oxalate in Plants: Formation and Function." *Annual Review of Plant Biology* 56: 41–71. https://doi.org/10.1146/annurev.arplant.56.032604.144106.

Gadd, G. M. 2004. "Mycotransformation of Organic and Inorganic Substrates." *Mycologist* 18: 60–70.

Gadd, G. M., and J. A. Sayer. 2000. "Influence of Fungi on the Environmental Mobility of Metals and Metalloids." In *Environmental Microbe-Metal Interactions*, edited by D. R. Lovley: ASM (American Society of Microbiology) Press. 237–56. Washington, DC.

Genre, Andrea, Luisa Lanfranco, Silvia Perotto, and Paola Bonfante. 2020. "Unique and Common Traits in Mycorrhizal Symbioses." *Nature Reviews. Microbiology* 18, no. 11: 649–60. https://doi.org/10.1038/s41579-020-0402-3.

González-Guerrero, Manuel, Elodie Oger, Karim Benabdellah, Concepción Azcón-Aguilar, Luisa Lanfranco, and Nuria Ferrol. 2010. "Characterization of a CuZn Superoxide Dismutase Gene in the Arbuscular Mycorrhizal Fungus *Glomus Intraradices*." *Current Genetics* 56, no. 3: 265–74. https://doi.org/10.1007/s00294-010-0298-y.

Gupta, Pratima, and Batul Diwan. 2017. "Bacterial Exopolysaccharide Mediated Heavy Metal Removal: A Review on Biosynthesis, Mechanism and Remediation Strategies." *Biotechnology Reports* 13: 58–71. https://doi.org/10.1016/j.btre.2016.12.006.

Hall, J. L. 2002. "Cellular Mechanisms for Heavy Metal Detoxification and Tolerance." *Journal of Experimental Botany* 53, no. 366: 1–11. https://doi.org/10.1093/jexbot/53.366.1.

Hamba, Y., and M. Tamiru. 2016. "Mycoremediation of Heavy Metals and Hydrocarbons Contaminated Environment." *Asian Journal of Natural & Applied Sciences* 5: 2.

Hansda, A., P. C. Kisku, V. Kumar, and Anshumali. 2022. "Plant-Microbe Association to Improve Phytoremediation of Heavy Metal." In *Advances in Microbe-Assisted Phytoremediation of Polluted Sites*: 113–46. Amsterdam: Elsevier.

Hansda, Arti, Vipin Kumar, and Anshumali. 2016. "A Comparative Review Towards Potential of Microbial Cells for Heavy Metal Removal with Emphasis on Biosorption and Bioaccumulation." *World Journal of Microbiology & Biotechnology* 32, no. 10: 170. https://doi.org/10.1007/s11274-016-2117-1.

Harley, J. L., and S. E. Smith. 1983. *Mycorrhizal Symbiosis*. London: Academic Press.

Herath, B. M. M. D., K. W. A. Madushan, J. P. D. Lakmali, and P. N. Yapa. 2021. "Arbuscular Mycorrhizal Fungi as a Potential Tool for Bioremediation of Heavy Metals in Contaminated Soil." *World Journal of Advance Research & Review* 10, no. 3: 217–28.

Herren, T., and U. Feller. 1997. "Influence of Increased Zinc Levels on Phloem Transport in Wheat Shoots." *Journal of Plant Physiology* 150: 228–31.

Hildebrandt, Ulrich, Marjana Regvar, and Hermann Bothe. 2007. "Arbuscular Mycorrhiza and Heavy Metal Tolerance." *Phytochemistry* 68, no. 1: 139–46. https://doi.org/10.1016/j.phytochem.2006.09.023.

Hou, Dandi, Kai Wang, Ting Liu, Haixin Wang, Zhi Lin, Jie Qian, Lingli Lu, and Shengke Tian. 2017. "Unique Rhizosphere Micro-characteristics Facilitate Phytoextraction of Multiple Metals in Soil by the Hyperaccumulating Plant Sedum alfredii." *Environmental Science & Technology* 51, no. 10: 5675–84. https://doi.org/10.1021/acs.est.6b06531.

Iram, S., R. Shabbir, H. Zafar, and M. Javaid. 2015. "Biosorption and Bioaccumulation of Copper and Lead by Heavy Metal-Resistant Fungal Isolates." *Arabian Journal for Science & Engineering* 40, no. 7: 1867–73. https://doi.org/10.1007/s13369-015-1702-1.

Javanbakht, Vahid, Seyed Amir Alavi, and Hamid Zilouei. 2014. "Mechanisms of Heavy Metal Removal Using Microorganisms as Biosorbent." *Water Science & Technology* 69, no. 9: 1775–87. https://doi.org/10.2166/wst.2013.718.

Jennrich, P. 2013. "The Influence of Arsenic, Lead, and Mercury on the Development of Cardiovascular Diseases." *ISRN Hypertension* 2013: 1–15. https://doi.org/10.5402/2013/234034.

Jiang, Q. Y., S. Y. Tan, F. Zhuo, D. J. Yang, Z. H. Ye, and Y. X. Jing. 2016. "Effect of Funneliformis mosseae on the Growth, Cadmium Accumulation and Antioxidant Activities of Solanum nigrum." *Applied Soil Ecology* 98: 112–20. https://doi.org/10.1016/j.apsoil.2015.10.003.

Kapahi, Meena, and Sarita Sachdeva. 2017. "Mycoremediation Potential of Pleurotus species for Heavy Metals: A Review." *Bioresources & Bioprocessing* 4, no. 1: 32. https://doi.org/10.1186/s40643-017-0162-8.

Kaur, Harmanjit, and Neera Garg. 2021. "Zinc Toxicity in Plants: A Review." *Planta* 253, no. 6: 129. https://doi.org/10.1007/s00425-021-03642-z.

Keen, C. L., J. L. Ensunsa, and M. S. Clegg. 2000. "Manganese Metabolism in Animals and Humans Including the Toxicity of Manganese." *Metal Ions in Biological Systems* 37: 89–121.

Khan, Ibrar, Maryam Aftab, SajidUllah Shakir, Madiha Ali, Sadia Qayyum, Mujadda Ur Rehman, Kashif Syed Haleem, and Isfahan Touseef. 2019. "Mycoremediation of Heavy Metal (Cd and Cr)–Polluted Soil Through Indigenous Metallotolerant Fungal Isolates." *Environmental Monitoring & Assessment* 191, no. 9: 585. https://doi.org/10.1007/s10661-019-7769-5.

Koller, M., and H. M. Saleh. 2018. "Introductory Chapter: Introducing Heavy Metals." *Heavy Metals* 1: 3–11.

Kumar, M., D. P. S. Rathore, and A. K. Singh. 2000. "Amberlite XAD-2 Functionalized with O Aminophenol: Synthesis and Applications as Extractant for Copper(ii), Cobalt(ii), Cadmium(ii), Nickel(II), Zinc(ii) and Lead(II)." *Talanta* 51: 1187–96.

Kumar, Vinay, and Shiv Kumar Dwivedi. 2021. "Mycoremediation of Heavy Metals: Processes, Mechanisms, and Affecting Factors." *Environmental Science & Pollution Research International* 28, no. 9: 10375–412. https://doi.org/10.1007/s11356-020-11491-8.

López Errasquín, E. L., and C. Vázquez. 2003. "Tolerance and Uptake of Heavy Metals by *Trichoderma atroviride* Isolated from Sludge." *Chemosphere* 50, no. 1: 137–43. https://doi.org/10.1016/s0045-6535(02)00485-x.

Luo, J.-M., X. Xiao, and S.-L. Luo. 2010. "Biosorption of Cadmium(II) from Aqueous Solutions by Industrial Fungus Rhizopus Cohnii." *Transactions of Nonferrous Metals Society of China* 20, no. 6. https://doi.org/10.1016/S1003-6326(09)60264-8.

Lynch, J. M., and A. J. Moffat. 2005. "Bioremediation–Prospects for the Future Application of Innovative Applied Biological Research." *Annals of Applied Biology* 146, no. 2: 217–21. https://doi.org/10.1111/j.1744-7348.2005.040115.x.

Mahmood, A., and R. N. Malik. 2014. "Human Health Risk Assessment of Heavy Metals via Consumption of Contaminated Vegetables Collected from Different Irrigation Sources in Lahore, Pakistan." *Arabian Journal of Chemistry* 7, no. 1: 91–9. https://doi.org/10.1016/j.arabjc.2013.07.002.

Marchal, M., R. Briandet, S. Koechler, B. Kammerer, and P. N. Bertin. 2010. "Effect of Arsenite on Swimming Motility Delays Surface Colonization in *Herminiimonas arsenicoxydans*." *Microbiology/Mikrobiologiya* 156, no. 8: 2336–42. https://doi.org/10.1099/mic.0.039313-0.

Mason, R. P. 2012. "The Methylation of Metals and Metalloids in Aquatic Systems." In *Methylation—from DNA, RNA and Histones to Diseases and Treatment, Anica Dricu*: 71–301. Intech Open. http://dx.doi.org/10.5772/51774.

Meena, Kamlesh K., Ajay M. Sorty, Utkarsh M. Bitla, Khushboo Choudhary, Priyanka Gupta, Ashwani Pareek, Dhananjaya P. Singh, Ratna Prabha, Pramod K. Sahu, Vijai K. Gupta, Harikesh B. Singh, Kishor K. Krishanani, and Paramjit S. Minhas. 2017. "Abiotic Stress Responses and Microbe-Mediated Mitigation in Plants: The Omics Strategies." *Frontiers in Plant Science* 8: 172. https://doi.org/10.3389/fpls.2017.00172.

Mitra, S., A. J. Chakraborty, A. M. Tareq, T. B. Emran, F. Nainu, A. Khusro, A. M. Idris, M. U. Khandaker, H. Osman, F. A. Alhumaydhi, and J. Simal-Gandara. 2022. "Impact of Heavy Metals on the Environment and Human Health: Novel Therapeutic Insights to Counter the Toxicity." *Journal of King Saud University—Science* 34, no. 3: 101865. https://doi.org/10.1016/j.jksus.2022.101865.

Mohamadhasani, Fereshteh, and Mehdi Rahimi. 2022. "Growth Response and Mycoremediation of Heavy Metals by Fungus *Pleurotus* sp." *Scientific Reports* 12, no. 1: 19947. https://doi.org/10.1038/s41598-022-24349-5.

Mukhopadhyay, Rita, and Barry P. Rosen. 2002. "Arsenate Reductases in Prokaryotes and Eukaryotes." *Environmental Health Perspectives* 110 Suppl. 5: 745–8. https://doi.org/10.1289/ehp.02110s5745.

Nadeem, Sajid Mahmood, Maqshoof Ahmad, Zahir Ahmad Zahir, Arshad Javaid, and Muhammad Ashraf. 2014. "The Role of Mycorrhizae and Plant Growth Promoting Rhizobacteria (PGPR) in Improving Crop Productivity Under Stressful Environments." *Biotechnology Advances* 32, no. 2: 429–48. https://doi.org/10.1016/j.biotechadv.2013.12.005.

Nriagu, J. 2007. *Zinc Toxicity in Humans*: 1–7. Ann Arbor, MI: School of Public Health, University of Michigan, Elsevier B.V.

Oladipo, Oluwatosin Gbemisola, Olusegun Olufemi Awotoye, Akinyemi Olayinka, Cornelius Carlos Bezuidenhout, and Mark Steve Maboeta. 2018. "Heavy Metal Tolerance Traits of Filamentous Fungi Isolated from Gold and Gemstone Mining Sites." *Brazilian Journal of Microbiology* 49, no. 1: 29–37. https://doi.org/10.1016/j.bjm.2017.06.003.

Oves, M., M. Saghir Khan, A. Huda Qari, M. Nadeen Felemban, and T. Almeelbi. 2016. "Heavy Metals: Biological Importance and Detoxification Strategies." *Journal of Bioremediation & Biodegradegradation* 7, no. 2: 1–15.

Pal, T. K., B. Sauryya, and B. Arunabha. 2010. "Cellular Distribution of Bioaccumulated Toxic Heavy Metals in *Aspergillus niger* and *Rhizopus arrhizus*." *International Journal of Pharmacy & Biological Sciences* 1, no. 2.

Pearson, J. N., Z. Rengel, C. F. Jenner, and R. D. Graham. 1995. "Transport of Zinc and Manganese to Developing Wheat Grains." *Physiologia Plantarum* 95: 449–55.

Perfus-Barbeoch, Laetitia, Nathalie Leonhardt, Alain Vavasseur, and Cyrille Forestier. 2002. "Heavy Metal Toxicity: Cadmium Permeates Through Calcium Channels and Disturbs the Plant Water Status." *Plant Journal: For Cell & Molecular Biology* 32, no. 4: 539–48. https://doi.org/10.1046/j.1365-313x.2002.01442.x.

Prasad, A. A., G. Varatharaju, C. Anushri, and S. Dhivyasree. 2013. "Biosorption of Lead by *Pleurotus florida* and *Trichoderma viride*." *British Biotechnology Journal* 1: 66–78.

Prasad, Majeti Narasimha Vara N. V., and Rajendra Prasad. 2012. "Nature's Cure for Cleanup of Contaminated Environment–A Review of Bioremediation Strategies." *Reviews on Environmental Health* 27, no. 4: 181–9. https://doi.org/10.1515/reveh-2012-0028.

Priyadarshini, E., S. S. Priyadarshini, B. G. Cousins, and N. Pradhan. 2021. "Metal-Fungus Interaction: Review on Cellular Processes Underlying Heavy Metal Detoxification and Synthesis of Metal Nanoparticles." *C hemosphere* 274: 129976.

Rahman, Md Sayedur, and Kathiresan V. Sathasivam. 2015. "Heavy Metal Adsorption onto *Kappaphycus* sp. from Aqueous Solutions: The Use of Error Functions for Validation of Isotherm and Kinetics Models." *BioMed Research International* 2015: 126298. https://doi.org/10.1155/2015/126298.

Rajkumar, Mani, Noriharu Ae, Majeti Narasimha Vara N. V. Prasad, and Helena Freitas. 2010. "Potential of Siderophore-Producing Bacteria for Improving Heavy Metal Phytoextraction." *Trends in Biotechnology* 28, no. 3: 142–9. https://doi.org/10.1016/j.tibtech.2009.12.002.

Read, D. J., J. G. Ducket, R. Francis, R. Ligron, and A. Russell. 2000. "Symbiotic Fungal Associations in 'lower' Land Plants." *Philosophical Transactions of the Royal Society of London. Series B, Biological Sciences* 355, no. 1398: 815–30; discussion 830. https://doi.org/10.1098/rstb.2000.0617.

Samadi, S., K. Karimi, and S. Behnam. 2017. "Simultaneous Biosorption and Bioethanol Production from Lead-Contaminated Media by *Mucor indicus*." *Biofuel Research Journal* 4, no. 1: 545–50. https://doi.org/10.18331/BRJ2017.4.1.4.

Satarug, Soisungwan, Jason R. Baker, Supanee Urbenjapol, Melissa Haswell-Elkins, Paul E. B. Reilly, David J. Williams, and Michael R. Moore. 2003. "A Global Perspective on Cadmium Pollution and Toxicity in Non-occupationally Exposed Population." *Toxicology Letters* 137, no. 1–2: 65–83. https://doi.org/10.1016/s0378-4274(02)00381-8.

Say, R., N. Yilmaz, and A. Denizli. 2003. "Removal of Heavy Metal Ions Using the Fungus *Penicillium canescens*." *Adsorption Science & Technology* 21, no. 7: 643–50. https://doi.org/10.1260/026361703772776420.

Shahid, Muhammad, Bertrand Pourrut, Camille Dumat, Muhammad Nadeem, Muhammad Aslam, and Eric Pinelli. 2014. "Heavy-Metal-Induced Reactive Oxygen Species: Phytotoxicity and Physicochemical Changes in Plants." *Reviews of Environmental Contamination & Toxicology* 232: 1–44. https://doi.org/10.1007/978-3-319-06746-9_1.

Shahid, Muhammad, S. Shamshad, M. Rafiq, S. Khalid, I. Bibi, and N. K. Niazi. 2017. "Chromium Speciation, Bioavailability, Uptake, Toxicity and Detoxification in Soilplant System: A Review." *Chemosphere* 178: 513–33. https://doi.org/10.1016/j.chemosphere.2017.03.074.

Shanker, Arun K., Carlos Cervantes, Herminia Loza-Tavera, and S. Avudainayagam. 2005. "Chromium Toxicity in Plants." *Environment International* 31, no. 5: 739–53. https://doi.org/10.1016/j.envint.2005.02.003.

Sharma, P., and R. S. Dubey. 2005. "Lead Toxicity in Plants." *Brazilian Journal of Plant Physiology* 17, no. 1: 35–52. https://doi.org/10.1590/S1677-04202005000100004.

Singh, A. L., R. S. Jat, V. Chaudhari, H. Bariya, and S. J. Sharma. 2010. "Toxicities and Tolerance of Mineral Elements Boron, Cobalt, Molybdenum and Nickel in Crop Plants. *P Stress*" 4: 31–56.

Singh, A., and S. M. Prasad. 2015. "Remediation of Heavy Metal Contaminated Ecosystem: An Overview on Technology Advancement." *International Journal of Environmental Science & Technology* 12, no. 1: 353–66. https://doi.org/10.1007/s13762-014-0542-y.

Singh, D., and N. Sharma. 2017. "Effect of Chromium on Seed Germination and Seedling Growth of Green Garm (*Phaseols aureus* L.) and Chickpea (*Cicer arietinum* L.)." *International Journal of Applied & Natural Science* 6: 37–46.

Srivastava, Pankaj Kumar, Aradhana Vaish, Sanjay Dwivedi, Debasis Chakrabarty, Nandita Singh, and Rudra Deo Tripathi. 2011. "Biological Removal of Arsenic Pollution by Soil Fungi." *Science of the Total Environment* 409, no. 12: 2430–42. https://doi.org/10.1016/j.scitotenv.2011.03.002.

Srivastava, Saumya, and Yogesh Kumar Sharma. 2013. "Impact of Arsenic Toxicity on Black Gram and Its Amelioration Using Phosphate." *ISRN Toxicology* 2013: 340925. https://doi.org/10.1155/2013/340925.

Stamets, P. 1999. "Helping the Ecosystem Through Mushroom Cultivation. In *Growing Gourmet and Medicinal Mushroom*, edited by A. Batellet: 452. Berkeley, CA: Ten Speed Press.

Sylvia, D. M. 1994. "Vesicular-Arbuscular Mycorrhizal Fungi." In *Methods of Soil Analysis*, edited by R. W. Weaver, S. Angle, P. Bottomley, D. Bezdicek, S. Smith, A. Tabatabai, and A. Wollum. Madison, WI: Soil Science Society of America. https://doi.org/10.2136/sssabookser5.2.c18.

Thippeswamy, B., C. K. Shivakumar, and M. Krishnappa. 2012. "Accumulation Potency of Heavy Metals by *Saccharomyces* sp. Indigenous to Paper Mill Effluent." *Journal of Environmental Research & Development* 6, no. 3: 439–45.

Tripathi, P., P. C. Singh, A. Mishra, S. Srivastava, R. Chauhan, S. Awasthi, S. Mishra, S. Dwivedi, P. Tripathi, A. Kalra, and R. D. Tripathi. 2017. "Arsenic Tolerant *Trichoderma* sp. Reduces Arsenic Induced Stress in Chickpea (*Cicer arietinum*)." *Environmental Pollution* 223: 137–45.

Vassilev, A., A. Nikolova, L. Koleva, and F. Lidon. 2011. "Effects of Excess Zn on Growth and Photosynthetic Performance of Young Bean Plants." *Journal of Phytology* 3, no. 6.

Veglio', F., and F. Beolchini. 1997. "Removal of Metals by Biosorption: A Review." *Hydrometallurgy* 44, no. 3: 301–16. https://doi.org/10.1016/S0304-386X(96)00059-X.

WHO, Fact Sheet. 2017. Accessed April 5, 2024. https://www.who.int/news-room/fact-sheets/detail/lead-poisoning-and-health.

Wierzba, S. 2017. "Biosorption of Nickel (II) and Zinc (II) from Aqueous Solutions by the Biomass of Yeast Yarrowia Lipolytica." *Polish Journal of Chemical Technology* 19, no. 1: 1–10. https://doi.org/10.1515/pjct-2017-0001.

Wuerfel, O., T. Frank, S. Marcel, H. Rainer, and D.-B. Roland. 2012. "Mechanism of Multi-Metal(loid) Methylation and Hydride Generation by Methylcobalamin and Cob(I)alamin: A Side Reaction of Methanogenesis." *Applied Organometallic Chemistry* 26. https://doi.org/10.1002/aoc.2821.

Yruela, I. 2005. "Copper in Plants." *Brazilian Journal of Plant Physiology* 17, no. 1: 145–56. https://doi.org/10.1590/S1677-04202005000100012.

Yusuf, M., F. Fariduddin, H. Shamsul, and A. Ahmad. 2011. "Nickel: An Overview of Uptake, Essentiality and Toxicity in Plants." *Bulletin of Environmental Contamination and Toxicology* 86: 1–17. https://doi.org/10.1007/s00128-010-0171-1.

Zeng, X., S. Su, X. Jiang, L. Li, L. Bai, and Y. Zhang. 2010. "Capability of Pentavalent Arsenic Bioaccumulation and Biovolatilization of Three Fungal Strains Under Laboratory Conditions." *CLEAN—Soil, Air, Water* 38, no. 3: 238–41. https://doi.org/10.1002/clen.200900282.

Zhang, X., Huanhuan Yang, and Zhaojie Cui. 2017. "*Mucor circinelloides*: Efficiency of Bioremediation Response to Heavy Metal Pollution." *Toxicology Research* 6, no. 4: 442–7. https://doi.org/10.1039/c7tx00110j.

4 Biological Solutions for Metal-Contaminated Environments
Role of Engineered Microbes and Plants

Pooja Sharma, Ambreen Bano, and Surendra Pratap Singh

4.1 INTRODUCTION

In order to reduce the adverse effects of heavy metal (HM) pollution on ecosystems and human health, biological solutions play a crucial role. The use of plants and microbes engineered to perform specific functions has emerged as an important tool in this endeavor. Microbes engineered to tolerate metals and detoxify them are referred to as bioaugmentation agents. Toxic metals can be sequestered, immobilized, or even transformed into less harmful forms with the introduction of these microorganisms into contaminated sites. In some bacteria, soluble metal ions can be converted into insoluble precipitates, reducing their bioavailability. Phytoremediation, on the other hand, uses plants. Plant species have a natural ability to accumulate metals in their tissues without showing toxic effects. Hyperaccumulators can be planted in contaminated areas to absorb and store metals in their roots, stems, and leaves. In addition to cleaning the soil, this approach provides the possibility of recovering metals through harvesting. Rhizoremediation, which combines the strengths of engineered microbes and plants, creates a synergistic effect. Due to uncontrolled urban expansion, industrial development, population expansion, and inadequate management policies, the levels of heavy metal contamination in soil, groundwater, and different water sources have exceeded established acceptable thresholds in numerous nations. In this situation, there is a hidden risk to the human food supply (Rakib *et al.*, 2022; Sarker *et al.*, 2022). Based on their requirement for plant nutrition, HMs can be classified into two groups: plant-essential and non-essential. The toxicity levels of non-essential metals can be quite high, posing a significant risk to human and plant physiology (Samreen *et al.*, 2017). To put it succinctly, HMs are metallic elements that have the potential to be harmful to living organisms, including humans. Among the major toxic effects are reproductive system abnormalities, kidney dysfunction, respiratory issues, high blood pressure, cellular mutations, lung cancer, and abdominal damage (Rehman *et al.*, 2018; Frisbie and Mitchell, 2022). Among the more widely recognized elements in this category are copper, chromium, cadmium, zinc, cobalt, iron, mercury, nickel, arsenic, lead, silver, and platinum. Cadmium, arsenic, mercury, and lead are classified as non-essential elements since they do not serve any biological function in the human body (Ahmed *et al.*, 2018). Environmental authorities around the world have designated these elements as "priority pollutants" due to their significant risks to human health and the environment (Stadlmair *et al.*, 2018). The removal of HMs from contaminated environments is therefore crucial to preserving both human health and the environment. The bioremediation process offers a means to transform toxic substances into non-toxic forms, and numerous bacterial species, including *Achromobacter, Pseudomonas, Dehalococcoides, Rhodococcus, Comamonas, Burkholderia,*

66

DOI: 10.1201/9781003423393-4

Biological Solutions for Metal-Contaminated Environments

Alcaligenes, Bacillus subtillus, Aspergillus niger, Deinococcus radiodurans, Acidithiobacillus ferrooxidans, Mesorhizobium huakuii, Pseudomonas K-62, *Ralstonia, Rhodopseudomonas palustris,* and *Sphingomonas,* have been identified for their proficiency in this regard (Zhang et al., 2020). A number of bacteria have been developed with enhanced pollutant-degrading capabilities as a result of genetic engineering and recombinant DNA technology (Bodor et al., 2020). Beneficial microbes flourish in the root zones of plants, enhancing metal uptake and transformation. In summary, biological solutions involving engineered microbes and plants offer sustainable and environmentally friendly methods to combat metal contamination. In metal-polluted environments, these approaches hold great promise for restoring ecosystems and safeguarding human well-being. This chapter comprehensively outlines the potential and current research status of genetically engineered microbes in the bioremediation of toxic HMs. Additionally, it delves into challenges, determinants, influencing factors, and mechanisms governing the construction of genetically modified organisms (GMOs) for pollutant removal. The passage of metals through GMOs via the biosorption process is thoroughly discussed, as is the construction of GMOs. Detailed insights into various molecular tools employed for the design and development of genetically engineered microbes are also provided to enhance comprehension.

4.2 HM DECONTAMINATION: ALARM FOR HUMAN AND ANIMAL HEALTH

Industrial technology's rapid expansion has led to the alarming release of HM contaminants into the atmosphere, posing a global menace to humanity's health and survival (Sharma et al., 2022a). Numerous reports (Sankhla et al., 2017; Mitra et al., 2022; Karahan et al., 2023) have underscored the serious threats posed by HM contamination, with As, Cr, Cd, Cu, Zn, Mn, Ni, Pb, and Al being major contributors to harmful health impacts in both humans and ecosystems. In the food chain, the accumulation of HMs has resulted in severe health hazards (as shown in Figure 4.1 and Table 4.1). For instance, Cr(VI) or its reduced form Cr(III) can result in the damage of DNA when accumulated in cells, while other HMs have shown similar harmful health consequences (Sharma et al., 2022b; Rasheed et al., 2018a, 2018b). This mounting evidence highlights the urgent need for proactive measures to address HM pollution and its potential catastrophic consequences on the environment and

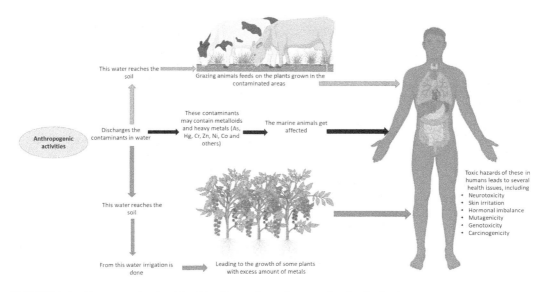

FIGURE 4.1 Human and animal health effects of metal contamination in food.

68 Eco-Restoration of the Polluted Environment

public health. Implementing strict regulations and adopting cleaner technologies can help mitigate the risks associated with HM contamination and safeguard the well-being of present and future generations (Sharma *et al.*, 2022c). The conventional techniques for HM removal, such as chemical precipitation, coagulation, reverse osmosis, biosorption, flocculation, filtration, ion exchange, and membrane processes, have been deemed uneconomical and have resulted in substantial generation of secondary wastes. In contrast, the utilization of biological agents, particularly microbes, presents an environmentally friendly and efficient alternative for eliminating HMs from various environmental media. Microorganisms possess inherent bioremediation attributes that stem from their self-protection mechanisms, including the secretion of enzymes and cellular morphological alterations. These exceptional properties of microbial isolates have been effectively harnessed for the degradation or reduction of the harmful contaminants, leading to the remediation of polluted environments (Jacob *et al.*, 2018).

Recently, significant research focus has been directed toward the use of genetically engineered bacteria for HM disposal due to their high treatment efficiency and strong adaptability. HMs can exert several physiological, biochemical, or genotoxic effects on microbes when present at elevated concentrations, inhibiting their metabolic functions and damaging genetic materials or transcription processes (Rasheed *et al.*, 2019; Aslam *et al.*, 2021; Verma *et al.*, 2021; Bajaj *et al.*, 2023). However, through biological detoxification, some microbes can develop resistance to HM toxicity. This resistance is primarily attributed to the presence of metal-binding proteins like metallothionein, which form complexes with high binding capacities for metal ions, and specific metal ion transport systems that facilitate the efflux of HM ions from the cell, reducing their uptake (Rasheed *et al.*, 2019). Numerous genetically engineered bacteria and biosensors have been created to detect and eliminate environmental pollutants. For instance, a genetically engineered strain of *E. coli* was created by expressing metallothionein from mammals and yeast, along with the exo-protein Lamb, resulting in a strain with 15–20 times higher cadmium ion (Cd^{2+}) enrichment capacity compared

TABLE 4.1
Genetic Introduction of Genes into Plants and Its Effects on HM Tolerance and Gene Expression

Genes	Bacteria	Target Plant	Enzymes Expressed	Beneficial Effects	References
CePCS and AtPCS1	*Caenorhabditis elegans*	*Nicotiana tabacum*	Phytochelatin synthase	As accumulation increased	Wojas *et al.*, 2010
AsPCS1 and gsh1	*Saccharomyces cerevisiae*	*Arabidopsis thaliana*	Phytochelatin synthase and glutathione synthase	As and Cd tolerance enhanced	Guo *et al.*, 2008
ECS genes	*Escherichia coli*	*Transgenic line of Populus tremula × P. alba* (Aiton) Smith	G-glutamylcysteine synthetase	Improved Cd fix and detoxification	He *et al.*, 2015
STGCS-GS	*Streptococcus thermophilus*	*Beta vulgaris*	Glutathione synthetase and g-glutamylcysteine synthetase	Increases the tolerance of Zn, Cu, and Cd	Liu *et al.*, 2015
ScMTII gene	*Saccharomyces cerevisiae*	*Nicotiana tabacum*	Metallothionein transporter	Zn and Cd accumulation enhanced	Daghan *et al.*, 2013
SaNramp1	*Pseudomonas fluorescens*	*Sedum alfredii*	Iron-regulated transporter	Improved uptake of Zn	Wang *et al.*, 2013

to the parent strain (Sousa *et al.*, 1998). Zhao *et al.* (2005) demonstrated that a recombinant *E. coli* JM109 expressing the transport protein (merT–merP) and metallothionein exhibited a remarkable potential for Hg bioaccumulation and cell propagation rate in LB medium with a higher concentration of Hg^{2+} up to 7.402 mg/L, surpassing the capabilities of the parent strain. In another approach, Kuroda *et al.* (2001) demonstrated the use of engineered bacterial oligopeptides for HM biosorption. Kuroda *et al.* (2001) expressed a histidine oligopeptide (hexa-His) with divalent metal chelating ability in *Saccharomyces cerevisiae*, resulting in a surface-engineered yeast with eight times the enrichment capacity of Cu^{2+} compared to the native counterpart. For the removal of Hg^{2+} from wastewater, Liang *et al.* (2015) developed a modified BL21-7 strain containing the synthetic operon P16S-g10-merT-merP-merB1-merB2-ppk-rpsT, which successfully eliminated approximately 43.7% of Hg^{2+} from the wastewater. Xiao-long *et al.* (2015) achieved enhanced HM ion accumulation and tolerance in recombinant *E. coli* by overexpressing phytochelatin synthase from *Pyruscalleryana*, which led to the synthesis of metal-binding cysteine-rich peptides known as phytochelatins. These studies demonstrate the potential of genetically engineered bacteria and engineered bacterial oligopeptides as effective and promising approaches for HM bioaccumulation and removal, offering valuable contributions to environmental remediation efforts.

4.3 NEED FOR ENGINEERED ORGANISMS

Efforts have been directed towards enhancing the pollutant removal capabilities of native microbes, which often exhibit limited or inadequate capacity to eliminate toxic compounds. This has been achieved by employing interdisciplinary approaches that encompass molecular techniques, bioinformatics, genomics, and microbiology (Figure 4.2 and Table 4.2). One effective strategy involves the genetic modification of these native microbes, leading to the creation of GMOs (Wu *et al.*, 2021; Pant *et al.*, 2021). The success of utilizing GMOs hinges on their ability to thrive under ecological stress conditions, and once their intended purpose is fulfilled, it is imperative that they possess mechanisms to be effectively removed from the target environment (Tran *et al.*, 2021).

The development of modified strains involves a series of steps, including mutagenesis-based screening and the implementation of various strategies. These strategies encompass:

- Identifying and isolating genes with high pollutant removal capabilities.
- Enhancing enzyme expression to facilitate increased secretion.

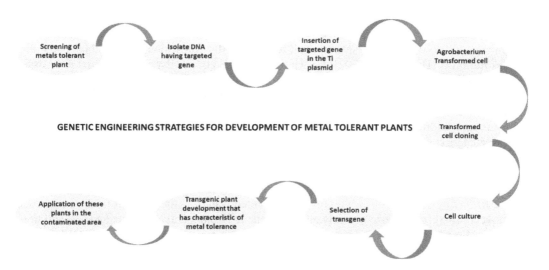

FIGURE 4.2 Genomic engineering strategies for metal-tolerant plant design.

TABLE 4.2
Utilizing Genetically Engineered Microorganisms for Detoxification of Hazardous Heavy Metals

Microbes	Heavy Metal	Genetically Modified Microbes	Removal Mechanism	References
Deinococcus radiodurans	Hg	MerH; ion transporter	Degrades marinum strain and produces resistance against mercury	Meruvu, 2021
Escherichia coli	As(III)	*E. coli* BL21	Surface MT fusion	Ma *et al.*, 2011
Escherichia coli	Cd(II)	*E. coli* MT3 and MT2	Eukaryotic metal-binding proteins	Uckun *et al.*, 2021
Ensifer medicae	Cu	MA11-copAB	copAB	Delgadillo *et al.*, 2014
Escherichia coli	Pb(II)	*E. coli* pBLP1	Surface MT fusion	Nguyen *et al.*, 2013
Escherichia coli	Cd(II)	SUMO-ShMT3	Modified proteins	Li *et al.*, 2021
Ensifermedicae	Cu	copAB	—	Pérez-Palacios *et al.*, 2017
Acidithiobacillus ferrooxidans	Hg	Mer C	Ion transporter, removal of Hg	Arshadi and Yaghmaei, 2020
E. coli	Ni	SE5000 strain	From aqueous system removal of Ni	Tsyganov *et al.*, 2020
Pseudomonas putida	Cd	X3 strain	Degradation of Cd and methyl parathion	Ayogu *et al.*, 2020
Ralstonia eutropha	HMs	AUM-01	Removes HMs	Sharma *et al.*, 2021
Pseudomonas	Hg	K-62	Hg degradation	Krout *et al.*, 2022

- Amplifying gene expression for various pollutant types to engineer specialized microbes.
- Employing protoplast fusion to combine advantages from different microbes for efficient contaminant removal.

Synergistic interactions among pollutant-remitting bacteria are commonplace, with microbial communities exhibiting mutualism. A deeper understanding of these microbial communities aids in deciphering contaminant occurrence in ecosystems and facilitating the development of novel mechanisms and tools for pollutant removal, as well as the cultivation of microbial populations with desired traits (Huang *et al.*, 2021; Guarin and Pagilla, 2021). Given the potential risks associated with releasing GMOs into the environment, the establishment of a biological control system becomes crucial. This system ensures the controlled elimination of the modified microorganisms (Wang *et al.*, 2019). Engineered organisms are needed to address complex challenges that natural organisms can't address. There is a pressing need for sustainable food sources, novel medicines, and alternative energy as the global population grows exponentially. By making production processes more efficient and sustainable, engineered organisms can produce specific proteins, biofuels, or nutrients. In regions plagued by famine, genetically modified crops can yield more food per acre, resist pests, and withstand harsh conditions, thus ensuring food security. Engineered bacteria can produce lifesaving drugs or absorb toxins from polluted waters in the medical field. A sustainable alternative to fossil fuels is also urgently needed as the environmental crisis escalates. In addition to generating biofuels, engineered algae and bacteria can also break down plastic, providing environmentally friendly solutions. Though controversial for some, these endeavors demonstrate the potential of biotechnology. Engineered organisms can play a crucial role in creating a more sustainable and healthier future through careful research and ethical considerations.

Biological Solutions for Metal-Contaminated Environments 71

4.4 ENGINEERED MICROBES

Utilizing microorganisms for bioremediation has been shown to be a successful method for reducing soil and water contamination brought on by HMs (Pande *et al.*, 2022). Microbes have the ability to decompose complex substances into simpler chemicals, providing a sustainable solution for HM contamination. Recent advancements in genetic engineering have facilitated the creation of biocatalysts or GMOs, surpassing natural microbes in their capacity to eliminate persistent compounds from the environment (Liu *et al.*, 2022). Various genetic and metabolic engineering strategies, including pathway construction, single-gene editing, and alterations to the coding and regulatory sequences of genes already in existence, have been employed to produce extremely efficient engineered microorganisms (Saravanan *et al.*, 2022). These modifications primarily focus on optimizing the rate-limiting stages of metabolic processes. Genetically modified bacteria have demonstrated efficient removal of HMs (Rafeeq *et al.*, 2023). Enzyme catalytic efficiency present within the cells or stimulated to act on specific substrates plays a crucial role in the speed of degradation. In genetic engineering, the application of recombinant DNA technology has enabled the amplification of the genome of genetically engineered microorganisms by mixing foreign genes from other species (Khan *et al.*, 2016). In microbial cells, this has enhanced the development of genetically engineered metabolic pathways in the processes of bioremediation. Genetically modified microbes (GEMs) have garnered public attention due to their advanced technological nature (Liu *et al.*, 2011). Metal-regulating genes have been incorporated into bacteria, which has aided the transformation of HMs from harmful forms to less toxic ones (Bondarenko *et al.*, 2008). Moreover, GEMs expressing metallothioneins (MTs) have shown increased HMS accumulation (Mishra *et al.*, 2017). Studies have extensively evaluated the use of GEMs and plants in the bioremediation of settings contaminated with HMs and other organic pollutants (Azad *et al.*, 2014).

Additionally, the *mer* genes now have the ability to reduce Hg^{2+} pollution at high temperatures thanks to the incorporation of the *mer* operon from bacterial resistance to Hg, which codes for that decrease (Nascimento and Chartone-Souza, 2003). Genes conferring resistance to HMs, such as *ndoB, nidA, alkB, xylE, alkM, alkB1*, and *alkB2*, have been found on plasmids within n-alkane-degrading microorganisms. These genes, which are widely utilized as indicators of microbial biodegradation, are known to be transferred horizontally (De Groote *et al.*, 2009). The genetic modification of microbial membrane transporters is a significant approach within environmental bioremediation to enhance the removal of hazardous metals (Chen and Wilson, 1997). In the remediation of these HMs, transporters and binding mechanisms play crucial roles. Transporter systems are typically categorized into three main groups: channels, secondary carriers, and primary active transporters. Channel transporters, like Mer T/P, Fps, and GlpF, enable the molecules to pass across the inner lipid membrane (Banerjee *et al.*, 2018; DalCorso *et al.*, 2019). Secondary carriers like Pho84, Hxt7, and NixA are also found in the inner lipid membrane, and primary active transporters such as MntA, cdtB/Ip_3327, TcHMSA3, and CopA are located there as well. Some transporters, like porin channels, can also be present in the outer lipid membrane (Kotrba *et al.*, 2009; Pande *et al.*, 2022). Once hazardous metals infiltrate the cell, various compounds like metallothioneins, phytochelatins, and polyphosphates work together to sequester the metals (Boechat *et al.*, 2016). Modifying the hazardous metal import-storage systems of microbes can possibly enhance their proficiency in extracting hazardous metals, aiding in the restoration process (Zhu *et al.*, 2020). GEMs have emerged as indispensable tools in tackling environmental pollutants effectively, efficiently degrading and neutralizing toxic substances and mitigating their impact on the ecosystem (Sharma *et al.*, 2021).

Recent research, aided by next-generation and high-throughput sequencing techniques, has identified genetic elements crucial for the degradation of the hazardous contaminant broad spectrum (Qin *et al.*, 2021). Innovative technologies, like CRISPR-Cas gene-editing tools, offer opportunities for optimizing bioremediation efforts. CRISPR technology allows for the transfer of HM remediation genes to plants or bacteria, facilitating HM detoxification and enhancing the synthesis of metallothioneins and phytochelatins (Basharat *et al.*, 2018). Numerous studies have shown that CRISPR-Cas9 gene-editing technology can improve metal remediation by transferring specific bacterial and plant genes. For instance, in *Arabidopsis* and tobacco plants, the *NAS1* gene, encoding the nicotianamine (NA) synthase enzyme, has improved tolerance to harmful metals (Kim *et al.*, 2005). By genetically modifying microorganisms with genes involved in breaking down stubborn substances, bioremediation processes can be significantly enhanced (Thai *et al.*, 2023). Bacteria have a high oxidation potential for contaminants, and modifying their metabolic pathways can efficiently eliminate different pollutants (Maqsood *et al.*, 2022). The bacterial community possesses exceptional virulent factors, including resistance to high Hg toxicity. The Hg-resistant gene *mer* in an operon aids bacterial cells in detoxifying Hg^{2+} to volatile Hg through the enzyme Hg reductase. The bacterial cell's resistance to Hg removes S and facilitates the transport of Hg^{2+} into the cell. Inside the cell, ionic Hg is transformed to elemental mercury, which is less toxic, by functional genes. Several forms of Hg resistance genes, such as *mer H, mer E, mer P, mer R, mer T, mer D, mer B,* and *mer A*, are involved in this process. Genetically modified *E. coli* containing *mer T–mer P* genes has shown the ability to adsorb Hg from wastewater. Metal-binding peptides like phytochelatins and metallothioneins can further improve the adsorption and removal of HM ions by these engineered bacteria, which have higher metal affinity and accumulation capacity (Wei *et al.*, 2020; Fathollahi *et al.*, 2021; Yu *et al.*, 2022). Modified *E. coli* has been utilized for the removal of Cd ions from contaminated sites. Other bacteria, such as *Caulobacter crescentus, Mesorhizobium huakuii,* and *Pseudomonas putida,* can also be engineered to remove Cd ions through the expression of phytochelatins and peptides. When compared to control strains, these recombinant engineered bacteria have been found to accumulate metal ions at higher concentrations (Appukuttan *et al.*, 2010).

Fungi can thrive in various climatic conditions and can disperse through spores in the air, contributing to the balance of the ecosystem (Zhou *et al.*, 2022). They have also been reported to survive in industrial effluent treatment plants treating various wastewaters. The diversity of habitats and the ability to secrete numerous enzymes make fungi potential candidates for bioremediation at various sites (Akerman-Sanchez and Rojas-Jimenez, 2021). Genetic engineering techniques can be utilized to modify fungi with desired characteristics. Fungal genes play a key role in mycotransformation during the remediation process (Sharma *et al.*, 2021; Saravanan *et al.*, 2022;). It is possible to create mutants of different fungus species that release unique enzymes for use in wastewater treatment. Protoplasts from fungi are a part of the mycoremediation process. Recent advancements have led to the discovery of fungal genes that are contaminant mineralizing (Garrigues *et al.*, 2022). Fungi demonstrate high resistance against HMs such as Ni, Cd, and Cu, making them well suited for removing contaminants from soil and wastewater (Oladipo *et al.*, 2018; Liaquat *et al.*, 2021). Fungal species like *Aspergillus niger* and *Aspergillus flavus* have been shown to aid in decreasing the toxicity of Cr ions from Cr^{6+} to Cr^{3+}. Similarly, algae have numerous mechanisms to protect against metal toxicity, including metal immobilization, gene regulation, chelation, and antioxidant production to counter the redox activity of HMs (Leong *et al.*, 2020; Manikandan *et al.*, 2022]). Microalgae produce antioxidant enzymes such as superoxide mutase, catalase, peroxidase, and glutathione reductase, which help neutralize the free radicals released by HMs during the adsorption process (Zeraatkar *et al.*, 2016; Al Ketife *et al.*, 2020). Through modification, microalgae can form complexes that enable them to withstand ecological stress conditions (Sarojini *et al.*, 2022). Certain algal species, such as *Chlorella* sp. and *Chlamydomonas* sp., have been engineered to possess high efficiency for pollutant removal, a process referred to as phycoremediation (Danouche *et al.*, 2021). Among the commonly used microalgal species for genetic engineering, Chlorophyta *C. reinhardtii* stands out. In this organism's genome, metallothionein genes from another algal species, *Festuca rubra*, have been transferred and cloned, resulting in a mutant algal strain with the desired characteristics (Qin *et al.*, 2012; Sproles *et al.*, 2021).

Biological Solutions for Metal-Contaminated Environments 73

4.5 ENGINEERED PLANTS

The primary objective of the transgenic approach is to develop plants with high capacity to tolerate, accumulate, or degrade HMs (da Conceição Gomes *et al.*, 2016). This approach also aims to obtain plants with favorable agronomic properties, such as high green biomass, deep root systems, and fast growth in diverse pedo-climatic conditions (Rai *et al.*, 2020). To enhance Hg detoxification, two bacterial gene enzymes, organomercuriallyase (merB) and mercuric ion reductase (merA), are involved. MerB converts the organic form of Hg into less toxic ionic form (Hg^{2+}) through protonolysis, while merA reduces Hg^{2+} to volatilize it into Hg0 (Bizily *et al.*, 2003). Various genes have been found to confer tolerance to HMs and modulate reactive oxygen species (ROS) scavenging in transgenic plants. For example, the *SbMT-2* gene isolated from *Salicornia brachiata* enhances Zn, Cd, and Cu tolerance in transgenic *Nicotiana tabacum* (Chaturvedi *et al.*, 2014). Overexpressed *SaNramp6* from *Sedum alfredii* significantly improves Cd accumulation in transgenic *Arabidopsis thaliana* (Chen *et al.*, 2017). The *OsMTP1* protein gene from *Oryza sativa* cv. IR64 is coupled with transgenic tobacco to significantly increase its tolerance to Cd (Das *et al.*, 2016). Similarly, the *AtACR2* gene (arsenic reductase 2) from *Arabidopsis thaliana* shows promise in As decontamination in transgenic tobacco (Nahar *et al.*, 2017). Vacuolar compartmentation plays a major role in HMs detoxification, involving numerous protein transporters like VIT-transporter, COPT5-transporter, H+-ATPase, and Na+/H+ antiporter (Zhang *et al.*, 2018). Hyper-accumulator *Sedum plumbizincicola* (Crassulaceae) overexpresses P1B-type ATPase (HMSA4 and HMSA2) transporters under Cd stress (Peng *et al.*, 2017). Tomato (*Solanum lycopersicum*) responds to diverse concentrations of $CdCl_2$, with the majority of Cd being compartmentalized in vacuoles primarily due to the high activity of the NRAMP3 transporter protein (Meena *et al.*, 2018). Phytochelatins (PCs), metallothioneins (MTs), and glutathione (GSH) play essential roles in HM detoxification by chelating toxic metals and transferring them to high-molecular-weight (HMW) or low-molecular-weight (LMW) compounds (Tan *et al.*, 2019). The overexpression of genes encoding enzymes like phytochelatin synthase (PCS) and glutathione synthase (GS) involved in PC biosynthetic pathways enhances tolerance and detoxification of HMs (Shukla *et al.*, 2012). Overexpressing the *Arabidopsis* ATP sulfurylase gene and transferring it to *Medicago sativa* enhances its tolerance against Cd (Kumar *et al.*, 2019).

Plants' ability to tolerate, extract, and accumulate HMs varies considerably among diverse plant species and is influenced by genetic factors and the environment (Gavrilescu, 2022; Singh *et al.*, 2023). For effective phytoremediation of contaminated soils, it is essential to identify metal-tolerant high-accumulating plants, known as hyperaccumulators, and understand the underlying mechanisms regulating HM hyperaccumulation. Many studies have reported HM hyperaccumulators in plants and successfully utilized them for reclaiming contaminated soils (Rascio and Navari-Izzo, 2011; Asad *et al.*, 2019). For instance, *Sedum plumbizincicola* and *Solanumnigrum* have been shown to remove Cd, while *Silenevis cidula* removes Cd, Zn, and Pb (Luo *et al.*, 2011; Liu *et al.*, 2017). Plant breeding and biotechnology methods can be used to increase hyperaccumulation and tolerance to HMs. Transgenic plants with more HM accumulation or remediation ability than natural plants have been successfully created using genetic engineering technologies, making them better options for phytoremediation or phytomining (Duan *et al.*, 2011). Overexpression of genes like phytochelatins, metallothioneins, antioxidant enzymes, and transporter proteins in plants (sweet potato, mustard, tobacco, and *Arabidopsis*) has shown improved phytoremediation ability (Balzano *et al.*, 2020). The uptake of HMs is enhanced by overexpressing plasma membrane and tonoplast transporter proteins involved in their uptake and mobilization, like NRAMP, HMSA4, aquaporin, ZIP4, YCF1, CAX4, and IRT1 (Lee *et al.*, 2010).

Arabidopsis cyclic nucleotide-gated ion channel AtCNGC1 and *Nicotiana tabacum* plasma membrane channel protein NtCBP4 are crucial for Pb^{2+} transport and play a role in Pb tolerance regulation (Sunkar *et al.*, 2000). The transfer of glutathione synthesis enzymes encoded by *GSH1* and *GSH2* genes from *E. coli* to *Brassica juncea* and *Populus canescens* led to Cd and Zn tolerance

(Nawaz *et al*., 2019). Overexpression of antioxidant defense enzymes like catalase, superoxide dismutase, and peroxidase has been shown to improve plant growth (Naing *et al*., 2016; Chiappero *et al*., 2019). Overexpression of the mercuric reductase genes *MerA* and *MerB* in sunflower callus provided tolerance even after HM accumulation (Jan *et al*., 2016). While transgenic technology shows promise in phytoremediation and soil reclamation, ethical considerations associated with such plants need to be addressed. Conducting risk analysis studies is crucial to prevent technology failure in the field (Linacre *et al*., 2003).

4.6 RISKS RELATED TO GENETICALLY MODIFIED MICROBES

Introducing foreign genes into the genetic makeup of other organisms can lead to a spectrum of outcomes, and the expression of the introduced gene may be altered upon integration. The functionality of the gene in the donor organism is intricately orchestrated, and its role within the recipient microbe might not align seamlessly with its original function (Jong *et al*., 2020). During the process, gene copies could undergo mutations such as insertions, deletions, integrations, or deletions, resulting in loss of specific functions or instability. This unpredictability gives rise to both defined and undefined risks associated with GMOs in open ecosystems. As each gene contributes to multiple functions within an organism, any single modification, whether an addition or deletion, might trigger unforeseen side effects. GMOs have the potential to disrupt native or wild strains, as they can outpace their growth rate, leading to competition. This competitive advantage could render GMOs invasive, causing ecological harm and economic instability. This intensified competition puts pressure on both genetically modified and unmodified organisms to adapt to changing environmental conditions, potentially leading to the emergence of distinct resistant communities (Peter *et al*., 2011). These changes within individual organisms reverberate across the ecosystem. Once GMOs become established in the environment, their removal becomes challenging, perpetuating problems (Jaiswal and Shukla, 2020).

Despite the incurred risks, GMOs are not unequivocally safe or beneficial. A particularly significant concern pertains to the risk of horizontal gene transfer (Gogarten and Townsend, 2005). This process involves microorganisms acquiring genes through mechanisms such as conjugation, transduction, or transformation in response to ecological shifts. These transferred genes could originate from GMOs and may pose health risks to living organisms. In the absence of proper precautions, these genes might propagate among native strains. The environmental ramifications of GMOs encompass potential pathogenicity, the dissemination of new diseases, and the proliferation of unwanted plants or pests if the affected species is a plant. Moreover, GMOs can alter the structure and function of invaded organisms, including the potential disruption of endogenous genes. Consequently, the consequences of introducing a novel gene are not limited to its addition but might also involve unintended modifications of existing genetic elements.

4.7 MANAGING RISKS FROM GMOS

After conducting a thorough risk evaluation, it becomes imperative to implement appropriate management strategies to control adverse effects stemming from GMOs. The decision to accept or reject the risks associated with GMOs lies within the view of governmental authorities. This decision-making process must consider ethical and socioeconomic dimensions while also employing effective techniques to minimize biologically identified risks (Renn, 2008). Risk assessment methodologies often segregate risk evaluation from risk management. Some techniques primarily focus on monitoring risks, whereas others encompass risk mitigation. However, a comprehensive understanding of risks should encompass the impact of any mitigation strategies that aim to minimize potential harm (Myhr and Traavik, 2002). It is crucial to recognize the intricate interconnection between risk assessment and risk management, as making conclusions based on insufficient information can lead to unpredictability. This underscores the importance of carefully analyzing

Biological Solutions for Metal-Contaminated Environments

the potential effects before arriving at decisions. A regulatory framework for biosafety serves as a vital methodology to ensure the proper use of biological outcomes and provide safety. Risk analysis demonstrates that the introduction of GMOs into the environment yields both positive and negative effects. It necessitates a combination of legal and practical measures for risk management to prevent harmful impacts (Devos *et al.*, 2014). Risk management involves various activities, and three central components stand out: assessing the impact, raising public awareness and participation, and establishing regulatory systems. The concepts underpinning these components are crucial in governing GMOs. While public awareness and participation are essential, governmental decisions hold more weight in designing safety measures for risk management.

The primary assessment should be substantiated by effective risk management and mitigation strategies. The nature of risks associated with GMOs varies based on parameters like the GMO's characteristics, its intended application, and the ecosystem's response. With each release into the environment, a GMO exhibits distinct traits. Obtaining proper authorization from authorities before conducting trials and releasing GMOs for commercial purposes is crucial (Schermer and Hoppichler, 2004). This step ensures that a GMO's potential to survive, reproduce, and persist within the environmental conditions of release is assessed. GMOs must demonstrate their ability to persist; thrive in the ecosystem; confer necessary benefits; transfer genes effectively; coexist harmoniously with native strains; exhibit positive effects on plants, animals, and humans; and fulfill their intended function. Small-scale trials may require a limited number of GMOs, yielding specific insights, while large-scale field trials within open ecosystems necessitate careful monitoring of the interaction between GMOs and the microbial community (Scipioni *et al.*, 2005).

4.8 SUGGESTIONS AND FUTURE RECOMMENDATIONS

In HM decontamination, investigate the potential for synergistic interactions between engineered microbes and plants. To enhance metal removal, immobilization, and overall remediation efficiency, further research could explore how different organisms can work together.

- To assess the real-world efficacy and practicality of engineered organisms for HM cleanup, move from laboratory experiments to larger-scale field trials. The performance, adaptability, and sustainability of these solutions will be analyzed over an extended period of time in diverse environments.
- Analyze the potential ecological consequences of introducing GMOs into natural ecosystems through comprehensive ecological risk assessments. The importance of understanding the unintended effects of a given technology on non-target organisms and ecosystem dynamics cannot be overstated.
- Develop clear guidelines for the deployment of engineered microbes and plants in environmental remediation in collaboration with regulatory bodies. Public acceptance and successful implementation require a framework that ensures safety, risk assessment, and responsible oversight.
- Promote understanding and acceptance of the benefits and possible risks of using GMOs for bioremediation by engaging local communities, stakeholders, and the general public. Building support and alleviating concerns can be achieved through transparent communication.
- Improve the performance of engineered organisms under varying soil and climate conditions by further refining genetic modifications. Bioremediation strategies that target genetic traits that enhance metal uptake, accumulation, and stress tolerance will be more efficient and reliable.
- Investigate the potential of metal-tolerant and hyperaccumulator plants and microbes for HM decontamination. Genetic diversity can be harnessed to develop sustainable solutions that minimize the need for extensive genetic engineering.

- Develop long-term monitoring programs to assess the persistence and stability of introduced organisms, as well as the long-term impact of HM removal on soil quality, biodiversity, and ecosystem functionality.
- Explore emerging technologies such as CRISPR-Cas gene editing and synthetic biology to create more precise and targeted genetic modifications that optimize HM remediation traits.
- Analyze the cost-effectiveness of using engineered organisms compared to traditional remediation methods. Bioremediation strategies must demonstrate economic feasibility before they can be widely adopted.

4.9 CONCLUSIONS

Engineering microbes and plants to decontaminate HM-polluted environments could be a promising approach. Throughout this chapter, we observed significant advances in understanding the genetic mechanisms underlying metal resistance and accumulation in different organisms. As a result of these advancements, tailored genetic modifications have been developed that enhance the efficiency and specificity of bioremediation processes. The expression of metal-binding proteins and transporters by engineered microbes, such as bacteria and fungi, has demonstrated remarkable potential in bioaccumulating and immobilizing heavy metals. The ability of genetically modified plants to uptake, translocate, and sequester metals also shows promise for phytoextraction and phytostabilization. A multifaceted approach maximizes the remediation potential of these organisms through the synergistic interactions between engineered microbes and plants. However, several challenges remain on the road to practical implementation. A careful balance must be struck between regulatory concerns, potential ecological impacts, and public acceptance of GMOs. Moreover, fine-tuning the genetic modifications to achieve optimal performance under diverse environmental conditions remains a priority. Bioremediation strategies have revolutionized HM decontamination by integrating genetic engineering techniques. The use of engineered microbes and plants in the remediation of metal-contaminated sites contributes to the restoration of ecological balance and protects human health. In order to realize the full potential of these innovative approaches and move toward a cleaner and healthier environment, ongoing research, collaboration between scientific communities, and proactive engagement with stakeholders are essential.

REFERENCES

Ahmed, T., M. Shahid, F. Azeem, I. Rasul, A. A. Shah, and M. Noman. 2018. "Biodegradation of Plastics: Current Scenario and Future Prospects for Environmental Safety." *Environmental Science & Pollution Control Series* 25, no. 8: 7287–98.

Akerman-Sanchez, G., and K. Rojas-Jimenez. 2021. "Fungi for the Bioremediation of Pharmaceutical-Derived Pollutants: A Bioengineering Approach to Water Treatment." *Environmental Advances* 4: 100071. https://doi.org/10.1016/j.envadv.2021.100071.

Al Ketife, A. M. D., F. Al Momani, and S. Judd. 2020. "A Bioassimilation and Bioaccumulation Model for the Removal of Heavy Metals from Wastewater Using Algae: New Strategy." *Process Safety & Environmental Protection* 144: 52–64. https://doi.org/10.1016/j.psep.2020.07.018.

Appukuttan, D., K. S. Nilgiriwala, C. Misra, and S. K. Apte. 2010. "Natural and Recombinant Bacteria for Bioremediation of Uranium from Acidic/Alkaline Aqueous Solutions in High Radiation Environment." *Journal of Biotechnology* 150: 53. https://doi.org/10.1016/j.jbiotec.2010.08.140.

Arshadi, M., and S. Yaghmaei. 2020. "Advances in Bioleaching of Copper and Nickel from Electronic Waste Using *Acidithiobacillus ferrooxidans*: Evaluating Daily pH Adjustment." *Chemical Papers* 74, no. 7: 2211–27. https://doi.org/10.1007/s11696-020-01055-y.

Asad, Saeed Ahmad, Muhammad Farooq, Aftab Afzal, and Helen West. 2019. "Integrated Phytobial Heavy Metal Remediation Strategies for a Sustainable Clean Environment-A Review." *Chemosphere* 217: 925–41. https://doi.org/10.1016/j.chemosphere.2018.11.021.

Aslam, M., A. Aslam, M. Sheraz, B. Ali, Z. Ulhassan, and N. Ullah. 2021. "Lead Toxicity in Cereals: Mechanistic Insight into Toxicity, Mode of Action and Management." *Frontiers in Plant Science* 11. https://doi.org/10.3389/fpls.2020.587785.

Ayogu, C. V., V. O. Ifeanyi, and N. P. Obasi. 2020. "Monitoring of Metabolic Compounds from Degradation of Petrochemicals Using Indigenous Consortium of *Pseudomonas* Strains." *Nigerian Journal of Microbiology* 34, no. 2: 5221–38.

Azad, M. A. K., L. Amin, and N. M. Sidik. 2014. "Genetically Engineered Organisms for Bioremediation of Pollutants in Contaminated Sites." *Chinese Science Bulletin* 59, no. 8: 703–14. https://doi.org/10.1007/s11434-013-0058-8.

Bajaj, T., H. Alim, A. Ali, and N. Patel. 2023. "Phytotoxicity Responses and Defence Mechanisms of Heavy Metal and Metal-Based Nanoparticles." In *Nanomaterials and Nanocomposites Exposures to Plants: Response, Interaction, Phytotoxicity and Defense Mechanisms*: 59–96. Singapore: Springer Nature.

Balzano, Sergio, Angela Sardo, Martina Blasio, Tamara Bou Chahine, Filippo Dell'Anno, Clementina Sansone, and Christophe Brunet. 2020. "Microalgal Metallothioneins and Phytochelatins and Their Potential Use in Bioremediation." *Frontiers in Microbiology* 11: 517. https://doi.org/10.3389/fmicb.2020.00517.

Banerjee, A., M. K. Jhariya, D. K. Yadav, and A. Raj. 2018. "Micro-remediation of Metals: A New Frontier in Bioremediation." *Handbook of Environmental Materials Management*: 1–36.

Basharat, Zarrin, Luís A. B. Novo, and Azra Yasmin. 2018. "Genome Editing Weds CRISPR: What Is in It for Phytoremediation?" *Plants* 7, no. 3: 51. https://doi.org/10.3390/plants7030051.

Bizily, Scott P., Tehryung Kim, Muthugapatti K. Kandasamy, and Richard B. Meagher. 2003. "Subcellular Targeting of Methylmercury Lyase Enhances Its Specific Activity for Organic Mercury Detoxification in Plants." *Plant Physiology* 131, no. 2: 463–71. https://doi.org/10.1104/pp.010124.

Bodor, A., N. Bounedjoum, G. E. Vincze, A. Erdein'e Kis, K. Laczi, and G. Bende. 2020. "Challenges of Unculturable Bacteria: Environmental Perspectives." *Reviews in Environment Science & Biotechnology* 19, no. 1: 1–22.

Boechat, Cácio Luiz, Vítor Caçula Pistóia, Clésio Gianelo, and F. A. D. O. Camargo. 2016. "Accumulation and Translocation of Heavy Metal by Spontaneous Plants Growing on Multi-metal-Contaminated Site in the Southeast of Rio Grande do Sul State, Brazil." *Environmental Science & Pollution Research International* 23, no. 3: 2371–80. https://doi.org/10.1007/s11356-015-5342-5.

Bondarenko, Olesja, Taisia Rõlova, Anne Kahru, and Angela Ivask. 2008. "Bioavailability of Cd, Zn and Hg in Soil to Nine Recombinant Luminescent Metal Sensor Bacteria." *Sensors* 8, no. 11: 6899–923. https://doi.org/10.3390/s8116899.

Chaturvedi, Amit Kumar, Manish Kumar Patel, Avinash Mishra, Vivekanand Tiwari, and Bhavanath Jha. 2014. "The SbMT-2 Gene from a Halophyte Confers Abiotic Stress Tolerance and Modulates ROS Scavenging in Transgenic Tobacco." *PLOS ONE* 9, no. 10: E111379. https://doi.org/10.1371/journal.pone.0111379.

Chen, S., and D. B. Wilson. 1997. "Genetic Engineering of Bacteria and Their Potential for Hg2+Bioremediation." *Biodegradation* 8, no. 2: 97–103. https://doi.org/10.1023/a:1008233704719.

Chen, Shuangshuang, Xiaojiao Han, Jie Fang, Zhuchou Lu, Wenmin Qiu, Mingying Liu, Jian Sang, Jing Jiang, and Renying Zhuo. 2017. "Sedum Alfredii SaNramp6 Metal Transporter Contributes to Cadmium Accumulation in Transgenic Arabidopsis thaliana." *Scientific Reports* 7, no. 1: 13318. https://doi.org/10.1038/s41598-017-13463-4.

Chiappero, J., L. Cappellari, L. G. S. Sosa Alderete, T. B. Palermo, and E. Banchio. 2019. "Plant Growth Promoting Rhizobacteria Improve the Antioxidant Status in Mentha piperita Grown Under Drought Stress Leading to an Enhancement of Plant Growth and Total Phenolic Content." *Industrial Crops & Products* 139: 111553. https://doi.org/10.1016/j.indcrop.2019.111553.

Daghan, H., M. Arslan, V. Uygur, and N. Koleli 2013. "Transformation of Tobacco with *ScMTII* Gene-Enhanced Cadmium and Zinc Accumulation." *Clean Soil Air Water* 41: 503–9. https://doi.org/10.1002/clen.201200298.

Dal Corso, Giovanni, Elisa Fasani, Anna Manara, Giovanna Visioli, and Antonella Furini. 2019. "Heavy Metal Pollutions: State of the Art and Innovation in Phytoremediation." *International Journal of Molecular Sciences* 20, no. 14: 3412. https://doi.org/10.3390/ijms20143412.

Danouche, Mohammed, Naïma El Ghachtouli, and Hicham El Arroussi. 2021. "Phycoremediation Mechanisms of Heavy Metals Using Living Green Microalgae: Physicochemical and Molecular Approaches for Enhancing Selectivity and Removal Capacity." *Heliyon* 7, no. 7: E07609. https://doi.org/10.1016/j.heliyon.2021.e07609.

Das, Natasha, Surajit Bhattacharya, and Mrinal K. Maiti. 2016. "Enhanced Cadmium Accumulation and Tolerance in Transgenic Tobacco over Expressing Rice Metal Tolerance Protein Gene OsMTP1

Is Promising for Phytoremediation." *Plant Physiology & Biochemistry* 105: 297–309. https://doi.org/10.1016/j.plaphy.2016.04.049.

De Groote, Valerie N., Natalie Verstraeten, Maarten Fauvart, Cyrielle I. Kint, Aline M. Verbeeck, Serge Beullens, Pierre Cornelis, and Jan Michiels. 2009. "Novel Persistence Genes in Pseudomonas Aeruginosa Identified by High-Throughput Screening." *FEMS Microbiology Letters* 297, no. 1: 73–9. https://doi.org/10.1111/j.1574-6968.2009.01657.x.

Delgadillo, J., A. Lafuente, B. Doukkali, S. Redondo-Gómez, E. Mateos-Naranjo, M. A. Caviedes, E. Pajuelo, and I. D. Rodríguez-Llorente. 2014. "Improving Legume Nodulation and Cu Rhizostabilization Using a Genetically Modified Rhizobia." *Environmental Technology* 36: 1–28. https://doi.org/10.1080/09593330.2014.983990.

Devos, G., M. Tuytens, and H. Hester. 2014. "Teachers' Organizational Commitment: Examining the Mediating Effects of Distributed Leadership." *American Journal of Education.* 120: 205–31. https://doi.org/10.1086/674370.

Duan, G. L., Y. Hu, W. J. Liu, R. Kneer, F. J. Zhao, and Y. G. Zhu. 2011. "Evidence for a Role of Phytochelatins in Regulating Arsenic Accumulation in Rice Grain." *Environmental & Experimental Botany* 71, no. 3: 416–21. https://doi.org/10.1016/j.envexpbot.2011.02.016.

Fathollahi, Alireza, Nazanin Khasteganan, Stephen J. Coupe, and Alan P. Newman. 2021. "A Meta-analysis of Metal Biosorption by Suspended Bacteria from Three Phyla." *Chemosphere* 268: 129290. https://doi.org/10.1016/j.chemosphere.2020.129290.

Frisbie, Seth H., and Erika J. Mitchell. 2022. "Arsenic in Drinking Water: An Analysis of Global Drinking Water Regulations and Recommendations for Updates to Protect Public Health." *PLOS ONE* 17, no. 4: E0263505. https://doi.org/10.1371/journal.pone.0263505.

Gogarten, J. Peter, and Jeffrey P. Townsend. 2005. "Horizontal Gene Transfer, Genome Innovation and Evolution." *Nature Reviews. Microbiology* 3, no. 9: 679–87. https://doi.org/10.1038/nrmicro1204.

da Conceição Gomes, M. A., R. A. Hauser-Davis, A. N. de Souza, and A. P. Vitória. 2016. "Metal Phytoremediation: General Strategies, Genetically Modified Plants and Applications in Metal Nanoparticle Contamination." *Ecotoxicology & Environmental Safety* 134: 133–47. https://doi.org/10.1016/j.ecoenv.2016.08.024.

Garrigues, S., R.S. Kun, M. Peng, D. Bauer, K. Keymanesh, A. Lipzen, V. Ng, I.V. Grigoriev, R.P. de Vries. 2022. "Unraveling the regulation of sugar beet pulp utilization in the industrially relevant fungus Aspergillus niger" *Science*, 25 :104065.

Gavrilescu, M. 2022. "Enhancing Phytoremediation of Soils Polluted with Heavy Metals." *Current Opinion in Biotechnology* 74: 21–31.

Guarin, Tatiana C., and Krishna R. Pagilla. 2021. "Microbial Community in Biofilters for Water Reuse Applications: A Critical Review." *Science of the Total Environment* 773: 145655. https://doi.org/10.1016/j.scitotenv.2021.145655.

Guo, Jiangbo, Xiaojing Dai, Wenzhong Xu, and M. Ma. 2008. "Overexpressing gsh1 and AsPCS1 Simultaneously Increases the Tolerance and Accumulation of Cadmium and Arsenic in *Arabidopsis thaliana*." *Chemosphere* 72, no. 7: 1020–6. https://doi.org/10.1016/j.chemosphere.2008.04.018.

He, J., H. Li, C. Ma, Y. Zhang, A. Polle, H. Rennenberg, X. Cheng, and Z. Luo. 2015. "Overexpression of Bacterial g-Glutamylcysteine Synthetase Mediates Changes in Cadmium Fux, Allocation and Detoxification in Poplar." *New Phytology* 205: 240–54.

Huang, Shan, Arianna Sherman, Chen Chen, and Peter R. Jaffé. 2021. "Tropical Cyclone Effects on Water and Sediment Chemistry and the Microbial Community in Estuarine Ecosystems." *Environmental Pollution* 286: 117228. https://doi.org/10.1016/j.envpol.2021.117228.

Jacob, Jaya Mary, Chinnannan Karthik, Rijuta Ganesh Saratale, Smita S. Kumar, Desika Prabakar, K. Kadirvelu, and Arivalagan Pugazhendhi. 2018. "Biological Approaches to Tackle Heavy Metal Pollution: A Survey of Literature." *Journal of Environmental Management* 217: 56–70. https://doi.org/10.1016/j.jenvman.2018.03.077.

Jaiswal, Shweta, and Pratyoosh Shukla. 2020. "Alternative Strategies for Microbial Remediation of Pollutants via Synthetic Biology." *Frontiers in Microbiology* 11: 808. https://doi.org/10.3389/fmicb.2020.00808.

Jan, S., B. Rashid, M. M. Azooz, M. A. Hossain, and P. Ahmad. 2016. "Genetic Strategies for Advancing Phytoremediation Potential in Plants: A Recent Update." *Plant Metal Interaction*: 431–54.

Jong, Mui-Choo, Colin R. Harwood, Adrian Blackburn, Jason R. Snape, and David W. Graham. 2020. "Impact of Redox Conditions on Antibiotic Resistance Conjugative Gene Transfer Frequency and Plasmid Fate in Wastewater Ecosystems." *Environmental Science & Technology* 54, no. 23: 14984–93. https://doi.org/10.1021/acs.est.0c03714.

Karahan, Faruk. 2023. "Evaluation of Trace Element and Heavy Metal Levels of Some Ethnobotanically Important Medicinal Plants Used as Remedies in Southern Turkey in Terms of Human Health Risk." *Biological Trace Element Research* 201, no. 1: 493–513. https://doi.org/10.1007/s12011-022-03299-z.

Khan, Suliman, Muhammad Wajid Ullah, Rabeea Siddique, Ghulam Nabi, Sehrish Manan, Muhammad Yousaf, and Hongwei Hou. 2016. "Role of Recombinant DNA Technology to Improve Life." *International Journal of Genomics* 2016: 2405954. https://doi.org/10.1155/2016/2405954.

Kim, Suyeon, Michiko Takahashi, Kyoko Higuchi, Kyoko Tsunoda, Hiromi Nakanishi, Etsuro Yoshimura, Satoshi Mori, and Naoko K. Nishizawa. 2005. "Increased Nicotianamine Biosynthesis Confers Enhanced Tolerance of High Levels of Metals, in Particular Nickel, to Plants." *Plant & Cell Physiology* 46, no. 11: 1809–18. https://doi.org/10.1093/pcp/pci196.

Kotrba, Pavel, Jitka Najmanova, Tomas Macek, Tomas Ruml, and Martina Mackova. 2009. "Genetically Modified Plants in Phytoremediation of Heavy Metal and Metalloid Soil and Sediment Pollution." *Biotechnology Advances* 27, no. 6: 799–810. https://doi.org/10.1016/j.biotechadv.2009.06.003.

Krout, Ian N., Thomas Scrimale, Daria Vorojeikina, Eric S. Boyd, and Matthew D. Rand. 2022. "Organomercuriallyase (merB)-Mediated Demethylation Decreases Bacterial Methylmercury Resistance in the Absence of Mercuric Reductase (merA)." *Applied & Environmental Microbiology* 88, no. 6: E0001022. https://doi.org/10.1128/aem.00010-22.

Kumar, V., S. Al Momin, A. Al-Shatti, H. Al-Aqeel, F. Al-Salameen, A. B. Shajan, and S. M. Nair. 2019. "Enhancement of Heavy Metal Tolerance and Accumulation Efficiency by Expressing Arabidopsis ATP Sulfurylase Gene in Alfalfa." *International Journal of Phytoremediation* 21, no. 11: 1112–21. https://doi.org/10.1080/15226514.2019.1606784.

Kuroda, K., S. Shibasaki, M. Ueda, and A. Tanaka. 2001. "Cell Surface-Engineered Yeast Displaying a Histidine Oligopeptide (Hexa-His) Has Enhanced Adsorption of and Tolerance to Heavy Metal Ions." *Applied Microbiology & Biotechnology* 57, no. 5–6: 697–701. https://doi.org/10.1007/s002530100813.

Lee, Kyunghee, Dong Won Bae, Sun Ho Kim, Hay Ju Han, Xiaomin Liu, Hyeong Cheol Park, Chae Oh Lim, Sang Yeol Lee, and Woo Sik Chung. 2010. "Comparative Proteomic Analysis of the Short-Term Responses of Rice Roots and Leaves to Cadmium." *Journal of Plant Physiology* 167, no. 3: 161–8. https://doi.org/10.1016/j.jplph.2009.09.006.

Leong, Yoong Kit, and Jo-Shu Chang. 2020. "Bioremediation of Heavy Metals Using Microalgae: Recent Advances and Mechanisms." *Bioresource Technology* 303: 122886. https://doi.org/10.1016/j.biortech.2020.122886.

Li, X., Z. Ren, M. J. C. Crabbe, L. Wang, and W. Ma. 2021. "Genetic Modifications of Metallothione in Enhance the Tolerance and Bioaccumulation of Heavy Metals in Escherichia coli." *Ecotoxicology & Environmental Safety* 222: 112512. https://doi.org/10.1016/j. ecoenv.2021.112512.

Liang, Xiao-Long, Feng Zhao, Rong-Jiu Shi, Yun-He Ban, Ji-Dong Zhou, Si-Qin Han, Ying Zhang. 2015. "Construction and Evaluation of an Engineered Bacterial Strain for Producing Lipopeptide Under Anoxic Conditions." *Ying Yong Sheng Tai Xue Bao* 26, no. 8: 2553–60.

Liaquat, F., U. Haroon, M. F. H. Munis, S. Arif, M. Khizar, W. Ali, C. Shengquan, and L. Qunlu. 2021. "Efficient Recovery of Metal Tolerant Fungi from the Soil of Industrial Area and Determination of Their Biosorption Capacity." *Environmental Technology & Innovation* 21: 101237. https://doi.org/10.1016/j. eti.2020.101237.

Linacre, Nicholas A., Steven N. Whiting, Alan J. Baker, J. Scott Angle, and Peter K. Ades. 2003. "Transgenics and Phytoremediation: The Need for an Integrated Risk Assessment, Management, and Communication Strategy." *International Journal of Phytoremediation* 5, no. 2: 181–5. https://doi.org/10.1080/713610179.

Liu, Dali, Zhigang An, Zijun Mao, Longbiao Ma, and Zhenqiang Lu. 2015. "Enhanced Heavy Metal Tolerance and Accumulation by Transgenic Sugar Beets Expressing Streptococcus thermophilus StGCS-GS in the Presence of Cd, Zn and Cu Alone or in Combination." *PLOS ONE* 10, no. 6: e0128824. https://doi.org/10.1371/journal.pone.0128824.

Liu, Huan, Haixia Zhao, Longhua Wu, Anna Liu, Fang-Jie Zhao, and Wenzhong Xu. 2017. "Heavy Metal ATPase 3 (HMA3) Confers Cadmium Hypertolerance on the Cadmium/Zinc Hyperaccumulator Sedum Plumbizincicola." *New Phytologist* 215, no. 2: 687–98. https://doi.org/10.1111/nph.14622.

Liu, Shuang, Fan Zhang, Jian Chen, and Guoxin Sun. 2011. "Arsenic Removal from Contaminated Soil via Biovolatilization by Genetically Engineered Bacteria Under Laboratory Conditions." *Journal of Environmental Sciences* 23, no. 9: 1544–50. https://doi.org/10.1016/s1001-0742(10)60570-0.

Liu, Yiting, Jing Feng, Hangcheng Pan, Xiuwei Zhang, and Yunlei Zhang. 2022. "Genetically Engineered Bacterium: Principles, Practices, and Prospects." *Frontiers in Microbiology* 13: 997587. https://doi.org/10.3389/fmicb.2022.997587.

Luo, S. L., Liang Chen, J. L. Chen, Xiao Xiao, T. Y. Xu, Yong Wan, Chan Rao, C. B. Liu, Y. T. Liu, Cui Lai, and G. M. Zeng. 2011. "Analysis and Characterization of Cultivable Heavy Metal-Resistant Bacterial Endophytes Isolated from Cd-Hyperaccumulator *Solanum nigrum* L. and Their Potential Use for Phytoremediation." *Chemosphere* 85, no. 7: 1130–8. https://doi.org/10.1016/j.chemosphere.2011.07.053.

Ma, Yao, Jianqun Lin, Chengjia Zhang, Yilin Ren, and Jianqiang Lin. 2011. "Cd(II) and As(III) Bioaccumulation by Recombinant *Escherichia coli* Expressing Oligomeric Human Metallothioneins." *Journal of Hazardous Materials* 185, no. 2–3: 1605–8. https://doi.org/10.1016/j.jhazmat.2010.10.051.

Manikandan, Arumugam, Palanisamy Suresh Babu, Shanmugasundaram Shyamalagowri, Murugesan Kamaraj, Peraman Muthukumaran, and Jeyaseelan Aravind. 2022. "Emerging Role of Microalgae in Heavy Metal Bioremediation." *Journal of Basic Microbiology* 62, no. 3–4: 330–47. https://doi.org/10.1002/jobm.202100363.

Maqsood, Quratulain, Nazim Hussain, Mehvish Mumtaz, Muhammad Bilal, and Hafiz M. N. Iqbal. 2022. "Novel Strategies and Advancement in Reducing Heavy Metals from the Contaminated Environment." *Archives of Microbiology* 204, no. 8: 478. https://doi.org/10.1007/s00203-022-03087-2.

Meena, M., M. Aamir, V. Kumar, P. Swapnil, and R. S. Upadhyay. 2018. "Evaluation of Morpho-physiological Growth Parameters of Tomato in Response to Cd Induced Toxicity and Characterization of Metal Sensitive NRAMP3 Transporter Protein." *Environmental & Experimental Botany* 148: 144–67. https://doi.org/10.1016/j.envexpbot.2018.01.007.

Meruvu, H. 2021. "Bacterial Bioremediation of Heavy Metals from Polluted Wastewaters." In *New Trends in Removal of Heavy Metals from Industrial Wastewater*: 105–14. Amsterdam: Elsevier.

Mishra, Jitendra, Rachna Singh, and Naveen K. Arora. 2017. "Alleviation of Heavy Metal Stress in Plants and Remediation of Soil by Rhizosphere Microorganisms." *Frontiers in Microbiology* 8: 1706. https://doi.org/10.3389/fmicb.2017.01706.

Mitra, S., A. J. Chakraborty, A. M. Tareq, T. B. Emran, F. Nainu, A. Khusro, A. M. Idris, M. U. Khandaker, H. Osman, F. A. Alhumaydhi, and J. Simal-Gandara. 2022. "Impact of Heavy Metals on the Environment and Human Health: Novel Therapeutic Insights to Counter the Toxicity." *Journal of King Saud University—Science* 34, no. 3: 101865. https://doi.org/10.1016/j.jksus.2022.101865.

Myhr, A. I., and T. Traavik. 2002. "The Precautionary Principle: Scientific Uncertainty and Omitted Research in the Context of GMO Use and Release." *Journal of Agricultural & Environmental Ethics* 15, no. 1: 73–86. https://doi.org/10.1023/A:1013814108502.

Nahar, Noor, Aminur Rahman, Neelu N. Nawani, Sibdas Ghosh, and Abul Mandal. 2017. "Phytoremediation of Arsenic from the Contaminated Soil Using Transgenic Tobacco Plants Expressing ACR2 Gene of Arabidopsis thaliana." *Journal of Plant Physiology* 218: 121–6. https://doi.org/10.1016/j.jplph.2017.08.001.

Naing, A. H., K. Il Park, M. Y. Chung, K. B. Lim, and C. K. Kim. 2016. "Optimization of Factors Affecting Efficient Shoot Regeneration in Chrysanthemum cv. Shinma." *Brazilian Journal of Botany* 39, no. 4: 975–84. https://doi.org/10.1007/s40415-015-0143-0.

Nascimento, Andréa M. A., and Edmar Chartone-Souza. 2003. "Operon Mer: Bacterial Resistance to Mercury and Potential for Bioremediation of Contaminated Environments." *Genetics & Molecular Research* 2, no. 1: 92–101.

Nawaz, Zarqa, Kaleem U. Kakar, Raqeeb Ullah, Shizou Yu, Jie Zhang, Qing-Yao Shu, and Xue-Liang Ren. 2019. "Genome-Wide Identification, Evolution and Expression Analysis of Cyclic Nucleotide-Gated Channels in Tobacco (*Nicotiana tabacum* L.)." *Genomics* 111, no. 2: 142–58. https://doi.org/10.1016/j.ygeno.2018.01.010.

Nguyen, Thuong T. L. T. L., Hae Ryong Lee, Soon Ho Hong, Ji-Ryang Jang, Woo-Seok Choe, and Ik-Keun Yoo. 2013. "Selective Lead Adsorption by Recombinant Escherichia coli Displaying a Lead-Binding Peptide." *Applied Biochemistry & Biotechnology* 169, no. 4: 1188–96. https://doi.org/10.1007/s12010–012-0073–2.

Oladipo, Oluwatosin Gbemisola, Olusegun Olufemi Awotoye, Akinyemi Olayinka, Cornelius Carlos Bezuidenhout, and Mark Steve Maboeta. 2018. "Heavy Metal Tolerance Traits of Filamentous Fungi Isolated from Gold and Gemstone Mining Sites." *Brazilian Journal of Microbiology* 49, no. 1: 29–37. https://doi.org/10.1016/j.bjm.2017.06.003.

Pande, Veni, Satish Chandra Pandey, Diksha Sati, Pankaj Bhatt, and Mukesh Samant. 2022. "Microbial Interventions in Bioremediation of Heavy Metal Contaminants in Agroecosystem." *Frontiers in Microbiology* 13: 824084. https://doi.org/10.3389/fmicb.2022.824084.

Pant, Gaurav, Deviram Garlapati, Urvashi Agrawal, R. Gyana Prasuna, Thangavel Mathimani, and Arivalagan Pugazhendhi. 2021. "Biological Approaches Practised Using Genetically Engineered Microbes for a Sustainable Environment: A Review." *Journal of Hazardous Materials* 405: 124631. https://doi.org/10.1016/j.jhazmat.2020.124631.

Peng, Jia-Shi, Yue-Jun Wang, G. Ding, Hai-Ling Ma, Yi-Jing Zhang, and Ji-Ming Gong. 2017. "A Pivotal Role of Cell Wall in Cadmium Accumulation in the Crassulaceae Hyperaccumulator Sedum Plumbizincicola." *Molecular Plant* 10, no. 5: 771–4. https://doi.org/10.1016/j.molp.2016.12.007.

Biological Solutions for Metal-Contaminated Environments

Pérez-Palacios, P., A. Romero-Aguilar, J. Delgadillo, B. Doukkali, M. A. Caviedes, and I. D. Rodríguez-Llorente. 2017. "Double Genetically Modified Symbiotic System for Improved Cu Phytostabilization in Legume Roots." *Environmental Science & Pollution Research Internation* 24: 14910–23. https://doi.org/10.1007/s11356-017-9092-4

Peter, R., J. Mojca, and P. Primož. 2011. "Genetically Modified Organisms (GMOs)." In *Encyclopedia of Environmental Health*, edited by J. O. Nriagu: 879–88. Amsterdam: Elsevier.

Qin, M., C. Y. Chen, B. Song, M. C. Shen, W. Cao, H. L. Yang, G. M. Zeng, and J. L. Gong. 2021. "A Review of Biodegradable Plastics to Biodegradable Microplastics: Another Ecological Threat to Soil Environments?" *Journal of Clean Production* 312: 127816.

Qin, Song, Hanzhi Lin, and Peng Jiang. 2012. "Advances in Genetic Engineering of Marine Algae." *Biotechnology Advances* 30, no. 6: 1602–13. https://doi.org/10.1016/j.biotechadv.2012.05.004.

Rafeeq, Hamza, Nadia Afsheen, Sadia Rafique, Arooj Arshad, Maham Intisar, Asim Hussain, Muhammad Bilal, and Hafiz M. N. Iqbal. 2023. "Genetically Engineered Microorganisms for Environmental Remediation." *Chemosphere* 310: 136751. https://doi.org/10.1016/j.chemosphere.2022.136751.

Rai, Prabhat Kumar, Ki-Hyun Kim, Sang Soo Lee, and Jin-Hong Lee. 2020. "Molecular Mechanisms in Phytoremediation of Environmental Contaminants and Prospects of Engineered Transgenic Plants/Microbes." *Science of the Total Environment* 705: 135858. https://doi.org/10.1016/j.scitotenv.2019.135858.

Rakib, Md Refat Jahan, Md Asrafur Rahman, Amarachi Paschaline Onyena, Rakesh Kumar, Aniruddha Sarker, M. Belal Hossain, Abu Reza Md Towfiqul Islam *et al.* 2022. "A Comprehensive Review of Heavy Metal Pollution in the Coastal Areas of Bangladesh: Abundance, Bioaccumulation, Health Implications, and Challenges." *Environmental Science & Pollution Research International* 29, no. 45: 67532–58. https://doi.org/10.1007/s11356-022-22122-9.

Rascio, Nicoletta, and Flavia Navari-Izzo. 2011. "Heavy Metal Hyperaccumulating Plants: How and Why Do They Do It? And What Makes Them so Interesting?." *Plant Science* 180, no. 2: 169–81. https://doi.org/10.1016/j.plantsci.2010.08.016.

Rasheed, T., F. Nabeel, M. Adeel, M. Bilal, and H. M. N. Iqbal. 2019. " 'Turn-on Fluorescent Sensor-Based Probing of Toxic." *Biocatalysis & Agricultural Biotechnology* 17: 696–701. https://doi.org/10.1016/j.bcab.2019.01.032.

Rasheed, Tahir, Muhammad Bilal, Faran Nabeel, Hafiz M. N. Iqbal, Chuanlong Li, and Yongfeng Zhou.2018a. "Fluorescent Sensor Based Models for the Detection of Environmentally Related Toxic Heavy Metals." *Science of the Total Environment* 615: 476–85. https://doi.org/10.1016/j.scitotenv.2017.09.126.

Rasheed, Tahir, Chuanlong Li, Muhammad Bilal, Chunyang Yu, and Hafiz M. N. Iqbal.2018b. "Potentially Toxic Elements and Environmentally Related Pollutants Recognition Using Colorimetric and Ratiometric Fluorescent Probes." *Science of the Total Environment* 640–1: 174–93. https://doi.org/10.1016/j.scitotenv.2018.05.232.

Rehman, K., F. Fatima, I. Waheed, and M. S. H. Akash. 2018. "Prevalence of Exposure of Heavy Metals and Their Impact on Health Consequences." *Journal of Cellular Biochemistry* 119: 157–184.

Renn, O. 2008. *Risk Governance: Coping with Uncertainty in a Complex World*. London, UK: Earthscan.

Samreen, T., H. U. Humaira, H. U. Shah, S. Ullah, and M. Javid. 2017. "Zinc Effect on Growth Rate, Chlorophyll, Protein and Mineral Contents of Hydroponically Grown Mungbeans Plant (Vigna radiata)." *Arabian Journal of Chemistry* 10: S1802–7. https://doi.org/10.1016/j.arabjc.2013.07.005.

Sankhla, M. S., K. Sharma, and R. Kumar. 2017. "Heavy Metal Causing Neurotoxicity in Human Health." *International Journal of Innovative Research in Science, Engineering & Technology* 6, no. 5.

Saravanan, A., P. SenthilKumar, B. Ramesh, and S. Srinivasan. 2022. "Removal of Toxic Heavy Metals Using Genetically Engineered Microbes: Molecular Tools, Risk Assessment and Management Strategies." *Chemosphere* 298: 134341. https://doi.org/10.1016/j.chemosphere.2022.134341.

Sarker, Aniruddha, Jang-Eok Kim, Abu Reza Md Towfiqul Islam, Muhammad Bilal, Md Refat Jahan Rakib, Rakhi Nandi, Mohammed M. Rahman, and Tofazzal Islam. 2022. "Heavy Metal Contamination and Associated Health Risks in Food Webs—A Review Focuses on Food Safety and Environmental Sustainability in Bangladesh." *Environmental Science & Pollution Research International* 29, no. 3: 3230–45. https://doi.org/10.1007/s11356-021-17153-7.

Sarojini, G., S. Venkatesh Babu, N. Rajamohan, M. Rajasimman, and Arivalagan Pugazhendhi. 2022. "Application of a Polymer-Magnetic-Algae Based Nanocomposite for the Removal of Methylene Blue–Characterization, Parametric and Kinetic Studies." *Environmental Pollution* 292, no. B: 118376. https://doi.org/10.1016/j.envpol.2021.118376.

Schermer, M., and J. Hoppichler. 2004. "GMO and Sustainable Development in Less Favoured Regions—the Need for Alternative Paths of Development." *Journal of Cleaner Production* 12, no. 5: 479–89. https://doi.org/10.1016/S0959-6526(03)00110-0.

Scipioni, A., G. Saccarola, F. Arena, and S. Alberto. 2005. "Strategies to Assure the Absence of GMO in FoodProductsApplicationProcess in a ConfectioneryFirm." *Food Control* 16, no. 7: 569–78. https://doi.org/10.1016/j.foodcont.2004.06.018.

Sharma, Pooja, Ambreen Bano, Surendra Pratap Singh, Swati Sharma, Changlei Xia, Ashok Kumar Nadda, Su Shiung Lam, and Yen Wah Tong. 2022a. "Engineered Microbes as Effective Tools for the Remediation of Polyaromatic Aromatic Hydrocarbons and Heavy Metals." *Chemosphere* 306: 135538. https://doi.org/10.1016/j.chemosphere.2022.135538.

Sharma, Pooja, Ranjna Sirohi, Yen Wah Tong, Sang Hyoun Kim, and Ashok Pandey. 2021. "Metal and Metal(Loids) Removal Efficiency Using Genetically Engineered Microbes: Applications and Challenges." *Journal of Hazardous Materials* 416: 125855. https://doi.org/10.1016/j.jhazmat.2021.125855.

Sharma, Pooja, Sheetal Kishor Parakh, Surendra Pratap Singh, Roberto Parra-Saldívar, Sang-Hyoun Kim, Sunita Varjani, and Yen Wah Tong. 2022b. "A Critical Review on Microbes-Based Treatment Strategies for Mitigation of Toxic Pollutants." *Science of the Total Environment* 834: 155444. https://doi.org/10.1016/j.scitotenv.2022.155444.

Sharma, Pooja, Surendra Pratap Singh, Sheetal Kishor Parakh, and Yen Wah Tong. 2022c. "Health Hazards of Hexavalent Chromium (Cr (VI)) and Its Microbial Reduction." *Bioengineered* 13, no. 3: 4923–38. https://doi.org/10.1080/21655979.2022.2037273.

Sharma, Pooja, R. Sirohi, Y. W. Tong, S. H. Kim, and A. Pandey. 2021. "Metal and Metal(loids) Removal Efficiency Using Genetically Engineered Microbes: Applications and Challenges." *Journal of Hazardous Materials* 416: 125855.

Shukla, Devesh, Ravi Kesari, Seema Mishra, Sanjay Dwivedi, Rudra Deo Tripathi, Pravendra Nath, and Prabodh Kumar Trivedi. 2012. "Expression of Phytochelatin Synthase from Aquatic Macrophyte *Ceratophyllum demersum* L. Enhances Cadmium and Arsenic Accumulation in Tobacco." *Plant Cell Reports* 31, no. 9: 1687–99. https://doi.org/10.1007/s00299-012-1283-3.

Singh, Akshay Kumar, Manoj Kumar, Kuldeep Bauddh, Ajai Singh, Pardeep Singh, Sughosh Madhav, and Sushil Kumar Shukla. 2023. "Environmental Impacts of Air Pollution and Its Abatement by Plant Species: A Comprehensive Review." *Environmental Science & Pollution Research International* 30, no. 33: 79587–616. https://doi.org/10.1007/s11356-023-28164-x.

Sousa, C., P. Kotrba, T. Ruml, A. Cebolla, and V. De Lorenzo. 1998. "Metalloadsorption by Escherichia coli Cells Displaying Yeast and Mammalian Metallothionein Anchored to the Outer Membrane Protein Lam B." *Journal of Bacteriology* 180, no. 9: 2280–4. https://doi.org/10.1128/JB.180.9.2280-2284.1998.

Sproles, A. E., F. J. Fields, T. N. Smalley, C. H. Le, A. Badary, and S. P. Mayfield. 2021. "Recent Advancements in the Genetic Engineering of Microalgae." *Algal Research* 53: 102158. https://doi.org/10.1016/j.algal.2020.102158.

Stadlmair, L. F., T. Letzel, J. E. Drewes, and J. Grassmann. 2018. "Enzymes in Removal of Pharmaceuticals from Wastewater: A Critical Review of Challenges, Applications and Screening Methods for Their Selection." *Chemosphere* 205: 649–61.

Sunkar, R., B. Kaplan, N. Bouché, T. Arazi, D. Dolev, I. N. Talke, F. J. Maathuis, D. Sanders, D. Bouchez, and H. Fromm. 2000. "Expression of a Truncated Tobacco NtCBP4 Channel in Transgenic Plants and Disruption of the Homologous Arabidopsis CNGC1 GeneConferPb2+Tolerance." *Plant Journal: For Cell & Molecular Biology* 24, no. 4: 533–42. https://doi.org/10.1046/j.1365-313x.2000.00901.x.

Tan, Xiaona, Kaixia Li, Zheng Wang, Keming Zhu, Xiaoli Tan, and Jun Cao. 2019. "A Review of Plant Vacuoles: Formation, Located Proteins, and Functions." *Plants* 8, no. 9: 327. https://doi.org/10.3390/plants8090327.

Thai, Thi Duc, Wonseop Lim, and Dokyun Na. 2023. "Synthetic Bacteria for the Detection and Bioremediation of Heavy Metals." *Frontiers in Bioengineering & Biotechnology* 11: 1178680. https://doi.org/10.3389/fbioe.2023.1178680.

Tran, Kha Mong, Hyang-Mi Lee, Thi Duc Thai, Junhao Shen, Seong-Il Eyun, and Dokyun Na. 2021. "Synthetically Engineered Microbial Scavengers for Enhanced Bioremediation." *Journal of Hazardous Materials* 419: 126516. https://doi.org/10.1016/j.jhazmat.2021.126516.

Tsyganov, Viktor E., Anna V. Tsyganova, Artemii P. Gorshkov, Elena V. Seliverstova, Viktoria E. Kim, Elena P. Chizhevskaya, Andrey A. Belimov, Tatiana A. Serova, Kira A. Ivanova, Olga A. Kulaeva, Pyotr G. Kusakin, Anna B. Kitaeva, and Igor A. Tikhonovich. 2020. "Efficacy of a Plant-Microbe System: Pisum sativum (L.) Cadmium-Tolerant Mutant and *Rhizobium leguminosarum* Strains, Expressing Pea Metallothionein Genes PsMT1 and PsMT2, for Cadmium Phytoremediation." *Frontiers in Microbiology* 11: 15. https://doi.org/10.3389/fmicb.2020.00015.

Uçkun, A. A., M. Uçkun and A. Şeyma. 2021. "Efficiency of Escherichia coli Jm109 and genetical engineering strains (E. coli MT2, E. coli MT3) in cadmium removal from aqueous solutions". *Environmental Technology & Innovation*. 24: 102024.

Verma, S., P. Bhatt, A. Verma, H. Mudila, P. Prasher, and E. R. Rene. 2021. "Microbial Technologies for Heavy Metal Remediation: Effect of Process Conditions and Current Practices." *Clean Technologies & Environmental Policy*: 1–23.

Wang, M., W. Wilhelm, and N. Bhullar. 2013. "Nicotianamine Synthase Overexpression Positively Modulates Iron Homeostasis-Related Genes in High Iron Rice." *Frontiers in Plant Science* 4: 156. https://doi.org/10.3389/fpls.2013.00156.

Wang, Qiong, Jiayuan Ye, Yingjie Wu, Sha Luo, Bao Chen, Luyao Ma, Fengshan Pan, Ying Feng, and Xiaoe Yang. 2019. "Promotion of the Root Development and Zn Uptake of Sedum alfredii Was Achieved by an Endophytic Bacterium Sasm05." *Ecotoxicology & Environmental Safety* 172: 97–104. https://doi.org/10.1016/j.ecoenv.2019.01.009.

Wei, Yuquan, Yue Zhao, Xinyu Zhao, Xintong Gao, Yansi Zheng, Huiduan Zuo, and Zimin Wei. 2020. "Roles of Different Humin and Heavy-Metal Resistant Bacteria from Composting on Heavy Metal Removal." *Bioresource Technology* 296: 122375. https://doi.org/10.1016/j.biortech.2019.122375.

Wojas, Sylwia, Stephan Clemens, Aleksandra Skłodowska, and Danuta Maria Antosiewicz. 2010. "Arsenic Response of AtPCS1- and CePCS-Expressing Plants-Effects of External As(V) Concentration on As-Accumulation Pattern and NPT Metabolism." *Journal of Plant Physiology* 167, no. 3: 169–75. https://doi.org/10.1016/j.jplph.2009.07.017.

Wu, Chen, Feng Li, Shengwei Yi, and Fei Ge. 2021. "Genetically Engineered Microbial Remediation of Soils Co-contaminated by Heavy Metals and Polycyclic Aromatic Hydrocarbons: Advances and Ecological Risk Assessment." *Journal of Environmental Management* 296: 113185. https://doi.org/10.1016/j.jenvman.2021.113185.

Yu, Ying, Kaixiang Shi, Xuexue Li, Xiong Luo, Mengjie Wang, Lin Li, Gejiao Wang, and Mingshun Li. 2022. "Reducing Cadmium in Rice Using Metallothionein Surface-Engineered Bacteria WH16–1-MT." *Environmental Research* 203: 111801. https://doi.org/10.1016/j.envres.2021.111801.

Zeraatkar, Amin Keyvan, Hossein Ahmad Zadeh, Ahmad Farhad Talebi, Navid R. Moheimani, and Mark P. McHenry. 2016. "Potential Use of Algae for Heavy Metal Bioremediation, a CriticalReview." *Journal of Environmental Management* 181: 817–31. https://doi.org/10.1016/j.jenvman.2016.06.059.

Zhang, Jie, Enrico Martinoia, and Youngsook Lee. 2018. "Vacuolar Transporters for Cadmium and Arsenic in Plants and Their Applications in Phytoremediation and Crop Development." *Plant & Cell Physiology* 59, no. 7: 1317–25. https://doi.org/10.1093/pcp/pcy006.

Zhang, W., Z. Lin, S. Pang, P. Bhatt, and S. Chen. 2020. "Insights into the Biodegradation of Lindane (γ-hexachlorocyclohexane) Using a Microbial System." *Frontiers in Microbiology* 11: 522.

Zhao, X. W., M. H. Zhou, Q. B. Li, Y. H. Lu, N. He, D. H. Sun, and X. Deng. 2005. "Simultaneous Mercury Bioaccumulation and Cell Propagation by Genetically Engineered Escherichia coli." *Process Biochemistry* 40, no. 5: 1611–6. https://doi.org/10.1016/j.procbio.2004.06.014.

Zhou, Jiaxi, Y. Cheng, Lifei Yu, Jian Zhang, and Xiao Zou. 2022. "Characteristics of Fungal Communities and the Sources of Mold Contamination in MildewedTobaccoLeavesStoredUnderDifferentClimaticConditions." *Applied Microbiology & Biotechnology* 106, no. 1: 131–44. https://doi.org/10.1007/s00253-021-11703-2.

Zhu, Nali, Bing Zhang, and Qilin Yu. 2020. "Genetic Engineering-Facilitated Co Assembly of Synthetic Bacterial Cells and Magnetic Nanoparticles for Efficient Heavy Metal Removal." *ACS Applied Materials & Interfaces* 12, no. 20: 22948–57. https://doi.org/10.1021/acsami.0c04512.

5 Phytoremediation of Organic Contaminants

*Idris Abdullahi Dabban, Oluwafemi Adebayo Oyewole,
Bello Aisha Bisola, Abdullahi Muhammad Asma'u,
Bawa Muhammed Muhammed, and Alfa Suleiman*

5.1 INTRODUCTION

With the arrival of the industrial age, the environment has constantly been under pressure due to the increase in human population, urbanization and subsequent boost in worldwide human and industrial activities, which has led to a buildup in biological and chemical pollutants in soils, water bodies and various ecological niches of the environment, hence posing serious potential threats to the stability of the environment. Research has revealed that the majority of these pollutants are toxic to the environment, with tendencies to move between mediums and accrue in living organisms such as plants, fishes, insects, birds and mammals (Bernard *et al.*, 1999; Kozak *et al.*, 2001). Pharmaceutical, agricultural, medical, municipal and industrial activities are the chief sources of organic contaminants of soil, water and air. Pharmaceutical sources of these pollutants include waste materials discarded as by-products of pharmaceutical manufacturing processes as well as by-products of numerous drugs excreted in urine, which accumulate in the environment (Kozak *et al.*, 2001).

Major agricultural pollutants of the environment include organic waste materials such as dead decaying plant and/or animals; animal waste matter; chemicals such as herbicides, organic fertilizers, insecticides, pesticides, salts and other trace elements from irrigation residues; and microorganisms present in the soil (Ansari *et al.*, 2018; Borah, 2020). The increasing and continuous application of pesticides and fertilizers has amplified health and environmental hazards since pesticides have been implicated as the cause of numerous human and animal illnesses, while chemicals from agricultural practices have reduced the quality of soil, air and water bodies of several environments (Andrews *et al.*, 2004; Mostafalou and Abdollahi, 2012; Borah, 2020).

A current major worldwide predicament is the contamination of soil by recalcitrant organic compounds, including petroleum hydrocarbons (e.g. benzopyrene, organochlorine pesticides (OCPs) and other phenolic compounds), chlorinated solvents such as polychlorinated biphenyls (PCBs), explosives (e.g. trinitrotoluene and other nitroaromatics), volatile organic carbons, linear halogenated hydrocarbons (e.g. trichloroethylene) and polycyclic aromatic hydrocarbons (PAHs) (Sosa *et al.*, 2017; Chandra and Kumar, 2018). The introduction of these compounds into the soil due to activities from industries and the military and incessant use of insecticides and fertilizers has led to a decline in soil health. Also, the organization of the ecosystem has been disrupted since these soil contaminants can pass into the food chain through the primary producers of the first trophic level (Rogimon *et al.*, 2022).

The removal of organic contaminants from the environment is tedious. However, one of the most efficient and cost-effective methods for removal of these pollutants includes the use of plants (Tokala *et al.*, 2002). Though still largely at the experimental stage, it is a sustainable and ecologically friendly strategy for remediation of contaminated environments and bringing them back to creative uses. Certain plants utilize a range of enzymes over and above physiological and cellular machinery that allows them to withstand elevated levels and high concentrations of organic contaminants

Phytoremediation of Organic Contaminants

without damaging the plant physiology (Rogimon *et al.*, 2022). These plants often achieve this by gathering and modifying these contaminants into less toxic or completely harmless compounds using specialized parts of the plants (Dixit *et al.*, 2015). The eradication of ecological contaminants via plant species through direct uptake, absorption, accumulation, extraction, sequestration, transport, transformation and degradation is generally termed phytoremediation (Chowdhury *et al.*, 2016, Tripathi *et al.*, 2020). Other terms such as vegetative remediation, agroremediation, green remediation and botanoremediation have also been used to describe cleanup of contaminants by plants (Chaney *et al.*, 1997; Chandra and Kumar, 2018).

Elimination of environmental contaminants by plants can be endogenous or exogenous, where various plants are used for extraction, containment, immobilization and degradation of various organic noxious wastes. Also, the degree of remediation of organic contaminants by plants through one or a combination of approaches such as phytouptake, phytoextraction, rhizofiltration, phytostabilization, phytodegradation, phytoimmobilization, phytotransformation, phytomineralization and phytovolatilization is majorly dependent on factors such as environment, plant type, genetic make-up of the plant and structure of the contaminating compound (Wiszniewska *et al.*, 2016). Herbs, shrubs, trees and various green plants found in both aquatic and terrestrial habitats with the potential for remediation of contaminated environments have been reported (Yoon *et al.*, 2006; Gupta and Sinha, 2007; Qixing *et al.*, 2011). Compared to various physical and chemical removal techniques, phytoremediation is a widely accepted, suitable, eco-friendly, safe and globally accepted choice for removal of environmental contaminants (Jan and Parray, 2016). Several groups of wild and recombinant plants have been effectively exploited for efficient removal of recalcitrant organic pollutants (Isiuku and Ebere, 2019). In view of this, this chapter focuses on the classification of organic pollutants, current phytoremediation techniques and plants used in phytoremediation. Advantages, limitations and future prospects of phytoremediation are also discussed.

5.2   ORGANIC CONTAMINANTS

Over the years, the health consequences of pollution of the environment on inhabitants have been an increasing basis for concern all over the earth. The World Health Organization (WHO) mentioned that 1/3 of the ailments that afflict humankind are attributable to extensive exposure to pollution. After the Second World War, researchers recognized numerous chemical pollutants which are noxious, unrelenting in the atmosphere, bioaccumulative and capable of long-term aerial transboundary movement and dumping that can be predicted to cause severe outcomes of health issues for human beings, fauna, and marine habitats far from as well as close to their emission source. These toxins are chemical contaminants, well known as the dirty dozen (Ashraf *et al.*, 2015). More contaminants have been revealed; the major concern is over the unique 12. These pollutants are ten intentionally formed chemicals: endrin, aldrin, dieldrin, chlordane, DDT, mirex, heptachlor, toxaphene, hexachlorobenzene (HCB) and PCBs, and two unintentionally formed substances, polychlorinated dibenzofurans (PCDFs) and polychlorinated dibenzo-p-dioxins (PCDDs) (Kendrovski *et al.*, 2001). Another kind of contaminant that has currently been noticed and categorized as a persistent organic compound is polycyclic aromatic hydrocarbons (PAHs), which are accidentally created from combustion in addition to the burning of organic compounds (Fiedler, 2003).

5.2.1   INTENTIONALLY PERSISTENT ORGANIC POLLUTANTS

Intentionally manufactured pollutants present or available on one occasion are utilized in manufacturing, agriculture, industrial processes or control of disease. These intentional persistent organic pollutants (POPs) compounds are manufactured as required products by diverse chemical reactions which use chlorine. These are organic molecules with associated atoms of chlorine, elevated lipophilicity and typically elevated neurotoxicity, and they are known as compounds of organochlorine. A number of the well-known instances of organochlorine compounds are the chlorinated

FIGURE 5.1 Industrial wastes.

insecticides, for example, polychlorinated biphenyls and dichlorodiphenyltrichloroethane. They have numerous compounds that can be divided based on sources into industrial, agricultural, biomedical, municipal and electronic wastes (Harner et al., 2006).

5.2.1.1 Industrial Chemicals

Polychlorinated biphenyls are remarkably stable mixtures that are resistant to extreme pressure and temperature. They are used in the collection of synthetic pollutants, fatty solids or liquids that range in color from yellow to transparent and have no taste or smell. They are mostly found in water, bird tissue, sediments and marine life tissue over the earth. These chemicals' structure is an important subset of particular wastes. PCBs are a grouping of chemical compounds wherein the biphenyl molecule has 2–10 chlorine atoms connected to it. By-products of the processes of manufacturing like mill mining and factories are also termed industrial waste materials (Figure 5.1). This waste furthermore includes metals, radioactive wastes, chemicals, paints and sandpapers in addition to paper products. Wastes from industrial supplies are potentially poisonous pollutants that require comprehensive treatment ahead of release into the environment (Maczulak, 2010).

5.2.1.2 Agricultural Waste

The function of agriculture cannot be overstated in economic and human development with the increasing human population, technological improvement towards green revolution and soil expansion for agriculture production resulting in greater generation of waste, which might amount to a severe public health challenge in the course of pollution (Adejumo et al., 2020). Agricultural waste (Figure 5.2) is generated from the cultivating and processing of raw farm products, resulting in by-products that could be valuable but have less cost-effective value and elevated management cost. Agro waste is made up of food and animal waste and dangerous and poisonous agricultural waste (insecticides, herbicides, and pesticides). The concentration of agriculture in emergent countries will possibly add to the increased production of agro waste worldwide, with around 998 million tons of agro waste produced annually (Agamuthu, 2009; Obi et al., 2016). Agricultural wastes are not appropriately handled because very little is recognized about the possible dangers and advantages related to appropriate management (Adejumo et al., 2020).

Agricultural waste can be exploited through anaerobic digestion, application of fertilizer, an absorbent in heavy metal removal, pyrolysis, direct combustion and animal feed. In order to prevent

Phytoremediation of Organic Contaminants

FIGURE 5.2 Agricultural wastes.

contamination of the soil, air and water, agricultural waste management requires that trash be viewed more voluntarily as likely resources than as undesirable materials. Inadequate handling of these wastes may also result in the establishment of breeding grounds for disease-carrying insects, deterioration of the soil's phosphorus loading, bad odors and gas emissions such as ammonia and methane (Obi *et al.*, 2016).

Organochlorine (OC) pesticides are a major agricultural contaminant, synthetic in nature and extensively utilized the world over. They are one of the chlorinated hydrocarbon offshoots that have enormous use in agriculture and the chemical industry. Pesticides are a group of chemicals utilized in killing insects, bacteria, fungi and weeds, as well as other organisms. Among the terms used to describe them are herbicides, insecticides, bactericides, fungicides and rodenticides. Most pesticides are capable of eliminating a wide range of weeds and pests, but some are targeted specifically at certain diseases or pests. Even though these substances are normally artificial, derivatives of plants and naturally occurring inorganic minerals are examples of exceptions that crop up naturally (Obi *et al.*, 2016).

5.2.1.3 Biomedical Waste

Biomedical wastes (Figure 5.3) are produced from healthcare institutions, for example, radioactive materials, non-sharp in addition to sharp objects, blood, chemicals and pharmaceutical products (Nwachukwu *et al.*, 2013). Approximately 85% of waste produced in health care is harmless, and the residual 15% is harmful and may also cause disease and be poisonous to the environment or toxic (radioactive) (WHO, 2018). WHO (2018) stated approximately 16 million injections are administered each year worldwide, causing the inappropriate dumping of syringes after use. Waste generated from healthcare centers exposes patients, care givers and waste handlers to possible injuries, infection and toxic materials, simultaneously polluting the environment. Such waste includes pharmaceutical wastes, radioactive materials, pathological waste, toxic waste and non-hazardous waste (Nwachukwu *et al.*, 2013). To sufficiently control health care waste, severance, suitable handling and safe disposal are imperative to facilitate appropriate disposal and recycling (Nwachukwu *et al.*, 2013). Burning these wastes might lead to the liberation of lethal chemicals and particles that cause pollution. Therefore, appropriate measures should be taken to ensure environmental safety

FIGURE 5.3 Biomedical wastes

in addition to health management being established to avoid severe health and environmental consequences, such as accidental release of biological and chemical threats, including drug-resistant microbes (WHO, 2018).

5.2.1.4 Municipal Waste

Municipal solid waste (MSW) (Figure 5.4) is refuse waste gathered from households, markets, schools, gardens, streets, malls and litter containers (Buragohain *et al.*, 2020; OECD, 2021). The rising quantity of MSW production from migration, industrialization, urbanization and inappropriate dumping of food waste causes a severe worldwide challenge (Rajan *et al.*, 2018). It is produced from diverse sources where human activities happen. Developing countries produce around 55–80% of household waste and 10–30% of market and commercial waste, which is from industries, streets, institutions and several others (Nabegu, 2017). Therefore, appropriate measures should be taken to ensure environmental safety in addition to health management being established to avoid severe health and environmental consequences, such as accidental release of biological and chemical threats, including drug-resistant microbes (WHO, 2018). Kaza *et al.* (2018) anticipated that worldwide, by the year 2050, the production of MSW will increase to 3.40 billion tons.

Because of inappropriate methods of disposing of waste, such as dumping in open spaces and burning solid waste, there are more health risks associated with waste in low-income nations than in high-income ones (Ferronato and Torretta, 2019). Restricted dumping sites, land-filling, incineration, composites, anaerobic digestion and recycling are a number of techniques of waste management and disposal (Kaza *et al.*, 2018; Vinti *et al.*, 2021).

5.2.1.5 Electronic Waste (E-Waste)

Electronic waste (e-waste) (Figure 5.5) consists of discarded electronic or electrical devices. E-waste is one of the kinds of waste infecting the globe presently. Used electronics which are intended for reuse, renovation, recovery and recycling through material recovery or dumping are also deemed e-waste (Buragohain *et al.*, 2020).

The casual disposal of e-waste in developing countries can have unfavorable effects on human health and initiate pollution of the environment. Scrap components of electronics, like

Phytoremediation of Organic Contaminants 89

FIGURE 5.4 Municipal waste site.

FIGURE 5.5 Electronic waste (e-waste) site.

central processing units (CPUs), have possible detrimental materials like lead, cadmium, beryllium and brominated flame retardants. E-waste recycling and disposal can pose considerable dangers to the health of workers and society in developing and developed countries (Buragohain *et al.*, 2020).

5.2.2 Unintentionally POPs

Unintentionally manufactured pollutants bring about consequences of the incineration of clinical disposal, embodiment and a number of industrial stages. They are subdivided into three categories, polycyclic aromatic hydrocarbons and furan and dioxin compounds (Abdel-Shafy and Mansour, 2016).

5.2.2.1 Polycyclic Aromatic Hydrocarbons

PAHs (Figure 5.6) are ever-present assemblages of many hundreds of chemicals that encompass two or more complex rings of benzene in linear, angular or group arrangements, holding only carbon

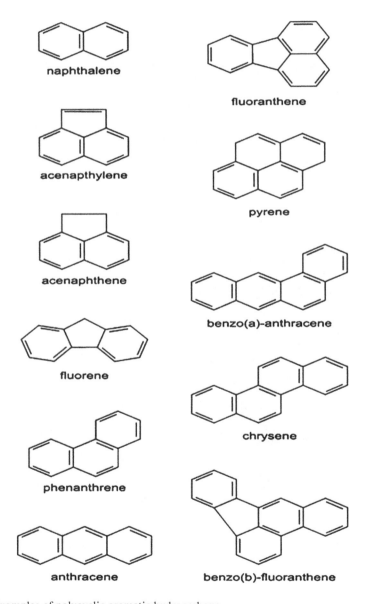

FIGURE 5.6 Examples of polycyclic aromatic hydrocarbons.

and hydrogen. The middle molecular arrangement is held mutually by steady carbon–carbon bonds. They are typically caused by partial burning of human-made or natural fuels like wood and coal, in addition to vehicular pollutants and smoke from cigarettes (Abdel-Shafy and Mansour, 2016). Dietary exposure accounts for over 70% of the exposure of humans in non-smokers (Martorell et al., 2010).

In accordance with a dietary study carried out in the United Kingdom, oils/fats and cereals make up a noteworthy source of intake of PAH (Dennis et al., 1983). Archetypal PAH contamination takes place when food is put through products of combustion in technical procedures like direct fire drying (Dennis et al., 1991). High concentrations of PAH in charcoal barbecued/grilled foods might also result from conventional home methods of cooking, for example, roasting, grilling, smoking and frying (Domingo and Nadal, 2015).

5.2.2.2 Dioxins and Dibenzofurans

Polychlorinated dibenzo-p-dioxins (PCDDs), dibenzofurans (PCDFs) and polychlorinated biphenyls (PCBs) (Figure 5.7) make up three groups of pertinent POPs with improved persistent toxicity. PCDD/Fs are released by a diversity of human actions and industrial processes and can be referred to as unwanted by-products. PCBs are omnipresent environmental pollutants on account of their significant manufacture until the end of the 1980s and their sustained usage (Grande et al., 1994). PCDD/Fs and PCBs can likewise be free from stationary sources such as waste burning, combustion of biomass and fossil fuel. PCDD/Fs and PCBs can be regarded as environmental indicators of anthropogenic activities with regard to this information, as their incidence is usually associated with human activities. PCDDs and PCDFs, usually known as "dioxins", are two classes of "quasi-planar" tricyclic aromatic ethers with 210 dissimilar compounds (congeners) in entirety (Grande et al., 1994).

5.3 PLANTS USED FOR PHYTOREMEDIATION

Several plants have been reported to have varying degrees of remediation potential (Figure 5.8). *Brassica juncea* (Indian mustard) is a high-biomass plant that has the ability to bioaccumulate lead,

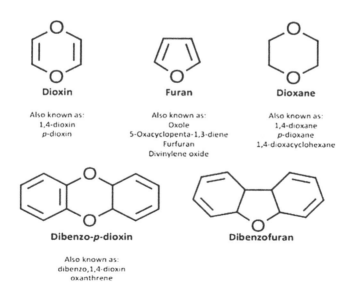

FIGURE 5.7 Dioxins, dibenzofurans and derivatives.

FIGURE 5.8   Some plants used in phytoremediation.

cadmium, copper, zinc, boron, chromium, nickel and selenium (Raskin *et al.*, 1994; Kumar *et al.*, 1995; Salt *et al.*, 1997). Owing to its large biomass, *B. juncea* can accumulate metals and transport lead (Pb) to its shoots, which accumulates more than 1.8% dry weight. In addition, the plant species is also reported to accumulate 0.82 to 10.9% of lead in its roots. *B. juncea* cultivars vary a lot in their ability to accumulate Pb, with some ranging between 0.04% and 3.5% Pb accumulation in their shoots, with 7 and 19% in their roots (Kumar *et al.*, 1995). The plant *Thlaspi caerulescens* (Alpine pennycress) has been reported to be used in the remediation of zinc and nickel (Brown *et al.*, 1994). Another plant, *Thlaspi rotundifolium* ssp. Cepaeifolium, is a non-crop brassica and one of the few accumulators of lead (Pb) that was reported (Kumar *et al.*, 1995). Another important plant used for remediation of nickel is *Alyssum wulfenianum* (Reeves and Brooks, 1983).

Hybrid poplar trees were reported to have been used in a field study in remediating wastes contaminated with cadmium and arsenic (Pierzynski *et al.*, 1994). Lambsquarter leaves have a higher arsenic concentration (14 mg/kg As) than other native plant and poplar leaves (8 mg/kg) in mine-tailing wastes (Pierzynski *et al.*, 1994). Baker (1995) found about 80 species of nickel-accumulating

Phytoremediation of Organic Contaminants

plants in the *Buxaceae* and *Euphoribiaceae* families. Some euphorbs have the ability to accumulate about 5% of their dry weight in nickel. The canola plant (*Brassica napus*) has been reported to accumulate boron and selenium. Tall fescue (*Festuca arundinacea*) and kenaf (*Hibiscus cannabinus*) also take up selenium, but not as much as canola (Bañuelos *et al.*, 1997). Sunflower is also known to take up cesium and strontium, with strontium moving into the shoots and cesium remaining in the roots (Adler, 1996). Plants such as sorghum, corn and alfalfa, which are metal accumulators, may be more effective than hyper-accumulators and remove a greater mass of metals due to their larger biomass and faster growth rate.

5.4 CURRENT PHYTOREMEDIATION TECHNIQUES

The overall mechanisms of phytoremediation processes involved in the remediation of toxic wastes from water and soil are phytostabilization, phytoextraction and phytovolatilization (Chandra and Kumar, 2017).

5.4.1 PHYTOSTABILIZATION

Phytostabilization, also called *in situ* or in place inactivation or phytoimmobilization, is a technique that involves immobilization of contaminants or heavy metals from water and soil through plant root accumulation and absorption, thereby reducing the bioavailability of the contaminants and preventing migration into the environment such as wind, soil and underground water (Marques *et al.*, 2009; Mirza *et al.*, 2014; Nikolic and Stevovic, 2015; Yan *et al.*, 2020). Reduction in bioavailability and mobility of contaminants by stabilizing plant root activity results in lower toxic effects. These toxic contaminants are bound within the plant as residues or in non-poisonous form that cannot released after accumulation. The rhizosphere stores the contaminants, embedding them in the root surface through transport proteins or in the root vacuole through a cellular process (Interstate Technology & Regulatory Council, 2009).

Biological contaminants and heavy metals can be converted to non-hazardous substances by forming complexes with amino acid derivatives, sugar and proteins within the rhizosphere. For instance, arsenic is immobilized by the epidermis of plant roots through binding with ferric sulfate in the vacuoles, resulting in As-tris-thiolate trivalent complex in the rhizosphere (Hammond *et al.*, 2018; Kafle *et al.*, 2022). In places with little vegetation due to high pollution levels, phytostabilization can be utilized to re-establish vegetation cover. The bacteria and mycorrhiza that are found in the rhizosphere of these plants also play a significant part in these processes since they actively contribute to the change in trace element speciation as well as helping the plants overcome phytotoxicity, which promotes revegetation (Mastretta *et al.*, 2006; Chandra and Kumar, 2017). The procedure is used when there are possible risks to human health and contaminants can lower exposure to dangerous compounds to safe levels. Utilizing this process can reduce exposure to hazardous substances to safe levels in situations where there may be health concerns to humans.

Before human activity may restart on the treated ground, phytostabilization calls for the establishment of a robust and healthy layer of plants. However, since some organic pollutants and their metabolic derivatives can adhere to or be integrated into plant components like lignins, there may be a chance for phytostabilization of organic pollutants. Phytolignification is the name given to this type of phytostabilization. However, one distinction between the two approaches is that, although phytostabilization of organic pollutants by phytolignification takes place above ground, phytostabilization of metals often occurs in the soil. Unlike phytoextraction or the excessive accumulation of metals in the shoots or roots, phytostabilization largely pays attention to the accumulation of metals in the rhizosphere rather than in plant tissues (Chandra and Kumar, 2017). Several plant species have been reported to have potential in the phytostabilization of heavy metal (Table 5.1). For effective phytostabilization, selecting the appropriate plant species is important. Other requirements

TABLE 5.1
Some Plants Capable of Phytostabilization of Heavy Metals

Plant Species	Metals	References
Atriplex canescens (Pursh) Nutt	Pb, As, Mn, Hg	Rosario *et al.* (2007)
Bidens humilis	Ag, As, Cd, Cu, Pb, Zn	Bech *et al.* (2002)
Dalea bicolor Humb. and Bonpl. Ex Willd.	Mn, Cd, Pb, Cu, Zn	Gonzalez and Gonzalez-Chavez (2006)
Euphorbia sp.	Zn, Cd, Mn, Cu, Pb	Gonzalez and Gonzalez-Chavez (2006)
Isocoma veneta (Kunth) Greene and *Teloxys graveolens*	Pb, Cu, Cd, Zn, Mn	Gonzalez and Gonzalez-Chavez (2006)
Lygeum spartum L.	Zn, Cu, Pb	Conesa *et al.* (2006)

include the ability to produce a large quantity of biomass, a dense rooting system and fast growth and easy maintenance in field conditions. An advantage of phytostabilization is that the disposal of toxic biomass is not a requirement of the process compared to phytoextraction (Wuana and Okieimen, 2011; Yan *et al.*, 2020).

5.4.2 PHYTOEXTRACTION

Phytoextraction, also known as phytoaccumulation, is a process of using plants to transport and accumulate pollutants from soil and water in their superterranean biomass (Jacob *et al.*, 2018). Plant roots absorb harmful metals, move them up shoots and then deposit them in the cell membrane, vacuoles and cell walls and in other plant tissues that are metabolically inert. Shoots and roots of hyperaccumulators known for their ability to eliminate heavy metals also have the ability to accumulate a complex level of harmful metals. The general pathways for hazardous heavy metal accumulation include metal cation absorption, followed by the production of metal-phytochelatin complexes (M-PCs) or metal-ligand complexes in plant cells (Lajayer *et al.*, 2019). These complexes are transported and stored in the vacuole of the plant (Yadav, 2010). The concentration of heavy metals in the superterranean plant biomass and tissues is a vital factor affecting the ability of a species to be extracted (Li *et al.*, 2010). Therefore, species suitable for phytoremediation that have the ability to resist and absorb heavy metals successfully must also produce a lot of biomass, grow quickly and be profitable (Bian *et al.*, 2020).

Plant species suitable for phytoextraction (Table 5.2) during phytoremediation of particular metals are, for example, *Sesbania drummondii* for lead (Pb), *Pteris vittata* for arsenic (As), *Sedum alfredii* for cadmium (Cd) and *Commoelina communis* for copper (Cu). Phytoaccumulation of arsenic (As) in *Arundo donax* is found to increase as a result of some plant growth-promoting bacteria, such as *Agrobacterium* sp. and *Stenotrophomonas maltophilia* (Guarino *et al.*, 2020). For effective phytoextraction process, selection of plant species is important, and the plants should possess the following characteristics: (i) resistance to the lethal side effects of HMs, (ii) capable of high extraction, (iii) high level of heavy metal accumulation in the superterranean parts, (iv) ample root and shoot system, (v) good environmental adaptabilities, and (vi) high resistance to pathogenic organisms and pests (Seth, 2012; Ali *et al.*, 2013). The most important variables that affect a plant species' ability to extract metals are its above-ground biomass and metal-accumulating capacity. Therefore, separate methods for plant selections are used: (i) using plants with high superterranean biomass with a likelihood of lower metal accumulation capacities but overall heavy metal accumulation compared to hyperaccumulator plants and (ii) using hyperaccumulator plants (Ali *et al.*, 2013).

Phytoremediation of Organic Contaminants

TABLE 5.2
Phytoremediation Plants Known to Utilize Phytoextraction

Common Name	Scientific Name	Contaminants	Remarks	Citations
Lettuce	*Lactuca sativa*	Ni, Co, Fe	Greater capacity for absorption and decreased intrinsic velocity	Hernández *et al.* (2019)
Tobacco	*Nicotiana tabacum*	Cd	Increased levels of Cd in stems and leaves	Yang *et al.* (2019b)
Indian mustard	*Brassica juncea*	Pb	To mobilize Pb and facilitate extraction, chelating compounds like ethylene diamine tetra-acetate (EDTA) are applied to soil	Salido *et al.* (2003)
Alpine pennygrass	*Thlaspi caerulescens*	Cd, Zn	Transport of Cd via the Zn and Ca pathway	Cosio *et al.* (2004)
Rattlebush	*Sesbania drummondii*	Pb	EDTA increases Pb absorption and accumulation	Barlow *et al.* (2000)
Common cocklebur	*Xanthium strumarium*	Cd, Pb, Ni, Zn	Possesses the highest BCF for Pb-1.048, Cd-1.574 and Ni-1.651.	Khalid *et al.* (2019)
Geranium	*Pelargonium hortorum*	Pb	Bacteria tolerant to Pb used	Manzoor *et al.* (2019)

One benefit of phytoextraction is that it is less expensive than the traditional techniques. Additionally, because vegetation covers the soil, less erosion occurs (Adhikary, 2015). Conversely, the degree of metal phytoextraction is decreased by a number of factors, including resistance to harmful metals, the availability of metals in plant rhizosphere cells and tissue and the influence of clay content and soil pH (Sinha *et al.*, 2007; Sharma and Pandey, 2014). Sunflower (*Helianthus annuus*) is a common species used in phytoextraction due to its quick growth, high biomass and successful remediation of radionuclides and some heavy metals (Lee, 2013; Adhikary, 2015). However, a number of factors typically affect this plant-based technology's effectiveness: Site selection is crucial, with the following requirements: (i) favorability of the site, (ii) the ability to amass and sift large quantities of HMs, (iii) plants should grow quickly and produce a large quantity of biomass, (iv) heavy metals must be bioavailable for easy absorption by the plants, (v) reasonable agronomic procedures and suitable land topography, and (vi) the ability to withstand difficult soil conditions like pH and salinity. Root depth also affects how well a phytoextraction process works (Malik *et al.*, 2014).

5.4.3 Phytovolatalization

Phytovolatilization is a phytoremediation technique where plants absorb pollutants from the soil, transform them into harmless and volatile forms and then, through the process of transpiration, which occurs through the leaves or foliage system, release the compounds into the atmosphere (Figure 5.9) (Kumari *et al.*, 2020). Mercury, selenium and arsenic are just a few of the toxins that can be detoxified with this technique (Kumar Yadav *et al.*, 2018; Bortoloti and Baron, 2022). In addition to metals, phytovolatilization has been established for organic contaminants. For instance, it has been demonstrated that the poplar tree (*Liriodendron* sp.), which is a phytovolatizer, may volatilize up to 90% of the trichloroethane absorbed from polluted soil. (McGrath and Zhao, 2003). Most organic contaminants respond favorably to the volatilization approach (Limmer and Burken, 2016), provided that the inorganic contaminants do not form methyl or hydride derivatives. While

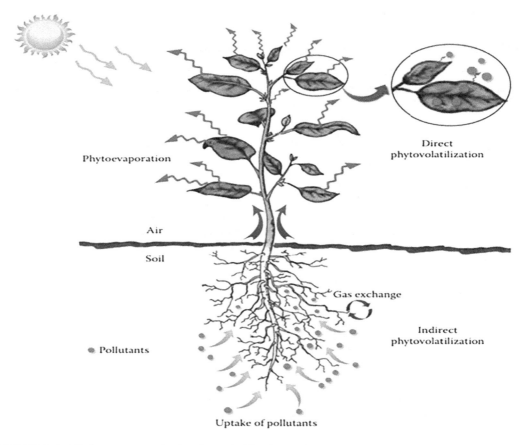

FIGURE 5.9 Various processes in the phytovolatilization method.

certain substances can volatilize directly from stems and leaves, others can do so because of the interaction between roots and soil. The plant's hydrophobic defenses include cuts, the epidermis, the suberin and other dermal layers, which allow the hydrophobic organic molecules to exit the stem and the leaves. Additionally, during the transpiration process, some compounds can be lost into the atmosphere as they migrate upward in the plant's system through the transpiration stream. Several plants have the ability to release benign forms of harmful organic pollutants into the environment through the volatilization mechanism (Table 5.3) (Kafle *et al.*, 2022).

Effective volatilizers include *Brassica juncea* and other members of the Brassicaceae family. The advantage of phytovolatilization over traditional phytoremediation techniques is that heavy metal (metalloid) contaminants are removed from the site and disseminated as gaseous compounds without the need for plant harvesting and disposal. However, some pollutants are not removed by phytovolatilization, and they remain in the environment. It only transports pollutants out of the soil and into the atmosphere, where they contaminate the air with their hazardous volatile compounds. Additionally, rainfall may re-deposit them in the soil (Vangronsveld *et al.*, 2009). Therefore, a risk analysis is required prior to its application in the field.

5.5 MERITS OF PHYTOREMEDIATION

In comparison to excavation and treatment, plants are solar powered and more environmentally benign because they allow the soil's natural habitat and texture to be preserved. Plant roots have

Phytoremediation of Organic Contaminants 97

TABLE 5.3

Plants Known to Utilize Phytovolatilization during Phytoremediation

Common Name	Scientific Name	Contaminants	Remarks	References
Broad leaf cattail	*Typha latifolia*	Se	Wetlands plants	Pilon-Smits *et al.* (1999)
Saltmarsh bulrush	*Scirpus robustus*	Se	Wetlands plants	Arthur *et al.* (2005)
Perennial reed grass	*Phragmites australis*	Organo-chlorines (OCs)	Volatilization occurs with 1,4-dichlorobenzene (DCB), 1,2,4-trichlorobenzene (TCB) and hexachlorocyclohexane (HCH)	San Miguel *et al.* (2013)
Parrot's feather	*Myriophyllum brasiliense*	Se	Wetlands plants	Pilon-Smits *et al.* (1999)
Rabbit foot grass	*Polypogon monspeliensis*	As	Pentamethylarsine (As(CH3)5) and dimethylchloroarsine (AsCl(CH3)2) volatilize. As organic forms are not volatile	Ruppert *et al.* (2013)
Hybrid poplar tree	*Populus trichocarpa* χ *Populus deltoïdes*, designated HI 1	TCE, PCE	Hybrid poplar tree HII varieties are popular for TCE phytovolatilization	Gordon *et al.* (1998)
Common rush	*Juncus effusus*		The release of ammonium methane	Wiessner *et al.* (2013)
Broccoli	*Brassica oleracea* var. italica	Se	Chelating agents are added, which encourages the volatilization of methyl selenate	Raskin *et al.* (1997)
Willow	*Salix* spp.	Tricholoro-ethylene (TCE) and tretracholoro ethylene (PCE)	Due to its large biomass and capacity for utilizing a variety of phytoremediation mechanisms, willow is a significant phytoremediating plant	Gordon *et al.* (1998)

the ability to find, alter or transport materials and molecules in the face of large chemical gradients (Ekta and Modi, 2018). Varieties of organic and inorganic substances are responsive to the phytoremediation process, and it is cost-effective, which is a significant benefit of phytoremediation. It is a green technology that, when properly applied, is both aesthetically pleasant to the general public and environmentally friendly. It is simple to apply and does not call for expensive equipment or skilled employees. With minimum interruption to the environment, its large-scale application is practical for clean-up activities (Chandra and Kumar, 2017).

The plants used in this procedure serve as soil stabilizers and can reduce the amount of contaminated dust that might be released into the atmosphere. It may be adaptable enough to handle a variety of dangerous environmental pollutants. Contrary to microbe-based bioremediation, phytoremediation is readily apparent; the health of the plants can be observed visually, and pollutant levels in samples of plant tissue may be checked over time. The potential for a valuable product, such as wood, pulp or bioenergy sources, that could help fund the cleaning is one of the advantages of phytoremediation over engineering or bioremediation techniques (Ekta and Modi, 2018).

5.6 LIMITATIONS/FACTORS THAT AFFECT PHYTOREMEDIATION

5.6.1 SITE CHARACTERISTICS

The characteristics of a contaminated site, especially the accessibility to materials that are contaminated, their mobility on the site, and some information that would be collected in the assessment report, will frequently dictate the technology choice (Cole, 1994). Some important elements in this include subsurface description, site description, and flow direction. Selby (1991) provided a detailed description of some methods for site assessment.

5.6.2 SOIL TYPE

Sands, gravels, clays, silts and sediments are all examples of contaminated soil types. The efficiency of bioremediation and, consequently, the strategy will depend on the soil's porosity and permeability (Guerin and Boyd, 1992). A key factor in contaminant sorption is the size of the particle, which is determined by the surface area of the grain. The higher the surface area of the grain, the greater the absorption capacity for contaminant by the plant, and this in turn lowers the bioavailability of the contaminants. By entrapment, sequestration and sorption, pollutants can attach to soil particles, and they can disperse into tiny spaces that, especially during the day, are inaccessible to bacteria. This prevents biological breakdown from occurring (Rieger and Knackmuss, 1995). Hydrophobicity of the soil may also play an important role in reducing bioavailability of the contaminant (Nam and Alexander, 1998). Also, the tight associations of contaminant and soil, for example, those occurring with PARs and TNT, often produce residues that are no longer extractable and result in them not being properly analyzed, a phenomenon described as humification (Lenke *et al.*, 1997).

5.6.3 MOISTURE CONTENT

In a typical soil column, soil moisture content, which is described as the degree of water saturation, varies with the depth of soil, ranging from the upper unsaturated vadose zone (20–60%) to the capillary zone (60–80%), the water table fluctuation zone (60–100%) and the saturated groundwater zone (100%).

5.6.4 NATURE OF THE CONTAMINANT

A lot of molecular species have structures or physicochemical properties that are not often transformed. These molecules can exist in soil for years, even though same plants have the ability to degrade many organic compounds. It is said that they are recalcitrant. Certain of these molecules are persistent but not toxic, such as some polymers like plastics. Many recalcitrant molecules, though, are hazardous. Examples of persistent chemicals that have been found to last at least 15 years in soil (Alexander, 1999) are chlordane, highly chlorinated PCB congeners, heptachlor, DDT, dieldrin, lindane and simazene. Some other molecules may be biodegraded, but plants don't access them easily because of their tendency to bind to soil and low water solubility. More soluble substances are usually more biodegradable. Naphthalene, o-xylene, benzene, anthracene and benzo[a]pyrene are a few instances of pollutants with water solubilities that are as follows: 58, 175, 1780, 0.07 and 0.004 mg/L, respectively (Mahro, 2000). A generalized sequence of petroleum components in order of decreasing biodegradability is shown as follows:

n-alkanes> branched-chain alkanes > branched alkenes > low molecular weight n-alkyl aromatics > monoaromatics > cyclic alkanes > polynuclear aromatics > asphaltenes

Predictive models have been designed for estimating the extent of petroleum hydrocarbon biodegradation (Huesemann, 1995).

Some compounds and their non-degradable metabolites may be toxic to microbial populations, with a possible/negative impact on bioremediation potential. Such considerations may affect the decision as to whether bioremediation is a cleanup option or the choice of approach to bioremediation.

5.6.5 TIME CONSTRAINTS

In some instances, expedited remediation of polluted areas is necessary for short-term real estate development plans since the lengthy time scale needed for natural attenuation and traditional phytoremediation methods are not acceptable (Field *et al.*, 1992). Advances in the understanding of physiology, genetics, microbial biochemistry and recombinant technologies can be exploited to maximize the extents of phytodegradation. Accelerated phytoremediation often utilizes increasing levels of plant expertise.

5.6.6 VOLATILIZATION

Many organic compounds that are pollutants in the environment are flammable. The huge amounts of volatile organic carbons found in contaminated soil that are harmful to human health and are emitted into the atmosphere represent a significant drawback of various bioremediation procedures. They also contribute to the formation of tropospheric ozone (Moseley and Meyer, 1992; Field *et al.*, 1992; Coppock *et al.*, 1995). In the Exxon Valdez spill in the relatively cold Alaskan climate, about 15–20% of the oil was lost to the atmosphere as a result of volatilization (Cole, 1994; Field *et al.*, 1992). Some reports indicated that up to 60% of remediation may be as a result to volatilization (Salanitro, 2001).

5.6.7 PHYTOSTIMULATION

A hierarchy that displays the different phytoremediation technical options was described by Untermann *et al.* (2000). In cases where the degrading population already exists in the contaminated zone but where nutrients or other conditions are insufficient to promote microbial activity in the roots of the plants, additional nutrients, typically sources of phosphorus and nitrogen, speed up the process for the remediation of carbon contaminants. Added co-substrates may aid in promoting development when the contaminant is broken down by co-metabolism.

5.6.7.1 Oxygen/Electron Acceptors

In surface phytoremediation techniques, aeration is achieved by putting the soil medium into close contact with the atmosphere. Air is typically employed to give oxygen when polluted soil is present in an in situ unsaturated zone or even in an engineered soil pile. When soil porosity is low or it is preferred to keep gas flow rates in the soil low because of problems with contaminant volatilization, compressed oxygen may be used instead of air. In the latter case, where the contamination is in a water-saturated zone and oxygen is poorly soluble in water, hydrogen peroxide may be employed as a substitute supply of oxygen. The fact that H_2O_2 is highly soluble in water allows it to slowly decompose into free oxygen in the aquifer. However, when biomass is injected into the earth, issues with biomass clogging up to the injection point have been noted. Magnesium peroxide and other inorganic peroxide compositions with delayed release have proven effective (Prince, 1998). Other species that can use inorganic electron acceptors like nitrate, sulphate or CO_2 or organic materials as electron acceptors may replace aerobic microorganisms in situations with low oxygen levels. Some strains may be able to anaerobically oxidize some simple aromatics, such benzoate and phenol, by using ferric iron as an electron acceptor (Lovley, 2000). The rates of these anaerobic conversions are often modest, and the microbial population must be acclimated over extremely long periods of time.

5.6.7.2 Co-Metabolic Substrates

Co-metabolism is the conversion of organic molecules by plants that cannot use the substrate as a source of energy or for considerable nutritional advantage (Alexander, 1999). The substrate is normally digested in the presence of another substrate that can promote development; however, under some conditions, metabolic events might take place without the additional co-substance that promotes growth. A single enzyme reaction or a string of reactions are involved in co-metabolic transformations. The substrate specificities of the downstream metabolizing enzymes, the presence of a substituent that obstructs a forward reaction, or the reaction product's reassembly or other chemical reactions to form a non-metabolizable end product can all contribute to the "dead-end" product's inability to participate in additional metabolism.

Smaller and frequently more volatile hydrocarbons are typically more biodegradable and more supportive of microbial growth and cell energy provision than more complicated components, which may rely on co-metabolic activities, in a mixed contamination environment, such as in hydrocarbon oil bioremediation (Alexander, 1999).

Therefore, the growth of a microbial population on these substrates that might contain the potential early stage catabolic enzymes with relaxed substrate specificities to initiate transformation of the more resistant compounds may be prevented in prolonged hydrocarbon biodegradation processes where volatile materials are lost to the atmosphere. Diesel fuel or other hydrocarbons increased the co-metabolic mineralization of benzo[a]pyrene in culture and soil (Kanaly *et al.*, 2000). By adding specific organic acid and alcohol substrates, anaerobic biostimulation of chlorinated solvent breakdown in situ was accomplished (Sewell *et al.*, 1998).

5.6.7.3 Surfactants

Many organic pollutants are hydrophobic, which frequently inhibits their ability to be assimilated and degraded by plants, which typically live in water environments. Numerous biosurfactants are produced by plants and bacteria that break down hydrocarbons, either attached to the cell surface or released into the extracellular medium (Makkar and Cameotra, 2002). The use of commercially available biosurfactant products in phytoremediation is therefore possibly uneconomical because the manufacture of biosurfactants by fermentation, as a potential addition for bioremediation, looks to be more expensive. Consequently, chemical surfactants, which are made at a far lower cost, could play a part. Chemical surfactants can improve phytodegradation if used properly (Van Hamme and Ward, 1999).

However, it has been noted that plant degradation of hydrocarbons has both increased and been inhibited (Mulligan *et al.*, 2001). Surfactants may inhibit the degradation of some poorly soluble petroleum hydrocarbons due to the following factors: (i) toxicity from high surfactant or soluble hydrocarbon concentrations, (ii) preferential metabolism of the surfactant itself, (iii) interference with membrane uptake or (iv) decreased bioavailability of micellar hydrocarbons (Rouse *et al.*, 1994). It is always a concern that the surfactant might be used as a carbon source rather than a pollutant. It is vital to offer an opinion on whether or how surfactants may be used in petroleum hydrocarbon degradation processes in order to increase rates and extents of degradation. Chemical surfactants have properties that affect their performance, including charge (non-ionic, anionic or cationic); hydrophilic lipophilic balance (HLB, a gauge of surfactant lipophilicity); and critical micellar concentration (cmc), the concentration at which surface tension reaches a minimum and surfactant monomer aggregate into micelles). When creating phytodegradation protocols, it is helpful to understand how plants react differently to surfactant modifications in accessing intractable hydrophobic substrates and how these responses change in the presence of other species (Van Hamme and Ward, 2001).

5.6.7.4 Solubilizing Agents

More resistant and hydrophobic compounds can be helped to dissolve by lower-molecular-weight hydrocarbons, increasing their bioavailability to the microbial community. Jimenez and Bartha (1996) attributed the solubilizing and mass transfer activity of solvents like paraffin oil to their

Phytoremediation of Organic Contaminants

capacity to facilitate mineralization of pyrene. Vegetable oil was added to soil, which promoted PAH mobilization and sped up their breakdown (Pannu *et al.*, 2003).

5.6.7.5 Bulking Agents

Low-porosity soils, particularly clays and silts, are difficult to aerate due to their restricted permeability. In static processes like in situ or constructed soil pile operations, and even in some mixed solid phase processes, this is a challenge (Venkateswaran *et al.*, 1995). Bulking agents, such as sand or plant material, especially woody materials with large particle sizes, may be used to increase the porosity of the soil. The main drawback of adding bulking agents is that they might significantly increase the initial mass of the contaminated medium.

5.6.8 BIOAUGMENTATION

Several laboratory and field studies have shown that the choice and acclimation of plants can reduce the risk of augmentation (Mohandass *et al.*, 1997), which did not significantly enhance rates of oil degradation over that achieved by nutrient enrichment of the natural plant population (biostimulation) (Venosa *et al.*, 1992; Glaser, 1993). This experience has been viewed by many as a general rule that phytoaugmentation is ineffective in petroleum and other biodegradation processes. Hence, there are sufficient data to indicate that possible advantages of bioaugmentation are not easily predicted, and there are variations in the variety selected. The environment to be treated and critical physico-chemical parameters influencing remediation are not always understood (Vogel, 1996). The facts at the very least prompt legitimate inquiries about whether and under what conditions plant variety plays a part in the processes of petroleum degradation. If recombinant plants might be used in the practice of environmental bioremediation and treatment, that would be another intriguing subject. The application of phytoaugmentation in agriculture and wastewater treatment has been successful (Rittmann and Whiteman, 1994). There are examples of phytoremediation achievements where the addition of capable plants seemed to increase the remediation rate and where the native variety seemed incapable (Shin and Crawford, 1995). Using phytoaugmentation, resistant substances like 2,4,6-TNT, carbon tetrachloride and PCP have been successfully treated (Witt *et al.*, 1995). One crucial aspect of this strategy may be the significance of utilizing plants that have shown lengthy survival times in a certain soil environment (Vandepitte *et al.*, 1995; Pearce *et al.*, 1995).

5.7 SAMPLING AND MONITORING

A critical problem and a necessity is the deployment of proper sample and monitoring techniques. An important cost factor for in situ phytoremediation processes is the ability to alter the concentrations of contaminants, nutrients and oxygen/electron acceptors both horizontally and vertically across the "plume" depending on the soil type, water saturation level and other factors. Monitoring wells must be constructed in order to monitor the progress of pollutant removal in a three-dimensional grid (EPA, 1983). Accessibility concern for samplings is almost completely eliminated by surface corrections such coastal restorations and land cropping. In mixed systems like land-farming systems, the higher the degree of mixing, the more homogeneous the system and the fewer samples required for process characterization. This reaches a limit in well-mixed slurry systems when homogeneity is reached or attained, requiring only a single sample location. Keith (1988) and Roemer (2000) give sampling protocols and methodologies for soil remediation, and EPA publications provide analytical methods (EPA 1984).

5.8 CONCLUSION AND FUTURE PROSPECTS

Environmental organic pollution is a major global problem, especially when the use of soil is under a constant and increasing pressure for food, feed, material production and urbanization.

Plants contain enzymes and systems coupled with other natural attributes that aid degradation of complex organic contaminants, making them a useful tool for cleaning up contaminated environments. Their use for cleanup of the environment is a potential valuable solution. Plant remediation of organic contaminating compounds present in the environment is continuously and actively being investigated, as it seems to be a promising technique, especially for large areas with low and medium contamination where the use of usual remediation techniques would be unavoidably costly. However, the field of phytoremediation most certainly requires additional and significant experimental research to develop and engineer new techniques and improve existing phytoremediation methods. In addition, new plant varieties with organic and inorganic pollutant-degrading potential should be sought after, classified, identified and investigated extensively using advanced selection techniques, as this will allow for the integration and application of these new plants into contaminated environments. Genetic modification using gene materials from microorganisms or other viable plants with great biodegradation potential should be extensively studied. Consequently, recombinant plant varieties with the potential to efficiently biodegrade organic pollutants can be developed. The resulting recombinant plants exhibiting biodegradation traits of the parent microorganisms/plants will bring great possibilities of efficient, effective and environmentally friendly technology for cleaning up contaminated environments. Also, in a bid to further improve the phytoremediation agenda, investigations should be conducted where plants will be used in consortium with microorganisms such as rhizopheric and endophytic bacteria, fungi (yeast and molds) and algae for their potential to biodegrade harmful organic compounds in the environment.

REFERENCES

Abdel-Shafy, H. I., and M. S. M. Mansour. 2016. "A Review on Polycyclic Aromatic Hydrocarbons: Source, Environmental Impact, Effect on Human Health and Remediation." *Egyptian Journal of Petroleum* 25, no. 1: 107–23. https://doi.org/10.1016/j.ejpe.2015.03.011.

Adejumo, O., I. Adebiyi, and A. Olufemi. 2020. "Agricultural Solid Wastes: Causes, Effects, and Effective Management." In *Strategies of Sustainable Solid Waste Management*, edited by Hosam M. Saleh. London: IntechOpen.

Adhikary, P. S. 2015. "Sustainable Management of Mining Area Through Phytoremediation: An Overview." *International Journal of Current Microbiology & Applied Sciences* 4, no. 3: 745–51.

Adler, T. 1996. "Botanical Cleanup Crews." *Science News* 150, no. 3: 4243. https://doi.org/10.2307/3980349.

Agamuthu, P. 2009. *Challenges and Opportunities in Agro-Waste Management: An Asian Perspective.* Inaugural meeting of First Regional 3RForum in Asia. Tokyo, Japan.

Alexander, M. 1999. *Biodegradation and Bioremediation.* 2nd ed. San Diego, CA: Academic Press.

Ali, Hazrat, Ezzat Khan, and Muhammad Anwar Sajad. 2013. "Phytoremediation of Heavy Metals-Concepts and Applications." *Chemosphere* 91, no. 7: 869–81. https://doi.org/10.1016/j.chemosphere.2013.01.075.

Andrews, S. S., D. L. Karlen, and C. A. Cambardella. 2004. "The Soil Management Assessment Framework: A Quantitative Soil Quality Evaluation Method." *Soil Science Society of America Journal* 68, no. 6: 1945–62. https://doi.org/10.2136/sssaj2004.1945.

Ansari, A. A., M. Naeem, and S. S. Gill. 2018. "Contaminants in Agriculture: Threat to Soil Health and Productivity." *Agricultural Research & Technology* 16, no. 1: 555975.

Arthur, E. L., P. J. Rice, P. J. Rice, T. A. Anderson, S. M. Baladi, K. L. D. Henderson, and J. R. Coats. 2005. "Phytoremediation—An Overview." *Critical Reviews in Plant Sciences* 24, no. 2: 109–22. https://doi.org/10.1080/07352680590952496.

Ashraf, M. A., M. Sarfraz, and R. Naureen. 2015. *Handbook of Environmental Impacts of Metallic Elements: Speciation, Bioavailability and Remediation.* Singapore: Springer. https://doi.org/10.1007/978–981–287–293–7.

Baker, A. J. M. 1995. "Metal Hyperaccumulation by Plants: Our Present Knowledge of the Ecophysiological Phenomenon. In *Will Plants Have a Role in Bioremediation?" Proceedings/Abstracts of the Fourteenth Annual Symposium, 1995, Current Topics in Plant Biochemistry, Physiology, and Molecular Biology*, April 19–22, 1995, edited by Interdisciplinary Plant Group. Missouri, CO: University of Missouri-Columbia.

Bañuelos, G. S., H. A. Ajwa, B. Mackey, L. L. Wu, C. Cook, S. Akohoue, and S. Zambruzuski. 1997. "Evaluation of Different Plant Species Used for Phytoremediation of High Soil Selenium." *Journal of Environmental Quality* 26, no. 3: 639–46. https://doi.org/10.2134/jeq1997.00472425002600030008x.

Barlow, R., N. Bryant, J. Andersland, and S. Sahi. 2000. *Lead Hyperaccumulation by Sesbania Drummond*. Vol. 2000. Paper presented at the Proceedings of the Conference on Hazardous Waste Research, Bowling Green, KY.

Bech, J., C. Poschenrieder, J. Barcelo, and A. Lansac. 2002. "Plants from Mine Spoils in the South American Area as Potential Sources of Germplasm for Phyotremediation Technologies." *Acta Biotechnology* 1–2: 5–11.

Bernard, A., C. Hermans, F. Broeckaert, G. De Poorter, A. De Cock, and G. Houins. 1999. "Food Contamination by PCbs and Dioxins." *Nature* 401, no. 6750: 231–2. https://doi.org/10.1038/45717.

Bian, Fangyuan, Zheke Zhong, Xiaoping Zhang, Chuanbao Yang, and X. Gai. 2020. "Bamboo—An Untapped Plant Resource for the Phytoremediation of Heavy Metal Contaminated Soils." *Chemosphere* 246: 125750. https://doi.org/10.1016/j.chemosphere.2019.125750.

Borah, P., M. Kumar, and P. Devi. 2020. "Types of Inorganic Pollutants: Metals/Metalloids, Acids, and Organic Forms." In: *Inorganic Pollutants. Water*, edited by Pooja Devi, Pardeep Singh, and Sushil Kumar Kansal: 17–31. Cambridge, MA: Elsevier.

Bortoloti, G. A., and D. Baron. 2022. "Phytoremediation of Toxic Heavy Metals by Brassica Plants: A Biochemical and Physiological Approach." *Environmental Advances* 8: 100204. https://doi.org/10.1016/j.envadv.2022.100204.

Brown, S., R. Chaney, J. S. Angle, and A. Baker. 1994. "Phytoremediation Potential of Thlaspi Caerulescens and Bladder Campion for Zinc- and Cadmium-Contaminated Soil." *Journal of Environmental Quality* 23. https://doi.org/10.2134/jeq1994.00472425002300060004x.

Buragohain, P., V. Nath, and H. K. Sharma. 2020. "Microbial Degradation of Waste: A Review." *Current Trends in Pharmaceutical Research* 7, no. 1: 107–25.

Chandra, R., and V. Kumar. 2017. "Phytoremediation: A Green Sustainable Technology for Industrial Waste Management." In *Phytoremediation of Environmental Pollutants*, edited by Ram Chandra, N.K. Dubey, Vineet Kumar: 1–42. Boca Raton, FL: CRC Press.

Chandra, S., and K. N. Kumar. 2018. "Exploring Factors Influencing Organizational Adoption of Augmented Reality in E-Commerce: Empirical Analysis Using Technology–Organization–Environment Model." *Journal of Electronic Commerce Research* 19: 237.

Chaney, R. L., M. Malik, Y. M. Li, S. L. Brown, E. P. Brewer, J. S. Angle, and A. J. M. Baker. 1997. "Phytoremediation of Soil Metals." *Current Opinion in Biotechnology* 8, no. 3: 279–84. https://doi.org/10.1016/s0958-1669(97)80004-3.

Chowdhury, S., N. Khan, G. H. Kim, J. Harris, P. Longhurst, and N. S. Bolan. 2016. "Zeolite for Nutrient Stripping from Farm Effluents." In *Environmental Materials and Waste*, edited by M. N. V. Prasad and Kaimin Shih: 569–89. London: Academic Press.

Cole, G. M. 1994. *Assessment and Remediation of Petroleum Contaminated Sites*. Boca Raton, FL: CRC Press.

Conesa, Héctor M., Angel Faz, and Raquel Arnaldos. 2006. "Heavy Metal Accumulation and Tolerance in Plants from Mine Tailings of the Semiarid Cartagena-La Union Mining District (SE Spain)." *Science of the Total Environment* 366, no. 1: 1–11. https://doi.org/10.1016/j.scitotenv.2005.12.008.

Coppock, R. W., M. S. Mostrom, A. A. Khan, and S. S. Semalulu. 1995. "Toxicology of Oil-Field Pollutants in Cattle: A Review." *Veterinary & Human Toxicology* 37, no. 6: 569–76.

Cosio, Claudia, Enrico Martinoia, and Catherine Keller. 2004. "Hyperaccumulation of Cadmium and Zinc in *Thlaspi caerulescens* and *Arabidopsis halleri* at the Leaf Cellular Level." *Plant Physiology* 134, no. 2: 716–25. https://doi.org/10.1104/pp.103.031948.

Dennis, M. J., R. C. Massey, G. Cripps, I. Venn, N. Howarth, and G. Lee. 1991. "Factors Affecting the Polycyclic Aromatic Hydrocarbon Content of Cereals, Fats and Other Food Products." *Food Additives & Contaminants* 8, no. 4: 517–30. https://doi.org/10.1080/02652039109374004.

Dennis, M. J., R. C. Massey, D. J. McWeeny, M. E. Knowles, and D. Watson. 1983. "Analysis of Polycyclic Aromatic Hydrocarbons in UK Total Diets." *Food & Chemical Toxicology* 21, no. 5: 569–74. https://doi.org/10.1016/0278-6915(83)90142-4.

Dixit, R., D. Wasiullah, D. Malaviya, K. Pandiyan, U. Singh, A. Sahu, R. Shukla, B. Singh, J. Rai, P. Sharma, H. Lade, and D. Paul. 2015. "Bioremediation of Heavy Metals from Soil and Aquatic Environment: An Overview of Principles and Criteria of Fundamental Processes." *Sustainability* 7, no. 2: 2189–212. https://doi.org/10.3390/su7022189.

Domingo, José L., and Martí Nadal. 2015. "Human Dietary Exposure to Polycyclic Aromatic Hydrocarbons: A Review of the Scientific Literature." *Food & Chemical Toxicology* 86: 144–53. https://doi.org/10.1016/j.fct.2015.10.002.

Ekta, P., and N. R. Modi. 2018. "A Review of Phytoremediation." *Journal of Pharmacognosy & Phytochemistry* 7, no. 4: 1485–9.

EPA. 1983. *Methods for Chemical Analysis of Water and Wastes*. EP-600/4-79020. Cincinnati, OH: Environmental Monitoring and Support Laboratory, US-EPA.

EPA. 1984. *Test Methods for Evaluating Solid Waste: Physical/Chemical Methods*. 2nd ed.: EPA-864. Washington, DC: Office of Solid Waste and Emergency Response, US-EPA.

Ferronato, Navarro, and Vincenzo Torretta. 2019. "Waste Mismanagement in Developing Countries: A Review of Global Issues." *International Journal of Environmental Research & Public Health* 16, no. 6: 1060. https://doi.org/10.3390/ijerph16061060.

Fiedler, H. 2003. *The Hand Book of Environmental Chemistry, Persistent Organic Pollutants*. 3rd ed. Berlin, Heidelberg: Springer-Verlag.

Field, R. A., M. E. Goldstone, J. N. Lester, and R. Perry. 1992. "The Sources and Behavior of Tropospheric Anthropogenie Volatile Hydrocarbons." *Atmospheric Environment. Part A. General Topics* 26, no. 16: 2983–96. https://doi.org/10.1016/0960-1686(92)90290-2.

Glaser, J. A. 1993. "Engineering Approaches Using Bioremediation to Treat Crude Oil-Contaminated Shoreline Following the Exxon Valdez Accident in Alaska." In *Bioremediation: Field Experience*, edited by P. E. Flathman, D. E. Jerger, and J. H. Exner: 81–103. Boca Raton, FL: Lewis Publishers.

González, R. Carrillo, and M. C. A. González-Chávez. 2006. "Metal Accumulation in Wild Plants Surrounding Mining Wastes." *Environmental Pollution* 144, no. 1: 84–92. https://doi.org/10.1016/j.envpol.2006.01.006.

Gordon, M., N. Choe, J. Duffy, G. Ekuan, P. Heilman, I. Muiznieks, L. Newman, M. Ruszaj, B. B. Shurtleff, S. Strand, and J. Wilmoth. 1998. "Phytoremediation of Trichloroethylene with Hybrid Poplars." *Environmental Health Perspectives* 106(Suppl 4): 1001–4. https://doi.org/10.1289/ehp.98106s41001.

Grande, M., S. Andersen, and D. Berge. 1994. "Effects of Pesticides on Fish. Norwegian Journal of Agricultural Sciences." *Experimental & Field Studies* 13: 195–209.

Guarino, Francesco, Antonio Miranda, Stefano Castiglione, and AngelaCicatelli. 2020. "Arsenic Phytovolatilization and Epigenetic Modifications in *Arundo donax* L. Assisted by a PGPR Consortium." *Chemosphere* 251: 126310. https://doi.org/10.1016/j.chemosphere.2020.126310.

Guerin, W. F., and S. A. Boyd. 1992. "Differential Availability of Soil-Sorbed Naphthalene to Two Bacteria." *Applied & Environmental Microbiology* 58, no. 4: 1142–52. https://doi.org/10.1128/aem.58.4.1142-1152.1992.

Gupta, A. K., and S. Sinha. 2007. "Phytoextraction Capacity of the *Chenopodium album* L. Grown on Soil Amended with Tannery Sludge." *Bioresource Technology* 98, no. 2: 442–6. https://doi.org/10.1016/j.biortech.2006.01.015.

Hammond, Corin M., Robert A. Root, Raina M. Maier, and Jon Chorover. 2018. "Mechanisms of Arsenic Sequestration by *Prosopis juliflora* During the Phytostabilization of Metalliferous Mine Tailings." *Environmental Science & Technology* 52, no. 3: 1156–64. https://doi.org/10.1021/acs.est.7b04363.

Harner, T., K. Pozo, T. Gouin, A. Macdonald, H. Hung, and I. Cieny. 2006. "Global Pilot Study for Persistent Organic Pollutants (POPs) Using PUF Disk Passive Air Samplers." *Environmental Pollution* 144: 445–52. https://doi.org/10.1016/j. envpol.2005.12.053.

Hernández, A., Loera, N., M. Contreras, L. Fischer, and D. Sanchez. 2019. "Comparison Between *Lactuca sativa* L. and *Lolium perenne*: Phytoextraction Capacity of Ni, Fe, and Co from Galvanoplastic Industry." *Energy Technology* 137–47.

Huesemann, M. H. 1995. "Predictive Model for Estimating the Extent of Petroleum Hydrocarbon Biodegradation in Contaminated Soils." *Environmental Science & Technology* 29, no. 1: 7–18. https://doi.org/10.1021/es00001a002.

Interstate Technology & Regulatory Council. 2009. *Phytotechnology Technical and Regulatory Guidance and Decision Trees (Revised)*: 1–131. Washington, DC. Interstate Technology & Regulatory Council.

Isiuku, B. O., and E. C. Ebere. 2019. "Water Pollution by Heavy Metal and Organic Pollutants: Brief Review of Sources, Effects and Progress on Remediation with Aquatic Plants". *Analytical Methods in Environmental Chemistry Journal* 2, no. 3: 5–38.

Jacob, Jaya Mary, Chinnannan Karthik, Rijuta Ganesh Saratale, Smita S. Kumar, Desika Prabakar, K. Kadirvelu, and Arivalagan Pugazhendhi. 2018. "Biological Approaches to Tackle Heavy Metal Pollution: A Survey of Literature." *Journal of Environmental Management* 217: 56–70. https://doi.org/10.1016/j.jenvman.2018.03.077.

Jan, S., and J. A. Parray. 2016. *Approaches to Heavy Metal Tolerance in Plants*. New Delhi, India: Springer.

Jimenez, I. Y., and R. Bartha. 1996. "Solvent Augmented Mineralization of Pyrene by a *Mycobacterium* sp." *Applied & Environmental Microbiology* 62, no. 7: 2311–6. https://doi.org/10.1128/aem.62.7.2311-2316.1996.

Kafle, A., A. Timilsina, A. Gautam, K. Adhikari, A. Bhattarai, and N. Aryal. 2022. "Phytoremediation: Mechanisms, Plant Selection and Enhancement by Natural and Synthetic Agents." *Environmental Advances* 8: 100203. https://doi.org/10.1016/j.envadv.2022.100203.

Kanaly, R. A., R. Bartha, K. Watanabe, and S. Harayama. 2000. "Rapid Mineralization of Benzo[a]Pyrene by a Microbial Consortium Growing on Diesel Fuel." *Applied & Environmental Microbiology* 66, no. 10: 4205–11. https://doi.org/10.1128/AEM.66.10.4205-4211.2000.

Kaza, S., L. Yao, P. Bhada-Tata, and F. Van Woerden. 2018. *What a Waste: A Global Snapshot of Solid Waste Management to 2050.* Singapore: World Bank.

Keith, L. H. 1988. *Principles of Environmental Sampling.* Washington, DC: American Chemical Society.

Kendrovski, V., E. Stikova, and L. Kolevska. 2001. "Contamination of Food and Agro Products in the Republic of Macedonia." *Arhiv za Higijenu Rada i Toksikologiju* 52, no. 1: 69–73.

Khalid, N., A. Noman, M. Aqeel, A. Masood, and A. Tufail. 2019. "Phytoremediation Potential of Xanthium strumarium for Heavy Metals Contaminated Soils at Road sides." *International Journal of Environmental Science & Technology* 16, no. 4: 2091–100. https://doi.org/10.1007/s13762-018-1825-5.

Kozak, R., I. Dhaese, and W. Verstraete. 2001. "Pharmaceuticals in the Environment: Focuson 17 Alpha-Ethinyloestradiol." In *Pharmaceuticals in the Environment: Sources, Fate, Effects and Risks*, edited by K. Kummerer: 49–66. Berlin: Springer Verlag.

Kumar, P. B. A. N., V. Dushenkov, H. Motto, and I. Raskin. 1995. "Phytoextraction: The Use of Plants to Remove Heavy Metals from Soils." *Environmental Science & Technology* 29, no. 5: 1232–8. https://doi.org/10.1021/es00005a014.

Kumari, S., Y. Amit, R. Jamwal, N. Mishra, D. K. Singh. 2020. "Recent Developments in Environmental Mercury Bioremediation and Its Toxicity: A Review." *Environmental Nanotechnology, Monitoring Management* 13: 100283 https://doi.org/10.1016/j. enmm.2020.100283.

Kumar Yadav, K., N. Gupta, A. Kumar, L. M. Reece, N. Singh, S. Rezania, and S. Ahmad Khan. 2018. "Mechanistic Understanding and Holistic Approach of Phytoremediation: A Review on Application and Future Prospects." *Ecological Engineering* 120: 274–98. https://doi.org/10.1016/j.ecoleng.2018.05.039.

Lajayer, Asgari, Behnam, Nader Khadem Moghadam, Mohammad Reza Maghsoodi, Mansour Ghorbanpour, and Khalil Kariman. 2019. "Phytoextraction of Heavy Metals from Contaminated Soil, Water and Atmosphere Using Ornamental Plants: Mechanisms and Efficiency Improvement Strategies." *Environmental Science & Pollution Research International* 26, no. 9: 8468–84. https://doi.org/10.1007/s11356-019-04241-y.

Lee, J. H. 2013. "An Overview of Phytoremediation as a Potentially Promising Technology for Environmental Pollution Control." *Biotechnology & Bioprocess Engineering* 18, no. 3: 431–9. https://doi.org/10.1007/s12257-013-0193-8.

Lenke, H., J. Warrelmann, D. Daun, U. Walter, U. Sieglen, and H.-J. Knackmuss. 1997. "Bioremediation of TNT –Contaminated Soil by an Anaerobic/Aerobic Process." In *In Situ and On-Site Bioremediation.* Vol. 2, edited by B. C. Alleman, and A. Leeson: 1–2. Columbus, OH: Battelle Press.

Li, J. T., B. Liao, C. Y. Lan, Z. H. Ye, A. J. Baker, and W. S. Shu. 2010. "Cadmium Tolerance and Accumulation in Cultivars of a High-Biomass Tropical Tree (*Averrhoa carambola*) and Its Potential for Phytoextraction." *Journal of Environmental Quality* 39, no. 4: 1262–8. https://doi.org/10.2134/jeq2009.0195.

Limmer, Matt, and Joel Burken. 2016. "Phytovolatilization of Organic Contaminants." *Environmental Science & Technology* 50, no. 13: 6632–43. https://doi.org/10.1021/acs.est.5b04113.

Lovley, D. R. 2000. "Anaerobic Benzene Degradation." *Biodegradation* 11, no. 2–3: 107–16. https://doi.org/10.1023/a:1011191220463.

Maczulak, A. E. 2010. *Pollution: Treating*: 120. Environmental Toxin Series. New York: Info Base Publishing.

Mahro, B. 2000. "Bioavailability of Contaminants." In *Biotechnology.* 2nd ed. Vol. 11b, edited by H.-J. Rehm, and G. Reed: 61–88. Weinheim: Wiley-VCH Press.

Makkar, R. S., and S. S. Cameotra. 2002. "An Update on the Use of Unconventional Substrates for Biosurfactant Production and Their New Applications." *Applied Microbiology & Biotechnology* 58, no. 4: 428–34. https://doi.org/10.1007/s00253–001–0924–1.

Malik, B., T. B. Pirzadah, I. Tahir, T. H. Dar, and R. Rehman. 2014. "Recent Trends and Approaches in Phytoremediation." *Soil Remediation & Plants*: 131–46.

Manzoor, Maria, Iram Gul, Iftikhar Ahmed, Muhammad Zeeshan, Imran Hashmi, Bilal Ahmad Zafar Amin, Jean Kallerhoff, and Muhammad Arshad. 2019. "Metal Tolerant Bacteria Enhanced Phytoextraction of Lead by Two Accumulator Ornamental Species." *Chemosphere* 227: 561–9. https://doi.org/10.1016/j.chemosphere.2019.04.093.

Marques, A. P. G. C., A. O. S. S. Rangel, and P. M. L. Castro. 2009. "Remediation of Heavy Metal Contaminated Soils: Phytoremediation as a Potentially Promising Clean-Up Technology." *Critical*

Reviews in Environmental Science & Technology 39, no. 8: 622–54. https://doi.org/10.1080/10643380 701798272.

Martorell, Isabel, Gemma Perelló, Roser Martí-Cid, Victòria Castell, Juan M. Llobet, and José L. Domingo. 2010. "Polycyclic Aromatic Hydrocarbons (PAH) in Foods and Estimated PAH Intake by the Population of Catalonia, Spain: Temporal Trend." *Environment International* 36, no. 5: 424–32. https://doi.org/10.1016/j.envint.2010.03.003.

Mastretta, Chiara, Tanja Barac, Jaco Vangronsveld, Lee Newman, Safiyh Taghavi, and Danielvan der Lelie. 2006. "Endophytic Bacteria and Their Potential Application to Improve the Phytoremediation of Contaminated Environments." *Biotechnology & Genetic Engineering Reviews* 23: 175–207. https://doi.org/10.1080/02648725.2006.10648084.

McCarty, P. L., M. N. Goltz, G. D. Hopkins, M. E. Dolan, J. P. Allan, B. T. Kawakami, and T. J. Carrothers. 1998. "Full-Scale Evaluation of *In Situ* Cometabolic Degradation of Trichloroethylene in Groundwater Through Toluene Injection." *Environmental Science & Technology* 32, no. 1: 88–100. https://doi.org/10.1021/es970322b.

McGrath, Steve P., and Fang-Jie Zhao. 2003. "Phytoextraction of Metals and Metalloids from Contaminated Soils." *Current Opinion in Biotechnology* 14, no. 3: 277–82. https://doi.org/10.1016/s0958-1669(03)00060-0.

Mirza, N., A. Pervez, Q. Mahmood, S. Sultan, and M. M. Shah. 2014. "Plants as Useful Vectors to Reduce Environmental Toxic Arsenic Content." *Science World Journal* 2014: 1–11.

Mohandass, C., J. J. David, S. Nair, P. A. Loka Bharathi, and D. Chandramohan. 1997. "Behaviour of Marine Oil-Degrading Bacterial Populations in a Continuous Culture System." *Journal of Marine Biotechnology* 5: 168–71.

Moseley, C. L., and M. R. Meyer. 1992. "Petroleum Contamination of an Elementary School: A Case History Involving Air, Soil-Gas, and Groundwater Monitoring." *Environmental Science & Technology* 26, no. 1: 185–92. https://doi.org/10.1021/es00025a023.

Mostafalou, S., and M. Abdollahi. 2012. "Concerns of Environmental Persistence of Pesticides and Human Chronic Diseases." *Clinical & Experimental Pharmacology* 5: e002.

Mulligan, C. N., R. N. Yong, and B. F. Gibbs. 2001. "Surfactant-Enhanced Remediation of Contaminated Soil: A Review." *Engineering Geology* 60, no. 1–4: 371–80.https://doi.org/10.1016/S0013–7952(00)00117–4.

Nabegu, A. B. 2017. *An Analysis of Municipal Solid Waste in Kano Metropolis, Nigeria*. A Paper presented in a Workshop at Kano State University of Science and Technology, Wudil.

Nam, K., and M. Alexander. 1998. "Role of Nanoporosity and Hydrophobicity in Sequestration and Bioavailability Tests with Model Solids." *Environmental Science & Technology* 32, no. 1: 71–4. https://doi.org/10.1021/es9705304.

Nikolic, M., and S. Stevovic. 2015. "Family Asteraceae as Sustainable Planning Tool in Phytoremediation and Its Relevance in Urban Areas." *Urban Forestry & Urban Greening* 14. https://doi.org/10.1016/j.ufug.2015.08.002.

Nwachukwu, N. C., A. F. Orji, and O. C. Ugbogu. 2013. "Health Care Waste Management—Public Health Benefits, and the Need for Effective Environmental Regulatory Surveillance in Federal Republic of Nigeria." In *Current Topics in Public Health*. Intech Open Science.

Obi, F. O., B. O. Ugwuishiwu, and J. N. Nwakaire. 2016. "Agricultural Waste Concept, Generation, Utilization and Management." *Nigerian Journal of Technology* 35, no. 4: 957–64. https://doi.org/10.4314/njt.v35i4.34.

Organization for Economic Co-operation and Development. 2021. *Municipal Waste (Indicator)*. Paris: OECD.

Pannu, Jasvir K., Ajay Singh, and Owen P. Ward. 2003. "Influence of Peanut Oil on Microbial Degradation of Polycyclic Aromatic Hydrocarbons." *Canadian Journal of Microbiology* 49, no. 8: 508–13. https://doi.org/10.1139/w03-068.

Pearce, K., H. G. Snyman, R. A. Oellermann, and A. Gerber. 1995. "Bioremediation of Petroleum Contaminated Soil." In *Bioaugmentation for Site Remediation*, edited by R. E. Hinchee, J. Frederickson, and B. C. Alleman: 71–6. Columbus, OH: Batelle Press.

Pierzynski, G. M., J. L. Schnoor, M. K. Banks, J. C. Tracy, L. A. Licht, and L. E. Erickson. 1994. "Vegetative Remediation at Superfund Sites. Mining and Its Environmental Impact." *Royal Society of Chemical Issues in Environmental Science & Technology* 1: 49–69.

Pilon-Smits, E. A. H., M. P. De Souza, G. Hong, A. Amini, R. C. Bravo, S. T. Payabyab, and N. Terry. 1999. "Selenium Volatilization and Accumulation by Twenty Aquatic Plant Species." *Journal of Environmental Quality* 28, no. 3: 1011–8. https://doi.org/10.2134/jeq1999.00472425002800030035x.

Prince, R. C. 1998. "Bioremediation." In *Encyclopedia of Chemical Technology*. Supplement to 4th ed.: 48–89. New York: Wiley.

Qixing, Z., C. Zhang, Z. Zhineng, and L. Weitao. 2011. "Ecological Remediation of Hydrocarbon Contaminated Soils with Weed Plant." *Journal of Resources & Ecology* 2, no. 2: 97–105.

Rajan, R., D. T. Robin, and M. V. Robert. 2018. "Biomedical Waste Management in Ayurveda Hospitals—Current Practices and Future Prospectives." *Journal of Ayurveda and Integrative Medicine* 10, no. 3: 214–21. https://doi.org/10.1016/j.jaim.2017.07.011.

Raskin, I., P. B. A. Nanda Kumar, S. Dushenkov, M. J. Blaylock, and D. Salt. 1994. "Phytoremediation—Using Plants to Clean up Soils and Waters Contaminated with Toxic Metals." *Emerging Technologies in Hazardous Waste Management* VI. Industrial: ACS & Engineering Chemistry Division Special Symposium. Atlanta, GAI, no. Sept. 19–21.

Raskin, I., R. D. Smith, and D. E. Salt. 1997. "Phytoremediation of Metals: Using Plants to Remove Pollutants from the Environment." *Current Opinion in Biotechnology* 8, no. 2: 221–6. https://doi.org/10.1016/s0958-1669(97)80106-1.

Reeves, R. D., and R. R. Brooks. 1983. "Hyperaccumulation of Lead and Zinc by Two Metallophytes from Mining Areas of Central Europe." *Environmental Pollution Series A, Ecological & Biological* 31, no. 4: 277–85.https://doi.org/10.1016/0143–1471(83)90064–8.

Rieger, P. G., and H. J. Knackmuss. 1995. "Basic Knowledge and Perspectives on Biodegradation of 2,4,6-Trinitrotoluene and Related Nitroaromatic Compounds in Contaminated Soil." In *Biodegradation of Nitroaromatic Compounds*, edited by J. C. Spain: 1–18. New York: Plenum Press.

Rittmann, B. E., and R. Whiteman. 1994. "Bioaugmentation: A Coming of Age." *Water Quality International* 1: 12–6.

Roemer, M. 2000. "Sampling and Investigation of Soil Matter." In *Biotechnology*. 2nd ed. Vol. 11b, edited by H.-J. Rehm and G. Reed: 477–507. Weinheim: Wiley-VCH Press.

Rogimon, P. T., P. Joby, V. Vinod, and V. M. Kannan. 2022. "Organic Contaminants and Phytoremediation: A Critical Appraisal." In *Bioenergy Crops: A Sustainable Means of Phytoremediation*. Boca Raton, FL: CRC Press. https://doi.org/10.1201/9781003043522.

Rosario, Karyna, Sadie L. Iverson, David A. Henderson, Shawna Chartrand, Casey McKeon, Edward P. Glenn, and Raina M. Maier. 2007. "Bacterial Community Changes During Plant Establishment at the San Pedro River Mine Tailings Site." *Journal of Environmental Quality* 36, no. 5: 1249–59. https://doi.org/10.2134/jeq2006.0315.

Rouse, J. D., D. A. Sabatini, J. M. Suflita, and J. H. Harwell. 1994. "Influence of Surfactants on Microbial Degradation of Organic Compounds." *Critical Reviews in Environmental Science & Technology* 24, no. 4: 325–70. https://doi.org/10.1080/10643389409388471.

Ruppert, L., Z. Q. Lin, R. P. Dixon, and K. A. Johnson. 2013. "Assessment of Solid Phase Microfiber Extraction Fibers for the Monitoring of Volatile Organoarsinicals Emitted from a Plant–Soil System." *Journal of Hazardous Materials* 262: 1230–6. https://doi.org/10.1016/j.jhazmat.2012.06.046.

Salanitro, J. P. 2001. "Bioremediation of Petroleum Hydrocarbons in Soil." *Advances in Agronomy* 72: 53–105. https://doi.org/10.1016/S0065-2113(01)72011-1.

Salido, Arthur L., Kelly L. Hasty, Jae-Min Lim, and David J. Butcher. 2003. "Phytoremediation of Arsenic and Lead in Contaminated Soil Using Chinese Brake Ferns (*Pteris vittata*) and Indian Mustard (*Brassica juncea*)." *International Journal of Phytoremediation* 5, no. 2: 89–103. https://doi.org/10.1080/713610173.

Salt, D. E., I. J. Pickering, R. C. Prince, D. Gleba, S. Dushenkov, R. D. Smith, and I. Raskin. 1997. "Metal Accumulation by Aquacultured Seedlings of Indian Mustard." *Environmental Science & Technology* 31, no. 6: 1636–44. https://doi.org/10.1021/es960802n.

San Miguel, Angélique, Patrick Ravanel, and Muriel Raveton. 2013. "A Comparative Study on the Uptake and Translocation of Organochlorines by Phragmites Australis." *Journal of Hazardous Materials* 244–245: 60–9. https://doi.org/10.1016/j.jhazmat.2012.11.025.

Selby, D. A. 1991. "A Critical Review of Site Assessment Methodologies." In *Hydrocarbon Contaminated Soils and Groundwater*, edited by P. T. Kostecki and E. J. Calabrese: 149. Boca Raton, FL: Lewis Publishers.

Seth, C. S. 2012. "A Review on Mechanisms of Plant Tolerance and Role of Transgenic Plants in Environmental Clean-Up." *Botany Review* 78: 32–62. https://doi.org/10.1007/s12229- 011–9092-x.

Sewell, G. W., M. F. DeFlaun, N. H. Baek, E. Kutz, B. Weesner, and B. Mahaffey. 1998. "Performance Evaluation of an In Situ Anaerobic Biotreatment System for Chlorinated Solvents." In *Designing and Applying Treatment Technologies for Remediation of Chlorinated and Recalcitrant Compounds*, edited by G. B. Wickramanayak and R. E. Hinchee: 15–20. Columbus, OH: Batelle.

Sharma, P., and S. Pandey. 2014. "Status of Phytoremediation in World Scenario." *International Journal of Environmental Bioremediation & Biodegradation* 2, no. 4: 178–91.

Shin, C. Y., and D. L. Crawford. 1995. "Biodegradation of Trinitrotoluene (TNT) by a Strain of *Clostridium bifermentans.*" In *Bioaugmentation for Site Remediation*, edited by R. E. Hinchee, J. Frederickson, and B. C. Alleman: 57–69. Columbus, OH: Batelle Press.

Sinha, K. R., S. Heart, and K. P. Tandon. 2007. "Phytoremediation: Role of Plants in Contaminated Site Management." In *Environmental Bioremediation Technologies*, edited by N. S. Singh and D. R. Tripathi: 215–330. Berlin, Heidelberg, and New York: Springer-Verlag.

Sosa, Dayana, Isabel Hilber, Roberto Faure, Nora Bartolomé, Osvaldo Fonseca, Armin Keller, Peter Schwab, Arturo Escobar, and Thomas D. Bucheli. 2017. "Polycyclic Aromatic Hydrocarbons and Polychlorinated Biphenyls in Soils of Mayabeque, Cuba." *Environmental Science & Pollution Research International* 24, no. 14: 1286012870: 12860–70. https://doi.org/10.1007/s11356-017-8810-2.

Tokala, Ranjeet K., Janice L. Strap, Carina M. Jung, Don L. Crawford, Michelle Hamby Salove, Lee A. Deobald, J. Franklin Bailey, and M. J. Morra. 2002. "Novel Plant-Microbe Rhizosphere Interaction Involving *Streptomyces lydicus* WYEC108 and the Pea Plant (*Pisum sativum*)." *Applied & Environmental Microbiology* 68, no. 5: 2161–71. https://doi.org/10.1128/AEM.68.5.2161-2171.2002.

Tripathi, S., V. P. Singh, P. Srivastava, R. Singh, R. S. Devi, A. Kumar, and R. Bhadouria. 2020. "Chapter 4. Abatement of Environmental Pollutants Trends and Strategies." In *Phytoremediation of Organic Pollutants: Current Status and Future Directions*, edited by Pardeep Singh, Ajay Kumar, and Anwesha Borthakur: 81–105. Amsterdam: Elsevier.

Untermann, R., M. F. DeFlaun, and R. J. Steffan. 2000. "Advanced In Situ Bioremediation—A Hierarchy of Technology Choices." In *Biotechnology.* 2nd ed. Vol. 11b, edited by H.-J. Rehm, and G. Reed. Weinheim: Wiley-VCH Press.

Van Hamme, J. D., and O. P. Ward. 1999. "Influence of Chemical Surfactants on the Biodegradation of Crude Oil by a Mixed Bacterial Culture." *Canadian Journal of Microbiology* 45, no. 2: 130–7. https://doi.org/10.1139/w98-209.

Van Hamme, J. D., and O. P. Ward. 2001. "Volatile Hydrocarbon Biodegradation by a Mixed-Bacterial Culture During Growth on Crude Oil." *Journal of Industrial Microbiology & Biotechnology* 26, no. 6: 356–62. https://doi.org/10.1038/sj.jim.7000145.

Vandepitte, V., P. Quataert, H. de Rore, and W. Verstraete. 1995. "Evaluation of the Gompertz Function to Model Survival of Bacteria Introduced into Soils." *Soil Biology & Biochemistry* 27, no. 3: 365–72. https://doi.org/10.1016/0038-0717(94)00158-W.

Vangronsveld, Jaco, Rolf Herzig, Nele Weyens, Jana Boulet, Kristin Adriaensen, Ann Ruttens, Theo Thewys, Andon Vassilev, Erik Meers, Erika Nehnevajova, Danielvan der Lelie, and Michel Mench. 2009. "Phytoremediation of Contaminated Soils and Groundwater: Lessons from the Field." *Environmental Science & Pollution Research International* 16, no. 7: 765–94. https://doi.org/10.1007/s11356-009-0213-6.

Venkateswaran, K., T. Hoaki, M. Kato, and T. Maruyama. 1995. "Microbial Degradation of Resins Fractionated from Arabian Light Erude Oil." *Canadian Journal of Microbiology* 41, no. 4–5: 418–24. https://doi.org/10.1139/m95-055.

Venosa, A. D., J. R. Haines, W. Nisamaneepong, R. Govind, S. Pradhan, and B. Siddique. 1992. "Efficacy of Commercial Products in Enhancing Oil Biodegradation in Closed Laboratory Reactors." *Journal of Industrial Microbiology* 10, no. 1: 13–23. https://doi.org/10.1007/BF01583629.

Vinti, Giovanni, Valerie Bauza, Thomas Clasen, Kate Medlicott, Terry Tudor, Christian Zurbrügg, and Mentore Vaccari. 2021. "Municipal Solid Waste Management and Adverse Health Outcomes: A Systematic Review." *International Journal of Environmental Research & Public Health* 18, no. 8: 4331. https://doi.org/10.3390/ijerph18084331.

Vogel, T. M. 1996. "Bioaugmentation as a Soil Bioremediation Approach." *Current Opinion in Biotechnology* 7, no. 3: 311–6. https://doi.org/10.1016/s0958-1669(96)80036-x.

WHO. 2018. *Fact Sheets on Detail Health Care Waste*. Geneva: WHO.

Wiessner, A., U. Kappelmeyer, M. Kaestner, L. Schultze-Nobre, and P. Kuschk. 2013. "Response of Ammonium Removal to Growth and Transpiration of *Juncus effusus* During the Treatment of Artificial Sewage in Laboratory-Scale Wetlands." *Water Research* 47, no. 13: 4265–73. https://doi.org/10.1016/j.watres.2013.04.045.

Wiszniewska, A., E. Hanus-Fajerska, E. Muszyńska, and K. Ciarkowska. 2016. "Natural Organic Amendments for Improved Phytoremediation of Polluted Soils: A Review of Recent Progress." *Pedosphere* 26, no. 1: 1–12. https://doi.org/10.1016/S1002-0160(15)60017-0.

Witt, M. E., M. J. Dybas, R. L. Heine, S. Nair, C. S. Criddle, and D. C. Wiggert. 1995. "Bioaugmentation and Transformation of Carbon Tetrachloride in a Model Aquifer." In *Bioaugmentation for Site Remediation*, edited by R. E. Hinchee, J. Frederiekson, and B. C. Alleman: 221–7. Columbus, OH: Batelle Press.

Phytoremediation of Organic Contaminants

Wuana, R. A., and F. E. Okieimen. 2011. "Heavy Metals in Contaminated Soils: A Review of Sources, Chemistry, Risks and Best Available Strategies for Remediation." *ISRN Ecology* 2011: 1–20. https://doi.org/10.5402/2011/402647.

Yadav, S. K. 2010. "Heavy Metals Toxicity in Plants: An Overview on the Role of Glutathione and Phytochelatins in Heavy Metal Stress Tolerance of Plants." *South African Journal of Botany* 76, no. 2: 167–79. https://doi.org/10.1016/j.sajb.2009.10.007.

Yan, A., Yamin Wang, Swee Ngin Tan, Mohamed Lokman Mohd Yusof, Subhadip Ghosh, and Zhong Chen. 2020. "Phytoremediation: A Promising Approach for Revegetation of Heavy Metal-Polluted Land." *Frontiers in Plant Science* 11: 359. https://doi.org/10.3389/fpls.2020.00359.

Yang, Yang, Yichen Ge, Pengfei Tu, Hongyuan Zeng, Xihong Zhou, Dongsheng Zou, Kelin Wang, and Qingru Zeng. 2019b. "Phytoextraction of Cd from a Contaminated Soil by Tobacco and Safe Use of Its Metal enriched Biomass." *Journal of Hazardous Materials* 363: 385–93. https://doi.org/10.1016/j.jhazmat.2018.09.093.

Yoon, J., X. Cao, Q. Zhou, and L. Q. Ma. 2006. "Accumulation of Pb, Cu, and Zn in Native Plants Growing on a Contaminated Florida Site." *Science of the Total Environment* 368, no. 2–3: 456–64. https://doi.org/10.1016/j.scitotenv.2006.01.016.

6 Microbial and Phytoremediation of Crude Oil-Contaminated Soil

Odangowei Inetiminebi Ogidi and Udeme Monday Akpan

6.1 INTRODUCTION

Soil pollution is now an international phenomenon. Due to industrial, agricultural, and uncontrolled waste disposal practices, heavy metals and chemicals have polluted soil in many regions of the globe. The leakage of petroleum products, especially crude oil, into the environment is responsible for a significant fraction of chemically polluted soil (Liao *et al.*, 2016). The danger of spills and environmental contamination has grown due to the global intensification of oil and gas operations, including exploration, drilling, production, onshore storage, and transportation of petroleum (Njoku *et al.*, 2008). A complex combination of hydrocarbons and organic molecules make up crude oil, some of which, like benzene and poly-aromatic hydrocarbons, are known to be hazardous to the environment and human health (Ebadi *et al.*, 2018). Due to probable human exposure to the hazardous chemicals and bioaccumulation of the compounds in agricultural products, soil polluted by crude oil might make it unsuitable for habitation and agricultural activity (Hegazy *et al.*, 2015).

Petroleum enters the soil environment via extraction, processing, and transportation (pipe rupture) (Abbaspour *et al.*, 2020; Freedman, 1995). The most dangerous and poisonous aliphatic, cycloaliphatic, and aromatic hydrocarbons are the main pollutants in soil that has been polluted with petroleum (Suganthi *et al.*, 2018). They lessen the variety of soil plants and bacteria, reduce soil fertility, disturb the biological balance of the soil, and potentially endanger human health (Ansari *et al.*, 2018). When cultivated in high petroleum-contaminated soil, crop germination is delayed, the chlorophyll content is low, and some crops die (Ekundayo *et al.*, 2001).

Additionally, pollutants may enter the body by inhaling, skin contact, or eating food contaminated with petroleum, leading to gastrointestinal problems, contact dermatitis, auditory and visual hallucinations, and a significantly increased risk of leukemia in children. Unlike some low-molecular-weight hydrocarbon pollutants, which may weather and decay over time, high-molecular-weight hydrocarbon pollutants will stay in the soil for extremely long times due to their hydrophobicity, creating secondary pollution throughout the ecosystem (Kumari *et al.*, 2018; Ozigis *et al.*, 2020). Soil that has been polluted by petroleum must be restored or remediated.

The traditional approach to cleaning up soil polluted by crude oil is excavating the soil and then treating it chemically or physically (Kumari *et al.*, 2018). Leaching, incineration, landfills, and chemical oxidation treatment techniques are also often used to clean up soil polluted with petroleum. In a polluted environment, these technologies may extract, remove, convert, or mineralize petroleum pollutants, turning them into a less harmful, nontoxic, and stable form (Maletic *et al.*, 2013). Both chemical oxidation and incineration, which can get rid of 99.0 and 92.3% of petroleum hydrocarbons, respectively, have limitations that limit their use for environmental restoration (Liao *et al.*, 2019).

Toxic substances such as dioxins, furans, polychlorinated biphenyls, and volatile heavy metals are released into the atmosphere when petroleum is burned incompletely (Giannopoulos *et al.*, 2007). The loss of carbon and organic matter reduces the soil ecosystem's capacity to recover

Phytoremediation of Crude Oil-Contaminated Soil   111

from incomplete combustion of petroleum, which raises the covert risks to environmental safety. Oxidants will prevent soil microbes from proliferating. As a result, despite lowering the concentration of petroleum pollutants in the soil, it won't result in secondary contamination of the soil or the ecosystem, which has turned into the primary factor in choosing remediation solutions.

The full mineralization of organic pollutants into carbon dioxide, water, inorganic chemicals, and cell proteins is possible with affordable microbial remediation. It may also be used to break down complicated organic contaminants into other simpler organics (Das and Chandran, 2011). Organic contaminants in the soil may be degraded by microorganisms using them as their only source of carbon (Velacano *et al.*, 2014). In 150 to 270 days, microorganisms in the soil degraded 62–75% of the petroleum hydrocarbons there (Chaîneau *et al.*, 2005). In the immobilized system (sodium alginate-diatomite beads), petroleum decayed at a rate of 29.8%, while loose cells deteriorated at a rate of 21.2% over the course of 20 days (Wang *et al.*, 2012). Microbial remediation has problems, including lengthy remediation timeframes and poor free-microorganism effectiveness. The microbial combination method is used to improve the biodegradation efficiency of microorganisms, which is necessary to overcome the challenge of microbial remediation of petroleum in the soil.

Phytoremediation is an economically feasible alternative to soil remediation because it speeds up the elimination of toxins from soil via plant physiological processes and microbiological activity at the roots of plants (Cunningham *et al.*, 1996). Local plant species are chosen for use in phytoremediation because they adapt well to the local climate and soil conditions, increasing the likelihood that they will develop and spread over polluted soil (Anh *et al.*, 2017). Soil environment, soil contaminated with petroleum hydrocarbons, biological remediation of petroleum hydrocarbon–polluted soil, advancements in the use of microorganisms in petroleum hydrocarbon, and variables affecting petroleum hydrocarbon soil remediation are all covered in this chapter.

6.2 SOIL ENVIRONMENT

The soft substance that covers the earth's surface is known as soil. All soils initially consist of solid rock, which is weathered physically and chemically into smaller fragments (Parker, 2009). Living things directly and practically depend on soil. It promotes the development of plants, human life, and animal life. Chemical buffering capacity, which is the balance between molecules that are adsorbed to soil surfaces and discharged into the soil solution, biodiversity, water filtering, nutrient cycling, organic matter decomposition, and a host of other ecological functions are all aided by healthy soil (Whalen and Sampedro, 2010). Soil is an important environmental agent due to the interactions it has with chemicals already there or introduced to it (Duffy, 2011). These biological, chemical, and physical characteristics all play a role in this interaction.

6.2.1 Soil Physical Properties

The composition, texture, structure, and permeability of soil are its primary physical characteristics.

6.2.1.1 Soil Composition

Soil is composed of three distinct phases: the solid phase (50% mineral particles andi5% organic matter), the liquid phase (25% water), and the gas phase (20% air) (Paria, 2008). Soil organic matter refers to the decomposed and undecomposed remnants of plants, animals, and microorganisms in the soil (Whalen and Sampedro, 2010).

6.2.1.2 Soil Texture

Soil is made up of particles with various shapes and properties. Sand, silt, and clay make up the three portions of these particles. The quantity of these components is referred to as soil texture. How readily a material flows through soil is directly correlated to the gaps between soil particles and determines the texture of the soil (Parker, 2009). Quartz and feldspar-based sand is mostly inert, low nutrient, low water retention (but excellent drainage), and easily aerated (Duffy, 2011). Colloids are microscopic particles (clays and silts small) less than 0.01 mm in size that have a high exchange surface (McCauley *et al.*, 2005). The soil texture triangle determines the name of the soil's texture. While the center of the triangle's soils are optimal for plant development, the presence of clay makes them chemically reactive (Duffy, 2011).

6.2.1.3 Soil Structure

The arrangement of soil particles is referred to as structure. It affects air flow and water absorption. These aggregates boost soil fertility and carbon sequestration, stabilize the soil against soil erosion, control water flow, and preserve soil porosity (McCauley *et al.*, 2005). The amount of pore or emptiness in the soil is known as porosity. A structure should contain a considerable amount of big pores and a high percentage of medium-sized aggregates (Parker, 2009).

6.2.1.4 Permeability

The capacity of soil to carry water is known as permeability or hydraulic conductivity. A healthy soil structure increases permeability compared to an unstructured soil. The major elements that affect how pollutants and water move through soil are texture and structure (Glatstein and Francisca, 2014).

6.2.2 Soil Chemical Properties

Colloid soil surfaces experience the majority of chemical reactions because of their substantial surface exchange. The three main processes that cause soil–contaminant interactions are sorption, complexion, and precipitation (Tan, 2000). To deal with inorganic pollutants, complexion and precipitation are used. Some trace elements may interact chemically or physically with the native molecules of soils when they are introduced, immobilizing them or producing compounds with limited solubility. Polycyclic aromatic hydrocarbons (PHCs), for example, are organic pollutants that interact with the soil through a process called sorption. Contaminants interact with the interface of the solid soil particles via sorption. This interaction results in a variety of physical manifestations (Volkering *et al.*, 1997).

Physical adsorption takes place when pollutants are drawn to the surface of soil particles, as opposed to chemical adsorption, which is caused by chemical bonding. Physical forces acting on the hydrophobic surface cause PHC to be adsorbed. The primary chemical characteristics of the soil, such as cation exchange capacity, pH, the amount of total and bioavailable elements, and salinity, are all directly impacted by this interaction.

The soil's electrostatic ability to fix and exchange positive ions on the surface is known as its "cation-exchange capacity" (CEC) (Duffy, 2011). The colloids that perform this function do so because they are mostly negatively charged, allowing them to draw cations from the soil (McCauley *et al.*, 2005). The kind and quantity of organic materials and clay that are present affect the CEC. CEC rises with an increase in organic matter (Parker, 2009).

6.2.2.1 Soil pH

The hydrogen ions (H^+) in the soil are measured by the pH of the soil (McCauley *et al.*, 2005). The pH ranges from 0 to 14. Low pH and high H^+ concentration go hand in hand. Acidic soils are those that contain more H^+ than hydroxyl ions (OH^-) (pH less than 7). Alkaline soils have a pH over 7, whereas neutral soils have a pH of 7. By changing the surface charge of the colloids, soil pH may have an impact on the CEC. For instance, the negative charges of a colloid are neutralized by a high concentration of H^+ (McCauley *et al.*, 2005). Additionally, the pH is a crucial chemical characteristic because it affects (i) soil microbial development, (ii) the breakdown of organic pollutants, and (iii) the availability of nutrients and pollutants (Cao *et al.*, 2009).

6.2.2.2 Total and Bioavailable Elements

Typically, the total elements are first determined while studying a soil. The kind of rock that the soil is made of and the processes it has gone through over time both affect its composition (Whalen and Sampedro, 2010). Not all soil components may act as pollutants. A subset of the total soil components engaged in physicochemical and biological processes are accessible for use (Duffy, 2011). PHCs are a mixture of hundreds of distinct compounds, each with a unique availability. The nature of the soil's physico-chemical characteristics, as well as how long it has been polluted, affects the elements that are still available (Chigbo and Batty, 2013).

6.2.2.3 Soil Salinity

There are three main types of salt that may have an effect on soil: saline, sodic, and saline-sodic. Saline soils include a lot of soluble salts such calcium (Ca^{2+}), magnesium (Mg^{2+}), and potassium (K^+), while sodium soils are dominated by sodium ions (Na^+). There are large levels of sodium and salts in saline-sodic soils. These salts have an impact on the plant's structure, porosity, and water content, which lowers production (McCauley *et al.*, 2005). The activity of soil organisms is governed by all of these physicochemical characteristics.

6.2.3 SOIL BIOLOGICAL PROPERTIES

Numerous living forms, including plants (flora), animals (fauna), and microbes, all perform vital roles in the soil ecosystem.

6.2.3.1 Soil Flora

In the creation of soil, plants play the primary function. In terrestrial ecosystems, they are the main producers (Oleszczuk, 2008). Through their leftover leaves and roots, they enrich the soil's organic content, structure, and porosity. Water and air may travel through plant roots. By forming aggregates, they help to maintain the ground's structure and porosity (Duffy, 2011). The soil region with the highest biological activity is the root zone, which is where the roots of plants come into direct contact with the soil. For soil organisms, root exudates serve as a source of nutrients (McCauley *et al.*, 2005).

6.2.3.2 Soil Fauna

Insects, nematodes, arthropods, and earthworms make up the majority of soil organisms. The most significant organisms are earthworms because they create channels that enhance soil porosity, boost biotic activity by breaking down organic materials into minute fragments, and secrete nutrient-rich compounds (McCauley *et al.*, 2005). Animals control the population of bacteria and fungi in the soil, start the decomposition of dead plants and animals, mix the soil layers, and make them more agglomerative (Whalen and Sampedro, 2010).

6.2.3.3 Soil Microorganisms

This is the soil's biggest and most varied group. Bacteria, protozoa, algae, fungus, and actinomycetes are examples of soil microorganisms. The most varied class of soil microorganisms is bacteria. They play a crucial role in processing the aggregation of microscopic soil particles as well as the breakdown of organic substances and minerals. Since rhizobial bacteria are connected to pulse crops and facilitate nitrogen fixation, they are being investigated. In terms of numbers, fungi dominate all other soil microbes (Whalen and Sampedro, 2010). They play a crucial role in both the stability of the aggregates and the degradation of organic materials. Generally speaking, microbes improve soil structure by secreting organic substances, mostly sugars, that bind the soil particles together (McCauley *et al.*, 2005). The hydraulic conductivity is increased in particular by stimulating soil microorganisms by introducing a nutrition source (Glatstein and Francisca, 2014). The rate and destiny of contaminants like trace elements and petroleum hydrocarbons are influenced by all of these soil physicochemical and biological characteristics.

6.3 PETROLEUM-CONTAMINATED SOIL

6.3.1 SOURCES OF PETROLEUM POLLUTANTS

The main source of hydrocarbon pollution in the soil is oil spills. There are 600,000 metric tons of natural petroleum leaks recorded annually on the planet (Abioye, 2011). In Europe,3.5 million sites are thought to have been contaminated by petroleum (Adipah, 2018). About 4.8 million hectares of soil in China may contain more petroleum than is safe (Wang *et al.*, 2017). Different countries and regions have various sampling and transit methods, as well as various petroleum pollution sources and intensities. In addition, rainfall-induced washing and leaching of pollutants into the surface and deep soil in both horizontal and vertical orientations as well as into the groundwater system will lead to accumulation of pollutants.

Groundwater's high permeability and low-molecular-weight hydrocarbon penetration are affected by the soil's physical and chemical properties, as well as by the environment's climate and flora (Adipah, 2018). As the concentration of petroleum hydrocarbons rises, the half-life of natural decay increases (it is 217 days at a petroleum concentration of 250 mg/L) (Kachieng'a and Momba, 2017). Alkane and aromatic pollutants have a natural half-life that lengthens with increasing molecular weight. Phenanthrene, a three-ring molecule, has a half-life of 16 to 126 days under normal circumstances, while benzo[a]pyrene, a five-ring molecule, has a half-life of 229 to 1400 days (Sarma *et al.*, 2019). Some specialized bacteria in contaminated soil may biodegrade and biotransform these hydrocarbons, incorporating them into biomass in the soil, but this is still a challenge owing to the non-polarity and chemical inertness of contaminants (Brakstad *et al.*, 2018).

6.3.2 COMPOSITION OF PETROLEUM POLLUTANTS

Petroleum-polluted soil often contains traces of petroleum, water, and solid particles. Symbolically, oil spills are often represented as the combination of water and petroleum, or W/O. Hydrocarbons, which comprise most petroleum, are composed of carbon (83–87%), hydrogen (11–4%), sulfur (0.06–0.8%), nitrogen (0.02–1.7%), oxygen (0.08–1.82%), and trace metal components (nickel, vanadium, iron, antimony, etc.). The majority (95–99%) of petroleum is made up of hydrocarbons, which are compounds created when carbon and hydrogen combine. Alkanes, cycloalkanes, and aromatic hydrocarbons are three types of hydrocarbons that are categorized based on their structural makeup.

The primary ingredients of gasoline, diesel, and jet fuel are alkanes (Schirmer *et al.*, 2010). The molecular structure is cyclic, linear, and branching. C_nH_{2n+2}, C_nH_{2n+2} (n > 2), and C_nH_{2n} (n > 3) are the generic formulas for linear alkanes, branched alkanes, and cycloalkanes, respectively. Asphalt, tar, lubricants, kerosene, diesel, and lubricants all include aromatic compounds (Zakaria *et al.*, 2018). They share cycloalkanes' molecular makeup, but they also include at least one benzene ring (Sadeghbeigi, 2012). The general formula for aromatics is C_nH_{2n-6}.

Bitumen is refined into petroleum, but it is challenging to release the oil back into the reservoir because the asphaltene's heaviest and most polar molecules are bound to the rock. So, the most common types of hydrocarbons are the least polar ones, such as saturated hydrocarbons, followed by aromatics. The molecular weight of hydrocarbons affects how easily they break down. Low-molecular-weight hydrocarbons are more bioavailable than their high-molecular-weight counterparts (Vasconcelos *et al.*, 2011). When compared to aromatics of low molecular weight, cyclic alkanes, and branching alkanes, linear alkanes are frequently more easily broken down by microorganisms (Imam *et al.*, 2019).

6.4 TOXIC EFFECTS OF PETROLEUM ON THE ENVIRONMENT

Petroleum mostly contains saturates, aromatics, and other toxic and dangerous hydrocarbons (Wang *et al.*, 2017). On soil, plants, and people, highly toxic petroleum pollutants (polycyclic aromatic

Phytoremediation of Crude Oil-Contaminated Soil

hydrocarbons (PAHs) or benzene, toluene, ethylbenzene, and xylene (BTEX)) will have a negative impact. In terrestrial invertebrates, high soil concentrations of polycyclic aromatic hydrocarbons may cause tumors, issues with growth, reproduction, and the immune system (Abdel-Shafy and Mansour, 2016). The neurological system, liver, kidneys, and respiratory system of an individual may be harmed by BTEX (Fooladi *et al.*, 2019). Pollutants block soil pores, change the amount and composition of organic matter in the soil, reduce the diversity and activity of soil microorganisms, and risk human health via the food chain (Wei *et al.*, 2020). Gomez-Eyles *et al.* (2011) used deuterated PAH (dPAH) to assess the bioavailability of PAH in soil (Gomez-Eyles *et al.*, 2011). The dPAH:PAH ratio of benzo[a]pyrene in earthworm tissues is higher than the dPAH:PAH ratio determined by conventional chemical procedures, according to studies. As the amount of PAH increases, the percentage of extra dPAH that earthworms acquire also grows. This shows that animal toxicity to petroleum contamination is substantially greater than previously believed. Through migration and dispersion, the petroleum in the soil also contaminates the groundwater environment, posing a threat to many aspects of human existence.

6.4.1 TOXIC EFFECTS OF PETROLEUM ON SOIL

According to Kumari *et al.* (2018) and Otele and Ogidi (2018), petroleum has a negative impact on the ecological structure and function of soils, significantly altering soil moisture, pH, and enzyme activity (Polyak *et al.*, 2018; Wei and Li, 2018). The amount of clay in contaminated soil increases with increasing pollutant concentrations; while soil porosity decreases, impermeability and hydrophobicity increase (Xin *et al.*, 2012), restricting the development of plant roots and the quantity of bacteria in the soil. When the soil contained 7791 mg/kg of petroleum hydrocarbons, the root length of *Lepidium sativum*, *Sinapis alba*, and *Sorghum saccharatum* decreased by 65.1%, 42.3%, and 47.3%, respectively (Steliga and Kluk, 2020). The number of bacterial species is most influenced by straight-chain alkanes (Mangse *et al.*, 2020). Studies reveal that the main pollutant that contributes to soil salinization and acidification is benzo[a]pyrene, a substance found in petroleum (Buzmakov and Khotyanovskaya, 2020).

6.4.2 TOXIC EFFECTS OF PETROLEUM ON PLANTS

There is a chance that petrochemical pollutants will enter plants via their leaves and spread throughout the tissues and blood vessels. Plant roots may absorb petroleum pollutants from the soil, then deposit them in the plant's leaves, and fruits, and maybe even convey them back to the roots. Corn's germination rate, plant height, leaf area, and dry matter production were all drastically reduced due to oil pollution (Uzoho *et al.*, 2006; Okereke *et al.*, 2016). As a result of a deficiency of oxygen and nutrients in the polluted soil, plant growth is stifled, stem length and diameter are diminished, the length of aboveground tissue is shortened, and the length of the roots and the leaf area of the plant are altered (depending on the plant type) (Lorestani *et al.*, 2012). Low petroleum hydrocarbon concentrations (10 g/kg) were shown to promote plant root vitality, but medium concentrations (30 g/kg) and high concentrations (50 g/kg) were found to diminish it. In addition, the chlorophyll content of soil polluted with 50 g/kg of petroleum is over 60% lower than that of uncontaminated soil (Zhen *et al.*, 2019).

6.4.3 TOXIC EFFECTS OF PETROLEUM ON HUMAN HEALTH

Whether it's via direct contact (polluted air and skin) or indirect contact (swimming in contaminated water and eating contaminated food), exposure to petroleum and petroleum products may cause major health concerns for humans (Kuppusamy *et al.*, 2020; Okereke *et al.*, 2014). Benzene and other polycyclic aromatic hydrocarbons are only two examples of the many petroleum pollutants that are dangerous, mutagenic, and carcinogenic. Certain aromatic compounds have been linked to cancer and to impaired liver and kidney function (Haller and Jonsson, 2020). Furthermore, because of their great lipophilicity, PAHs are readily absorbed via the digestive system of animals (Abdel-Shafy and Mansour, 2016). According to O'Callaghan-Gordo *et al.* (2016), prolonged exposure to polluted environments might result in fatigue, respiratory issues, eye irritation, headaches, and a higher risk of spontaneous miscarriages in females. Many people's health is negatively impacted by home oil extraction, according to studies, particularly in poor and medium income nations. Six hundred and thirty-eight million people in low and middle-income countries live in rural areas close to conventional oil sources (O'Callaghan-Gordo *et al.*, 2016). The elderly, pregnant women, babies, children, and those with preexisting medical disorders are the groups most susceptible to pollution from oil, even though they are less likely to work in these areas and engage in daily activities like farming and bathing that could be affected. Parallel to this, the burning of natural gas in oil wells may result in the production of benzo[a]pyrene, polycyclic aromatic hydrocarbons, nitrogen dioxide (NO_2), sulfur dioxide (SO_2), volatile organic compounds (VOCs), and nitrogen dioxide (NO_2).

6.5 BIOLOGICAL REMEDIATION OF PETROLEUM HYDROCARBON–POLLUTED SOIL

Using living organisms (bacteria, fungus, and plants) to break down dangerous compounds found in the environment is a practice known as bioremediation. The removal of crude oil from the soil via bioremediation is a highly effective, affordable, and green approach. The hydrocarbon content, soil properties, and pollutant makeup all affect how successful this procedure is (Balba *et al.*, 1998). The most hazardous and resilient class of soil contaminants found in crude oil is the PAH family. After entering the soil, PAHs are retained by the soil matrix and get trapped in the soil pores, so it is really challenging to remove them from the soil (Safdari *et al.*, 2018). The best way to remove PAHs from soil is via bioremediation since bacteria and plant roots can readily enter these small holes.

6.5.1 MICROBIAL REMEDIATION

Due to the diversity of microbial populations that live there, soil is a complex ecosystem. Only a resistant consortium of microorganisms survives and actively contributes to the cleansing of contaminated soil because the composition of naturally existing bacteria changes with the content and concentration of toxins (Zhao *et al.*, 2018; Okore *et al.*, 2021). Numerous naturally occurring hydrocarbon-degrading microorganisms may be found in the polluted soil, where they employ their enzymatic systems to convert complex hydrocarbons into simpler forms. The metabolic adaptability of microorganisms is used in microbial remediation to break down harmful contaminants for the ecological recovery of petroleum waste-contaminated locations (Ogidi and Njoku, 2017). Bacteria are often chosen among the microorganisms because of their quick metabolic rates, wide variety of degradation routes, and capacity for genetic manipulation to enhance bioremediation (Prakash and Irfan, 2011).

Numerous bacteria, including Rhodococcus species, Pseudomonas species, and *Scedosporium boydii*, have been shown in studies to be able to breakdown petroleum pollutants (Pi *et al.*, 2018; Yuan *et al.*, 2018). Bacteria primarily break down hydrocarbons via aerobic routes (Wang *et al.*, 2010). Hydrocarbon catabolism is often enhanced when oxygen acts as an electron acceptor (Cao *et al.*, 2009). Degradation in aerobic mode is mediated by the processes of oxidation, reduction, hydroxylation, and dehydrogenation (Sirajuddin and Rosenzweig, 2015; Huang *et al.*, 2014).

In microorganisms, terminal oxidation, subterminal oxidation, ω-oxidation, and β-oxidation are the main routes for alkane and PAH metabolism. The most typical method by which alkanes are destroyed is via the terminal oxidation route. In order to oxidize terminal methyl and create alcohols, alkane hydroxylase adds molecular oxygen to hydrocarbons. These alcohols are next oxidized to create aldehydes and fatty acids, and finally, the -oxidation pathway produces carbon dioxide and water (Oyetibo *et al.*, 2017; Rojo, 2009). PAHs, on the other hand, have stable structural

Phytoremediation of Crude Oil-Contaminated Soil

characteristics that make them resistant to biodegradation. Through a mixed functional oxidase system mediated by the cytochrome P450 enzyme, PAHs are predominantly broken down, with oxidation or hydroxylation serving as the first step and the synthesis of diols serving as the intermediate result. Through ortho- or meta-cleavage routes, these intermediates are transformed into catechol intermediates, which are subsequently incorporated into the tricarboxylic acid cycle (TCA) (Xia *et al.*, 2005).

Additionally, it has been found that *Coriolus versicolor, Pleurotus ostreatus*, and *Phanerochaete chrysosporium* all degrade TPH (Yateem *et al.*, 1997). In addition to bacteria and fungus, it has also been claimed that green algae, cyanobacteria, and brown algae are involved in the breakdown of petroleum hydrocarbons (Cerniglia, 1993). *Prototheca zopfii* that could break down crude oil and several hydrocarbon substrates was identified by researchers. Depending on how readily available they are to microbes for their mineralization, hydrocarbons may be classified in the following order according to how differently they are attacked by the microorganism: Asphaltenes are followed by n-alkanes, branched alkanes, low-molecular-weight aromatics, cyclic alkanes, and high-molecular-weight aromatics (Cerniglia, 1993).

The eradication of environmental pollutants that endure is the most urgent problem facing the world. Due to their hydrophobicity, PAHs have become one of the most serious environmental pollutants (Premnath *et al.*, 2021).

Long-term exposure to low quantities of petroleum hydrocarbons requires PAHs to survive two to four times longer than they do in their native environment. Despite the identification of bacteria that can break down naphthalene, phenanthrene, and pyrene, it is still difficult to break down polycyclic aromatic hydrocarbons with high molecular weight by biodegradation. Most studies on the microbial decomposition of petroleum pollutants have been conducted in the lab, on mineral basal media (liquid), and have not been transferred to actual petroleum-contaminated soil. Although some studies have shown that a single strain may break down petroleum-contaminated soil, there are still issues with a single bioremediation technique, including lengthy repair periods, unstable microbial activity, and inadequate eradication of free microorganisms.

6.5.2 PHYTOREMEDIATION

The process of phytoremediation involves removing pollutants from the environment using plants, soil amendments, and the accompanying microbes (Wang *et al.*, 2012; Gerhardt *et al.*, 2009). By immobilizing, removing, and promoting organic chemicals and microbial decomposition, plants help pollutants dissipate (Megharaj *et al.*, 2011).

Some substances are taken up by plant roots and brought to the tissues of aboveground plants, where they are then collected (phytoextraction) (Pilon-Smits and Freeman, 2006). Organic pollutants may be partially or completely degraded by enzymes via phytodegradation (plant enzyme) or phytostimulation (microbial activity). Contaminants are absorbed by plant roots during phytovolatilization and subsequently released as volatile compounds. Finally, plant roots' phytostabilization prevents mobilization by allowing pollutants to accumulate or precipitate (Pilon-Smits and Freeman, 2006).

6.5.2.1 Phytoremediation of Oil Spillage in Soils

Initially, plants had a high tolerance for soil tainted with crude oil. For instance, the four o'clock flower (*Mirabilis jalapa* L.), which can withstand crude oil pollution, has been effectively used as a phytoremediator (Peng *et al.*, 2009). *M. jalapa* increased the removal effectiveness of total petroleum hydrocarbons (TPHs) during a 127-day period (Peng *et al.*, 2009).

Teak (*Tectona grandis*) and gmelina (*Gmelina arborea*) are examples of forest tree species that have shown adequate capacities to grow well in a polluted environment with crude oil levels up to 10% weight-weight of soil. However, with greater doses of oil treatments, the test plants' biomass and height were severely impacted (Agbogidi *et al.*, 2007). It has also been shown that the native

Brazilian tree known as the branquilho (*Sebastiania commersoniana*) can withstand soil pollution with petroleum. This tree was able to lower levels of petroleum hydrocarbons by up to 94% in soil that had previously been contaminated based on research by Ramos *et al.* (2009). Seven plant species had their early development and seed germination negatively affected by oil field generated water in experimental setups. Results showed that sorghum, okra, millet, and maize had a higher resistance to oil phytotoxicity than other plants (Pardue *et al.*, 2015). Additionally, two plant species, soybean (*G. max*) and corn (*Z. mays*), have shown tolerance to soils polluted with crude oil (Issoufi *et al.*, 2006).

Another trait that makes plants appropriate for phytoremediation is their capacity to bioaccumulate petroleum hydrocarbons. Boonsaner *et al.* reported finding significant levels of BTEX in canna lily (*Canna indica* L.) shoots (2011). In 21 days, canna eliminated 80% of BTEX from the soil in the root zone. Siam weed (*Chromolaena odorata* L.), a tropical ornamental shrub, has a significant capacity for phytoaccumulation in soils polluted with heavy metals and crude oil. These organisms cleaned up to 80% of crude oil from oil- and heavy metal–polluted soil (Atagana, 2011).

Some vegetation's roots have physical and morphological traits that enable them to draw additional microbes to the area surrounding them and promote hydrocarbon breakdown (Ansari *et al.*, 2014). It has been shown that the roots of mulberry (*Morus* spp.), apple (*Malus domestica*), and Osage orange (*Maclura pomifera*) trees emit phenolic and flavonoid chemicals that encourage PAH-degrading bacteria (Atagana, 2011). *Impatiens balsamina* L., sometimes known as garden balsam, has been suggested as a possible decorative plant for the efficient removal of oil from polluted soils. The number of live microorganisms around the plant root increased significantly over the course of the four-month culture period, and they were mostly responsible for the oil deterioration (Cai *et al.*, 2010). Hybrid poplar cuttings (*Populus deltoids Populus nigra*) planted in benzene-filled flow-through reactors increased benzene breakdown, volatilization, and mass reduction in effluents in a laboratory study on phytoremediation (Ojuederie and Babalola, 2017).

Plant species having tap roots, as opposed to surface roots, might influence toxins that are situated deeper in the soil or in the water table (Kang *et al.*, 2016). Trees with deep roots, such as poplars and willows, have been utilized to effectively extract water from groundwater contaminated with total petroleum hydrocarbons (Ferro *et al.*, 2013).

6.5.3 Rhizoremediation (Plant–Microbe Assisted Remediation: Recent Technology)

It has been shown that the ability of plants to clean up crude oil increases when soil microorganisms are present. In a mutually beneficial partnership, plants and soil microorganisms work to break down pollutants utilizing their enzyme systems in exchange for space and a hospitable habitat in the rhizosphere area. Rhizoremediation, a mix of bioremediation and phytoremediation, is the name for this symbiotic connection. Rhizoremediation is now the most efficient and economical method for eliminating all pollutants found in crude oil. This process occurs naturally, but it may also be accelerated by adding certain bacteria.

Rhizoremediation involves plants that can both thrive in soil that has been polluted with oil and provide a suitable environment for the microorganisms that break down contaminants by exudate secretion or aeration. The plant-microbe approach enhances the physical and chemical characteristics of the soil, increases microbial access to the soil's contaminants, and boosts not just the metabolic activity of rhizosphere bacteria (Kuiper *et al.*, 2004) but also soil microbial population.

In a controlled environment over the course of 175 days, the PAH-degrading bacterial strain *Rhodococcus ruber* Em1 demonstrated an increased degradation rate when coupled with *Orychophragmus violaceus* (mesocosms). In the mesocosoms, the expression of the genes linA and RHD that code for PAH-ring hydroxylating dioxygenase increased by three to five times (Kong *et al.*, 2018). When a maize plant received an indigenous microbial biomass inoculum, there was an increase in the amount of pollutants that the maize plant degraded (Segura and Luis, 2012).

Phytoremediation of Crude Oil-Contaminated Soil

One of the plants that has the potential to remediate hydrocarbons is the glycine max (soybean) plant. Research revealed that *G. max* and rhizospheric bacteria worked together to remediate crude oil in soybeans rather than phytoaccumulation. *G. max* growth in polluted soil was shown to have an impact on the overall number of bacteria, amount of water, pH, and amount of organic matter (Njoku *et al.*, 2016). According to a research on wheat plants cultivated in hydroponic environments, wheat seedlings remove more than 20% of the oil from the medium, but when grown in conjunction with Azospirillum, this remediating capacity increases to 29% (Muratova *et al.*, 2018).

Yellow alfalfa with *Acinetobacter* sp. strain S-33 enhanced the bioremediation effectiveness of oil-contaminated soil by 39% when compared to alfalfa alone (34%) and *Acinetobacter* sp. S-33 (35%). According to a fractional contaminants study, plant–microbe association is the most effective method for removing aromatic hydrocarbons from soil (Muratova *et al.*, 2018). Plant tolerance and resistance to soil pollutants are encouraged by bacteria that promote plant growth (PGPR). When 13% TPH concentration was employed, ryegrass cultivated with PGPR demonstrated improved hydrocarbon degradation to 61.5% for 3 years. Low concentration was shown to accelerate degradation and vice versa (Dos Santos and Maranho, 2018; Huang *et al.*, 2005).

After leaking, crude oil becomes physically entangled or stuck with soil particles; plant roots provide access to these tiny gaps. Microbes are found both on the surface and in the rhizosphere of plant roots. Therefore, the root creates a route via which these microorganisms may reach these toxins. GPR produces biosurfactants or adheres to the surface of the oil droplets to boost their solubility after it enters the soil micropores. By incorporating oxygen atoms into PHC, microbial surface membrane oxygenases then produce fatty acid mimics. Microbes continue to multiply and break down pollutants in this manner. Microbes reportedly need 150 mg of nitrogen and 30 mg of potassium to break down 1 g of PHC (Gerhardt *et al.*, 2009). Utilizing plants and microorganisms in tandem is a successful way to restore polluted soil. Although it may take a while, the technique is safer and more environmentally friendly.

6.5.4 Combined Microbial Method Remediation

The three main kinds of microbial combination approaches are microorganism-physical, microorganism-chemical, and microorganism-biology. A range of tools and techniques have been used in the microbial combination approach to break down soil that has been polluted with petroleum. Due to the hydrophobicity and fluidity of petroleum, the majority of remediation combination approaches aim to increase soil microbial activity and aeration. Petroleum-contaminated soil was treated with an electric field, fertilizers, biosurfactants, biocarrier, biochar, and plants in order to speed up the deterioration of the system (Mukome *et al.*, 2020). The effectiveness of microbial degradation of petroleum pollutants may be increased by combining microorganisms with physical or chemical methods. Electric fields, fertilizers, biosurfactants, and biochar may all be added to high-concentration petroleum-contaminated soil (>10,000 mg/kg) to increase the clearance rate of petroleum pollutants to over 60%. The best degradation result was achieved by ryegrass and mixed microbial strains within 162 days, with a breakdown rate of 58% (the original oil content was 6.19%). Alfalfa and microbes may work together to break down 63% of petroleum hydrocarbons in 60 days (the starting oil concentration is 12%) (Tahseen *et al.*, 2016).

6.5.4.1 Advantages and Challenges in Combined Microbial Method Application for Hydrocarbon Removal

The extraction of petroleum has significantly increased the release of harmful pollutants into the soil environment. Oily soil remediation research is now centered on bioremediation because of its benefits, including its simplicity, economic viability, and lack of secondary pollution (Tao *et al.*, 2016). Microorganisms' biological activity, stability, and capability to break down petroleum pollutants are all improved by the addition of charcoal, nutrients, and plants. The advantages of combining three microbiological techniques are as follows. Following restoration, these techniques will really

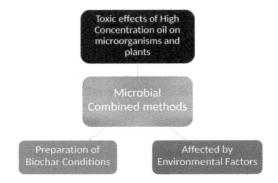

FIGURE 6.1 Some challenges of three microbial combined methods.

make the physical, chemical, and biological features of the soil ecosystem better. Additionally, without introducing secondary pollution, they may convert organic pollutants (such as carbon dioxide and water) into completely non-polluting inorganic compounds.

Combining oily soil remediation with microbial biomass and the quantity of PAH-degrading bacteria led to much higher levels of soil enzyme activity, microbial biomass, and PAH-degrading bacteria than did any of the other treatments studied (Li et al., 2021; Chakravarty and Deka, 2021). According to a Biology study, soil microbial communities' diversity, richness, and homogeneity have changed as a result of restoration (Dekang et al., 2017). According to a DNA sequencing research of soil microbial diversity, joint restoration has increased the genetic diversity of soil microbial populations (Iffis et al., 2017). Few strains are used in engineering repair, and the three repair methods are still just at the laboratory stage. There is still much to learn about many contributing factors and degradation processes; thus additional research is necessary. Some of the issues with the three integrated microbiological approaches are outlined in Figure 6.1.

The thermal degradation of biomass and the characteristics of biochar are most sensitive to the pyrolysis temperature. Biochar's elemental composition, pore structure, surface area, and functional groups are only some of the physicochemical features and structures that are influenced by the pyrolysis temperature (Yaashikaa et al., 2020). Biochar is rich in oxygen-containing functional groups when pyrolysis is performed between 300 and 500°C. There are fewer oxygen-containing functional groups, more minerals, and more micropores when the pyrolysis temperature is between 500 and 700°C (Zhang et al., 2019). These characteristics dictate how they interact with their surroundings. Additionally, the stability of the carbon in the biochar and carbon retention throughout the pyrolysis process are impacted by the pyrolysis temperature (Nan et al., 2021).

Past studies have shown that concentrations of petroleum pollution in soil lower than 5% are optimal for the majority of soil microbes and plants involved in soil regeneration. There have been many studies on this topic (Iqbal et al., 2019; Cai et al., 2016). Any increase (even by 5%) in the quantity of petroleum pollutants in the soil has an immediate negative effect on the capacity of microorganisms and plants to clean up the environment. *Festuca arundinacea* (TPH) was removed by 64.0% to 1.6% and biological flora by 54.6% to 1.3% after 90 days of restoration. At first, there was 1.21% oil present and after 70 days of restoration, tall fescue can remove 48.4% of oil pollution from soils with a 5% oil concentration (Tang et al., 2010). When compared to the control group, ryegrass had a lower stem and root biomass after 90 days of healing (Abedi-Koupai et al., 2007).

After 70 days of remediation, microorganisms may remove 15% of petroleum pollutants from soil with an oil content of 5.6% (Zhang et al., 2020). Plants and microorganisms are put at risk when soil oil concentrations are too high, since they become less effective at degrading petroleum contaminants and may potentially perish as a result. Soil temperatures below 10°C and pH values below 4 and above 9 might hinder the ability of microorganisms and plants to degrade petroleum

Phytoremediation of Crude Oil-Contaminated Soil   121

pollutants. Alterations in soil pH and either abiotic or biodegradation of the biochar will hasten the desorption of PAH from biochar into sediments. Despite having little secondary pollution and being inexpensive, microbial remediation still faces several difficulties.

6.6 FACTORS INFLUENCING PETROLEUM HYDROCARBON SOIL REMEDIATION

Numerous elements, which may be categorized into three types, influence the amount and pace of bioremediation.

   i. Physical and chemical factors (chemical composition, pollution concentrations, and the physical and chemical characteristics of the soil);
   ii. Environmental factors (temperature, pH, oxygen, and nutrient availability);
   iii. Microorganisms (genetic makeup, variety, and interactions of microorganisms) (Haritash and Kaushik, 2009; Singh, 2006).

As a result, bioremediation is limited by the heterogeneity of contaminants, their concentration, and the diversity of environmental circumstances. The next section discusses a few of the most crucial elements.

6.6.1 BIOLOGICAL FACTORS

Native species have a greater potential for degradation compared to plants and microorganisms, which can break down a wide range of contaminants. The selection of appropriate species that remove, stabilize, or collect contaminants and create substantial volumes of biomass is essential for the effectiveness of the bioremediation approach (Tangahu *et al.*, 2011). Genetic adaptation may emerge from the acclimatization of microbes as a consequence of exposure to increasing amounts of contaminants (Haritash and Kaushik, 2009). Hydrocarbon biodegradation is influenced by the enzymes released by microorganisms since most of them have substrate specificity and are active at various temperatures (Haritash and Kaushik, 2009).

6.6.2 PHYSICO-CHEMICAL FACTORS

Pollutant bioavailability is a crucial component in bioremediation. The equilibrium between the quantity of the pollutant adsorbed and the amount of dissolved organic matter determines bioavailability. In the case of PAH, the reduction in solubility and bioavailability is caused by an increase in molecular weight (Haritash and Kaushik, 2009). Instead of being related to the slow rate of microorganism destruction, the slow rate of remediation in soil is mostly caused by the slow rate of desorption of contaminants from the soil particles (Haritash and Kaushik, 2009). Pollutant concentration also affects the pace of biodegradation since able bacteria cannot produce degradation enzymes at low pollutant concentrations. On the other hand, organisms are poisonous to compounds in large quantities (Borja *et al.*, 2005).

Studies have shown that the presence of PAHs in a mixture results in interaction effects that might alter the pace at which certain PAHs degrade. For instance, phenanthrene prevented microorganisms from breaking down pyrene. The addition of fluoranthene also hindered Rhodococcus sp.'s ability to degrade anthracene (Dean-Ross *et al.*, 2002). Phenanthrene, on the other hand, encouraged *Pseudomonas putida* strain KBM-1 to degrade pyrene and phenanthrene. Researchers Yuan *et al.* (2002) found that the rate of biodegradation was faster when fluorene, phenanthrene, acenaphthene, anthracene, and pyrene were present together rather than individually.

Biodegradation of PAHs with more than three rings was impeded by a lack of oxygen (Wiegel and Wu, 2000). Furthermore, it is well known that the presence of nutrients as carbon sources in polluted soil accelerates the breakdown of hydrocarbons. To improve cleanup, agronomic procedures such oxygen and pH correction, chelator addition, and fertilizer use have been established. Compost, for instance, may be added to improve hydrocarbon decomposition by feeding the microbial population, transferring oxygen, preventing the production of extractable polar intermediates, and improving soil texture (Haritash and Kaushik, 2009). Because it affects the adsorption and desorption processes, pH is crucial for organic and inorganic cleanup (Borja *et al.*, 2005).

6.7 CONCLUSION

The majority of nations have access to crude oil, which is a rapid and convenient energy source. Because it includes mutagenic and carcinogenic substances, spillage during extraction and transportation has put the environment in peril. It is crucial to get rid of soil pollution brought on by crude oil leaks since it negatively affects human health and plant development. There are several techniques, including physical, chemical, thermal, and biological ones, that have been developed to extract crude oil from soil. In order to improve their effectiveness and lessen their drawbacks, physio-chemical and thermal procedures have undergone several changes and developments. However, these approaches are less favored by society and have several disadvantages. However, bioremediation techniques are chosen because of their effectiveness, affordability, and environmental friendliness.

REFERENCES

Abbaspour, Ali, Farshad Zohrabi, Vajiheh Dorostkar, Angel Faz, and Jose A. Acosta. 2020. "Remediation of an Oil-Contaminated Soil by Two Native Plants treated with Biochar and Mycorrhizae." *Journal of Environmental Management* 2, no. 54: 109755. https://doi.org/10.1016/j.jenvman.2019.109755.

Abdel-Shafy, H. I., and M. S. M. Mansour. 2016. "A Review on Polycyclic Aromatic Hydrocarbons: Source, Environmental Impact, Effect on human Health and Remediation." *Egyptian Journal of Petroleum* 25, no. 1: 107–23. https://doi.org/10.1016/j.ejpe.2015.03.011.

Abedi-Koupai, J., M. Vossoughi-Shavari, S. Yaghmaei, M. Borghei, and A. R. Ezzatian. 2007. "The Effects of Microbial Population on Phytoremediation of Petroleum Contaminated Soils Using Tall Fescue." *International Journal of Agriculture & Biology* 9: 242–6.

Abioye, O. P. 2011. "Biological Remediation of Hydrocarbon and Heavy Metals Contaminated Soil." *Journal of Soil Contamination* 7: 127–42.

Adipah, S. 2018. "Introduction of Petroleum Hydrocarbons Contaminants and Its Human Effects." *Journal of Environmental Science & Public Health* 03, no. 1: 001–9. https://doi.org/10.26502/jesph.96120043.

Agbogidi, M. O., D. E. Dolor, and M. E. Okechukwu. 2007. "Evaluation of *Tectona grandis* (Linn.) and *Gmelina arborea* (Roxb.) for Phytoremediation in Crude Oil Contaminated Soils." *Agriculture Conspectus Scientificus (ACS)*72, no. 2: 149–52.

Anh, B. T. K., N. T. H. Ha, L. T. Danh, V. Van-Minh, and D. D. Kim. 2017. "Phytoremediation Applications for Metal-Contaminated Soils Using Terrestrial Plants in Vietnam." *Phytoremediation*: 157–81.

Ansari, A. A., S. S. Gill, R. Gill, G. R. Lanza, and L. Newman. 2014. *Phytoremediation: Management of Environmental Contaminants*. Vol. 1. Cham, Heidelberg, New York, Dordrecht, and London: Springer.

Ansari, N., M. Hassanshahian, and H. Ravan. 2018. "Study the Microbial Communities' Changes in Desert and Farmland Soil After Crude Oil pollution." *International Journal of Environmental Research* 12, no. 3: 391–8. https://doi.org/10.1007/s41742-018-0099-6.

Atagana, H. I. 2011. "Bioremediation of Co-contamination of Crude oil and Heavy Metals in Soil by Phytoremediation Using *Chromolaena odorata* (L.) King and HE Robinson." *Water, Air, & Soil Pollution* 215, no. 1–4: 261–71. https://doi.org/10.1007/s11270-010-0476-z.

Balba, M. T., N. Al-Awadhi, and R. Al-Daher. 1998. "Bioremediation of Oil-Contaminated Soil: Microbiological Methods for Feasibility Assessment and Field Evaluation." *Journal of Microbiological Methods* 32, no. 2: 155–64. https://doi.org/10.1016/S0167-7012(98)00020-7.

Boonsaner, M., S. Borrirukwisitsak, and A. Boonsaner. 2011. "Phytoremediation of BTEX Contaminated Soil by *Canna generalis*." *Ecotoxicology & Environmental Safety* 74, no. 6: 1700–7. https://doi.org/10.1016/j.ecoenv.2011.04.011.

Borja, J., D. M. Taleon, J. Auresenia, and S. Gallardo. 2005. "Polychlorinated Biphenyls and Their Biodegradation." *Process Biochemistry* 40, no. 6: 1999–2013. https://doi.org/10.1016/j.procbio.2004.08.006.

Brakstad, Odd G., Deni Ribicic, Anika Winkler, and Roman Netzer. 2018. "Biodegradation of Dispersed Oil in Seawater Is Not Inhibited by a Commercial Oil Spill Dispersant." *Marine Pollution Bulletin* 129, no. 2: 555–61. https://doi.org/10.1016/j.marpolbul.2017.10.030.

Buzmakov, S. A., and Y. V. Khotyanovskaya. 2020. "Degradation and Pollution of Lands Under the Influence of Oil Resources Exploitation." *Applied Geochemistry* 113: 104443. https://doi.org/10.1016/j.apgeochem.2019.104443.

Cai, B., J. Ma, G. Yan, X. Dai, M. Li, and S. Guo. 2016. "Comparison of Phytoremediation, Bioaugmentation and Natural Attenuation for Remediating Saline Soil Contaminated by Heavy Crude Oil." *Biochemical Engineering Journal* 112: 170–7. https://doi.org/10.1016/j.bej.2016.04.018.

Cai, Zhang, Qixing Zhou, Shengwei Peng, and Kenan Li. 2010. "Promotion Biodegradation and Microbiological Effects of Petroleum Hydrocarbons by *Impatiens balsamina* L. with Strong Endurance." *Journal of Hazardous Materials* 183, no. 1–3: 731–7. https://doi.org/10.1016/j.jhazmat.2010.07.087.

Cao, Bin, Karthiga Nagarajan, and Kai-Chee Loh. 2009. "Biodegradation of Aromatic compounds: Current Status and Opportunities for Biomolecular Approaches." *Applied Microbiology & Biotechnology* 85, no. 2: 207–28. https://doi.org/10.1007/s00253-009-2192-4.

Cerniglia, C. E. 1993. "Biodegradation of Polycyclic Aromatic Hydrocarbons." *Current Opinion in Biotechnology* 4, no. 3: 331–8. https://doi.org/10.1016/0958-1669(93)90104-5.

Chaîneau, C. H., G. Rougeux, C. J. Yéprémian, and J. Oudot. 2005. "Effects of Nutrient Concentration on the Biodegradation of Crude Oil and Associated Microbial Populations in the Soil." *Soil Biology & Biochemistry* 37, no. 8: 1490–7. https://doi.org/10.1016/j.soilbio.2005.01.012.

Chakravarty, Paramita, and Hemen Deka. 2021. "Enzymatic Defense of Cyperus brevifolius in Hydrocarbons Stress Environment and Changes in Soil Properties." *Scientific Reports* 11, no. 1: 718. https://doi.org/10.1038/s41598-020-80854-5.

Chigbo, Chibuike, and Lesley Batty. 2013. "Phytoremediation Potential of *Brassica juncea* in Cu-Pyrene Co-contaminated Soil: Comparing Freshly Spiked Soil with Aged Soil." *Journal of Environmental Management* 129: 18–24. https://doi.org/10.1016/j.jenvman.2013.05.041.

Cunningham, S. D., T. A. Anderson, A. P. Paul Schwab, and F. C. Hsu. 1996. "Phytoremediation of Soils Contaminated with Organic Pollutants." *Advances in Agronomy* 56: 55–114. https://doi.org/10.1016/S0065-2113(08)60179-0.

Das, N., and P. Chandran. 2011. "Microbial Degradation of Petroleum Hydrocarbon Contaminants: An Overview." *Biotechnology Research International* 10: 1–13.

Dean-Ross, Deborah, Joanna Moody, and C. E. Cerniglia. 2002. "Utilization Of mixtures of Polycyclic Aromatic Hydrocarbons by Bacteria Isolated from Contaminated Sediment." *FEMS Microbiology Ecology* 41, no. 1: 1–7. https://doi.org/10.1111/j.1574-6941.2002.tb00960.x.

Dekang, K., W. Hongqi, L. Zili, X. Jie, and X. Ying. 2017. "Remediation of Petroleum Hydrocarbon Contaminated Soil by Plant-Microbe and the Change of Rhizosphere Microenvironment." *Asian Journal of Ecotoxicology* 12: 644–51.

Dos Santos, Jéssica Janzen, and Leila Teresinha Maranho. 2018. "Rhizospheric Microorganisms as a Solution for the Recovery of Soils Contaminated by Petroleum: A Review." *Journal of Environmental Management* 210: 104–13. https://doi.org/10.1016/j.jenvman.2018.01.015.

Duffy, S. J. 2011. *Environmental Chemistry: A Global Perspective*. Oxford: Oxford University Press.

Ebadi, A., N. A. K. Sima, M. Olamaee, M. Hashemi, and R. G. Nasrabadi. 2018. "Remediation of Saline Soils Contaminated with Crude Oil Using the Halophyte Salicornia persica in Conjunction with Hydrocarbon-Degrading Bacteria." *Journal of Environment Management* 219: 260–8.

Ekundayo, E. O., T. O. Emede, and D. I. Osayande. 2001. "Effects of Crude Oil Spillage on Growth and Yield of Maize (*Zea mays* L.) in Soils of Midwestern Nigeria." *Plant Foods for Human Nutrition* 56, no. 4: 313–24. https://doi.org/10.1023/a:1011806706658.

Ferro, Ari M., Tareq Adham, Brett Berra, and David Tsao. 2013. "Performance of Deep-Rooted Phreatophytic Trees at a Site Containing Total Petroleum Hydrocarbons." *International Journal of Phytoremediation* 15, no. 3: 232–44. https://doi.org/10.1080/15226514.2012.687195.

Fooladi, M., R. Moogouei, S. A. Jozi, F. Golbabaei, and G. Tajadod. 2019. "Phytoremediation of BTEX from Indoor Air by Hyrcanian Plants." *Environmental Health Engineering & Management* 6, no. 4: 233–40. https://doi.org/10.15171/EHEM.2019.26.

Freedman, B. 1995. "Oil Pollution." In *Environmental Ecology*. 2nd ed.: 159–88. San Diego, CA: Academic Press.

Gerhardt, K. E., X. D. Huang, B. R. Glick, and B. M. Greenberg. 2009. "Phytoremediation and Rhizoremediation of Organic Soil Contaminants: Potential and Challenges." *Plant Science* 176, no. 1: 20–30. https://doi.org/10.1016/j.plantsci.2008.09.014.

Giannopoulos, D., D. I. Kolaitis, A. Togkalidou, G. Skevis, and M. A. Founti. 2007. "Quantification of Emissions from the Co-incineration Of cutting Oil Emulsions in Cement Plants—Part II: Trace Species." *Fuel* 86, no. 16: 2491–501. https://doi.org/10.1016/j.fuel.2007.02.034.

Glatstein, Daniel Alejandro, and Franco Matías Francisca. 2014. "Hydraulic Conductivity of Compacted Soils Controlled by Microbial Activity." *Environmental Technology* 35, no. 13–16: 1886–92. https://doi.org/10.1080/09593330.2014.885583.

Gomez-Eyles, Jose L., Chris D. Collins, and Mark E. Hodson. 2011. "Using Deuterated PAH Amendments to Validate Chemical Extraction Methods To predict PAH Bioavailability in Soils." *Environmental Pollution* 159, no. 4: 918–23. https://doi.org/10.1016/j.envpol.2010.12.015.

Haller, Henrik, and Anders Jonsson. 2020. "Growing Food in Polluted Soils: A Review of Risks and Opportunities Associated with Combined Phytoremediation and Food Production (CPFP)." *Chemosphere* 254: 126826. https://doi.org/10.1016/j.chemosphere.2020.126826.

Haritash, A. K., and C. P. Kaushik. 2009. "Biodegradation Aspects of Polycyclic Aromatic Hydrocarbons (PAHs): A Review." *Journal of Hazardous Materials* 169, no. 1–3: 1–15. https://doi.org/10.1016/j.jhazmat.2009.03.137.

Hegazy, A. K., N. H. Mohamed, Y. M. Moustafa, and A. A. Hamad. 2015. "Phytoremediation of Soils Polluted with Crude Petroleum Oil Using *Bassia scoparia* and Its Associated Rhizosphere Microorganisms." *International Journal of Biodeterioration & Biodegradation* 98: 113–20.

Huang, H., D. Shen, N. Li, D. Shan, J. Shentu, and Y. Zhou. 2014. "Biodegradation of 1, 4-Dioxane by a Novel Strain and Its Biodegradation Pathway." *Water, Air, & Soil Pollution* 225, no. 9: 2135. https://doi.org/10.1007/s11270-014-2135-2.

Huang, X. D., Y. El-Alawi, J. Gurska, B. R. Glick, and B. M. Greenberg. 2005. "A Multi- Process Phytoremediation System for Decontamination of Persistent Total Petroleum Hydrocarbons (TPHs) from Soils." *Microchemical Journal* 81, no. 1: 139–47. https://doi.org/10.1016/j.microc.2005.01.009.

Iffis, Bachir, Marc St-Arnaud, and Mohamed Hijri. 2017. "Petroleum Contamination and Plant Identity Influence Soil and Root Microbial Communities While AMF Spores Retrieved from the Same Plants Possess Markedly Different Communities." *Frontiers in Plant Science* 8: 1381. https://doi.org/10.3389/fpls.2017.01381.

Imam, A., S. K. Suman, D. K. Ghosh, and P. K. Kanaujia. 2019. "Analytical Approaches Used in Monitoring the Bioremediation of Hydrocarbons in Petroleum-Contaminated Soil and Sludge." *TrAC Trends in Analytical Chemistry* 118: 50–64. https://doi.org/10.1016/j.trac.2019.05.023.

Iqbal, Aneela, Maitreyee Mukherjee, Jamshaid Rashid, Saud Ahmed Khan, Muhammad Arif Ali, and Muhammad Arshad. 2019. "Development of Plant-Microbe Phytoremediation System for Petroleum Hydrocarbon Degradation: An Insight from Alkb Gene Expression and Phytotoxicity Analysis." *Science of the Total Environment* 671: 696–704. https://doi.org/10.1016/j.scitotenv.2019.03.331.

Issoufi, I., R. L. Rhykerd, and K. D. Smiciklas. 2006. "Seedling Growth of Agronomic Crops in Crude Oil Contaminated Soil." *Journal of Agronomy & Crop Science* 192, no. 4: 310–7. https://doi.org/10.1111/j.1439-037X.2006.00212.x.

Kachieng'a, L., and M. N. B. Momba. 2017. "Kinetics of Petroleum Oil Biodegradation by a Consortium of Three Protozoan Isolates (*Aspidisca* sp., *Trachelophyllum* sp. and *Peranema* sp.)." *Biotechnology Reports* 15: 125–31. https://doi.org/10.1016/j.btre.2017.07.001.

Kang, C. H., Y. J. Kwon, and J. S. So. 2016. "Bioremediation of Heavy Metals by Using Bacterial Mixtures." *Ecological Engineering* 89: 64–9. https://doi.org/10.1016/j.ecoleng.2016.01.023.

Kong, Fan-Xin, Guang-Dong Sun, and Zhi-Pei Liu. 2018. "Degradation of Polycyclic Aromatic Hydrocarbons in Soil Mesocosms by Microbial/Plant Bioaugmentation: Performance and Mechanism." *Chemosphere* 198: 83–91. https://doi.org/10.1016/j.chemosphere.2018.01.097.

Kuiper, Irene, Ellen L. Lagendijk, Guido V. Bloemberg, and Ben J. J. Lugtenberg. 2004. "Rhizoremediation: A Beneficial Plant-Microbe Interaction." *Molecular Plant–Microbe Interactions* 17, no. 1: 6–15. https://doi.org/10.1094/MPMI.2004.17.1.6.

Kumari, Smita, Raj Kumar Regar, and Natesan Manickam. 2018. "Improved Polycyclic Aromatic Hydrocarbon Degradation in a Crude Oil by Individual and A consortium of Bacteria." *Bioresource Technology* 254: 174–9. https://doi.org/10.1016/j.biortech.2018.01.075.

Kuppusamy, S., M. N. Raju, M. Mallavarapu, and V. Kadiyala. 2020. "Impact of Total Petroleum Hydrocarbons on Human Health." In *Total Petroleum Hydrocarbons* 139–65. Cham: Springer.

Li, Xiaokang, Jinling Li, Chengtun Qu, Tao Yu, and Mingming Du. 2021. "Bioremediation of Clay with High Oil Content and Biological Response After Restoration." *Scientific Reports* 11, no. 1: 9725. https://doi.org/10.1038/s41598-021-88033-w.

Liao, C., W. Xu, G. Lu, F. Deng, X. Liang, C. Guo, and Z. Dang. 2016. "Biosurfactant-Enhanced Phytoremediation of Soils Contaminated by Crude Oil Using Maize (Zea Mays. L)." *Ecological Engineering* 92: 10–7. https://doi.org/10.1016/j.ecoleng.2016.03.041.

Liao, Xiaoyong, Zeying Wu, You Li, Hongying Cao, and Chunming Su. 2019. "Effect of Various Chemical Oxidation Reagents on Soil Indigenous Microbial Diversity in Remediation of Soil Contaminated by PAHs." *Chemosphere* 226: 483–91. https://doi.org/10.1016/j.chemosphere.2019.03.126.

Lorestani, B., N. Kolahchi, M. Ghasemi, M. Cheraghi, and N. Yousefi. 2012. "Survey the Effect of Oil Pollution on Morphological Characteristics in Faba Vulgaris and Vicia ervilia." *Journal of Chemical Health Risks* 26727: 2251.

Maletíc, S., B. Dalmacija, and S. Rončevíc. 2013. "Petroleum Hydrocarbon Biodegradability in Soil– Implications for Bioremediation." *Intech* 43: 43–64.

Mangse, George, David Werner, Paola Meynet, and Chukwuma C. Ogbaga. 2020. "Microbial Community Responses to Different Volatile Petroleum Hydrocarbon class Mixtures in an Aerobic Sandy Soil." *Environmental Pollution* 264: 114738. https://doi.org/10.1016/j.envpol.2020.114738.

McCauley, A., J. Clain, and J. Jeff. 2005. *Basic Soil Properties*. Bozeman: Montana State University Extension Services, Montana State University.

Megharaj, Mallavarapu, Balasubramanian Ramakrishnan, Kadiyala Venkateswarlu, Nambrattil Sethunathan, and Ravi Naidu. 2011. "Bioremediation Approaches for Organic Pollutants: A Critical Perspective." *Environment International* 37, no. 8: 1362–75. https://doi.org/10.1016/j.envint.2011.06.003.

Mukome, Fungai N. D., Maya C. Buelow, Junteng Shang, Juan Peng, Michael Rodriguez, Douglas M. Mackay, Joseph J. Pignatello, Natasha Sihota, Thomas P. Hoelen, and Sanjai J. Parikh. 2020. "Biochar Amendment as a Remediation Strategy for Surface Soils Impacted by Crude Oil." *Environmental Pollution* 265, no. B: 115006. https://doi.org/10.1016/j.envpol.2020.115006.

Muratova, A. Y., L. V. Panchenko, D. V. Semina, S. N. Golubev, and O. V. Turkovskaya. 2018. "New Strains of Oil- Degrading Microorganisms for Treating Contaminated Soils and Wastes." In *IOP Conference Series: Earth & Environmental Science* 107: 012–66. https://doi.org/10.1088/1755-1315/107/1/012066.

Nan, Hongyan, Jianxiang Yin, Fan Yang, Ying Luo, Ling Zhao, and Xinde Cao. 2021. "Pyrolysis Temperature-Dependent Carbon Retention and Stability of Biochar with Participation of Calcium: Implications to Carbon Sequestration." *Environmental Pollution* 287: 117566. https://doi.org/10.1016/j.envpol.2021.117566.

Njoku, K., M. Akinola and B. Oboh. 2016. Phytoremediation of Crude Oil Contaminated Soil Using Glycine max (Merril); Through Phytoaccumulation or Rhizosphere Effect? *Journal of Biological & Environmental Sciences* 10(30): 115–124.

Njoku, K. L., M. O. Akinola, and B. O. Oboh. 2008. "Does Crude Oil Affect the pH, Moisture and Organic Content of Soils?" *Ecology, Environment & Conservation* paperNumber: 14(04): 731–736.

O'Callaghan-Gordo, Cristina, Martí Orta-Martínez, and Manolis Kogevinas. 2016. "Health Effects of Non-occupational Exposure to Oil Extraction." *Environmental Health: A Global Access Science Source* 15: 56. https://doi.org/10.1186/s12940-016-0140-1.

Ogidi, O. I., and O. C. Njoku. 2017. "A Review on the Possibilities of the Application of Bioremediation Methods in the Oil Spill Clean-Up of Ogoni Land." *International Journal of Biological Sciences & Technology* 9, no. 6: 48–59.

Ojuederie, Omena Bernard, and Olubukola Oluranti Babalola. 2017. "Microbial and Plant-Assisted Bioremediation of Heavy Metal Polluted Environments: A Review." *International Journal of Environmental Research & Public Health* 14, no. 12: 1504. https://doi.org/10.3390/ijerph14121504.

Okereke, J. N., O. I. Ogidi, and A. A. Nwachukwu. 2014. "Environmental Challenges Associated with Oil Spillage and Gas Flaring in Nigeria: A Review." *International Journal for Environmental Health & Human Development* 15, no. 2: 1–11.

Okereke, J. N., A. C. Udebuani, A. A. Ukaoma, K. O. Obasi, O. I. Ogidi, and U. C. Onyekachi. 2016. "Performance of Zea mays On Soil Contaminated with Petroleum (Oily) Sludge." *International Journal of Innovative Research & Advanced Studies (IJIRAS)* 3, no. 10: 159–66.

Okore, C. C., T. E. Ogbulie, O. I. Ogidi, C. Ejiogud, P. Duruojinkeya, I. B. Ogbuka, and J. U. Ajoku. 2021. "Plasmid Profiles of Bacterial Isolates from Kerosene, Diesel and Crude Oil Polluted Soils." *Global Scientific Journal* 9, no. 7: 1994–2005.

Oleszczuk, Patryk. 2008. "Phytotoxicity of Municipal Sewage Sludge Composts Related to Physico-chemical Properties, PAHs and Heavy Metals." *Ecotoxicology & Environmental Safety* 69, no. 3: 496–505. https://doi.org/10.1016/j.ecoenv.2007.04.006.

Otele, A., and O. I. Ogidi. 2018. "Crude Oil Spillage and Its Effect on Soil Biodegradation." *International Journal of Innovative Environmental Studies Research* 6, no. 2: 17–23.

Oyetibo, G. O., M. F. Chien, W. Ikeda-Ohtsubo, H. Suzuki, O. S. Obayori, S. A. Adebusoye, M. O. Ilori, O. O. Amund, and G. Endo. 2017. "Biodegradation of Crude Oil and Phenanthrene by Heavy Metal Resistant *Bacillus subtilis* Isolated from a Multi-polluted Industrial Wastewater Creek." *International Biodeterioration & Biodegradation* 120: 143–51. https://doi.org/10.1016/j.ibiod.2017.02.021.

Ozigis, Mohammed S., Jorg D. Kaduk, Claire H. Jarvis, Polyannada Conceição Bispo, and Heiko Balzter. 2020. "Detection of Oil Pollution Impacts on Vegetation Using Multifrequency SAR, Multispectral images with Fuzzy Forest and Random Forest Methods." *Environmental Pollution* 256: 113360. https://doi.org/10.1016/j.envpol.2019.113360.

Pardue, Michael J., James W. Castle, John H. Rodgers, and George M. Huddleston. 2015. "Effects of Simulated Oil field Produced Water on Early Seedling Growth After Treatment in a Pilot-Scale Constructed Wetland System." *International Journal of Phytoremediation* 17, no. 1–6: 330–40. https://doi.org/10.1080/1522 6514.2014.910168.

Paria, Santanu. 2008. "Surfactant-Enhanced Remediation of Organic Contaminated Soil and Water." *Advances in Colloid & Interface Science* 138, no. 1: 24–58. https://doi.org/10.1016/j.cis.2007.11.001.

Parker, R. 2009. *Plant and Soil Science: Fundamentals and Applications.* Noida: Cengage Learning.

Peng, Shengwei, Qixing Zhou, Zhang Cai, and Zhineng Zhang. 2009. "Phytoremediation of Petroleum Contaminated Soils by *Mirabilis jalapa* L." *Journal of Hazardous Materials* 168, no. 2–3: 1490–6. https://doi.org/10.1016/j.jhazmat.2009.03.036.

Pi, Y., X. Li, Q. Xia, J. Wu, Y. Li, J. Xiao, and Z. Li. 2018. "Adsorptive and Photocatalytic Removal of Persistent Organic Pollutants (POPs) in Water by Metal-Organic Frameworks (MOFs)." *Chemical Engineering Journal* 337: 351–71. https://doi.org/10.1016/j.

Pilon-Smits, E. A. H., and J. L. Freeman. 2006. "Environmental Cleanup Using Plants: Biotechnological Advances and Ecological Considerations." *Frontiers in Ecology & the Environment* 4, no. 4: 203–10. https://doi.org/10.1890/1540-9295(2006)004[0203:ECUPBA]2.0.CO;2.

Polyak, Y. M., L. G. Bakina, M. V. Chugunova, N. V. Mayachkina, A. O. Gerasimov, and V. M. Bure. 2018. "Effect of Remediation Strategies on Biological Activity Of oil-Contaminated Soil—A Field Study." *International Biodeterioration & Biodegradation* 126: 57–68. https://doi.org/10.1016/j.ibiod.2017.10.004.

Prakash, B., and M. Irfan. 2011. "Pseudomonas aeruginosa Is Present in Crude Oil Contaminated Sites of Barmer Region (India)." *Journal of Bioremediation & Biodegradation* 2: 129.

Premnath, N., K. Mohanrasu, R. G. R. Guru Raj Rao, G. H. Dinesh, G. Siva Prakash, V. Ananthi, Kumar Ponnuchamy, Govarthanan Muthusamy, and A. Arun. 2021. "A Crucial Review on Polycyclic Aromatic Hydrocarbons–Environmental Occurrence and Strategies for Microbial Degradation." *Chemosphere* 280: 130608. https://doi.org/10.1016/j.chemosphere.2021.130608.

Ramos, D. T., L. T. Maranho, A. F. L. Godoi, M. A. da Silva Carvalho Filho, L. G. Lacerda, and E. C. de Vasconcelos. 2009. "Petroleum Hydrocarbons Rhizodegradation by *Sebastiania Commersoniana* (BAILL.) L. B. SM. and Downs." *Water, Air, & Soil Pollution* 9, no. 3–4: 293–302.

Rojo, Fernando. 2009. "Degradation of Alkanes by Bacteria." *Environmental Microbiology* 11, no. 10: 2477–90. https://doi.org/10.1111/j.1462-2920.2009.01948.x.

Sadeghbeigi, R. 2012. "FCC Feed Characterization." In *Fluid Catalytic Cracking Handbook.* 3rd ed. Oxford, UK: Butterworth-Heinemann.

Safdari, Mohammad-Saeed, Hamid-Reza Kariminia, Mahmood Rahmati, Farhad Fazlollahi, Alexandra Polasko, Shaily Mahendra, W. Vincent Wilding, and Thomas H. Fletcher. 2018. "Development of Bioreactors for Comparative Study of Natural Attenuation, Biostimulation, and Bioaugmentation of Petroleum-Hydrocarbon Contaminated Soil." *Journal of Hazardous Materials* 342: 270–8. https://doi.org/10.1016/j.jhazmat.2017.08.044.

Sarma, H., A. R. Nava, and M. N. V. Prasad. 2019. "Mechanistic Understanding and Future Prospect of Microbe-Enhanced Phytoremediation of Polycyclic Aromatic Hydrocarbons in Soil." *Environmental Technology & Innovation* 13: 318–30. https://doi.org/10.1016/j.eti.2018.12.004.

Schirmer, Andreas, Mathew A. Rude, Xuezhi Li, Emanuela Popova, and Stephen B. B. D. del Cardayre. 2010. "Microbial Biosynthesis of Alkanes." *Science* 329, no. 5991: 559–62. https://doi.org/10.1126/science.1187936.

Segura, A., and J. Luis Ramos. 2012. "Plant–Bacteria Interactions in the Removal of Pollutants." *Current Opinion in Biotechnology* 24, no. 3: 467–73.

Singh, H. 2006. *Mycoremediation: Fungal Bioremediation.* New York: Wiley Interscience.

Sirajuddin, Sarah, and Amy C. Rosenzweig. 2015. "Enzymatic Oxidation of Methane." *Biochemistry* 54, no. 14: 2283–94. https://doi.org/10.1021/acs.biochem.5b00198.

Steliga, Teresa, and Dorota Kluk. 2020. "Application of Festuca arundinacea in Phytoremediation of Soils Contaminated with Pb, Ni, Cd and Petroleum hydrocarbons." *Ecotoxicology & Environmental Safety* 194: 110409. https://doi.org/10.1016/j.ecoenv.2020.110409.

Suganthi, S. H., S. Murshid, and S. Sriram. 2018. "Enhanced Biodegradation of Hydrocarbons in Petroleum Tank Bottom Oil Sludge and Characterization of Biocatalysts and Biosurfactants." *Journal of Environment Management* 220: 87–95.

Tahseen, R., M. Afzal, S. Iqbal, G. Shabir, Q. M. Khan, Z. M. Khalid, and I. M. Banat. 2016. "Rhamnolipids and Nutrients Boost Remediation of Crude Oil-Contaminated Soil by Enhancing Bacterial Colonization and Metabolic Activities." *International Biodeterioration & Biodegradation* 115: 192–8. https://doi.org/10.1016/j.ibiod.2016.08.010.

Tan, K. H. 2000. *Environmental Soil Science*. 3rd ed. Boca Raton, FL: Taylor & Francis.

Tang, J., R. Wang, X. Niu, M. Wang, and Q. Zhou. 2010. "Characterization on the Rhizoremediation of Petroleum Contaminated Soil as Affected by Different Influencing Factors." *Biogeosciences Discussions* 7: 4665–88.

Tangahu, B. V., S. A. Siti-Rozaimah, B. Hassan, I. Mushrifah, A. Nurina, and M. Muhammad. 2011. "A Review on Heavy Metals (As, Pb, and Hg) Uptake by Plants Through Phytoremediation." *International Journal of Chemical Engineering* 8: 274–82.

Tao, Kaiyun, Xiaoyan Liu, Xueping Chen, Xiaoxin Hu, Liya Cao, and Xiaoyu Yuan. 2017. "Biodegradation of Crude Oil by a Defined Co-culture of Indigenous Bacterial Consortium and Exogenous *Bacillus subtilis*." *Bioresource Technology* 224: 327–32. https://doi.org/10.1016/j.biortech.2016.10.073.

Uzoho, B., N. Oti, and E. Onweremadu. 2006. "Effect of Crude Oil Pollution on Maize Growth and Soil Properties in Ihiagwa, Imo State, Nigeria." *International Journal of Agriculture & Rural Development* 5, no. 1: 91–100. https://doi.org/10.4314/ijard.v5i1.2568.

Vasconcelos, U., F. Pd. França, and F. J. S. Oliveira. 2011. "Removal of High-Molecular Weight Polycyclic Aromatic Hydrocarbons." *Química Nova* 34, no. 2: 218–21. https://doi.org/10.1590/S0100-4042201 1000200009.

Velacano, M., A. Castellanohinojosa, A. F. Vivas, and M. V. M. Toledo. 2014. "Effect of Heavy Metals on the Growth of Bacteria Isolated from Sewage Sludge Compost Tea." *Advances in Microbiology* 4: 644–55.

Volkering, F., A. M. Breure, and W. H. Rulkens.1997–1998. "Microbiological Aspects of Surfactant Use for Biological Soil Remediation." *Biodegradation* 8, no. 6: 401–17. https://doi.org/10.1023/a:1008291130109.

Wang, Liping, Wanpeng Wang, Qiliang Lai, and Zongze Shao. 2010. "Gene Diversity of CYP153A and AlkB Alkane Hydroxylases in Oil-Degrading Bacteria Isolated from the Atlantic Ocean." *Environmental Microbiology* 12, no. 5: 1230–42. https://doi.org/10.1111/j.1462-2920.2010.02165.x.

Wang, S., Y. Xu, Z. Lin, J. Zhang, N. Norbu, and W. Liu. 2017. "The Harm of Petroleum-Polluted Soil and Its Remediation Research." *AIP Conference Proceedings* 1864: 020222. https://doi.org/10.1063/1.4993039.

Wang, Z. Y., Y. Xu, H. Y. Wang, J. Zhao, D. M. Gao, F. M. Li, and B. Xing. 2012. "Biodegradation of Crude Oil in Contaminated Soils by Free and Immobilized Microorganisms." *Pedosphere* 22, no. 5: 717–25. https://doi.org/10.1016/S1002-0160(12)60057-5.

Wei, Y., and G. Li. 2018. "Effect of Oil Pollution on Water Characteristics of Loessial Soil.". *IOP Conference Series: Earth & Environmental Science*. IOP Conference Series Environmental Earth Sciences 170 170: 032154. https://doi.org/10.1088/1755-1315/170/3/032154.

Wei, Zhuo, Jim J. Wang, Yili Meng, Jiabing Li, Lewis A. Gaston, Lisa M. Fultz, and Ronald D. DeLaune. 2020. "Potential Use of Biochar and Rhamnolipid Biosurfactant for Remediation of Crude Oil-Contaminated Coastal Wet land Soil: Ecotoxicity Assessment." *Chemosphere* 253: 126617. https://doi.org/10.1016/j.chemosphere.2020.126617.

Whalen, J. K., and Luis Sampedro. 2010. *Soil Ecology and Management*: 320. Wallingford: CABI.

Wiegel, J., and Q. Wu. 2000. "Microbial Reductive Dehalogenation Of polychlorinated Biphenyls." *FEMS Microbiology Ecology* 32, no. 1: 1–15. https://doi.org/10.1111/j.1574-6941.2000.tb00693.x.

Xia, Ying, Hang Min, Gang Rao, Z. M. Lv, J. Liu, Y. F. Ye, and X. J. Duan. 2005. "Isolation and Characterization of Phenanthrene-Degrading Sphingomonas paucimobilis strain ZX4." *Biodegradation* 16, no. 5: 393–402. https://doi.org/10.1007/s10532-004-2412-7.

Xin, L., Z. Hui-hui, Y. Bing-bing, X. Nan, Z. Wen-xu, H. Ju-wei, and S. Guang-yu. 2012. "Effects of Festuca arundinacea on the Microbial community in Crude Oil-Contaminated Saline-Alkaline Soil." *Chinese Journal of Applied Ecology* 23: 3414–20.

Yaashikaa, P. R., P. Senthil Kumar, Sunita Varjani, and A. Saravanan. 2020. "A Critical Review on the Biochar Production Techniques, Characterization, Stability and Applications for Circular Bioeconomy." *Biotechnology Report s* 28: E00570. https://doi.org/10.1016/j.btre.2020.e00570.

Yateem, A., M. Balba, N. Al-Awadhi, and A. El-Nawawy. 1997. "White Rot Fungi and Their Role In remediating Oil-Contaminated Soil." *Environment International* 24: 181–7.

Yuan, S. Y., L. C. Shiung, and B. V. Chang. 2002. "Biodegradation of Polycyclic Aromatic Hydrocarbons by Inoculated Microorganisms in Soil." *Bulletin of Environmental Contamination & Toxicology* 69, no. 1: 66–73. https://doi.org/10.1007/s00128-002-0011-z.

Yuan, Xiaoyu, Xinying Zhang, Xueping Chen, Dewen Kong, Xiaoyan Liu, and Siyuan Shen. 2018. "Synergistic Degradation of Crude Oil by Indigenous Bacterial Consortium and T Exogenous Fungus Scedosporium boyd." *Bioresource Technology* 264: 190–7. https://doi.org/10.1016/j.biortech.2018.05.072.

Zakaria, M. P., C. W. Bong, and V. Vaezzadeh. 2018. "Fingerprinting of Petroleum Hydrocarbons in Malaysia Using Environmental Forensic Techniques: A 20-Year Field Data Review." In *Oil Spill Environmental Forensics Case Studies*: 345–72. Oxford, UK: Butterworth-Heinemann.

Zhang, Meng, Penghong Guo, B. Wu, and Shuhai Guo.2020. "Change in Soil Ion Content and Soil Water-Holding Capacity During Electro-bioremediation of Petroleum Contaminated Saline Soil." *Journal of Hazardous Materials* 387: 122003. https://doi.org/10.1016/j.jhazmat.2019.122003.

Zhang, Yue, Xiaoyun Xu, Pengyu Zhang, L. Zhao, Hao Qiu, and Xinde Cao. 2019. "Pyrolysis-Temperature Depended Quinone and Carbonyl Groups as the Electron Accepting Sites in Barley Grass Derived Biochar." *Chemosphere* 232: 273–80. https://doi.org/10.1016/j.chemosphere.2019.05.225.

Zhao, Xiaohui, Fuqiang Fan, Huaidong Zhou, Panwei Zhang, and Gaofeng Zhao. 2018. "Microbial Diversity and Activity of an Aged Soil Contaminated by Polycyclic Aromatic Hydrocarbons." *Bioprocess & Biosystems Engineering* 41, no. 6: 871–83. https://doi.org/10.1007/s00449-018-1921-4.

Zhen, Meinan, Hongkun Chen, Qinglong Liu, Benru Song, Yizhi Wang, and Jingchun Tang. 2019. "Combination of Rhamnolipid and Biochar in Assisting Phytoremediation of Petroleum Hydrocarbon Contaminated Soil Using Spartina anglica." *Journal of Environmental Sciences* 85: 107–18. https://doi.org/10.1016/j.jes.2019.05.013.

7 The Biological Remediation of Water and Wastewaters Using Different Treatment Techniques

Mohamed Saad Hellal, Aleksandra Ziembińska-Buczyńska, Mohamed Azab El-Liethy, and Joanna Surmacz-Górska

7.1 INTRODUCTION

Water scarcity will affect approximately 66% of the total population by 2025. Furthermore, food production should more than double by 2050 to feed 9 billion people. As a result, approximately 15 million m^3/day of untreated wastewater polluted with pathogens, heavy metals, and excess salts will be used for direct crop irrigation globally (Ungureanu *et al.*, 2020). Untreated wastewater contains high levels of toxic, organic, and inorganic pollutants, as well as heavy metals such as Fe, Mn, Zn, Co, and Cr. Furthermore, the presence of pathogenic microorganisms in untreated wastewater and surface water, such as *Salmonella* spp., *Shigella* spp., rotavirus, and adenovirus, has a direct impact on human health (Akram *et al.*, 2018). Water pollution is a global environmental issue that poses significant threats to ecosystems and human health. The contamination of water bodies, such as rivers, lakes and groundwater, with various pollutants, including heavy metals, organic compounds, and nutrients has become a pressing concern (WHO/UNICEF, 2021). As a result, untreated wastewater should be subjected to adequate treatment before being reused for crop irrigation. Furthermore, domestic and industrial wastewater remediation must be cost effective. Traditional water treatment methods, such as chemical and physical processes, have limitations in terms of cost, efficiency, and potential formation of toxic byproducts (Abou-Elela *et al.*, 2019a). As a result, there is a growing interest in the development and implementation of environmentally friendly and sustainable remediation technologies, such as biological remediation and phytoremediation. Biological remediation involves the use of microorganisms, such as bacteria, fungi, and algae, to degrade, transform, or immobilize contaminants in water and wastewaters. These microorganisms have various metabolic capabilities and enzymatic systems that enable them to metabolize or bind pollutants effectively (Abou-Elela *et al.*, 2019c). Biological remediation can occur naturally in aquatic environments, but it can also be enhanced through engineered systems. Various treatment techniques have been developed to optimize and accelerate the biological remediation process (Hellal *et al.*, 2021).

Phytoremediation, on the other hand, is a specific type of biological remediation that utilizes plants and their associated microorganisms to remediate water and wastewaters (Kurade *et al.*, 2021). Plants can absorb contaminants from water through their root systems and translocate them to their shoots and leaves. Once in the aerial parts of the plant, contaminants can be degraded or sequestered, thereby reducing their concentration in the water (Abou-Elela *et al.*, 2013). Phytoremediation is an attractive option for water treatment due to its low cost, aesthetic appeal, and potential for sustainable biomass production (Abou-Elela *et al.*, 2017).

Different treatment techniques have been used to enhance the efficiency of biological remediation and phytoremediation processes. Some of the common techniques include bioaugmentation,

DOI: 10.1201/9781003423393-7

biofilm reactors, constructed wetlands, and algal bioreactors. Bioaugmentation involves the addition of specific microbial strains or consortia to contaminated water to enhance the degradation of target pollutants (Muter, 2023). This approach has shown promising results in the treatment of various contaminants, including organic compounds, heavy metals, and recalcitrant pollutants (Herrero and Stuckey, 2015). The added microbial populations can possess unique metabolic capabilities and enzymatic systems that enable them to efficiently degrade the target pollutants. Furthermore, bioaugmentation can be combined with other treatment technologies to optimize the remediation process, such as the use of biofilm reactors or constructed wetlands (Bai *et al.*, 2011, 2010; Manirakiza and Sirotkin, 2021).

Constructed wetlands are engineered systems that mimic natural wetlands and utilize the combined actions of plants, microorganisms, and soil to remove pollutants from water (Abou-Elela and Hellal, 2012). Plants in constructed wetlands play a crucial role in pollutant uptake, filtration, and oxygenation of the system (Abou-Elela *et al.*, 2014). The root systems of plants facilitate the uptake of contaminants, which can be subsequently metabolized or stored within plant tissues (Abou-Elela *et al.*, 2019c). The soil and associated microbial communities contribute to the degradation and transformation of pollutants, enhancing the overall treatment efficiency.

Algal bioreactors employ photosynthetic microorganisms, such as microalgae, to assimilate nutrients and capture heavy metals from water through biosorption and bioaccumulation processes (Abd-Elmaksoud *et al.*, 2021). Microalgae can remove nutrients, such as nitrogen and phosphorus, through assimilation, thereby reducing the eutrophication potential of treated water. Additionally, microalgae can accumulate heavy metals through biosorption and bioaccumulation processes. Algal bioreactors offer several advantages, including high growth rates, rapid pollutant removal, and potential for biomass production, which can be used for biofuel or value-added products (Heidari *et al.*, 2022; Hoh *et al.*, 2016).

Numerous studies have demonstrated the effectiveness of biological remediation and phytoremediation techniques in treating various contaminants in water and wastewaters. For example, researchers have reported successful removal of heavy metals, such as lead, cadmium, and chromium, using bioaugmentation with metal-resistant bacteria or phytoremediation with metal-accumulating plants (Bhat *et al.*, 2022; Delgado-González *et al.*, 2021; Naeem *et al.*, 2023; Yaashikaa *et al.*, 2022). Similarly, organic compounds, including pesticides, pharmaceuticals, and hydrocarbons, have been effectively degraded by microbial consortia or metabolically active plants.

Despite the significant progress made in biological remediation and phytoremediation, several challenges must be addressed. These challenges include the optimization of treatment processes, selection of appropriate plant species or microbial strains, management of system operation and maintenance, and scaling up of these technologies for large-scale applications. Additionally, the potential release of contaminants or their metabolites into the environment and the long-term sustainability of the treatment systems should be carefully evaluated.

The objective of this chapter is to provide a comprehensive overview of biological remediation and phytoremediation techniques for water and wastewater treatment. The chapter aims to discuss the principles and mechanisms underlying these processes, including the role of microorganisms, plants, and associated microbial communities in pollutant removal and transformation. It explores different treatment techniques utilized in biological remediation and phytoremediation, such as bioaugmentation, biofilm reactors, constructed wetlands, and algal bioreactors, and highlights their advantages and limitations. The chapter summarizes recent advancements in the application of these techniques for the removal of various contaminants from water and wastewater, including heavy metals, organic compounds, and nutrients. It evaluates the efficiency, cost-effectiveness, and sustainability aspects of biological remediation and phytoremediation compared to conventional water treatment methods. In addition, the discussion of challenges and prospects associated with the implementation of these techniques includes optimization strategies, selection of appropriate plant species or microbial strains, operation and maintenance, and scalability for large-scale applications.

7.2 BIOLOGICAL WASTEWATER TREATMENT

7.2.1 IMPORTANCE OF WASTEWATER TREATMENT

Wastewater originating from industrial and domestic activities must meet specific criteria before being reintroduced into the natural ecosystem. The concentration of toxic substances and microorganisms in wastewater varies depending on the source, as natural environments are incapable of processing high volumes of pollutants. In some cases, municipal wastewater, which includes domestic and industrial waste, is untreated or inadequately treated before being released into bodies of water such as rivers, lakes, and oceans. Common components of municipal wastewater include human waste, food waste, soaps, detergents, and soil residues. For wastewater intended for irrigation purposes, there is no obligation to remove nutrients or harmless dissolved and suspended organic matter. However, toxic chemicals and microorganisms are undesirable constituents and should be removed (USEPA, 2012). In the case of wastewater discharge into the hydrosphere, not only should toxic chemicals and microorganisms be eliminated but also nutrients and dissolved organic matter. The water bodies maintain adequate levels of dissolved oxygen to oxidize animal and vegetative waste through aerobic microbial reactions. This process results in the conversion of dissolved oxygen into CO_2 (Vanloon *et al.*, 2005), which is subsequently used during photosynthesis to revert back to oxygen, ensuring a stable natural ecosystem. Excessive amounts of dissolved oxygen introduced through wastewater discharge disrupt this self-purification cycle (Sonune and Ghate, 2004). The excess degradable organic matter gives rise to anoxic conditions that hinder vital aerobic biological degradation processes. The total suspended solids (TSS) also contributes to increased turbidity that hampers photosynthesis.

Phosphorous, nitrogenous, and silicate nutrients are essential for plant development and reproduction in aquatic environments. However, an overabundance of nitrogen and phosphorus can adversely affect aquatic ecosystems by stimulating excessive plant and algae proliferation (Chopin *et al.*, 2012). The large-scale availability of inorganic nitrogen and phosphorus to phytoplankton contributes significantly to eutrophication within aquatic environments. Various forms of inorganic nitrogen, including ammonia nitrogen, nitrates, and nitrites, can be present in wastewater discharges and lead to oxidation of rivers and the mortality of numerous fish species (Wei *et al.*, 2017).

To assess the extent of pollution in wastewater, specific parameters such as biochemical oxygen demand (BOD5), chemical oxygen demand (COD), total suspended solids, total phosphorus (TP), and total nitrogen (TN) are measured (Abou-Taleb *et al.*, 2020). Therefore, it is imperative that wastewater discharged into the hydrosphere maintain low concentrations of toxic chemicals that pose a threat to the ecosystem, oxidizable components, nutrients that promote microbial growth, and pathogens (Abou-Elela *et al.*, 2018). Parameters such as BOD5, TSS, TP, and TN are of significant concern as they signify the pollutant concentrations that adversely impact the environment and disrupt aquatic life.

To protect water resources, governmental regulations delineate permissible levels of contaminants. Based on the intended end use, effluents from treatment facilities must adhere to defined standards. Consequently, a diverse assortment of wastewater treatment processes has been conceived to purify water and attain compliance with effluent regulations. Conventional treatments encompass primary and secondary processing stages during which suspended particles are eliminated and organic matter transformed into bacterial biomass, water, and carbon dioxide. Certain processes entail a tertiary treatment phase where disinfection is executed or specific contaminants like nitrogen and phosphorus are extracted (Hellal *et al.*, 2021). Such treatments involve aerobic techniques using the activated sludge process, chemical coagulants for reduction of turbidity, anaerobic processing, such as sludge digestion, and attached growth systems. These may integrate physical, chemical, and biological procedures. Several large-scale systems have been implemented capable of treating up to 10,000 m³ of municipal wastewater daily (Techobanoglous *et al.*, 2014). Continuous advancements or replacements are observed in these processes and designs.

7.2.2 Anaerobic Treatment

Anaerobic treatment is a complex particulate technique that requires strict anaerobic conditions and has been described as a multistep process of series and parallel reactions. It depends on the coordinated activity of a complex microbial association to transform organic material into mainly carbon dioxide (CO_2) and methane (CH_4). Several groups of bacteria catalyze the reactions taking place during the anaerobic process: (1) fermentative bacteria, (2) hydrogen-producing acetogenic bacteria, (3) hydrogen-consuming acetogenic bacteria, (4) carbon dioxide–reducing methanogens, and (5) acetolactic methanogens. One of the promising advanced techniques for wastewater treatment is the anaerobic packed bed reactor. In this reactor, microorganisms grow as a biofilm on the surface of inert carriers. Microorganisms exist not only in the spaces within the carriers but also attached to its surface; hence, a high-density microbial population is retained within the reactor, allowing a biomass retention time longer than the hydraulic retention time (HRT) (Gavrilescu, 2002). The anaerobic process has been successfully applied for domestic sewage treatment at 20°C or higher, with COD removal efficiencies in the range of 57% to 82% and HRT ranging from 5 to 15 hours (Abou-Elela *et al.*, 2015).

Advances in anaerobic treatment of domestic wastewater offer a few promising options, including biofilm reactors. These reactors can be assembled in a number of configurations including batch, continuous stirred tank reactor (CSTR; including agitating continuous reactors, and rotary continuous reactors) (Usack *et al.*, 2012), trickling bed (TBR) (Cheng *et al.*, 2022), fluidized bed (FBR) (Aslam *et al.*, 2017), anaerobic baffled reactor (ABR) (Liu *et al.*, 2010), upflow anaerobic sludge blanket (UASB) (Elmitwalli and Otterpohl, 2007), and expanded bed reactors (EGSB) (Wang *et al.*, 2015). The operation of these reactors changes from reactor to reactor. In a batch biofilm reactor, the immobilized cells have to be utilized for repeated batches. However, it is likely that during the late stationary phase of chemical production, the culture would experience inhibition, thus reducing productivity.

7.2.3 Aerobic Treatment

Aerobic treatment is based on the metabolic activity of aerobic microorganisms, such as bacteria and fungi, that thrive in the presence of oxygen (Abou-Elela *et al.*, 2019b). These microorganisms decompose and transform organic matter, converting it into simpler and more stable compounds through biochemical reactions (Hellal *et al.*, 2020). Oxygen acts as an electron acceptor during the breakdown of organic substances, facilitating the generation of energy required for the organisms of microbial growth and the degradation of pollutants. This aerobic degradation process plays a crucial role in wastewater treatment. Most successful biological treatment systems incorporate an aerobic process. Aerobic degradation has been found to effectively reduce the high COD, BOD, and NH_3-N concentrations in wastewater (Bengtsson *et al.*, 2018). Other advantages of aerobic treatment compared to anaerobic treatment include lower temperature requirements, better ammonia removal, and less sensitivity to acidic conditions (Oliveira *et al.*, 2009). Aerobic bacteria can also treat many trace organic compounds, often with complete or near-complete degradation. The amount of sludge that is produced under aerobic conditions will vary depending on sludge age (Mishra *et al.*, 2023). As bacteria grow, some of the organic compounds present in the leachate are used in cell synthesis.

7.2.4 Natural Treatment Systems

Natural wastewater treatment (NWT) systems are biological treatment systems that require no or very little electrical energy. They also called passive aerated treatment systems (Gad *et al.*, 2022). Constructed wetlands, waste stabilization ponds (WSPs), and natural ventilation biofilters are common components of passive systems that require minimal energy and chemical input and rely on biological processes to treat waste effluents (Hellal and Abou-Elela, 2021). This is in contrast to conventional treatment systems such as activated sludge. Instead they rely on entirely natural processes, principally biochemical and in particular photosynthetic reactions, to provide the energy required for wastewater treatment. Natural wastewater treatment systems are anaerobic or aerobic, and some have both aerobic and anaerobic zones. Because they are not energy-intensive processes, they require a larger volume or area to enable wastewater treatment to proceed to the required level. There is thus a trade-off: either more money is spent on land (either by purchase or by leasing) for NWT, or more money is spent on electromechanical equipment and electrical energy for conventional treatment processes such as activated sludge. Trickling filters with different configurations are considered natural ventilation biofilters.

7.3 BIOREMEDIATION TECHNIQUES IN WASTEWATER TREATMENT

7.3.1 WATER AND WASTEWATER BIOREMEDIATION DEFINITION

Bioremediation is considered a promising wastewater treatment technology that has recently been used to reduce the toxicity of wastewater effluent while also ensuring its long-term use (Mandeep *et al.*, 2020).

The primary distinction between biodegradation and bioremediation processes is that biodegradation occurs naturally in the environment. Bioremediation, on the other hand, is a human-engineered technique to clean up pollutants in the environment. Both processes are primarily governed by many microorganisms such as bacteria, fungi, and microalgae, which have the ability to degrade organic and inorganic pollutants in the environment. The main goal of bioremediation is to alter hazardous substances into non-toxic or less toxic substances through the use of biological agents such as microorganisms or plants (Yan *et al.*, 2022). The bioremediation technique is a waste management method that involves the use of biological agents to remove contaminants in the environment. Also, it is a planned process that involves human intervention. This is a quicker method. It is a controlled procedure. It always has a positive impact. This occurs at the contaminated site. This process must be designed and implemented by experts (Riser-Roberts, 2019). Besides that, bioremediation treatment has several advantages, including the fact that the process is completely natural and has no side effects, as well as low startup and maintenance costs, low energy requirements, and a high recycle rate for water and soil reuse, and most applications are completed in-situ, with no hazardous transport (Ihsanullah *et al.*, 2020).

There are two types of bioremediation: in-situ bioremediation and ex-situ bioremediation. In-situ bioremediation refers to the breakdown of contaminants within the polluted area. Ex-situ bioremediation, on the other hand, refers to biological treatment that takes place outside of the polluted area (Azubuike *et al.*, 2016; Boopathy, 2000; Wu *et al.*, 2014). Many studies have used in-situ bioremediation for organic and heavy metal pollutants using different methods such as compositing, bioreactors, biofilters, bioaugmentation, biostimulation, and phytoremediation (Abdelaal *et al.*, 2021; Chen *et al.*, 2015; El-Liethy *et al.*, 2022; Jing *et al.*, 2012; Wijekoon *et al.*, 2013). Ex-situ bioremediation for environmental pollutants has been also studied in many research works (Firmino *et al.*, 2015; Gomez and Sartaj, 2013; Hussien *et al.*, 2020). Despite the fact that in-situ bioremediation methods have been in use for two to three decades, they have yet to produce the expected results. Their limited success has been attributed to poor environmental sustainability. As a result, a stable microbial eco-system with a balanced structure and composition must be obtained during the implementation of a bioremediation engineering project (Pandey *et al.*, 2009).

Bacterial bioremediation is an extensively used method for industrial wastewater treatment. *Aeromonas* sp., *Staphylococcus aureus*, *Lactobacillus* sp., *Bacillus subtilis*, *Micrococcus* sp., *Bacillus* sp., *Bacillus megaterium*, and *E. coli* were used by Ali *et al.* (2009) to remediate heavy metal (Cd, Cu, Pb, Zn, Fe, Cr, Ni, and Mn) pollutants and dyes in textile wastewater effluent. Kamika and Momba (2013) investigated the bioremediation of heavy metals by selected protozoan and bacterial species in highly polluted industrial wastewater. The results found that *Pseudomonas*

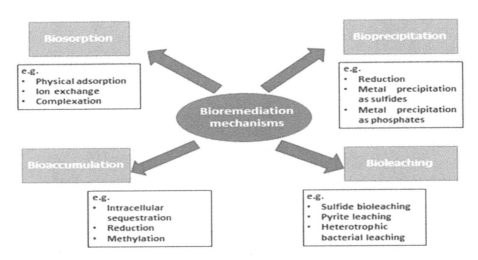

FIGURE 7.1 Different mechanisms of microbial bioremediation in wastewater (adapted from Sreedevi *et al.* 2022).

putida had the highest heavy metal removal rates for Co, Ni, Mn, V, Pb, Ti, and Cu with 71, 51, 45, 83, 96, 100, and 49%, respectively, then *Bacillus licheniformis* (Al 23% and Zn 53%). Many microorganisms have the ability to decolorize dyes and are capable of mineralizing several azo dyes under optimum environmental conditions. The microorganisms used include bacteria (Ikram *et al.*, 2022), fungi (Krishnamoorthy *et al.*, 2018), algae (Kapoor *et al.*, 2021; Wang *et al.*, 2016), actinomycetes (Mohamed *et al.*, 2016; Selvaraj *et al.*, 2021), and plants as phytoremediation (Abou-Elela *et al.*, 2019c; Hellal *et al.*, 2021). Ikram *et al.* (2022) investigated the ability of 11 different bacterial strains to degrade basic orange 2 dye. They discovered that *E. coli* could remediate the dye with 78.90% degradation activity. They concluded that *E. coli* was the most promising strain for the degradation of the chosen dye.

There are numerous mechanisms of bacterial bioremediation for organic and inorganic pollutants, particularly heavy metal removal (Figure 7.1). The first mechanism is the biosorption on bacterial cell walls, intracellular sequestration by accumulation, extracellular sequestration as insoluble compounds by precipitation, or production of metabolites that solubilize and chelate metal compounds, resulting in leaching (Sreedevi *et al.*, 2022).

Fungal bioremediation, also known as mycoremediation, is a bioremediation strategy that has seen tremendous success in recent decades (Rhodes, 2014). Fungal enzymes and their products have the potential to bioremediate pollutants in wastewater. Many fungal spores, including *Aspergillus niger, Aspergillus niveus*, and *Aspergillus fumigatus*, have bioremediation potential (Chaudhary *et al.*, 2022). The environment is teeming with fungal consortia that play an important role in maintaining the balance of various environmental cycles. Due to the good exchange of metabolites among its individuals and populations, organisms in consortium have a high tolerance (Zhang *et al.*, 2018). The fungal consortium, as a result of this property, is a reliable and promising agent for environmental remediation operations such as wastewater treatment. Several studies have been published (Chaudhary *et al.*, 2022). Microbial bioremediation employs indigenous microbial species to remove pollutants from the environment. The detoxification degree of pollutants is influenced by two main factors: the first is the biotic factor, which includes the composition of indigenous microbial species (bacteria, fungi, algae, and protozoa), and the second is the abiotic factor, which depends on the nature and amount of the pollutant, bioavailability, as well as environmental circumstances such as pH, oxygen, temperature, nutrients (carbon and nitrogen), salinity, and pressure (Ahmad *et al.*, 2018).

7.3.2 BIOAUGMENTATION AND BIOSTIMULATION

Indigenous microorganisms cannot completely remediate all heavy metals, organic pollutants, and microbial pathogens in contaminated industrial wastewater. In the case of complex pollutants, different microbial consortiums are required that differ from what naturally occurs in the polluted environment. Furthermore, the use of indigenous bacteria in bioremediation may be inhibited by abiotic factors such as pH and temperature (Harms and Zehnder, 1995). All of these contribute to the production of poor water quality; therefore, bioaugmentation methods could be beneficial in such cases (Al-Gheethi *et al.*, 2018). Bioaugmentation is a type of bioremediation treatment process that involves the addition of external microorganisms, plants, or their enzymes to a polluted environment (such as water, sediments, or soil) or any toxic substances. It was thought to be a simple and cost-effective technology that could be useful in biological wastewater treatment (Ibrahim *et al.*, 2020). The selected exogenous microorganisms should have high synergistic interactions with the indigenous microorganisms, easy recovery on cultural media, fast growth, high ability to remediate pollutants, ability to resist huge amounts of contaminants, and finally ability to survive in different environmental circumstances (Koul and Gauba, 2014).

The first type of microbial bioaugmented inoculum is allochthonous bioaugmentation, in which microbes are isolated from one site and added to another. The second type is autochthonous bioaugmentation, in which microorganisms are isolated from contaminated sites and re-inoculated in the same polluted sites, and the third type is gene bioaugmentation, in which genetically engineered microorganisms with genes encoding for enzymes capable of enhancing pollution and toxic substance bioremediation are used. These gene-encoding enzymes have the ability to transfer to indigenous microorganisms via horizontal gene transfer, allowing them to break down the target pollutants (Sun *et al.*, 2020). Many studies have used plasmid-mediated bioaugmentation for biodegradation of organic, heavy metal and pesticide pollutants; for instance, plasmid pDOD was used for dichlorodiphenyltrichloroethane (DDT) bioaugmentation, and plasmid pJP4 was used for 2, 4-dichlorophenoxyacetic acid (2,4-D) bioaugmentation. Moreover, plasmid pDOC was used for chlorpyrifos bioaugmentation (Gao *et al.*, 2015; Inoue *et al.*, 2012; Q. Zhang *et al.*, 2012). In order to break down hydrocarbons from oily wastewater, Mazumder *et al.* (2020) looked into the bioaugmentation potential of the isolated hydrocarbon oclastic bacterium *Rhodococcus pyridinivorans*. *Rhodococcus pyridinivorans* showed strong hydrocarbon tolerance at a hydrocarbon concentration of 8% (v/v). Moreover, response surface methodology (RSM), which was discovered and validated at the temperature of 37°C with a neutral pH for a wide range of salinity, was used also in this study to simulate the optimization of the growth conditions. When using an isolated strain in a free bacterial suspension state with both free and emulsified forms of oil, the maximum percentage of hydrocarbon degradation was 79%; however, when using a cell-entrapped alginate bead without the addition of activated carbon as a doping agent, the percentage of hydrocarbon degradation increased to 86%. The percentage of breakdown was further raised to 95% after alginate beads were used to dope activated carbon. The increased recyclability of the microbial cells in the bioaugmenting of the oil from oily wastewater was found to increase the sustainability of the technique with trapping. In another study, *Bacillus subtilis, Lactobacillus plantarum, Pediococcus acidilactici*, and *Pediococcus pentosaceus* were added as bioaugmented exogenous microorganisms to municipal wastewater supplemented with three-stage biofilters consisting of a sedimentation step followed by gravel biofiltration and then sand biofiltration at a lab scale. The removal efficiencies of H_2S, COD, BOD_5, total solids, total dissolved solids, total suspended solids, ammonia, nitrate, total phosphorus, and oil and grease reached 85, 93.4, 83.5, 37, 49.2, 93.4, 100, 55.7, 76.6, and 76.6%, respectively (Ibrahim *et al.*, 2020). In another study, total organic carbon (TOC) was removed from municipal wastewater through the bioaugmentation process using a consortium of commercial and local bacterial strains (Hesnawi *et al.*, 2014). Tuo *et al.* (2012) used *Bacillus* sp. isolated from petroleum oil–contaminated soil for quinolone bioaugmentation into wastewater treatment systems. A mixture of local bacteria to bioaugment a *Cupriavidus* sp. strain with ratio 10:3 was used for bioremediation of sulfolane. Oxygen as a biostimulant agent was added to enhance bioremediation of high concentrations of sulfolane pollutant. The byproducts of bioaugmentation of sulfolane were non-toxic (Chang *et al.*, 2021).

A more effective strategy to address some of these issues is biostimulation, in which nutrients are supplied during the bioaugmentation of organic and heavy metal pollutants to encourage the activity of native degrading bacteria (Varjani and Upasani, 2019) Microorganisms require additional nutrients than contaminants which considered as a source of nutrients throughout the biostimulation process in order to improve bioremediation. This raises the amount of microorganisms, which in turn accelerates their biodegradation. Additionally, separating promising degraders from their native microbial populations and cultivating them in a lab setting could speed up biodegradation (Bradford *et al.*, 2018) For the bioremediation of crude oil–contaminated soils, Yaman (2020) studied bioaugmentation, biostimulation, and natural attenuation processes. To achieve a C:N:P ratio of 100:5:1, ammonium chloride (NH_4Cl) and potassium phosphate (KH_2PO_4) were utilized as biostimulants. The reduction in total petroleum hydrocarbons for biostimulation and mixed bioaugmentation and biostimulation was reported at 66 and 74%, respectively. This result demonstrated the great significance of mixed bioaugmentation and biostimulation in crude oil–contaminated soil bioremediation. For natural attenuation and just bioaugmentation, lower percentages of 35 and 41% were noted, respectively. Ali *et al.* (2009) recently discovered that some recognized bacterial species could remove 53 to 63% of the oil from a desert soil sample that had been saturated with 17.3% w/w crude oil. According to Cai *et al.* (2020), adding nutrients like ammonium sulphate ((NH_4)2SO_4, 0.43 g/kg soil) and monopotassium dihydrogen phosphate (KH_2PO_4, 0.067 g/kg soil) to crude oil–contaminated soil stimulated a variety of bacteria that break down hydrocarbons and microorganisms that break down nitrogen. M. Wu *et al.* (2019) conducted a microcosm study to check the petroleum hydrocarbon degradation efficiency of an *Acinetobacter* strain for bioaugmentation and nitrogen and phosphorus nutrients for biostimulation in oil-contaminated soil. However, little research on the examination of process parameter influence and optimum conditions at which bioaugmentation and biostimulation of crude oil–polluted soil could be achieved has been conducted. This is necessary for industrial scale-up of the bioaugmentation-coupled biostimulation technique. In a separate study, sulfolane was efficiently bioremediated using a bioagumented *Cupriavidus* sp. strain. The stimulation process was carried out by oxygen addition. Sulfolane was effortlessly broken down into nontoxic substances (Chang *et al.*, 2021).

7.3.3 Biofilters and Bioreactors

In wastewater treatment, biofilters and bioreactors are commonly employed as biological treatment systems to remove pollutants through microbial degradation. These systems utilize the metabolic activities of microorganisms to break down organic matter and transform or remove contaminants (Abou-Elela *et al.*, 2015). Biofilters and bioreactors are versatile and can be designed for various types of wastewater, making them valuable tools in sustainable wastewater treatment. Recent studies have focused on enhancing the performance of biofilters and bioreactors through process optimization, microbial community manipulation, and the integration of advanced technologies. For example, the use of bioaugmentation, where specific microbial strains are added to enhance pollutant degradation, has shown promising results in improving treatment efficiency (Abou-Elela *et al.*, 2010). Additionally, advancements in molecular techniques, such as high-throughput sequencing and metagenomics, have provided insights into the microbial communities involved in wastewater treatment processes, aiding in system optimization and control (Ziembińska-Buczyńska and Surmacz-Górska, 2021). Furthermore, the integration of sensor technologies, automation, and artificial intelligence–based control systems has enabled real-time monitoring and optimization of biofilter and bioreactor performance. In this in-depth discussion, we will explore different types of biofilters and bioreactors used for wastewater treatment and their applications.

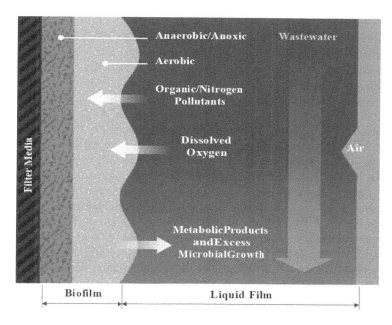

FIGURE 7.2 Cross-section schematic of the wastewater, biofilm, and media in attached growth of trickling filter.

7.3.3.1 Trickling Filters

Trickling filters are a type of biofilter that uses a solid support medium, such as rocks, plastic media, or wood chips, as a substrate for microbial attachment and growth (Zhu and Rothermel, 2014). Wastewater is distributed over the support medium, allowing the microorganisms to form a biofilm. As wastewater trickles down through the biofilm, organic pollutants are metabolized by the microbial community. Trickling filters are commonly used for the treatment of municipal wastewater and industrial effluents with moderate organic loads (Muralikrishna and Manickam, 2017). They offer advantages such as simplicity, low energy requirements, and robust performance. However, their efficiency may be limited for the removal of certain pollutants, such as nitrogen and phosphorus compounds. Trickling filters are essentially a solid-liquid-gas system in which the wastewater (liquid) flows over the biofilm (solids) and contacts air (gas), and at the same time, organic matter and nitrogen pollutants are absorbed and subsequently degraded by the microorganisms in the biofilm (Metcalf & Eddy, 2003), as shown in Figure 7.2. During the operation of a trickling filter, the biofilm tends to grow in thickness as a result of microbial growth. Portions of the biofilm will detach when they lose the ability to remain attached to the media, which is called "sloughing." In modern trickling filters with rotary distributor arms, biofilm sloughing can be effectively controlled through a flushing operation, during which the shear force of the wastewater flow refreshes the biofilm (Wang *et al.*, 1986). The sloughed biofilm together with the treated wastewater will drain to the filter bottom and then be conveyed to the downstream solid separation units such as clarifiers for settling. Advanced filtration such as sand filters may be used following secondary clarifiers or nitrifying trickling filters where higher effluent quality is desired.

7.3.3.2 Rotating Biological Contactors

Rotating biological contactors (RBCs) are another type of biofilter that consists of a series of rotating discs or cylinders partially submerged in wastewater. The rotating motion provides oxygen transfer and creates a thin biofilm on the surface of the discs or cylinders (Waqas *et al.*, 2021). As wastewater flows over the biofilm, microorganisms utilize the organic matter for growth and degradation. RBCs are effective in treating wastewater with moderate organic loads and can achieve high removal efficiencies for organic pollutants (Figure 7.3). They are commonly used in small to medium-sized wastewater treatment plants and can be easily expanded or upgraded based on demand (Hassard *et al.*, 2015). However, RBCs may require additional post-treatment processes for nutrient removal.

RBCs could be classified as single-stage or multi-stage RBCs. Single-stage RBCs consist of a single reactor unit with one or more rotating discs or cylinders. Wastewater is evenly distributed over the media surfaces using spray nozzles or distribution arms. As the media rotate, the biofilm comes into contact with the wastewater, facilitating the treatment process. Treated water is then separated from the biofilm and collected for further treatment or discharge (Cvetkovic *et al.*, 2014). Multi-stage RBCs involve a series of reactor units arranged in a sequence. Each stage operates independently, allowing for staged treatment and improved pollutant removal efficiency. Wastewater flows from one stage to another, increasing contact time with the biofilm and enhancing treatment performance (Buchanan and Leduc, 1994). Multi-stage RBCs are commonly used when higher treatment efficiency or specific pollutant removal is required. A study conducted by Mohamed *et al.* (2022) examined the performance of a pilot-scale RBC system treating domestic wastewater. The results showed high removal efficiencies for organic matter, with average BOD and COD removal rates of 94.4 and 86.9%, respectively. Ouyang (1981) investigated the sludge production and characteristics of a full-scale RBC system treating municipal wastewater were investigated. The results indicated that the RBC system generated lower sludge production compared to conventional activated sludge systems. The sludge production rate was found to be approximately 0.67 kg of dry solids per kg of BOD removed, highlighting the potential for reduced sludge handling and disposal costs. Also, Cabije *et al.* (2009) evaluated the performance of an RBC system in removing nitrogen and phosphorus from synthetic wastewater. The results showed that the RBC system achieved significant nutrient removal, with average TN and TP removal rates of 83 and 94%, respectively. The study concluded that RBCs can be an effective option for nutrient removal in wastewater treatment. A review study by Cortez *et al.* (2008) assessed the operation and maintenance aspects of RBC systems in a rural wastewater treatment plant. The results indicated that RBCs exhibited stable and reliable performance over a 2-year period. The study highlighted the ease of operation and

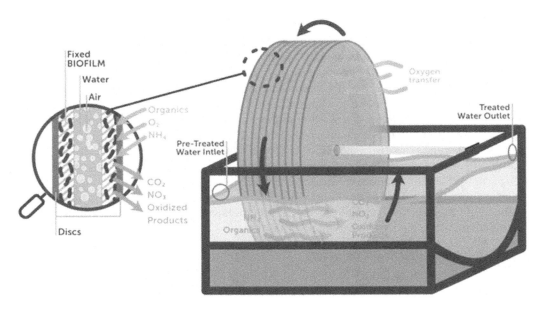

FIGURE 7.3 Mechanism of wastewater treatment in RBCs.

maintenance of RBC systems, with minimal need for chemical dosing and relatively low energy requirements. In Dutta *et al.* (2007), the performance of a laboratory-scale RBC system treating synthetic wastewater was investigated with varying operating parameters. The results demonstrated that optimal removal efficiencies were achieved at specific rotational speeds and hydraulic retention times. The study emphasized the importance of process optimization for maximizing treatment performance in RBC systems.

7.3.3.3 Membrane Bioreactors

Membrane bioreactors (MBRs) combine biological treatment with membrane filtration to achieve high-quality effluent. In MBRs, microorganisms biodegrade organic matter in wastewater, and the treated water is separated from biomass using a membrane filtration system, such as microfiltration or ultrafiltration (Malleviale *et al.*, 1996). The membrane acts as a physical barrier, effectively removing suspended solids, bacteria, and pathogens, resulting in a higher level of treated water quality. In recent years, MBRs have emerged as a promising technology that combines biological treatment and membrane filtration. Previous studies have yielded insightful findings, shedding light on the performance and potential of MBRs in wastewater treatment.

One notable area of research has focused on exploring novel membrane materials and configurations. For instance, hollow fiber membranes with modified surfaces have demonstrated enhanced fouling resistance and longer operational lifespans, ensuring sustained system performance (Razavi and Miri, 2015). Additionally, the use of immersed membranes with effective backwash mechanisms has proven successful in reducing fouling and the frequency of membrane replacements, thereby improving operational efficiency (Lee *et al.*, 2016). Process optimization and control strategies have also been extensively investigated. Advanced control algorithms, such as model predictive control and fuzzy logic control, have been implemented to enhance nutrient removal and increase energy efficiency in MBR systems. Al-Sayed *et al.* (2023b) conducted an in-depth examination of flat sheet submerged membrane bioreactor (FS-SMBR) process optimization, specifically focusing on hydraulic retention times (HRT) of 6 and 4 hours and a 40-day solid retention time using genuine municipal wastewater (Figure 7.4). By utilizing the GPS-X software simulator for validation and calibration, the researchers established complete alignment between the modeled data and experimental results under identical conditions. It was determined that FS-SMBR permeate adhered to the appropriate standards for reuse, achieving non-detectable fecal coliform levels at HRTs of 6 and 4 hours, accompanied by pore-blocking resistance (R_t) values of 28×10^{11} and 40×10^{11}/m, respectively. Furthermore, upon assessing sludge viscosity as another crucial factor, it was discovered that a 4-hour HRT yielded the most favorable outcomes for domestic wastewater treatment. Studies on the application of sludge disintegration techniques, such as ultrasound and ozonation, have shown promise in improving organic matter degradation rates and reducing sludge

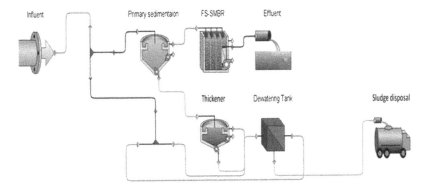

FIGURE 7.4 Process optimization of FS-SMBR using GPS-X modelling (Al-Sayed et al., 2023a).

production, thereby optimizing overall system performance (Wagner and Rosenwinkel, 2000; Yoon *et al.*, 2004). To further enhance treatment capabilities, researchers have explored the integration of supplementary treatment units. The coupling of adsorption or advanced oxidation processes with MBRs has resulted in improved micropollutant removal, ensuring the production of high-quality effluent (Boonnorat *et al.*, 2016; Tawfik *et al.*, 2022). Additionally, the incorporation of anaerobic digestion systems into MBR configurations has demonstrated the potential for energy recovery through biogas production, contributing to the overall sustainability of the wastewater treatment process (Annop *et al.*, 2014; Cho *et al.*, 2005). Despite these advancements, challenges persist in the widespread implementation of MBRs, and membrane fouling remains a primary concern, necessitating ongoing efforts to develop effective fouling control strategies (Chae *et al.*, 2006; Le-Clech *et al.*, 2006). High energy consumption and capital costs also pose obstacles to large-scale adoption (Huang *et al.*, 2020). However, these challenges have sparked further research aimed at optimizing operational parameters, exploring innovative membrane materials, and developing sustainable approaches to reduce energy consumption.

7.3.3.4 Anaerobic Bioreactors

Anaerobic bioreactors are specifically designed for the treatment of wastewater under anaerobic conditions, where microorganisms metabolize organic matter in the absence of oxygen. These bioreactors facilitate the conversion of organic pollutants into biogas, predominantly methane. Common types of anaerobic bioreactors include up-flow anaerobic sludge blanket (UASB) reactors, expanded granular sludge bed (EGSB) reactors, and modified anaerobic baffled reactors (MABRs). Anaerobic bioreactors are particularly suitable for high-strength organic wastewater, such as agro-industrial effluents and food processing wastewaters (Filer *et al.*, 2019). They offer advantages such as energy recovery from biogas production, reduced sludge generation, and lower operational costs (Hellal *et al.*, 2022). However, anaerobic bioreactors require careful control of operating conditions, such as temperature, pH, and organic loading rates, to maintain optimal microbial activity. Also, the effluent of anaerobic bioreactor mostly requires further treatment, and anaerobic bioreactors should be combined with other treatment steps (Lin *et al.*, 2010). Abou-Elela *et al.* (2015) studied the influence of seasonal variation on the performance of an integrated anaerobic-aerobic pilot-plant for wastewater treatment. The pilot plant was composed of a packed bed anaerobic sludge blanket, followed by a biological aerated filter. The packing material utilized in both bioreactors was a non-woven polyester textile. The apparatus functioned for an extended period exceeding 2 years at ambient temperature. The outcomes implied that the pilot plant's operation was exceptionally satisfactory throughout the entire seasons. Nevertheless, during the summer months, marginally superior results were attained in terms of COD, BOD, TSS, and TN due to the increase in temperature. The corresponding average removal percentages were 90, 91, 97, and 54%, respectively, as opposed to 88, 90, 91, and 46% during winter months. For the treatment of wastewater from poultry slaughterhouses, Debik and Coskun (2009) studied two different types of anaerobic reactors: one static granular bed reactor (SGBR), which uses a newly developed anaerobic process that is a fully anaerobic granule, and another SGBR that contains both anaerobic granular biomass and non-granular biomass. Average COD removal efficiencies were greater than 95% for both reactors. A nine-chambered modified anaerobic baffled reactor was developed by Bodkhe (2009) to evaluate its suitability for the treatment of municipal wastewater and biogas production and to establish an understanding of the relationship between reactor design and operational parameters. Reactor performance evaluation was carried out for 375 d at 11 different HRTs ranging from 6 d to 3 h. The HRT of 6 h was judged appropriate for this reactor configuration. At a HRT of 6 h, the efficiencies of reduction in TSS, BOD, and COD were found to be 86, 87, and 84%, respectively. Specific biogas yield and methane content were found to be 0.34 m^3 CH_4/Kg_{COD} and 67%, respectively.

7.3.3.5 Advanced Biofilters

Recent studies have explored the development of advanced biofilters to enhance pollutant removal and system performance. For instance, researchers have investigated the use of novel filter media

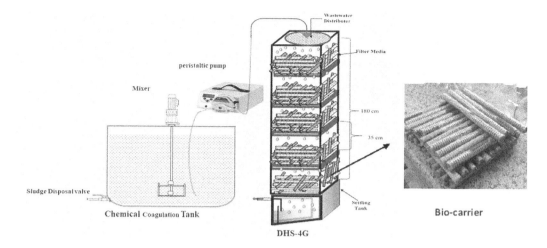

FIGURE 7.5 Combined DHS-4G reactor with chemical treatment of SWW (Hellal and Doma, 2022).

with high surface areas and enhanced microbial attachment, such as activated carbon, biochar, and zeolites, and even geotextile and concrete wastes. These media provide favorable conditions for microbial growth and promote the removal of specific contaminants, including heavy metals and emerging pollutants like pharmaceuticals and personal care products. The down-flow hanging sponge (DHS) reactor represents a cutting-edge biofilm system implemented in wastewater treatment processes. This innovative biotower-trickling filter system not only functions as post-treatment for other primary treatment outlets but also employs polyurethane packing material. Superiorities of the DHS reactor system include elevated biomass concentration with increased sludge residence time (SRT), reduced hydraulic retention time (HRT), a smaller spatial requirement in comparison to traditional treatment methods, and the elimination of external aeration (Nurmiyanto and Ohashi, 2019; Onodera *et al.*, 2014). In the down-flow hanging sponge system, the sponge is suspended naturally in the air, allowing wastewater to flow downwards and oxygen to dissolve (Tandukar *et al.*, 2005). Recently, DHS was employed for industrial wastewater treatment. Hellal and Doma (2022) utilized a pilot-scale fourth-generation DHS reactor (DHS-4G) for the treatment of chemically pretreated slaughterhouse wastewater (SWW). A DHS-4G reactor capacity of 100 liters was installed and operated under two different hydraulic retention times (HRTs), 12 and 8 h. A schematic representation of the DHS system can be observed in Figure 7.5. The integrated treatment system gave reasonable organic and nitrogen removal efficiencies. The removal efficiency was 91% and 87% for COD, 92% and 88% for BOD, and 94% and 84% for TSS at HRT 12 h and 8 h, respectively. Also, analysis of retained biomass indicated that sludge retention time was 27 days, which indicated a low sludge amount produced from the reactor.

Another type of advanced bioreactor that was recently developed is the passively aerated biological filter (PABF). Passive wastewater treatment systems rely on the attached growth process, employing tangential flow as opposed to the permeation mode associated with substrate–biofilm contact in granular filters (Abou-Elela *et al.*, 2019b). The configuration of these passive wastewater treatment systems consists of semi-closed basins filled with substantial amounts of crushed rock or plastic media. This design ensures an extensive surface area for biofilm attachment and continuous, sizable pores. During operational phases, wastewater is distributed over the top of the bed via a rotating spray apparatus. The wastewater subsequently percolates through the bed and across the biofilm-coated media. Autoclaved aerated concrete waste (AACW) from construction and demolition was used as a biocarrier in a passively aerated biological filter treating municipal wastewater in a study conducted by Abou-Elela *et al.* (2019a). The AACW had a rough surface with a high specific surface area of 42.8 m^2/g. The use of AACW as a biocarrier in a pilot-scale PABF reactor greatly improved the removal of organic and inorganic pollutants (Figure 7.6). Average removal values of

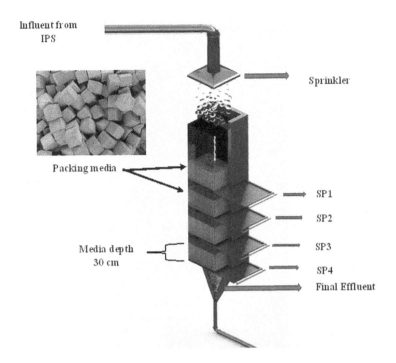

FIGURE 7.6 PABF packed with AACW for domestic wastewater treatment (Abou-Elela et al., 2019a).

COD, BOD, and TSS were 90%, 92%, and 89%. Results indicated that the potential of using AACW as a biocarrier in a PABF is recommended, and it provides a promising approach to utilize construction and demolition solid waste.

7.3.4 Factors Affecting Biofilter Performance

Several important parameters affect the efficiency of biological wastewater treatment: ammonia and nitrite concentrations, surface area of support media, and bed porosity of biofilters. Nitrifiers are very sensitive to operating conditions. For a given aquaculture recirculation system, the nitrification rate varies depending on the chemical and physical properties of the wastewater, the dimensional parameters of biofilters, the support media size, and the operating conditions of biofilters. Important properties of aquaculture wastewater include pH value, temperature, C:N ratio, and dissolved oxygen concentration, which affect nitrification efficiency. In addition, the density and size of support media, the wastewater loading rate, the mass transfer rate within the biofilm, the growth rate of the biofilm, and competition for space and oxygen between different kinds of bacteria also have a strong influence on nitrification efficiency.

7.3.4.1 Bed Porosity and Support Media Size

Bed porosity (void ratio of bed) is the ratio of the volume of liquid in a biofilter to the total volume of the support medium and liquid (Hellal *et al.*, 2020). A high bed porosity reduces clogging and consequently reduces power consumption. The selection of support media usually depends on cost, availability, the media size and specific surface area, and weight per unit volume. Supporting media, manufactured or naturally occurring, usually exhibit different sizes and shapes. The support medium is solid material placed into a nitrification filter to provide surface area on which nitrification bacteria grow. The voiding of support media is an important factor in the attachment of bacteria to the bedding materials. Media that have a high void space inside allow more microorganisms

Biological Remediation of Water and Wastewaters

to attach under low shear conditions, where a greater liquid flow rate can be reached (Sánchez Guillén *et al.*, 2015).

There are several ways to describe particle sizes, such as arithmetic mean diameter of particles, geometric mean diameter of particles, volume mean diameter of particles, and weight mean diameter of particles. For a biofilter, the surface area is more important, so the volume-surface diameter of particles is usually applied (Naz *et al.*, 2016). An upflow fluidized bed requires relatively small solid particles to provide high specific surface area for bacteria growth. The higher the surface area of the media, the more bacteria grow in the biofilters. Consequently, a greater ammonia removal rate can be achieved per unit volume of biofilter (Hellal and Abou-Elela, 2021). Sand is a good medium for an upflow biofilter, because it is cheap and easy to obtain. However, its relatively high density will introduce considerable head, loss indicating high energy consumption (Khan *et al.*, 2015). Plastic materials are light in weight, have a high void ratio, and will last almost indefinitely if not exposed to sunlight. They are ideal support media for downflow fluidized bed biofilters. Ben Rebah *et al.* (2010) indicated that plastic cross-flow media had an advantage over rock media that permitted higher volumetric loading because of their high specific surface area.

7.3.4.2 pH

The pH value in biofilters is one of the most important factors in avoiding inhibition caused by free ammonia accumulation, because ammonia and nitric acid are toxic to fish. A small change in pH will have a rather strong effect on the ammonia concentration (Nurmiyanto and Ohashi, 2019). Lydmark *et al.* (2006) reported that microorganisms and nitrifiers were able to function in a pH range of 6.0 to 8.5. Nitrifying bacteria usually prefer an alkaline environment with an optimal pH value ranging from 7.5 to 8.5. Jing *et al.* (2012) found that within the pH range of 5.0–9.0, a pH increase of one unit led to a 13% increase of organic removal and nitrification efficiency. Wąsik and Chmielowski (2017) presented a linear relationship between optimum pH value and the logarithm of total ammonia-N concentration, which indicated that the optimum pH value decreases with increasing ammonia concentration. The best pH value for biological growth is between 6.5 and 7.5, although bacteria can keep growing in the range of 4.0 to 9.5 (Biplob *et al.*, 2011). The pH level in a biofilter should be kept stable while operating. Rapid pH changes of more than 0.5 or 1.0 within a few minutes will greatly reduce the efficiency of bacteria until they adapt to the new operating condition.

7.3.4.3 Temperature

Temperature has a strong effect on bacterial metabolic activities. Microorganism bacteria prefer moderate temperatures that range from 20 to 30°C (Metcalf and Eddy, 2016). As temperature drops below 20°C, the metabolic activity of bacteria will decline. At 40°C or above, the activity of bacteria completely ceases. Yang *et al.* (2007) indicated that a linear relationship exists between ammonia oxidation rate and temperature within the 7 to 30°C temperature range, and the maximum nitrification rate occurred at approximately 30°C, while the ammonia removal rate dropped sharply by 50% below 15°C. Banach-Wiśniewska *et al.* (2021) also reported that it was necessary to increase temperature in order to maximize ammonia oxidizer activity.

7.3.4.4 Hydraulic Loading Rate

Hydraulic loading is a measure of the amount of wastewater pumped into the biofilter per unit area of biofilter top surface per unit time (Machdar, 2016). For biofilters, minimum hydraulic loading is the rate that keeps the minimum velocity of particles, while maximum hydraulic loading is the rate that prevents the scouring of bacteria off the support media, producing excessive head loss, or flowing support media out of the biofilter. In addition, biofilter height is an important consideration for pollutant removal, since the bacteria need sufficient time to convert ammonia to nitrite and nitrite to nitrate. Many factors should be considered for the design of biofilter height, such as support media size and hydraulic and organic loading rates (Wallace, 2013).

7.4 PHYTOREMEDIATION TECHNIQUES IN WASTEWATER TREATMENT

7.4.1 PHYTOEXTRACTION

Phytoextraction is a remediation method that facilitates the elimination of contaminants from soil, groundwater, or surface water by employing plants with a high propensity for accumulating toxic substances (Sessitsch *et al.*, 2013). Adaptability to high concentrations of heavy metals or organic compounds is essential for plant species utilized in this approach. Additionally, the selected plants should exhibit rapid growth and generate substantial biomass. Divided into two categories—continuous and induced phytoextraction processes—the former encompasses plants that accumulate elevated levels of toxic contaminants throughout their entire life cycle. In contrast, induced phytoextraction incorporates chelators during specific stages of plant development, thereby enhancing toxin accumulation in plant tissues (Bhargava *et al.*, 2012; Bian *et al.*, 2018; Suman *et al.*, 2018). Monferrán *et al.* (2012) examined *Potamogeton pusillus*'s tolerance to Cr and Cu, its bioaccumulation process, and its ability to eliminate these metals from aqueous solutions. The findings demonstrated that *P. pusillus* roots and leaves accumulated greater quantities of Cu and Cr compared to stems, leading to the conclusion that this plant species is suitable for heavy metal phytoextraction from contaminated waters.

7.4.2 PHYTODEGRADATION

Phytodegradation is a technique employing plants to generate enzymes that facilitate the breakdown of xenobiotic compounds (Newman and Reynolds, 2004). This degradation process can transpire both within and outside the plant, particularly when enzymes are excreted into the soil surrounding the root zone (García-Sánchez *et al.*, 2018). The method has been applied in treating soil, river sediments, and sludges, as well as groundwater and surface water (Awa and Hadibarata, 2020; Bhat *et al.*, 2022; Farraji *et al.*, 2020; Morita and Moreno, 2022). A study by Zazouli *et al.* (2014) examined *Azolla filiculoides*'s potential in eliminating bisphenol A (BPA) from aqueous solutions. The experiment involved culturing *A. filiculoides* in solutions containing varying BPA concentrations of 5, 10, 25, and 50 ppm. Results demonstrated that BPA degradation was contingent upon the *A. filiculoides* biomass quantity and BPA concentration. Notably, over 90% removal efficiency was achieved with a BPA concentration of 5 ppm and a biomass amount of 0.9 grams. Zazouli *et al.* (2014) deduced that *A. filiculoides* eliminates BPA from aqueous solutions through plant metabolic processes or via the breakdown of adjacent contaminants using enzymes produced and secreted by the plant itself.

7.4.3 PHYTOFILTRATION

Phytofiltration, also known as rhizofiltration, is a technique employed for treating surface wastewater generated from industrial and agricultural sources. This process involves either spraying wastewater onto plant roots or submerging the plants in the water requiring treatment (Odinga *et al.*, 2019). The choice of plants for this method necessitates high tolerance to toxic compounds; resistance to low oxygen concentrations; and the possession of an extensive, rapidly growing root system capable of generating substantial biomass (Banerjee and Roychoudhury, 2022). Rhizofiltration is particularly effective in the removal of heavy metals such as lead (Pb) and radioactive elements (Yadav *et al.*, 2011; Yang *et al.*, 2015). In a laboratory phytofiltration study conducted by Han *et al.* (2014), four plant species, lettuce, Chinese cabbage, radish sprouts and buttercup, were investigated for their ability to decrease uranium (U) concentrations in contaminated ground water. Their findings revealed a rapid uptake of U that resulted in up to an 83% reduction in water concentration. Consequently, the authors deduced that specific plant species could be effectively utilized for remediating water contaminated with uranium.

7.4.4 Phytostabilization

In the realm of soil remediation, the technique of phytostabilization employs plant roots to mitigate the spread of contaminants. By preventing contaminants from leaching into groundwater and migrating to surface soil through rainwater runoff, this process effectively curbs further environmental damage (Mendez and Maier, 2008). Ideal plants for phytostabilization exhibit a well-developed root system capable of adsorbing, absorbing, and accumulating contaminants in their tissues while converting them into less soluble compounds within the rhizosphere (Radziemska et al., 2017). Additionally, these plants ought to possess low pollutant accumulation capacity in aboveground tissues and exhibit high tolerance to variations in pH, salinity, and soil moisture (Saran et al., 2020). In a 2014 study, Polechońska and Klink explored the phytoremediation potential of red canary grass (*Phalaris arundinace*) regarding trace metals (Zn, Fe, Mn, Pb, Cu, Ni, Cd, Co, and Cr) found in water and sediments. Their findings demonstrated that different concentrations of trace metals were present in the distinct organs of the plant – roots contained the highest concentrations, while leaves exhibited the lowest. The authors posited that due to limited translocation of absorbed trace metals within red canary grass, this species holds promise for phytostabilization applications in sediments contaminated with metals, particularly Co and Cd (Polechońska and Klink, 2014).

7.5 BACTERIAL COMMUNITY CHARACTERISTICS AND ROLES IN VARIOUS SYSTEMS OF WASTEWATER BIOREMEDIATION AND PHYTOREMEDIATION

Biological wastewater treatment and pollutant removal is based on the performance of a variety of microbes, bacteria, archaea, and fungi. In both natural and engineered environments, microbes live in complex groups called bacterial communities or biocenoses (Stubbendieck et al., 2016; Ziembińska-Buczyńska and Surmacz-Górska, 2021). These communities vary in their size and complexity, from relatively small, multicellular aggregates to groups of billions or trillions of cells, and different microbial species creating communities are linked with ecophysiological interactions (Stubbendieck et al., 2016). The features of the communities influence the efficacy of their performance, which is important especially in the case of engineered environments because there a community of a particular functionality is required. The level of diversity is also vital for community performance efficacy (Xu et al., 2018). Biodiversity of the community is not directly linked with its performance efficacy. It is rather the point of its resistance to harmful environmental factors influencing the community performance. The higher the community biodiversity, the higher the resistance towards these factors. Species diversity in biocenosis, also known as alpha diversity, is a combination of observed richness (number of taxa present in the community) or evenness (the relative abundances of those taxa) of an average sample within a particular biotope (Walters and Martiny, 2020). It should be underlined that because of this indirect lineage between community diversity and its performance, it is important to measure and analyze biodiversity in the context of community functionality and effectiveness (Willis, 2019).

In case of wastewater treatment, microorganisms can occur in various morphological forms. They could be activated sludge, granular sludge, or biofilm (Ziembińska-Buczyńska and Surmacz-Górska, 2021). The first type of community is a flocculated mixture of bacteria, archaea, protozoa, and metazoa. In this system, 95% of sludge consists of bacteria (Xu et al., 2018). The second type is a regular granule in which oxic, anaerobic, and anoxic zones are distinctly marked. These two microbiological communities are suspended inside bioreactors, and they don't need any solid surface to grow on and perform. In the case of biofilm, which is a 3D community of microbes linked strongly with extracellular substances (EPSs) produced by its members, a solid basis for community growth is required (Ziembińska-Buczyńska and Surmacz-Górska, 2021). In terms of taxonomic composition, a single community is unique; however, the proportion of particular microbial groups can be similar, especially working under the same technological parameters. This is because microbial community composition depends heavily on the technological features of the performed process and the type of the bioreactors used. Among the most important community-shaping factors, temperature, pH, solid retention time (SRT), dissolved oxygen concentration, feeding medium composition, and seed sludge composition can be mentioned (Baquero-Rodríguez et al., 2022; Xu et al., 2018).

Up-to-date molecular tools, such as next-generation sequencing, are now used to present taxonomical composition and changeability together with diversity measurements in all sorts of microbial communities. On the basis of the results obtained on the basis of molecular tools, the community structure can be presented, analyzed, and compares. For example, it was stated that in case of full-scale activated sludge-based wastewater treatment plants, *Proteobacteria* are recognized as the most ubiquitous phylum (Gao et al., 2016; Wang et al., 2012; Xu et al., 2018), while representatives of phyla *Bacteroidetes, Acidobacteria*, and *Chloroflexi* are also known to be present in large quantities there (Xu et al., 2018; T. Zhang et al., 2012; Zhao et al., 2014). As the composition of the community depends heavily also on the feeding medium composition, the type of wastewater treated shapes the community, and different bacterial representatives will dominate. Interestingly, more important is the chemical composition of the feeding medium, which serves as a source of nutrition for the dominant microbes already present in the activated sludge. The medium will be less likely to support the growth of microbes which are carried with real wastewater to WWTP. This is clearly visible at the genus level in the case of Shchegolkova et al.'s (2016) research, which presented the composition of activated sludge treatment for municipal wastewater, and together with petroleum for slaughterhouse wastewater. In the first case, *Caldilinea* and *Prostheobacter* dominated in the activated sludge. In the second case, it was *Opitutus* together with *Caldilinea* and *Prostheobacter*. In the last case, *Flavobacterium* and *Luteolibacter* were the dominant genera. Interestingly, these bacteria were not dominant in incoming wastewater. It was *Acinetobacter* which was dominant in all wastewater types directed to bioreactors, together with *Akkermansia* and *Trichococcus* for municipal wastewater; *Trichococcus* and *Arcobacter* for petroleum; and *Cloacibacterim, Acidaminococcus*, and *Streptococcus* for slaughterhouse wastewater. None of these microbes dominating in feeding real wastewater gained dominance in activated sludge working in the bioreactors. These results underline that the dominant genera are present in the activated sludge already, and their abundance and/or dominance rely on the sort of food sources brought with the feeding medium and the environmental parameters created in the bioreactor to support their growth.

From the community composition point of view (at the level of genus rather than higher taxonomic levels), the bioreactor configuration and the aeration system used to shape the community heavily. It was reported by Begmatov et al. (2022) as a result of community analysis from nine full-scale activated sludge systems. Two plants used conventional aerobic processes, one was based on nitrification/denitrification, and six operated with anaerobic/anoxic/oxic systems. All systems were dominated by *Proteobacteria, Bacteroidota*, and *Actinobacteriota*, and at this taxonomic level, the differences were not strongly marked. But at the genus level, the systems varied from the point of view of the biochemical processes dominating the technological infrastructure. For example, the anaerobic/anoxic/oxic process was efficient not only in organic matter removal but also nitrogen and phosphorus, and these communities presented high content of ammonia-oxidizing *Nitrosomonas* sp. and phosphate-accumulating bacteria. In the case of nitrification-denitrification, both ammonia and nitrite oxidizers were present in high numbers, but phosphate-accumulating bacteria and other functional groups were in the minority because this system was not supporting them well enough.

Another example of a community shaped by bioreactor and aeration configuration is the wastewater technology used for ammonia-rich low-carbon wastewater treatment known as anaerobic ammonia oxidation (anammox). This process can be performed with floccular or granular sludge, as well as in biofilms (i.e. rotating biological contactor). Due to the fact that this process is performed by representatives of *Planctomycetes*, which are a relatively slowly growing bacterial group, they are observed as co-dominating in anammox systems, together with *Proteobacteria* (Y. Wu et al., 2019) or with *Proteobacteria, Chlorobi*, and *Chloroflexi* (Chen et al., 2017). To date, anammox bacteria represent six Candidatus genera: *Jettenia, Bocadia, Scalindua, Anammoxoglobus*,

Biological Remediation of Water and Wastewaters

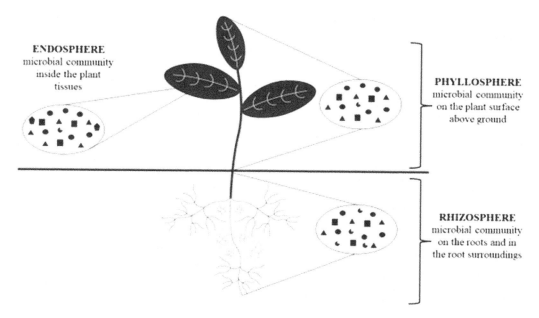

FIGURE 7.7 A scheme of plant microbiota.

Kuenenia, and *Anammoximicrobium* (Zhang and Okabe, 2020). In wastewater treatment systems, *Candidatus brocadia* and *Candidatus kuenenia* are commonly described (Chen *et al.*, 2017; Hu *et al.*, 2010). Also, *Candidatus jettenia* is detected in anammox–upflow anaerobic sludge blankets (Chen *et al.*, 2017) and in systems with decreased C/N ratios, which seem to promote *Candidatus jettenia* growth (Wang *et al.*, 2022). However, there is evidence of anammox process performance by unidentified *Planctomycetes* dominating in anammox performing systems with no particular anammox genus dominance (Tomaszewski *et al.*, 2019). These results show that there is a possibility of a wider group of anammox-performing bacteria presence in *Planctomycetes* that do not belong to any of the six known genera.

Bacteria are important not only in case of wastewater treatment systems based on activated sludge or biofilm but also in the plant-based biotechnological processes microbes play crucial role. Depending on the type of biotechnological processes in which the plant microbiome is used, one of three microbiota parts can be used: rhizosphere, phyllosphere, or bacterial endosphere (Figure 7.7) (Compant *et al.*, 2019).

Functionally, this microbiome consists of three groups of microbes: neutral, beneficial, or pathogenic for plants (Compant *et al.*, 2019), and all three functional parts are highly variable and diverse. Plant-related microbes play a vital role not only in plant growth, health, and development but also in biotechnological processes in which plants are widely used. In the case of wastewater treatment, the most important for process efficacy are rhizospheric microbiota and endophytes, which are also useful in soil phytoremediation.

As Edwards *et al.* (2015) described, plants can create unique rhizosphere microbial communities, with tremendous importance for plant nutrition and health. It should be also mentioned that rhizosphere microbial communities respond differently to soil properties and plant performance in terms of community diversity and structure and in their interactions with their host plants (Jia *et al.*, 2022). As the rhizosphere is also useful in plant-based technologies for wastewater treatment, gaining new knowledge in the field of plant–microbiome relationships is crucial for understanding this collaboration and improving biotechnological recalcitrant removal processes. In the case of wastewater, the pollutants removed in plant-based technologies are usually heavy metals, dyes, and organic or inorganic compounds (pharmaceuticals, hydrocarbons, or pesticides) (Khan *et al.*, 2022).

The composition of the rhizospheric or endophitic communities also depends not only on the type of the host plant used but also on the type of wastewater and other technological parameters which were used to obtain particular recalcitrant removal. At the level of phyla, the community structure does not differ for plant-based and activated sludge-based (conventional) technologies. In constructed wetlands, the dominating phyla include *Proteobacteria, Bacteroidetes, Actinobacteria*, and *Firmicutes* (Wang *et al.*, 2022). As Wang *et al.* (2022) reported, the dominance of a particular genus is shaped by the dominating biochemical process performed by the community in the researched system. Thus it could be stated that the mechanisms shaping the communities during wastewater treatment don't differ much for plant-based and activated sludge–based systems. For example, in case of heavy metal removal for Cd and Zn biosorption, *Serratia* and *Pseudomonas* dominated (Wang *et al.*, 2022; Yu *et al.*, 2020). In Fe biooxidation, *Thiomonas* and *Sideroxydans* were identified as main players in the community (Chen *et al.*, 2021; Wang *et al.*, 2022) In the case of antibiotic removal, the dominant genera depends on the type of pharmaceutical removed from wastewater. For sulfonamide degradation under aerobic conditions, *Bacillus* and *Geobacter* were reported as dominating (Chen *et al.*, 2020; Wang *et al.*, 2022), but the first one is linked with sulfamethoxazole and the second one with sulfadiazine degradation. Particular microbes' presence and dominance in the case of antibiotics removal are probably based on the presence of degradation genes (Liu *et al.*, 2019; Wang *et al.*, 2022).

Interestingly, in case of plant-related microbes used for pollutant removal, it is often observed that genera dominating in the community and performing the recalcitrant removal effectively also play other important roles in plant–microbe stimulation. Di Gregorio *et al.* (2015) showed that the plant growth-promoting rhizobacteria (PGPR) *Stenotrophomonas* and *Sphingobium* were identified as effectively removing nonylphenol, mono-ethoxylated nonylphenols, and di-ethoxylated nonylphenols in *Phragmites australis*-based phytoremediation. It might be said that the adaptation processes of such factors as the feeding medium and other technological parameters play key roles in the composition and changeability of the microbial population utilized in various sorts of wastewater treatment. Each community, however, is a unique microbial group despite the high resemblance of communities treating particular types of wastewater.

7.6 OPPORTUNITIES AND CHALLENGES OF BIOREMEDIATION AND PHYTOREMEDIATION

7.6.1 OPPORTUNITIES OF BIOREMEDIATION AND PHYTOREMEDIATION

Bioremediation, including microorganisms (microbial remediation) and plants (phytoremediation), provides a viable and suitable technique with high performance, low maintenance costs, and excellent selectivity and has the ability to reduce pollutants in highly polluted environments (water, sediment, and soil). Bioremediation is usually used to eradicate toxic materials such as pesticides, fertilizers, and heavy metals from the environment. Bioremediation and phytoremediation techniques align with the principles of sustainability by utilizing natural processes to treat wastewater. They offer significant environmental benefits by minimizing the use of chemicals and energy-intensive operations associated with traditional treatment methods (Schwitzguébel *et al.*, 2011). Additionally, these techniques often result in lower carbon footprints and reduced greenhouse gas emissions compared to conventional treatment approaches (D'Orazio *et al.*, 2013). Bioremediation and phytoremediation techniques are versatile and can be tailored to address specific contaminants in wastewater. Recent studies have focused on the development of microbial strains and genetically modified plants with enhanced pollutant removal capabilities (Walters and Martiny, 2020). Microorganisms can be engineered to express specific enzymes or pathways for the degradation of targeted pollutants, while plants can be selected based on their ability to accumulate

Biological Remediation of Water and Wastewaters

or transform specific contaminants. This versatility allows for the design of treatment systems that can effectively target a wide range of pollutants and adapt to different wastewater compositions. Bioremediation and phytoremediation techniques offer the potential for resource recovery from wastewater, such as the production of biofuels, bioplastics, and other value-added products. Additionally, the recovery of nutrients, such as nitrogen and phosphorus, through phytoremediation can contribute to their recycling and reduce the reliance on synthetic fertilizers. This circular economy approach enhances the economic viability and sustainability of wastewater treatment processes. Also, these techniques generally have lower operation and maintenance costs compared to conventional treatment methods. The reliance on natural processes and the reduced need for chemical additives result in cost savings.

7.6.2 CHALLENGES OF BIOREMEDIATION AND PHYTOREMEDIATION

Environmental factors, such as temperature, pH, and oxygen availability, significantly influence the performance of bioremediation and phytoremediation techniques. Recent studies have highlighted the importance of understanding microbial and plant responses to varying environmental conditions (Pearce and Jarvis, 2011). Strategies such as temperature control, pH adjustment, and aeration optimization can help mitigate the impact of environmental variability and maintain consistent treatment efficiency. Wastewater often contains complex mixtures of pollutants, including organic compounds, heavy metals, and emerging contaminants. Recent studies have shown that the presence of co-contaminants can influence the overall efficiency of bioremediation and phytoremediation processes. Interactions among contaminants may lead to competitive inhibition or toxic effects, affecting microbial activity or plant health (Xia *et al.*, 2010). Also, the selection of suitable microbial strains or plant species is critical for the success of bioremediation and phytoremediation techniques. The studies presented in Section 7.5 explored the potential of microbial consortia or synthetic microbial communities to enhance pollutant removal efficiency. The use of plant-associated bacteria or mycorrhizal fungi has also shown promising results in improving plant growth and pollutant uptake. Advancements in molecular techniques, such as high-throughput sequencing and metagenomics, have facilitated the identification and characterization of microbial communities in different treatment systems (Ziembińska-Buczyńska and Surmacz-Górska, 2021). Integrating omics approaches with functional assays can help optimize microbial and plant selection for enhanced treatment performance. Scaling up bioremediation and phytoremediation techniques from laboratory-scale to full-scale applications presents challenges in terms of on optimizing reactor design, hydraulic retention time, and biomass retention to ensure consistent treatment performance. Computational modeling and simulation tools have been used to aid in the design and optimization of treatment systems (Al-Sayed *et al.*, 2023a; Hellal and Abou-Elela, 2021). Additionally, pilot-scale and full-scale demonstrations are necessary to validate the feasibility and performance of these techniques under real-world conditions. Continuous monitoring and control of bioremediation and phytoremediation systems are essential for their efficient operation. Ensuring the long-term sustainability and safety of bioremediation and phytoremediation techniques requires comprehensive risk assessments and environmental monitoring. Life cycle assessment (LCA) studies have been conducted to evaluate the overall environmental impacts and compare the sustainability of different treatment options (Abello-Passteni *et al.*, 2020). Integrating risk assessment and LCA approaches can provide insights into the potential trade-offs and benefits associated with bioremediation and phytoremediation techniques.

7.7 CONCLUSION

Bioremediation and phytoremediation techniques offer significant opportunities for sustainable wastewater treatment. Their versatility, resource recovery potential, and low operation and maintenance costs make them attractive alternatives to conventional treatment methods. However, challenges such as environmental variability, complex pollutants, microbial and plant selection, system optimization, monitoring and control, and long-term sustainability must be addressed to maximize their effectiveness. Recent studies have provided valuable insights into these challenges and proposed innovative solutions to overcome them. Continued research, technological advancements, and real-world demonstrations are crucial to further develop and optimize bioremediation and phytoremediation techniques for effective and sustainable wastewater treatment.

REFERENCES

Abdelaal, Mohamed, Ibrahim A. Mashaly, Dina S. Srour, Mohammed A. Dakhil, Mohamed Azab El-Liethy, Ali El-Keblawy, Reham F. El-Barougy, Marwa Waseem A. Halmy, and Ghada A. El-Sherbeny. 2021. "Phytoremediation Perspectives of Seven Aquatic Macrophytes for Removal of Heavy Metals from Polluted Drains in the Nile Delta of Egypt." *Biology* 10, no. 6. https://doi.org/10.3390/biology 10060560.

Abd-Elmaksoud, S., S. M. Abdo, M. Gad, A. Hu, M. A. El-Liethy, N. Rizk, M. A. Marouf, I. A. Hamza, and H. S. Doma. 2021. "Pathogens Removal in a Sustainable and Economic High-Rate Algal Pond Wastewater Treatment System." *Sustainability* 13, no. 23: 1–13. https://doi.org/10.3390/su132313232.

Abello-Passteni, V., E. M. Muñoz Alvear, S. Lira, and E. Garrido-Ramírez. 2020. "Eco-efficiency Assessment of Domestic Wastewater Treatment Technologies Used in Chile." *Tecnologia y Ciencias del Agua* 11, no. 2: 190–228. https://doi.org/10.24850/j-tyca-2020-02-05.

Abou-Elela, S. I., S. A. Abo-El-Enein, and M. S. Hellal. 2019a. "Utilization of Autoclaved Aerated Concrete Solid Waste as a Bio-carrier in Immobilized Bioreactor for Municipal Wastewater Treatment." *Desalination & Water Treatment* 168: 108–16. https://doi.org/10.5004/dwt.2019.24640.

Abou-Elela, Sohair I., Mohamed A. Elekhnawy, Magdy T. Khalil, and Mohamed S. Hellal. 2017. "Factors Affecting the Performance of Horizontal Flow Constructed Treatment Wetland Vegetated with Cyperus Papyrus for Municipal Wastewater Treatment." *International Journal of Phytoremediation* 19, no. 11: 1023–8. https://doi.org/10.1080/15226514.2017.1319327.

Abou-Elela, S. I., S. A. El-Shafai, M. E. Fawzy, M. S. Hellal, and O. Kamal. 2018. "Management of Shock Loads Wastewater Produced from Water Heaters Industry." *International Journal of Environmental Science & Technology* 15, no. 4: 743–54. https://doi.org/10.1007/s13762-017-1433-9.

Abou-Elela, Sohair I., G. Golinelli, Abdou Saad El-Tabl, and Mohammed S. Hellal. 2014. "Treatment of Municipal Wastewater Using Horizontal Flow Constructed Wetlands in Egypt." *Water Science & Technology* 69, no. 1: 38–47. https://doi.org/10.2166/wst.2013.530.

Abou-Elela, S. I., G. Golinielli, E. M. Abou-Taleb, and M. S. Hellal. 2013. "Municipal Wastewater Treatment in Horizontal and Vertical Flows Constructed Wetlands." *Ecological Engineering* 61: 460–8. https://doi.org/10.1016/j.ecoleng.2013.10.010.

Abou-Elela, S. I., and M. S. Hellal. 2012. "Municipal Wastewater Treatment Using Vertical Flow Constructed Wetlands Planted with Canna, Phragmites and Cyprus." *Ecological Engineering* 47: 209–13. https://doi.org/10.1016/j.ecoleng.2012.06.044.

Abou-Elela, Sohair I., Mohamed S. Hellal, Olfat H. Aly, and Salah A. Abo-Elenin.2019b. "Decentralized Wastewater Treatment Using Passively Aerated Biological Filter." *Environmental Technology* 40, no. 2: 250–60. https://doi.org/10.1080/09593330.2017.1385648.

Abou-Elela, S. I., M. S. Hellal, and M. A. Elekhnawy.2019c. "Phytoremediation of Municipal Wastewater for Reuse Using Three Pilot-Scale HFCW Under Different HLR, HRT and Vegetation: A Case Study from Egypt." *Desalination & Water Treatment* 140: 80–90. https://doi.org/10.5004/dwt.2019.23362.

Abou-Elela, S. I., M. S. Hellal, and A. H. Harb. 2015. "Assessment of Seasonal Variations on the Performance of P-UASB/BAF for Municipal Wastewater Treatment." *Desalination & Water Treatment* 57: 1–8. https://doi.org/10.1080/19443994.2015.1103308.

Abou-Elela, S. I., M. M. Kamel, and M. E. Fawzy. 2010. "Biological Treatment of Saline Wastewater Using a Salt-Tolerant Microorganism." *Desalination* 250, no. 1: 1–5. https://doi.org/10.1016/j.desal.2009.03.022.

Abou-Taleb, E. M., M. E. M. Ali, M. S. Hellal, K. H. Kamal, S. M. Abdel Moniem, N. S. Ammar, and H. S. Ibrahim. 2020. "Sustainable Solutions for Urban Wastewater Management and Remediation." *Egyptian Journal of Chemistry* 63, no. 2: 405–15. https://doi.org/10.21608/ejchem.2019.13605.1840.

Ahmad, Maqshoof, Lisa Pataczek, Thomas H. Hilger, Zahir Ahmad Zahir, Azhar Hussain, Frank Rasche, Roland Schafleitner, and Svein Ø. Solberg. 2018. "Perspectives of Microbial Inoculation for Sustainable Development and Environmental Management." *Frontiers in Microbiology* 9: 2992. https://doi.org/10.3389/fmicb.2018.02992.

Akram, R., V. Turan, H. M. Hammad, S. Ahmad, S. Hussain, A. Hasnain, M. M. Maqbool, M. I. A. Rehmani, A. Rasool, N. Masood, F. Mahmood, M. Mubeen, S. R. Sultana, S. Fahad, K. Amanet, M. Saleem, Y. Abbas, H. M. Akhtar, S. Hussain, F. Waseem, R. Murtaza, A. Amin, S. A. Zahoor, Samiul Din, M., and W. Nasim. 2018. *Fate of Organic and Inorganic Pollutants in Paddy Soils*: 197–214. https://doi.org/10.1007/978-3-319-93671-0_13.

Al-Gheethi, A. A., A. N. Efaq, J. D. Bala, I. Norli, M. O. Abdel-Monem, and Ab. 2018. "Removal of Pathogenic Bacteria from Sewage-Treated Effluent and Biosolids for Agricultural Purposes." *Applied Water Science* 8, no. 2. https://doi.org/10.1007/s13201-018-0698-6.

Ali, Naeem, Abdul Hameed, and Safia Ahmed. 2009. "Physicochemical Characterization and Bioremediation Perspective of Textile Effluent, Dyes and Metals by Indigenous Bacteria." *Journal of Hazardous Materials* 164, no. 1: 322–8. https://doi.org/10.1016/j.jhazmat.2008.08.006.

Al-Sayed, A., G. K. Hassan, M. T. Al-Shemy, and F. A. El-Gohary. 2023a. "Effect of Organic Loading Rates on the Performance of Membrane Bioreactor for Wastewater Treatment Behaviours, Fouling, and Economic Cost." *Scientific Reports* 13, no. 1: 15601.

Al-Sayed, A., M. S. Hellal, M. T. Al-Shemy, and G. K. Hassan. 2023b. "Performance Evaluation of Submerged Membrane Bioreactor for Municipal Wastewater Treatment: Experimental Study and Model Validation with GPS-X Software Simulator." *Water & Environment Journal* 37, no. 3: 480–92. https://doi.org/10.1111/wej.12852.

Annop, S., P. Sridang, U. Puetpaiboon, and A. Grasmick. 2014. "Effect of Solids Retention Time on Membrane Fouling Intensity in Two-Stage Submerged Anaerobic Membrane Bioreactors Treating Palm Oil Mill Effluent." *Environmental Technology* 35, no. 17–20: 2634–42. https://doi.org/10.1080/09593330.2014.914575.

Aslam, Muhammad, Perry L. Mc Carty, Chungheon Shin, Jaeho Bae, and Jeonghwan Kim. 2017. "Low Energy Single-Staged Anaerobic Fluidized Bed Ceramic Membrane Bioreactor (AFCMBR) for Wastewater Treatment." *Bioresource Technology* 240: 33–41. https://doi.org/10.1016/j.biortech.2017.03.017.

Awa, S. H., and T. Hadibarata. 2020. "Removal of Heavy Metals in Contaminated Soil by Phytoremediation Mechanism: A Review." *Water, Air, & Soil Pollution* 231, no. 2: 47. https://doi.org/10.1007/s11270-020-4426-0.

Azubuike, Christopher Chibueze, Chioma Blaise Chikere, and Gideon Chijioke Okpokwasili. 2016. "Bioremediation Techniques–Classification Based on Site of Application: Principles, Advantages, Limitations and Prospects." *World Journal of Microbiology & Biotechnology* 32, no. 11: 180. https://doi.org/10.1007/s11274-016-2137-x.

Bai, Yaohui, Qinghua Sun, Renhua Sun, Donghui Wen, and Xiaoyan Tang. 2011. "Bioaugmentation and Adsorption Treatment of Coking Wastewater Containing Pyridine and Quinoline Using Zeolite-Biological Aerated Filters." *Environmental Science & Technology* 45, no. 5: 1940–8. https://doi.org/10.1021/es103150v.

Bai, Yaohui, Qinghua Sun, Cui Zhao, Donghui Wen, and Xiaoyan Tang. 2010. "Bioaugmentation Treatment for Coking Wastewater Containing Pyridine and Quinoline in a Sequencing Batch Reactor." *Applied Microbiology & Biotechnology* 87, no. 5: 1943–51. https://doi.org/10.1007/s00253-010-2670-8.

Banach-Wiśniewska, Anna, Mariusz Tomaszewski, Mohamed S. Hellal, and Aleksandra Ziembińska-Buczyńska. 2021. "Effect of Biomass Immobilization and Reduced Graphene Oxide on the Microbial Community Changes and Nitrogen Removal at Low Temperatures." *Scientific Reports* 11, no. 1: 840. https://doi.org/10.1038/s41598-020-80747-7.

Banerjee, A., and A. Roychoudhury. 2022. "Assessing the Rhizofiltration Potential of Three Aquatic Plants Exposed to Fluoride and Multiple Heavy Metal Polluted Water." *Vegetos* 35, no. 4: 1158–64. https://doi.org/10.1007/s42535-022-00405-3.

Baquero-Rodríguez, Gustavo Andrés, Sandra Martínez, Julián Acuña, Daniel Nolasco, and Diego Rosso. 2022. "How Elevation Dictates Technology Selection in Biological Wastewater Treatment." *Journal of Environmental Management* 307: 114588. https://doi.org/10.1016/j.jenvman.2022.114588.

Begmatov, Shahjahon, Alexander G. Dorofeev, Vitaly V. Kadnikov, Alexey V. Beletsky, Nikolai V. Pimenov, Nikolai V. Ravin, and Andrey V. Mardanov. 2022. "The Structure of Microbial Communities of Activated Sludge of Large-Scale Wastewater Treatment Plants in the City of Moscow." *Scientific Reports* 12, no. 1: 3458. https://doi.org/10.1038/s41598-022-07132-4.

Ben Rebah, F. B., A. Kantardjieff, A. Yezza, and J. P. Jones. 2010. "Performance of Two Combined Anaerobic-Aerobic Biofilters Packed with Clay or Plastic Media for the Treatment of Highly Concentrated Effluent." *Desalination* 253, no. 1–3: 141–6. https://doi.org/10.1016/j.desal.2009.11.018.

Bengtsson, S., M. de Blois, B. M. Wilén, and D. Gustavsson. 2018. "Treatment of Municipal Wastewater with Aerobic Granular Sludge." *Critical Reviews in Environmental Science & Technology* 48, no. 2: 119–66. https://doi.org/10.1080/10643389.2018.1439653.

Bhargava, Atul, Francisco F. Carmona, Meenakshi Bhargava, and Shilpi Srivastava. 2012. "Approaches for Enhanced Phytoextraction of Heavy Metals." *Journal of Environmental Management* 105: 103–20. https://doi.org/10.1016/j.jenvman.2012.04.002.

Bhat, Shakeel Ahmad, Omar Bashir, Syed Anam Ul Haq, Tawheed Amin, Asif Rafiq, Mudasir Ali, Juliana Heloisa Pinê Américo-Pinheiro, and Farooq Sher. 2022. "Phytoremediation of Heavy Metals in Soil and Water: An Eco-friendly, Sustainable and Multidisciplinary Approach." *Chemosphere* 303, no. 1: 134788. https://doi.org/10.1016/j.chemosphere.2022.134788.

Bian, X., J. Cui, B. Tang, and L. Yang. 2018. "Chelant-Induced Phytoextraction of Heavy Metals from Contaminated Soils: A Review." *Polish Journal of Environmental Studies* 27, no. 6: 2417–24. https://doi.org/10.15244/pjoes/81207.

Biplob, P., S. Fatihah, Z. Shahrom, and E. Ahmed. 2011. "Nitrogen-Removal Efficiency in an Upflow Partially Packed Biological Aerated Filter (BAF) Without Backwashing Process." *Journal of Water Reuse & Desalination* 1, no. 1: 27–35. https://doi.org/10.2166/wrd.2011.008.

Bodkhe, S. Y. 2009. "A Modified Anaerobic Baffled Reactor for Municipal Wastewater Treatment." *Journal of Environmental Management* 90, no. 8: 2488–93. https://doi.org/10.1016/j.jenvman.2009.01.007.

Boonnorat, Jarungwit, Somkiet Techkarnjanaruk, Ryo Honda, and Pradthana Prachanurak. 2016. "Effects of Hydraulic Retention Time and Carbon to Nitrogen Ratio on Micro-pollutant Biodegradation in Membrane Bioreactor for Leachate Treatment." *Bioresource Technology* 219: 53–63. https://doi.org/10.1016/j.biortech.2016.07.094.

Boopathy, R. 2000. "Factors Limiting Bioremediation Technologies." *Bioresource Technology* 74, no. 1: 63–7. https://doi.org/10.1016/S0960-8524(99)00144-3.

Bradford, Lauren M., Gisle Vestergaard, András Táncsics, Baoli Zhu, Michael Schloter, and Tillmann Lueders. 2018. "Transcriptome-Stable Isotope Probing Provides Targeted Functional and Taxonomic Insights into Microaerobic Pollutant-Degrading Aquifer Microbiota." *Frontiers in Microbiology* 9: 2696. https://doi.org/10.3389/fmicb.2018.02696.

Buchanan, I., and R. Leduc. 1994. *Probabilistic Design of Multi-stage Rotating Biological Contactors*: 113–25. https://doi.org/10.1007/978-94-017-3081-5_9.

Cabije, A. H., R. C. Agapay, and M. V. Tampus. 2009. "Carbon-Nitrogen-Phosphorus Removal and Biofilm Growth Characteristics in an Integrated Wastewater Treatment System Involving a Rotating Biological Contactor." *Asia-Pacific Journal of Chemical Engineering* 4, no. 5: 735–43. https://doi.org/10.1002/apj.329.

Cai, P., Z. Ning, Y. Liu, Z. He, J. Shi, and M. Niu. 2020. "Diagnosing Bioremediation of Crude Oil-Contaminated Soil and Related Geochemical Processes at the Field Scale Through Microbial Community and Functional Genes." *Annals of Microbiology* 70, no. 1. https://doi.org/10.1186/s13213-020-01580-x.

Chae, S. R., Y. T. Ahn, S. T. Kang, and H. S. Shin. 2006. "Mitigated Membrane Fouling in a Vertical Submerged Membrane Bioreactor (VSMBR)." *Journal of Membrane Science* 280, no. 1–2: 572–81. https://doi.org/10.1016/j.memsci.2006.02.015.

Chang, Shih-Hsien, Cheng-Fang Wu, Chu-Fang Yang, and Chi-Wen Lin. 2021. "Evaluation Use of Bioaugmentation and Biostimulation to Improve Degradation of Sulfolane in Artificial Groundwater." *Chemosphere* 263: 127919. https://doi.org/10.1016/j.chemosphere.2020.127919.

Chaudhary, P., V. Beniwal, P. Sharma, S. Goyal, R. Kumar, A. A. M. Alkhanjaf, and A. Umar. 2022. "Unloading of Hazardous Cr and Tannic Acid from Real and Synthetic Waste Water by Novel Fungal Consortia." *Environmental Technology & Innovation* 26. https://doi.org/10.1016/j.eti.2021.102230.

Chen, C., C. Li, G. Reniers, and F. Yang. 2021. "Safety and Security of Oil and Gas Pipeline Transportation: A Systematic Analysis of Research Trends and Future Needs Using WoS." *Journal of Cleaner Production* 279: 123583. https://doi.org/10.1016/j.jclepro.2020.123583.

Chen, Jianfei, Tianli Tong, Xinshu Jiang, and Shuguang Xie. 2020. "Biodegradation of Sulfonamides in Both Oxic and Anoxic Zones of Vertical Flow Constructed Wetland and the Potential Degraders." *Environmental Pollution* 265, no. B: 115040. https://doi.org/10.1016/j.envpol.2020.115040.

Chen, Ming, Piao Xu, Guangming Zeng, Chunping Yang, Danlian Huang, and Jiachao Zhang. 2015. "Bioremediation of Soils Contaminated with Polycyclic Aromatic Hydrocarbons, Petroleum, Pesticides, Chlorophenols and Heavy Metals by Composting: Applications, Microbes and Future Research Needs." *Biotechnology Advances* 33, no. 6 Pt. 1: 745–55. https://doi.org/10.1016/j.biotechadv.2015.05.003.

Chen, W., X. Dai, D. Cao, X. Hu, W. Liu, and D. Yang. 2017. "Characterization of a Microbial Community in an Anammox Process Using Stored Anammox Sludge." *Water* 9, no. 11: 829. https://doi.org/10.3390/w9110829.

Cheng, George, Florian Gabler, Leticia Pizzul, Henrik Olsson, Åke Nordberg, and Anna Schnürer. 2022. "Microbial Community Development During Syngas Methanation in a Trickle Bed Reactor with

Various Nutrient Sources." *Applied Microbiology & Biotechnology* 106, no. 13–16: 5317–33. https://doi.org/10.1007/s00253-022-12035-5.

Cho, J., K. G. Song, S. Hyup Lee, and K. H. Ahn. 2005. "Sequencing Anoxic/Anaerobic Membrane Bioreactor (SAM) Pilot Plant for Advanced Wastewater Treatment." *Desalination* 178, no. 1–3: 219–25. https://doi.org/10.1016/j.desal.2004.12.018.

Chopin, T., J. A. Cooper, G. Reid, S. Cross, and C. Moore. 2012. "Open-Water Integrated Multi-trophic Aquaculture: Environmental Biomitigation and Economic Diversification of Fed Aquaculture by Extractive Aquaculture." *Reviews in Aquaculture* 4, no. 4: 209–20. https://doi.org/10.1111/j.1753-5131.2012.01074.x.

Compant, Stéphane, Abdul Samad, Hanna Faist, and Angela Sessitsch. 2019. "A Review on the Plant Microbiome: Ecology, Functions, and Emerging Trends in Microbial Application." *Journal of Advanced Research* 19: 29–37. https://doi.org/10.1016/j.jare.2019.03.004.

Cortez, S., P. Teixeira, R. Oliveira, and M. Mota. 2008. "Rotating Biological Contactors: A Review on Main Factors Affecting Performance." *Reviews in Environmental Science & Bio/Technology* 7, no. 2: 155–72. https://doi.org/10.1007/s11157-008-9127-x.

Cvetkovic, D., V. Susterstic, D. Gordic, M. Bojic, and S. Stosic. 2014. "Performance of Single-Stage Rotating Biological Contactor with Supplemental Aeration." *Environmental Engineering & Management Journal* 13, no. 3: 681–8. https://doi.org/10.30638/eemj.2014.072.

D'Orazio, V., A. Ghanem, and N. Senesi. 2013. "Phytoremediation of Pyrene Contaminated Soils by Different Plant Species." *CLEAN—Soil, Air, Water* 41, no. 4: 377–82. https://doi.org/10.1002/clen.201100653.

Debik, E., and T. Coskun. 2009. "Use of the Static Granular Bed Reactor (SGBR) with Anaerobic Sludge to Treat Poultry Slaughterhouse Wastewater and Kinetic Modeling." *Bioresource Technology* 100, no. 11: 2777–82. https://doi.org/10.1016/j.biortech.2008.12.058.

Delgado-González, Cristián Raziel, Alfredo Madariaga-Navarrete, José Miguel Fernández-Cortés, Margarita Islas-Pelcastre, Goldie Oza, Hafiz M. N. Iqbal, and Ashutosh Sharma. 2021. "Advances and Applications of Water Phytoremediation: A Potential Biotechnological Approach for the Treatment of Heavy Metals from Contaminated Water." *International Journal of Environmental Research & Public Health* 18, no. 10: 5215. https://doi.org/10.3390/ijerph18105215.

Di Gregorio, S., L. Giorgetti, M. Ruffini Castiglione, L. Mariotti, and R. Lorenzi. 2015. "Phytoremediation for Improving the Quality of Effluents from a Conventional Tannery Wastewater Treatment Plant." *International Journal of Environmental Science & Technology* 12, no. 4: 1387–400. https://doi.org/10.1007/s13762-014-0522-2.

Dutta, S., E. Hoffmann, and H. H. Hahn. 2007. "Study of Rotating Biological Contactor Performance in Wastewater Treatment Using Multi-culture Biofilm Model." *Water Science & Technology* 55, no. 8–9: 345–53. https://doi.org/10.2166/wst.2007.276.

Edwards, Joseph, Cameron Johnson, Christian Santos-Medellín, Eugene Lurie, Natraj Kumar Podishetty, Srijak Bhatnagar, Jonathan A. Eisen, Venkatesan Sundaresan, and L. D. Jeffery. 2015. "Structure, Variation, and Assembly of the Root-Associated Microbiomes of Rice." *Proceedings of the National Academy of Sciences of the United States of America* 112, no. 8: E911–20. https://doi.org/10.1073/pnas.1414592112.

El-Liethy, Mohamed Azab, Mohammed A. Dakhil, AliEl-Keblawy, Mohamed Abdelaal, Marwa Waseem A. Halmy, Abdelbaky Hossam Elgarhy, Ilunga Kamika, Ghada A. El-Sherbeny, and Mai Ali Mwaheb. 2022. "Temporal Phytoremediation Potential for Heavy Metals and Bacterial Abundance in Drainage Water." *Scientific Reports* 12, no. 1: 8223. https://doi.org/10.1038/s41598-022-11951-w.

Elmitwalli, Tarek A., and Ralf Otterpohl. 2007. "Anaerobic Biodegradability and Treatment of Grey Water in Upflow Anaerobic Sludge Blanket (UASB) Reactor." *Water Research* 41, no. 6: 1379–87. https://doi.org/10.1016/j.watres.2006.12.016.

Farraji, H., B. Robinson, P. Mohajeri, and T. Abedi. 2020. "Phytoremediation: Green Technology for Improving Aquatic and Terrestrial Environments." *Nippon Journal of Environmental Science* 1, no. 1. https://doi.org/10.46266/njes.1002.

Filer, J., H. H. Ding, and S. Chang. 2019. "Biochemical Methane Potential (BMP) Assay Method for Anaerobic Digestion Research." *Water* 11, no. 5: 923. https://doi.org/10.3390/w11050921.

Firmino, P. I. M., R. S. Farias, A. N. Barros, P. M. C. Buarque, E. Rodríguez, A. C. Lopes, and A. B. dos Santos. 2015. "Understanding the Anaerobic BTEX Removal in Continuous-Flow Bioreactors for Ex Situ Bioremediation Purposes." *Chemical Engineering Journal* 281: 272–80. https://doi.org/10.1016/j.cej.2015.06.106.

Gad, M., S. M. Abdo, A. Hu, M. A. El-Liethy, M. S. Hellal, H. S. Doma, and G. H. Ali. 2022. "Performance Assessment of Natural Wastewater Treatment Plants by Multivariate Statistical Models: A Case Study." *Sustainability* 14, no. 13: 7658. https://doi.org/10.3390/su14137658.

Gao, Chunming, Xiangxiang Jin, Jingbei Ren, Hua Fang, and Yunlong Yu. 2015. "Bioaugmentation of DDT-Contaminated Soil by Dissemination of the Catabolic Plasmid pDOD." *Journal of Environmental Sciences (China)* 27: 42–50. https://doi.org/10.1016/j.jes.2014.05.045.

Gao, Pin, Wenli Xu, Philip Sontag, Xiang Li, Gang Xue, Tong Liu, and Weimin Sun. 2016. "Correlating Microbial Community Compositions with Environmental Factors in Activated Sludge from Four Full-Scale Municipal Wastewater Treatment Plants in Shanghai, China." *Applied Microbiology & Biotechnology* 100, no. 10: 4663–73. https://doi.org/10.1007/s00253-016-7307-0.

García-Sánchez, Mercedes, Zdeněk Košnář, Filip Mercl, Elisabet Aranda, and Pavel Tlustoš. 2018. "A Comparative Study to Evaluate Natural Attenuation, Mycoaugmentation, Phytoremediation, and Microbial-Assisted Phytoremediation Strategies for the Bioremediation of an Aged PAH-Polluted Soil." *Ecotoxicology & Environmental Safety* 147: 165–74. https://doi.org/10.1016/j.ecoenv.2017.08.012.

Gavrilescu, M. 2002. "Engineering Concerns and New Developments in Anaerobic Waste-Water Treatment." *Clean Technologies & Environmental Policy* 3, no. 4: 346–62. https://doi.org/10.1007/s10098-001-0123-x.

Giloteaux, Ludovic, Dawn E. Holmes, Kenneth H. Williams, Kelly C. Wrighton, Michael J. Wilkins, Alison P. Montgomery, Jessica A. Smith, Roberto Orellana, Courtney A. Thompson, Thomas J. Roper, Philip E. Long, and Derek R. Lovley. 2013. "Characterization and Transcription of Arsenic Respiration and Resistance Genes During *In Situ* Uranium Bioremediation." *ISME Journal* 7, no. 2: 370–83. https://doi.org/10.1038/ismej.2012.109.

Gomez, F., and M. Sartaj. 2013. "Field Scale Ex-Situ Bioremediation of Petroleum Contaminated Soil Under Cold Climate Conditions." *International Biodeterioration & Biodegradation* 85: 375–82. https://doi.org/10.1016/j.ibiod.2013.08.003.

Han, Y., S. Kim, H. Heo, and M. Lee. 2014. "Application of Rhizofiltration Using Lettuce, Chinese Cabbage, Radish Sprouts and Buttercup for the Remediation of Uranium Contaminated Groundwater." *Journal of Soil & Groundwater Environment* 19, no. 6: 37–48. https://doi.org/10.7857/JSGE.2014.19.6.037.

Harms, H., and A. J. B. Zehnder. 1995. "Bioavailability of Sorbed 3-Chlorodibenzofuran." *Applied & Environmental Microbiology* 61, no. 1: 27–33. https://doi.org/10.1128/aem.61.1.27-33.1995.

Hassard, F., J. Biddle, E. Cartmell, B. Jefferson, S. Tyrrel, and T. Stephenson. 2015. "Rotating Biological Contactors for Wastewater Treatment—A Review." *Process Safety & Environmental Protection* 94: 285–306. https://doi.org/10.1016/j.psep.2014.07.003.

Heidari, S., D. A. Wood, and A. F. Ismail. 2022. "Algal-Based Membrane Bioreactor for Wastewater Treatment." In *Bioresource Technology: Concept, Tools and Experiences*, edited by Tanveer Bilal Pirzadah, Bisma Malik, Rouf Ahmad Bhat, and Khalid Rehman Hakeem: 347–72. John Wiley & Sons Ltd. https://doi.org/10.1002/9781119789444.ch11.

Hellal, M. S., and S. I. Abou-Elela. 2021. "Simulation of a Passively Aerated Biological Filter (PABF) Immobilized with Non-woven Polyester Fabric (NWPF) for Wastewater Treatment Using GPS-X." *Water & Environment Journal* 35, no. 4: 1192–203. https://doi.org/10.1111/wej.12709.

Hellal, M. S., S. I. Abou-Elela, and O. H. Aly. 2020. "Potential of Using Nonwoven Polyester Fabric (NWPF) as a Packing Media in Multistage Passively Aerated Biological Filter for Municipal Wastewater Treatment." *Water & Environment Journal* 34, no. 2: 247–58. https://doi.org/10.1111/wej.12458.

Hellal, M. S., E. M. Abou-Taleb, A. M. Rashad, and G. K. Hassan. 2022. "Boosting Biohydrogen Production from Dairy Wastewater via Sludge Immobilized Beads Incorporated with Polyaniline Nanoparticles." *Biomass & Bioenergy* 162: 106499. https://doi.org/10.1016/j.biombioe.2022.106499.

Hellal, M. S., A. Al-Sayed, M. A. El-Liethy, and G. K. Hassan. 2021. "Technologies for Wastewater Treatment and Reuse in Egypt: Prospectives and Future Challenges." In *Handbook of Advanced Approaches Towards Pollution Prevention and Control*: 275–310. Elsevier. https://doi.org/10.1016/B978-0-12-822134-1.00010-5.

Hellal, M. S., and H. S. Doma. 2022. "Combined Slaughterhouse Wastewater Treatment via Pilot Plant Chemical Coagulation Followed by 4th Generation Downflow Hanging Sponge (DHS-4G)." *Journal of Environmental Science & Health—Part A* 0: 1–11. https://doi.org/10.1080/10934529.2022.2130634.

Herrero, M., and D. C. Stuckey. 2015. "Bioaugmentation and Its Application in Wastewater Treatment: A Review." *Chemosphere* 140: 119–28. https://doi.org/10.1016/j.chemosphere.2014.10.033.

Hesnawi, R., K. Dahmani, A. Al-Swayah, S. Mohamed, and S. A. Mohammed. 2014. "Biodegradation of Municipal Wastewater with Local and Commercial Bacteria." In *Procedia Engineering* 70: 810–4. https://doi.org/10.1016/j.proeng.2014.02.088.

Hoh, D., S. Watson, and E. Kan. 2016. "Algal Biofilm Reactors for Integrated Wastewater Treatment and Biofuel Production: A Review." *Chemical Engineering Journal* 287: 466–73. https://doi.org/10.1016/j.cej.2015.11.062.

Biological Remediation of Water and Wastewaters

Hu, B. L., Ping Zheng, C. J. Tang, J. W. Chen, Erwinvan der Biezen, Lei Zhang, B. J. Ni, Mike S. M. Jetten, Jia Yan, Han-Qing Yu, and Boran Kartal. 2010. "Identification and Quantification of Anammox Bacteria in Eight Nitrogen Removal Reactors." *Water Research* 44, no. 17: 5014–20. https://doi.org/10.1016/j.watres.2010.07.021.

Huang, Shujuan, Ching Kwek Pooi, Xueqing Shi, Sunita Varjani, and How Yong Ng. 2020. "Performance and Process Simulation of Membrane Bioreactor (MBR) Treating Petrochemical Wastewater." *Science of the Total Environment* 747: 141311. https://doi.org/10.1016/j.scitotenv.2020.141311.

Hussien, M. T. M., M. A. El-Liethy, A. L. K. Abia, and M. A. Dakhil. 2020. "Low-Cost Technology for the Purification of Wastewater Contaminated with Pathogenic Bacteria and Heavy Metals." *Water, Air, & Soil Pollution* 231, no. 8: 400. https://doi.org/10.1007/s11270-020-04766-w.

Ibrahim, Salma, Mohamed Azab El-Liethy, Akebe Luther King Abia, Mohammed Abdel-Gabbar, Ali Mahmoud Al Zanaty, and Mohamed Mohamed Kamel. 2020. "Design of a Bioaugmented Multistage Biofilter for Accelerated Municipal Wastewater Treatment and Deactivation of Pathogenic Microorganisms." *Science of the Total Environment* 703: 134786. https://doi.org/10.1016/j.scitotenv.2019.134786.

Ihsanullah, I., A. Jamal, M. Ilyas, M. Zubair, G. Khan, and M. A. Atieh. 2020. "Bioremediation of Dyes: Current Status and Prospects." *Journal of Water Process Engineering* 38: 101680. https://doi.org/10.1016/j.jwpe.2020.101680.

Ikram, M., M. Naeem, M. Zahoor, M. M. Hanafiah, A. A. Oyekanmi, R. Ullah, D. A. A. Farraj, M. S. Elshikh, I. Zekker, and N. Gulfam. 2022. "Biological Degradation of the Azo Dye Basic Orange 2 by Escherichia coli: A Sustainable and Ecofriendly Approach for the Treatment of Textile Wastewater." *Water* 14, no. 13: 2063. https://doi.org/10.3390/w14132063.

Inoue, Daisuke, Yuji Yamazaki, Hirofumi Tsutsui, Kazunari Sei, Satoshi Soda, Masanori Fujita, and Michihiko Ike. 2012. "Impacts of Gene Bioaugmentation with pJP4-Harboring Bacteria of 2,4-D-Contaminated Soil Slurry on the Indigenous Microbial Community." *Biodegradation* 23, no. 2: 263–76. https://doi.org/10.1007/s10532-011-9505-x.

Jia, P., Fenglin Li, Shengchang Zhang, Guanxiong Wu, Yutao Wang, and Jin-Tian Li. 2022. "Microbial Community Composition in the Rhizosphere of Pteris vittata and Its Effects on Arsenic Phytoremediation Under a Natural Arsenic Contamination Gradient." *Frontiers in Microbiology* 13: 989272. https://doi.org/10.3389/fmicb.2022.989272.

Jing, Zhaoqian, Yu-You Li, Shiwei Cao, and Yuyu Liu. 2012. "Performance of Double-Layer Biofilter Packed with Coal Fly Ash Ceramic Granules in Treating Highly Polluted River Water." *Bioresource Technology* 120: 212–7. https://doi.org/10.1016/j.biortech.2012.06.069.

Kamika, Ilunga, and Maggy N. B. Momba. 2013. "Assessing the Resistance and Bioremediation Ability of Selected Bacterial and Protozoan Species to Heavy Metals in Metal-Rich Industrial Wastewater." *BMC Microbiology* 13: 28. https://doi.org/10.1186/1471-2180-13-28.

Kapoor, R. T., M. Danish, R. S. Singh, M. Rafatullah, and A. K. Abdul. 2021. "Exploiting Microbial Biomass in Treating Azo Dyes Contaminated Wastewater: Mechanism of Degradation and Factors Affecting Microbial Efficiency." *Journal of Water Process Engineering* 43: 102255. https://doi.org/10.1016/j.jwpe.2021.102255.

Khan, Atta Ullah, Allah Nawaz Khan, Abdul Waris, Muhammad Ilyas, and Doaa Zamel. 2022. "Phytoremediation of Pollutants from Wastewater: A Concise Review." *Open Life Sciences* 17, no. 1: 488–96. https://doi.org/10.1515/biol-2022-0056.

Khan, Z. U., I. Naz, A. Rehman, M. Rafiq, N. Ali, and S. Ahmed. 2015. "Performance Efficiency of an Integrated Stone Media Fixed Biofilm Reactor and Sand Filter for Sewage Treatment." *Desalination & Water Treatment* 54, no. 10: 2638–47. https://doi.org/10.1080/19443994.2014.903521.

Koul, S., and P. Gauba. 2014. "Bioaugmentation-A Strategy for Cleaning Up Soil." *Journal of Civil Engineering & Environmental Technology* 1, no. 5. ISSN 2349-8404.

Krishnamoorthy, R., P. A. Jose, M. Ranjith, R. Anandham, K. Suganya, J. Prabhakaran, S. Thiyageshwari, J. Johnson, N. O. Gopal, and K. Kumutha. 2018. "Decolourisation and Degradation of Azo Dyes by Mixed Fungal Culture Consisted of Dichotomomyces Cejpii MRCH 1–2 and Phoma tropica MRCH 1–3." *Journal of Environmental Chemical Engineering* 6, no. 1: 588–95. https://doi.org/10.1016/j.jece.2017.12.035.

Kurade, M. B., Y. H. Ha, J. Q. Xiong, S. P. Govindwar, M. Jang, and B. H. Jeon. 2021. "Phytoremediation as a Green Biotechnology Tool for Emerging Environmental Pollution: A Step Forward Towards Sustainable Rehabilitation of the Environment." *Chemical Engineering Journal* 415: 129040. https://doi.org/10.1016/j.cej.2021.129040.

Le-Clech, P., V. Chen, and T. A. G. Fane. 2006. "Fouling in Membrane Bioreactors Used in Wastewater Treatment." *Journal of Membrane Science* 284, no. 1–2: 17–53. https://doi.org/10.1016/j.memsci.2006.08.019.

Lee, E. J., A. K. J. An, P. Hadi, and D. Y. S. Yan. 2016. "Characterizing Flat Sheet Membrane Resistance Fraction of Chemically Enhanced Backflush." *Chemical Engineering Journal* 284: 61–7. https://doi.org/10.1016/j.cej.2015.08.136.

Lin, Jun, Xingwang Zhang, Zhongjian Li, and Lecheng Lei. 2010. "Biodegradation of Reactive Blue 13 in a Two-Stage Anaerobic/Aerobic Fluidized Beds System with a *Pseudomonas* sp. Isolate." *Bioresource Technology* 101, no. 1: 34–40. https://doi.org/10.1016/j.biortech.2009.07.037.

Liu, R., Q. Tian, and J. Chen. 2010. "The Developments of Anaerobic Baffled Reactor for Wastewater Treatment: A Review". *African Journal of Biotechnology* 9, no. 11: 1535–42.

Liu, Xiaohui, Xiaochun Guo, Ying Liu, Shaoyong Lu, Beidou Xi, Jian Zhang, Zhi Wang, and Bin Bi. 2019. "A Review on Removing Antibiotics and Antibiotic Resistance Genes from Wastewater by Constructed Wetlands: Performance and Microbial Response." *Environmental Pollution* 254, no. A: 112996. https://doi.org/10.1016/j.envpol.2019.112996.

Lydmark, Pär, Magnus Lind, Fred Sörensson, and Malte Hermansson. 2006. "Vertical Distribution of Nitrifying Populations in Bacterial Biofilms from a Full-Scale Nitrifying Trickling Filter." *Environmental Microbiology* 8, no. 11: 2036–49. https://doi.org/10.1111/j.1462-2920.2006.01085.x.

Machdar, I. 2016. "Hydraulic Behavior in the Downflow Hanging Sponge Bioreactor." *Jurnal Litbang Industri* 6, no. 2: 83. https://doi.org/10.24960/jli.v6i2.1679.83-88.

Mallevialle, J., P. E. Odendaal, and M. R. Wiesner. 1996. *Water Treatment Membrane Processes*. Denver, CO: American Water Works Association.

Mandeep, Kumar Gupta, Guddu Kumar Gupta, and Pratyoosh Shukla. 2020. "Insights into the Resources Generation from Pulp and Paper Industry Wastes: Challenges, Perspectives and Innovations." *Bioresource Technology* 297: 122496. https://doi.org/10.1016/j.biortech.2019.122496.

Manirakiza, B., and A. C. Sirotkin. 2021. "Bioaugmentation of Nitrifying Bacteria in Up-Flow Biological Aerated Filter's Microbial Community for Wastewater Treatment and Analysis of Its Microbial Community." *Scientific African* 14: E00981. https://doi.org/10.1016/j.sciaf.2021.e00981.

Mazumder, A., S. Das, D. Sen, and C. Bhattacharjee. 2020. "Kinetic Analysis and Parametric Optimization for Bioaugmentation of Oil from Oily Wastewater with Hydrocarbonoclastic Rhodococcus pyridinivorans F5 Strain." *Environmental Technology & Innovation* 17. https://doi.org/10.1016/j.eti.2020.100630.

Mendez, Monica O., and Raina M. Maier. 2008. "Phytostabilization of Mine Tailings in Arid and Semiarid Environments—An Emerging Remediation Technology." *Environmental Health Perspectives* 116, no. 3: 278–83. https://doi.org/10.1289/ehp.10608.

Metcalf & Eddy. 2003. "Trickling Filters. Wastewater Eng." *Treatment & Reuse* 8: 890–929.

Metcalf and Eddy. 2016. *Metcalf and Eddy, Inc. Wastewater Engineering Treatment and Reuse*. New York: McGraw-Hill.

Mishra, Saurabh, Virender Singh, Banu Ormeci, Abid Hussain, Liu Cheng, and Kaushik Venkiteshwaran. 2023. "Anaerobic–Aerobic Treatment of Wastewater and Leachate: A Review of Process Integration, System Design, Performance and Associated Energy Revenue." *Journal of Environmental Management* 327: 116898. https://doi.org/10.1016/j.jenvman.2022.116898.

Mohamed, M., M. Raja, A. Raja, S. M. Salique, and P. Gajalakshmi. 2016. "Studies on Effect of Marine Actinomycetes on Amido Black (Azo Dye) Decolorization." *Journal of Chemical & Pharmaceutical Research* 8: 640–4.

Mohamed, M. A., H. A. Fouad, and R. M. Hefny. 2022. "Rotating Biological Contactor Wastewater Treatment Using Geotextiles, Sugarcane Straw and Steel Cylinder for Green Areas Irrigation." *Egyptian Journal of Chemistry* 65: 59–72. https://doi.org/10.21608/EJCHEM.2021.82581.4065.

Monferrán, Magdalena V., María L. Pignata, and Daniel A. Wunderlin. 2012. "Enhanced Phytoextraction of Chromium by the Aquatic Macrophyte Potamogeton pusillus in Presence of Copper." *Environmental Pollution* 161: 15–22. https://doi.org/10.1016/j.envpol.2011.09.032.

Morita, A. K. M., and F. N. Moreno. 2022. "Phytoremediation Applied to Urban Solid Waste Disposal Sites." *Engenharia Sanitaria e Ambiental* 27, no. 2: 377–84. https://doi.org/10.1590/S1413-415220210105.

Muralikrishna, I. V., and V. Manickam. 2017. "Wastewater Treatment Technologies." In *Environmental Management*. Elsevier: 249–63. https://doi.org/10.1016/B978-0-12-811989-1.00012-9.

Muter, Olga. 2023. "Current Trends in Bioaugmentation Tools for Bioremediation: A Critical Review of Advances and Knowledge Gaps." *Microorganisms* 11, no. 3: 710. https://doi.org/10.3390/microorganisms11030710.

Naeem, M., K. Khadeeja, A. Salam, A. Maria, U. Iftikhar, A. Anwar, M. Saleem, M. W. Mazhar, and I. Tariq. 2023. "Phytoremediation of Heavy Metals from Irrigation Water, Faisalabad, Pakistan." *Global Academic Journal of Pharmacy & Drug Research* 5, no. 1: 10–5. https://doi.org/10.36348/gajpdr.2023.v05i01.003.

Naz, I., W. Ullah, S. Sehar, A. Rehman, Z. U. Khan, N. Ali, and S. Ahmed. 2016. "Performance Evaluation of Stone-Media Pro-type Pilot-Scale Trickling Biofilter System for Municipal Wastewater Treatment." *Desalination & Water Treatment* 57, no. 34: 15792–805. https://doi.org/10.1080/19443994.2015.1081111.

Newman, Lee A., and Charles M. Reynolds. 2004. "Phytodegradation of Organic Compounds." *Current Opinion in Biotechnology* 15, no. 3: 225–30. https://doi.org/10.1016/j.copbio.2004.04.006.

Nurmiyanto, A., and A. Ohashi. 2019. "Downflow Hanging Sponge (DHS) Reactor for Wastewater Treatment—A Short Review.". *MATEC Web of Conferences* 280: 05004. https://doi.org/10.1051/matecconf/201928005004.

Odinga, C. A., A. Kumar, M. S. Mthembu, F. Bux, and F. M. Swalaha. 2019. "Rhizofiltration System Consisting of *Phragmites australis* and *Kyllinga nemoralis*: Evaluation of Efficient Removal of Metals and Pathogenic Microorganisms." *Desalination & Water Treatment* 169: 120–32. https://doi.org/10.5004/dwt.2019.24428.

Oliveira, M., C. Queda, and E. Duarte. 2009. "Aerobic Treatment of Winery Wastewater with the Aim of Water Reuse." *Water Science & Technology* 60, no. 5: 1217–23. https://doi.org/10.2166/wst.2009.558.

Onodera, Takashi, Madan Tandukar, Doni Sugiyana, Shigeki Uemura, Akiyoshi Ohashi, and Hideki Harada. 2014. "Development of a Sixth-Generation Down-Flow Hanging Sponge (DHS) Reactor Using Rigid Sponge Media for Post-treatment of UASB Treating Municipal Sewage." *Bioresource Technology* 152: 93–100. https://doi.org/10.1016/j.biortech.2013.10.106.

Ouyang, C. F. 1981. "Characteristics of Rotating Biological Contactor (RBC) Sludges." *Journal of the Chinese Institute of Engineers* 4, no. 2: 53–60. https://doi.org/10.1080/02533839.1981.9676669.

Pandey, Janmejay, Archana Chauhan, and Rakesh K. Jain. 2009. "Integrative Approaches for Assessing the Ecological Sustainability of InSitu Bioremediation." *FEMS Microbiology Reviews* 33, no. 2: 324–75. https://doi.org/10.1111/j.1574-6976.2008.00133.x.

Pearce, P., and S. Jarvis. 2011. "Operational Experiences with Structured Plastic Media Filters: 10 Years On." *Water & Environment Journal* 25, no. 2: 200–7. https://doi.org/10.1111/j.1747-6593.2009.00210.x.

Polechońska, L., and A. Klink. 2014. "Trace Metal Bioindication and Phytoremediation Potentialities of *Phalaris arundinacea* L. (Reed Canary Grass)." *Journal of Geochemical Exploration* 146: 27–33. https://doi.org/10.1016/j.gexplo.2014.07.012.

Radziemska, Maja, Magdalena D. Vaverková, and Anna Baryła. 2017. "Phytostabilization-Management Strategy for Stabilizing Trace Elements in Contaminated Soils." *International Journal of Environmental Research & Public Health* 14, no. 9: 928. https://doi.org/10.3390/ijerph14090958.

Razavi, S. M. R., and T. Miri. 2015. "A Real Petroleum Refinery Wastewater Treatment Using Hollow Fiber Membrane Bioreactor (HF-MBR)." *Journal of Water Process Engineering* 8: 136–41. https://doi.org/10.1016/j.jwpe.2015.09.011.

Rhodes, C. J. 2014. "Mycoremediation (bioremediation with fungi) – growing mushrooms to clean the earth, Chemical Speciation & Bioavailability". vol. 26 no. 3: 196–198.

Riser-Roberts, E. 2019. *Remediation of Petroleum Contaminated Soils: Biological, Physical, and [WWW Document]*. Boca Raton, FL: CRC Press.

Sánchez Guillén, J. A., P. R. Cuéllar Guardado, C. M. Lopez Vazquez, L. M. de Oliveira Cruz, D. Brdjanovic, and J. B. van Lier. 2015. "Anammox Cultivation in a Closed Sponge-Bed Trickling Filter." *Bioresource Technology* 186: 252–60. https://doi.org/10.1016/j.biortech.2015.03.073.

Saran, A., L. Fernandez, F. Cora, M. Savio, S. Thijs, J. Vangronsveld, and L. J. Merini. 2020. "Phytostabilization of Pb and Cd Polluted Soils Using *Helianthus petiolaris* as Pioneer Aromatic Plant Species." *International Journal of Phytoremediation* 22, no. 5: 459–67. https://doi.org/10.1080/15226514.2019.1675140.

Schwitzguébel, Jean-Paul, Elena Comino, Nadia Plata, and Mohammadali Khalvati. 2011. "Is Phytoremediation a Sustainable and Reliable Approach to Clean-Up Contaminated Water and Soil in Alpine Areas?" *Environmental Science & Pollution Research International* 18, no. 6: 842–56. https://doi.org/10.1007/s11356-011-0498-0.

Selvaraj, V., T. Swarna Karthika, C. Mansiya, and M. Alagar. 2021. "An over Review on Recently Developed Techniques, Mechanisms and Intermediate Involved in the Advanced Azo Dye Degradation for Industrial Applications." *Journal of Molecular Structure* 1224: 129195. https://doi.org/10.1016/j.molstruc.2020.129195.

Sessitsch, Angela, Melanie Kuffner, Petra Kidd, Jaco Vangronsveld, Walter W. Wenzel, Katharina Fallmann, and Markus Puschenreiter. 2013. "The Role of Plant-Associated Bacteria in the Mobilization and Phytoextraction of Trace Elements in Contaminated Soils." *Soil Biology & Biochemistry* 60, no. 100: 182–94. https://doi.org/10.1016/j.soilbio.2013.01.012.

Shchegolkova, Nataliya M., George S. Krasnov, Anastasia A. Belova, Alexey A. Dmitriev, Sergey L. Kharitonov, Kseniya M. Klimina, Nataliya V. Melnikova, and Anna V. Kudryavtseva. 2016. "Microbial Community Structure of Activated Sludge in Treatment Plants with Different Wastewater Compositions." *Frontiers in Microbiology* 7: 90. https://doi.org/10.3389/fmicb.2016.00090.

Sonune, A., and R. Ghate. 2004. "Developments in Wastewater Treatment Methods." *Desalination* 167: 55–63. https://doi.org/10.1016/j.desal.2004.06.113.

Sreedevi, P. R., K. Suresh, and G. Jiang. 2022. "Bacterial Bioremediation of Heavy Metals in Wastewater: A Review of Processes and Applications." *Journal of Water Process Engineering* 48: 102884. https://doi.org/10.1016/j.jwpe.2022.102884.

Stubbendieck, Reed M., Carol Vargas-Bautista, and Paul D. Straight. 2016. "Bacterial Communities: Interactions to Scale." *Frontiers in Microbiology* 7: 1234. https://doi.org/10.3389/fmicb.2016.01234.

Suman, Jachym, Ondrej Uhlik, Jitka Viktorova, and Tomas Macek. 2018. "Phytoextraction of Heavy Metals: A Promising Tool for Clean-Up of Polluted Environment?" *Frontiers in Plant Science* 9: 1476. https://doi.org/10.3389/fpls.2018.01476.

Sun, Y., M. Kumar, L. Wang, J. Gupta, and D. C. W. Tsang. 2020. "Biotechnology for Soil Decontamination: Opportunity, Challenges, and Prospects for Pesticide Biodegradation." In *Bio-based Materials and Biotechnologies for Eco-efficient Construction. Ernando Pacheco-Torgal*, edited by Volodymyr Ivanov, and Daniel C. W. Tsang: 261–83. Woodhead Publishing. https://doi.org/10.1016/B978-0-12-819481-2.00013-1.

Tandukar, M., S. Uemura, I. Machdar, A. Ohashi, and H. Harada. 2005. "A Low-Cost Municipal Sewage Treatment System with a Combination of UASB and the 'Fourth-Generation' Downflow Hanging Sponge Reactors." *Water Science & Technology* 52, no. 1–2: 323–9. https://doi.org/10.2166/wst.2005.0534.

Tawfik, Ahmed, Aly Al-Sayed, Gamal K. Hassan, Mahmoud Nasr, Saber A. El-Shafai, Nawaf S. Alhajeri, Mohd Shariq Khan, Muhammad Saeed Akhtar, Zubair Ahmad, Patricia Rojas, and Jose L. Sanz. 2022. "Electron Donor Addition for Stimulating the Microbial Degradation of 1,4 Dioxane by Sequential Batch Membrane Bioreactor: A Techno-Economic Approach." *Chemosphere* 306: 135580. https://doi.org/10.1016/j.chemosphere.2022.135580.

Techobanoglous, G., F. L. Burton, H. D., and S. 2014. *Wastewater Engineering: Treatment and Reuse*. 5th ed. Metcalf and Eddy. https://doi.org/10.1016/0309–1708(80)90067–6.

Tomaszewski, Mariusz, Grzegorz Cema, Slawomir Ciesielski, Dariusz Łukowiec, and Aleksandra Ziembińska-Buczyńska. 2019. "Cold Anammox Process and Reduced Graphene Oxide—Varieties of Effects During Long-Term Interaction." *Water Research* 156: 71–81. https://doi.org/10.1016/j.watres.2019.03.006.

Tuo, B. H., J. B. Yan, B. A. Fan, Z. H. Yang, and J. Z. Liu. 2012. "Biodegradation Characteristics and Bioaugmentation Potential of a Novel Quinoline-Degrading Strain of Bacillus sp. Isolated from Petroleum-Contaminated Soil." *Bioresource Technology* 107: 55–60. https://doi.org/10.1016/j.biortech.2011.12.114.

Ungureanu, N., V. Vlăduț, and G. Voicu. 2020. "Water Scarcity and Wastewater Reuse in Crop Irrigation." *Sustainability* 12, no. 21: 9055. https://doi.org/10.3390/su12219055.

Usack, Joseph G., Catherine M. Spirito, and Largus T. Angenent. 2012. "Continuously-Stirred Anaerobic Digester to Convert Organic Wastes into Biogas: System Setup and Basic Operation." *Journal of Visualized Experiments: JoVE* 13, no. 65: E3978. https://doi.org/10.3791/3978.

USEPA. 2012. *Guidelines for Water Reuse*. EPA/600/R-12/618. EPA/625/R-04/108. https://doi.org/EPA-600/R-12/618 vol. 643.

Vanloon, G. W., S. J. Duffy, and J. Wiley. 2005. *Environmental Chemistry: A Global Perspective*. Oxford: Oxford University Press.

Varjani, Sunita, and Vivek N. Upasani. 2019. "Influence of Abiotic Factors, Natural Attenuation, Bioaugmentation and Nutrient Supplementation on Bioremediation of Petroleum Crude Contaminated Agricultural Soil." *Journal of Environmental Management* 245: 358–66. https://doi.org/10.1016/j.jenvman.2019.05.070.

Wagner, J., and K.-H. Rosenwinkel. 2000. "Sludge Production in Membrane Bioreactors Under Different Conditions." *Water Science & Technology* 41, no. 10–11: 251–8. https://doi.org/10.2166/wst.2000.0655.

Wallace, J. 2013. *Evaluating the Performance and Water Chemistry Dynamics of Passive Systems Treating Municipal Wastewater and Landfill Leachate*. A thesis submitted to Queen's University Kingston, Ontario, Canada.

Walters, Kendra E., and Jennifer B. H. Martiny. 2020. "Alpha-, Beta-, and Gamma-Diversity of Bacteria Varies Across Habitats." *PLOSONE* 15, no. 9: E0233872. https://doi.org/10.1371/journal.pone.0233872.

Wang, Jianwu, Yuannan Long, Guanlong Yu, Guoliang Wang, Zhenyu Zhou, Peiyuan Li, Yameng Zhang, Kai Yang, and Shitao Wang. 2022. "A Review on Microorganisms in Constructed Wetlands for Typical

Pollutant Removal: Species, Function, and Diversity." *Frontiers in Microbiology* 13: 845725. https://doi.org/10.3389/fmicb.2022.845725.

Wang, Jin, Qaisar Mahmood, Jiang-Ping Qiu, Yin-Sheng Li, Yoon-Seong Chang, and Xu-Dong Li. 2015. "Anaerobic Treatment of Palm Oil Mill Effluent in Pilot-Scale Anaerobic EGSB Reactor." *BioMed Research International* 2015: 398028. https://doi.org/10.1155/2015/398028.

Wang, L. K., M. H. S. Wang, and C. P. C. Poon. 1986. "Trickling Filters." In *Biological Treatment Processes*: 361–425. Totowa, NJ: Humana Press. https://doi.org/10.1007/978-1-4612-4820-0_8.

Wang, Xiaohui, Man Hu, Y. Xia, Xianghua Wen, and Kun Ding. 2012. "Pyrosequencing Analysis of Bacterial Diversity in 14 Wastewater Treatment Systems in China." *Applied & Environmental Microbiology* 78, no. 19: 7042–7. https://doi.org/10.1128/AEM.01617-12.

Wang, Yue, Shih-Hsin Ho, Chieh-Lun Cheng, Wan-Qian Guo, Dillirani Nagarajan, Nan-Qi Ren, Duu-Jong Lee, and Jo-Shu Chang. 2016. "Perspectives on the Feasibility of Using Microalgae for Industrial Wastewater Treatment." *Bioresource Technology* 222: 485–97. https://doi.org/10.1016/j.biortech.2016.09.106.

Waqas, S., M. R. Bilad, Z. B. Man, H. Suleman, N. A. Hadi Nordin, J. Jaafar, M. H. Dzarfan Othman, and M. Elma. 2021. "An Energy-Efficient Membrane Rotating Biological Contactor for Wastewater Treatment." *Journal of Cleaner Production* 282: 124544. https://doi.org/10.1016/j.jclepro.2020.124544.

Wąsik, E., and K. Chmielowski. 2017. "Ammonia and Indicator Bacteria Removal from Domestic Sewage in a Vertical Flow Filter Filled with Plastic Material." *Ecological Engineering* 106: 378–84. https://doi.org/10.1016/j.ecoleng.2017.05.015.

Wei, Zhangliang, Jiaguo You, Hailong Wu, Fangfang Yang, Lijuan Long, Qiao Liu, Yuanzi Huo, and Peimin He. 2017. "Bioremediation Using Gracilaria lemaneiformis to Manage the Nitrogen and Phosphorous Balance in an Integrated Multi-trophic Aquaculture System in Yantian Bay, China." *Marine Pollution Bulletin* 121, no. 1–2: 313–9. https://doi.org/10.1016/j.marpolbul.2017.04.034.

WHO/UNICEF, Sdg6 Program Reports: Imi-Sdg6. 2021. *WHO/UNICEF Joint Monitoring Program for Water Supply, Sanitation and Hygiene (JMP)—Progress on Household Drinking Water, Sanitation and Hygiene 2000–2020*. WWW Document. https://www.unwater.org/publications/who/unicef-joint-monitoring-program-water-supply-sanitation-and-hygiene-jmp-progress-0.

Wijekoon, Kaushalya C., Takahiro Fujioka, James A. McDonald, Stuart J. Khan, Faisal I. Hai, William E. Price, and Long D. Nghiem. 2013. "Removal of N-Nitrosamines by an Aerobic Membrane Bioreactor." *Bioresource Technology* 141: 41–5. https://doi.org/10.1016/j.biortech.2013.01.057.

Willis, Amy D. 2019. "Rarefaction, Alpha Diversity, and Statistics." *Frontiers in Microbiology* 10: 2407. https://doi.org/10.3389/fmicb.2019.02407.

Wu, Manli, Jialuo Wu, Xiaohui Zhang, and Xiqiong Ye. 2019. "Effect of Bioaugmentation and Biostimulation on Hydrocarbon Degradation and Microbial Community Composition in Petroleum-Contaminated Loessal Soil." *Chemosphere* 237: 124456. https://doi.org/10.1016/j.chemosphere.2019.124456.

Wu, Yang, Yuexing Wang, Yashika G. De Costa, Zhida Tong, Jay J. Cheng, Lijie Zhou, Wei-Qin Zhuang, and K. Yu. 2019. "The co-Existence of Anammox Genera in an Expanded Granular Sludge Bed Reactor with Biomass Carriers for Nitrogen Removal." *Applied Microbiology & Biotechnology* 103, no. 3: 1231–42. https://doi.org/10.1007/s00253-018-9494-3.

Wu, Yonghong, Lizhong Xia, Zhiqiang Yu, Sadaf Shabbir, and Philip G. Kerr. 2014. "In Situ Bioremediation of Surface Waters by Periphytons." *Bioresource Technology* 151: 367–72. https://doi.org/10.1016/j.biortech.2013.10.088.

Xia, Siqing, Jixiang Li, Shuying He, Kang Xie, Xiaojia Wang, Yanhao Zhang, Liang Duan, and Zhiqiang Zhang. 2010. "The Effect of Organic Loading on Bacterial Community Composition of Membrane Biofilms in a Submerged Polyvinyl Chloride Membrane Bioreactor." *Bioresource Technology* 101, no. 17: 6601–9. https://doi.org/10.1016/j.biortech.2010.03.082.

Xu, Shuang, Junqin Yao, Meihaguli Ainiwaer, Ying Hong, and Yanjiang Zhang. 2018. "Analysis of Bacterial Community Structure of Activated Sludge from Wastewater Treatment Plants in Winter." *BioMed Research International* 2018: 8278970. https://doi.org/10.1155/2018/8278970.

Yaashikaa, P. R., P. Senthil Kumar, S. Jeevanantham, and R. Saravanan. 2022. "A Review on Bioremediation Approach for Heavy Metal Detoxification and Accumulation in Plants." *Environmental Pollution* 301: 119035. https://doi.org/10.1016/j.envpol.2022.119035.

Yadav, B. K., M. A. Siebel, and J. J. A. van Bruggen. 2011. "Rhizofiltration of a Heavy Metal (Lead) Containing Wastewater Using the Wetland Plant Carex pendula." *CLEAN—Soil, Air, Water* 39, no. 5: 467–74. https://doi.org/10.1002/clen.201000385.

Yaman, C. 2020. "Performance and Kinetics of Bioaugmentation, Biostimulation, and Natural Attenuation Processes for Bioremediation of Crude Oil-Contaminated Soils." *Processes* 8, no. 8: 883. https://doi.org/10.3390/PR8080883.

Yan, Chicheng, Zhengzhe Qu, Jieni Wang, Leichang Cao, and Qiuxia Han. 2022. "Microalgal Bioremediation of Heavy Metal Pollution in Water: Recent Advances, Challenges, and Prospects." *Chemosphere* 286, no. 3: 131870. https://doi.org/10.1016/j.chemosphere.2021.131870.

Yang, Minjune, James W. Jawitz, and Minhee Lee. 2015. "Uranium and Cesium Accumulation in Bean (Phaseolus vulgaris L. var. vulgaris) and Its Potential for Uranium Rhizofiltration." *Journal of Environmental Radioactivity* 140: 42–9. https://doi.org/10.1016/j.jenvrad.2014.10.015.

Yang, Qing, Yongzhen Peng, Xiuhong Liu, Wei Zeng, Takashi Mino, and Hiroyasu Satoh. 2007. "Nitrogen Removal via Nitrite from Municipal Wastewater at Low Temperatures Using Real-Time Control to Optimize Nitrifying Communities." *Environmental Science & Technology* 41, no. 23: 8159–64. https://doi.org/10.1021/es070850f.

Yoon, Seong-Hoon, Hyung-Soo Kim, and Ik-Tae Yeom. 2004. "The Optimum Operational Condition of Membrane Bioreactor (MBR): Cost Estimation of Aeration and Sludge Treatment." *Water Research* 38, no. 1: 37–46. https://doi.org/10.1016/j.watres.2003.09.001.

Yu, Guanlong, Guoliang Wang, Jianbing Li, Tianying Chi, Shitao Wang, Haiyuan Peng, Hong Chen, Chunyan Du, Changbo Jiang, Yuanyuan Liu, L. Zhou, and Haipeng Wu. 2020. "Enhanced Cd^{2+} and Zn^{2+} Removal from Heavy Metal Wastewater in Constructed Wetlands with Resistant Microorganisms." *Bioresource Technology* 316: 123898. https://doi.org/10.1016/j.biortech.2020.123898.

Zazouli, Mohammad Ali, Yousef Mahdavi, Edris Bazrafshan, and Davoud Balarak. 2014. "Phytodegradation Potential of Bisphenol A from Aqueous Solution by *Azolla filiculoides*." *Journal of Environmental Health Science & Engineering* 12: 66. https://doi.org/10.1186/2052-336X-12-66.

Zhang, Lei, and Satoshi Okabe. 2020. "Ecological Niche Differentiation Among Anammox Bacteria." *Water Research* 171, no. 171: 115468. https://doi.org/10.1016/j.watres.2020.115468.

Zhang, Qun, Baichuan Wang, Zhengya Cao, and Yunlong Yu. 2012. "Plasmid-Mediated Bioaugmentation for the Degradation of Chlorpyrifos in Soil." *Journal of Hazardous Materials* 221–222: 178–84. https://doi.org/10.1016/j.jhazmat.2012.04.024.

Zhang, Shu, Nancy Merino, Akihiro Okamoto, and Phillip Gedalanga. 2018. "Interkingdom Microbial Consortia Mechanisms to Guide Biotechnological Applications." *Microbial Biotechnology* 11, no. 5: 833–47. https://doi.org/10.1111/1751-7915.13300.

Zhang, Tong, Ming-Fei Shao, and Lin Ye. 2012. "454 Pyrosequencing Reveals Bacterial Diversity of Activated Sludge from 14 Sewage Treatment Plants." *ISME Journal* 6, no. 6: 1137–47. https://doi.org/10.1038/ismej.2011.188.

Zhao, Dayong, Rui Huang, Jin Zeng, Zhongbo Yu, Peng Liu, Shupei Cheng, and Qinglong L. Wu. 2014. "Pyrosequencing Analysis of Bacterial Community and Assembly in Activated Sludge Samples from Different Geographic Regions in China." *Applied Microbiology & Biotechnology* 98, no. 21: 9119–28. https://doi.org/10.1007/s00253-014-5920-3.

Zhu, J., and B. Rothermel. 2014. "Everything You Need to Know About Trickling Filters." *Clear Waters* 44: 16–19.

Ziembińska-Buczyńska, A., and J. Surmacz-Górska. 2021. "Nitrogen Removal Bacterial Communities Characteristics and Dynamics at Lab-Scale Reactors." In *Wastewater Treatment Reactors: Microbial Community Structure*, edited by Maulin P. Shah, and Susana Rodriguez-Couto: 39–63. Amsterdam: Elsevier. https://doi.org/10.1016/B978-0-12-823991-9.00023-X.

8 Effective Microorganisms
A Microbial Technology to Improve Polluted Water Quality

Mohit Yadav

8.1 INTRODUCTION

Microorganisms are important factors and have a crucial impact on the environment. The unpredictable nature, biosynthetic capabilities, and uniqueness of microorganisms also depend on environmental conditions and make them problem-solving elements in various fields of science. All healthy ecosystems have their communities of microorganisms that decompose biological matter. However, anthropogenic activities (for example, mining, farming, manufacturing, etc.) are involved in the accumulation of a wide range of chemicals and other contaminants in aquatic environments (Carpenter *et al.*, 2011). Sewage and human waste contaminations can disrupt the natural balance of different bacteria and affect aquatic ecosystems. Contamination in aquatic ecosystems such as rivers, groundwater, lakes, oceans, and reservoirs not only influences the local community but also impact the entire biosphere. For many years, microorganisms have been used successfully to promote environmental protection through the treatment of wastewater, and agricultural and municipal waste. Nowadays such advantageous microorganisms are involved in numerous applications (Aracic *et al.*, 2015). For example, these microorganisms are involved in adjusting the population of algae in water bodies to prevent the deterioration of the water quality, strengthen the immune system of aquatic animals, and promote their disease resistance. Microorganisms also assist in controlling the challenges of dioxin and polychlorinated biphenyl, prevent the progression of diseases in fishes, and inhibit the decomposition of organic matter in some aquatic plants (Zhou *et al.*, 2009).

A major problem facing municipalities throughout the world is the treatment, disposal, and recycling of sewage sludge. Generally, sludge from municipal waste consists mainly of biodegradable organic materials with a significant amount of inorganic matter (Comesaña *et al.*, 2018). However, sludge exhibits wide variations in its physical, chemical, and biological properties (Comesaña *et al.*, 2018). At present, several methods are being used to dispose of sewage sludge from disposal to landfill to land application, although with these methods, there are numerous concerns raised regarding the presence of constituents including heavy metals, pathogens, and other toxic substances. This requires the selection of the correct disposal method focusing on efficient and environmentally safe disposal. New technologies are being produced to assist in the treatment and disposal of sewage sludge, conforming to strict environmental regulations. One of the new technologies that are being promoted is the application of effective microorganisms (EM).

At the Institute of Ryukyus in Okinawa, Japan, Professor Teruo Higa created the EM technology in the 1970s (Higa and Parr, 1994). A multi-culture of coexisting anaerobic and aerobic beneficial microorganisms has been described as an effective microorganism. Effective microorganisms are a combination of groups of organisms that have a revitalizing effect on people, animals, and the natural environment (Higa and Parr, 1994). According to studies, EM may be used in a variety of industries, including farming, raising livestock, gardening and landscaping, composting, bioremediation, septic tank cleaning, algal control, and household uses (Safwat and Matta, 2021; Abd El-Mageed

DOI: 10.1201/9781003423393-8

et al., 2022; Hidalgo, Corona, and Martín-Marroquín, 2022). The following species are among the major players in EM (Figure 8.1 and Table 8.1):

- Lactic acid bacteria—*Lactobacillus plantarum, L. casei, Streptococcus lactis*
- Photosynthetic bacteria—*Rhodopseudomonas palustris, Rhodobacter sphaeroides*
- Yeasts—*Saccharomyces cerevisiae, Candida utilis*
- Actinomycetes—*Streptomyces albus, S. griseus*
- Fermenting fungi—*Aspergillus oryzae, Mucor hiemalis.*

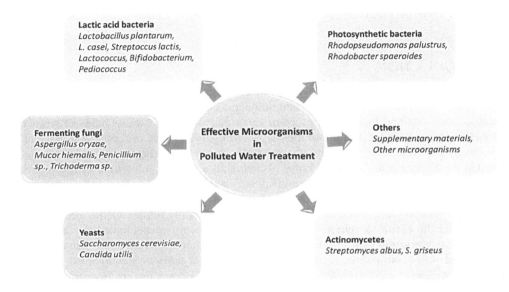

FIGURE 8.1 Outline of the main components of EM technology.

TABLE 8.1
Primary Microbial Components of EM

Microbial Components	Approach to Nutrition	Examples
Lactic acid bacteria (LAB)	Utilize sugar as the carbon source, and the LAB rely exclusively on it through homolactic fermentation, which also results in the production of lactate.	*Lactobacillus casei, L. plantarum, Streptococcus lactis*
Photosynthetic bacteria	Reduces the production of sludge and breaks down difficult biodegradables.	*Rhodobacter Sphaeroides, Rhodopseudomonas palustris*
Yeast	In the presence of free molecular oxygen, yeast can break down organic materials into carbon dioxide and water, and in the absence of it, it can turn sugars into ethanol.	*Candida utilis, Saccharomyces cerevisiae*
Actinomycetes	With sufficient hydraulic retention time, the activity of actinomycetes causes the breakdown of organic phosphorus compounds and the release of orthophosphate.	*Streptomyces griseus, S. albus*
Fungi	Alcohol, esters, and antimicrobial substances are produced when fungi break down organic matter, which lessens the smell of waste.	*Mucor hiemalis, Aspergillus oryzae*

Effective Microorganisms **163**

8.1.1 Lactic Acid Bacteria

Lactic acid bacteria (LAB) are acid-tolerant, low GC-containing, Gram-positive bacteria that generally grow at a pH between 4 and 4.5. This bacterium can thrive in conditions that other bacteria cannot because of its ability to tolerate acid (de Souza *et al.*, 2015). LAB is composed of different bacterial genera such as *Lactococcus, Lactobacillus, Bifidobacterium, Leuconostoc,* and *Streptococcus* and can be extracted from the intestinal tract of animals or fermented foods. LAB is microaerophilic and can survive in a 5% CO_2-containing atmosphere. These bacteria show a slow growth pattern that depends on temperature (optimum temperature is 30°C) (Londoño *et al.*, 2015). Because LAB can produce lactic acid in an acidic environment, it outcompetes other bacteria, inhibits their growth, and breaks down organic materials in contaminated water (Soto *et al.*, 2017; Xu *et al.*, 2018). These bacteria also increase the fermentation and breakdown of different organic materials, which generally take a longer time (Xu *et al.*, 2018). LAB exhibits antagonistic activity to other pathogens in polluted water by decreasing pH and producing bacteriocins such as nisin which have antimicrobial properties. LAB also retards the growth of some harmful fungi such as *Fusarium* (Timmusk *et al.*, 2020).

8.1.2 Photosynthetic Bacteria

Photosynthetic bacteria such as *Rhodopseudomonas palustris, Rhodobacter sphaeroides,* and other facultative autotrophic bacteria are major components of EM. These bacteria are capable of synthesizing essential molecules, for example, nucleic acids, sugars, and amino acids by utilizing contaminants from water, and these molecules are consumed as a substrate by other heterotrophs to increase their populations (Su *et al.*, 2017). *R. palustris* is a Gram-negative, purple, nonsulfur, rod-shaped bacterium and also produces the previously mentioned essential molecules that are utilized by other heterotrophs (MA Luna, 2016). *R. sphaeroides* is a purple, photosynthetic, rod-shaped, Gram-negative bacterium that is involved in nitrogen fixation (Morocho and Agrícola, 2019). Besides nitrogen fixation and photosynthesis, these bacteria have diverse metabolic activity to grow heterotrophically and are involved in aerobic and anaerobic respiration (Morocho and Agrícola, 2019). Under oxygen deprivation, for growth, *R. sphaeroides* acquires energy by light through photosynthesis. However, in the presence of oxygen, they grow by utilizing toxic and nontoxic carbon compounds from contaminants in water.

8.1.3 Yeasts

Yeasts are also an important component of EM solution and utilize various carbon sources present in polluted water such as glucose, galactose, hydrolyzed whey, sucrose, maltose, fructose, and alcohol. Yeasts utilize various nitrogen sources in the form of urea, ammonia, or its salts and a complex of amino acids, and these microorganisms are incapable of assimilating nitrites or nitrates (Fayemi and Ojokoh, 2014). Some other nutrients needed for yeast growth are magnesium, iron, phosphorus, copper, calcium, and zinc (Meena and Meena, 2017). Numerous species from the genus *Saccharomyces* are the main constituent of this microbial community, while *Candida utilis* and *Saccharomyces cerevisiae* predominate. Through fermentative metabolism, these microorganisms produce a high concentration of ethanol that also works as an antimicrobial substance against some pathogenic microorganisms (Farrag and Bakr, 2021). Apart from ethanol, yeasts also synthesize other antimicrobial substances and bioactive substances such as enzymes and hormones. These by-products also support the growth of other beneficial microorganisms present in the EM solution.

8.1.4 Actinomycetes

Actinomycetes have some characteristics similar to fungi, such as their growth showing branched mycelia and belonging to filamentous bacteria. Many Gram-positive bacteria belong to

Actinomycetes and are found in both aquatic as well as terrestrial environments (Servin *et al.*, 2008). Various species from actinomycetes such as Streptomyces bacteria have efficient biological control activity. Some examples of actinomycete species used in EM technology are *Streptomyces griseus* and *Streptomyces albus* (Vurukonda *et al.*, 2018). These bacteria have antagonistic activity to pathogenic fungi due to the synthesis of antifungal compounds. These antifungal compounds include chitinases, β-1,3-glucanase, and other extracellular hydrolytic enzymes that inhibit the mycelial growth by degrading the cell walls of different pathogenic fungi such as *Sclerotium rolfsii*, *Fusarium oxysporum*, and *Sclerotinia minor* (Chaurasia *et al.*, 2018). Actinomycetes are complementary to photosynthetic bacteria and can improve wastewater treatment aided by EM technology with the advantage of increased antimicrobial properties.

8.1.5 FERMENTING FUNGI

Fungi are one of the primary decomposers in the ecosystem and are involved in the breakdown of pollutants to synthesize esters and alcohols and reduce odors and antimicrobial substances. Fungi also have an antagonistic feature to other pathogens present in contaminated water and are involved in the mineralization process of pollutants. This microorganism has a special feature, as it grows both sexually and asexually. In favorable conditions, by utilizing acidic and carbon-rich substrates, it multiplies asexually, while under unfavorable conditions, it forms spores sexually. The primary choices of fungi used in EM are *Aspergillus oryzae*, *Trichoderma* species, *Penicillium* species, and *Mucor hiemalis*. These microorganisms have lower nitrogen requirements and provide an advantage in decomposing organic materials (Yang *et al.*, 2017). *A. oryzae* is a filamentous and aerobic microorganism that is involved in fermentative as well as cellulolytic activity (Morocho and Agrícola, 2019). *Penicillium* species are known for their rapid growth, adaptability to acidic and water-stressed environments, and production of extracellular enzymes that help break down contaminants in water (El-Gendy *et al.*, 2017), while *Mucor hiemalis* and *Trichoderma* species can both degrade the organic material found in wastewater and use a variety of substrates. These microbes are also facultative anaerobes that enable them to survive in different environments such as in all latitudes, from the polar zones to the equatorial, and give greater ecological plasticity. *Trichoderma* species have wide distribution, compete for nutrients and space, and have different biocontrol features such as antibiosis, mycoparasitism, and resistance induction (Horwath, 2017). All the mentioned EMs here exhibit some beneficial characteristics such as fermenting properties, synthesis of bioactive compounds, antagonist activity to pathogens, coexistence with other beneficial microorganisms, and exerting a positive impact on the ecosystem (López and Ambiental, 2017). Another theory put forth by some researchers was that EMs could be used to magnify the individual actions of each microorganism and provide a collective application (Ramirez Martinez, 2006). Lactic acid bacteria, phototropic bacteria, yeasts, actinomycetes, and fungi are just a few examples of the representative microbes that can be found in an EM environment. Each of these organisms has a particular job to do and, through coexistence and co-prosperity, enhances the activities of the others.

8.2 ACTIVATION OF THE EM

Before use, EM must be activated because it is available in a dormant state. Usually, seven days before application, almost 3 liters of dormant EM are activated by mixing it with 8 liters of chlorine-free water and almost 2 kilograms of brown sugar (Figure 8.2). These elements are combined and stored in a big container in an area with little temperature variation. The survival of microorganisms is significantly influenced by the environment's temperature, but there are additional factors, such as pH, that can also have an impact. It is suggested that the pH of the EM should range from 3 to 4.5 (Szymanski and Patterson, 2003).

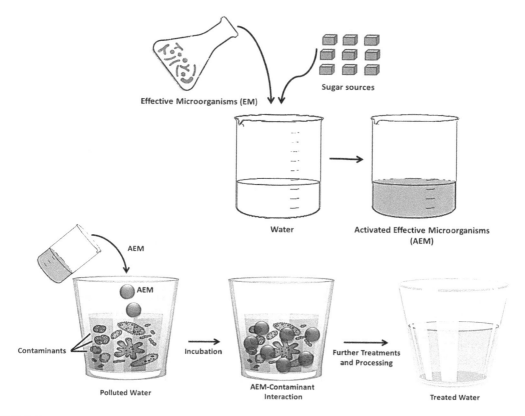

FIGURE 8.2 Overview of activation of EMs and contaminated water treatment.

8.3 APPROACH TO USING EM IN WASTEWATER IMPROVEMENT

Three steps make up the polluted water improvement process: primary, secondary, and tertiary. The second stage of improving contaminated water employs effective microorganisms. Figure 8.3 depicts the secondary treatment's general layout. Similar to the activated sludge process, the microbes may be suspended in this method. The benefit is that since microorganisms are suspended, they can easily access nutrients and get enough oxygen. As a result, the process switches to aerobic.

The purpose of the secondary treatment is to eliminate soluble organics from polluted water. The microbial process used in secondary treatment, secondary settling, is intended to significantly reduce the biological content of the sewage, including that found in soaps, detergents, food waste, and human waste. Most municipal and industrial wastewater treatment facilities use aerobic microbial processes to treat the settled sewage liquor. The microorganisms need oxygen and a substrate to live on for this to work. This can be accomplished in a variety of ways. The biodegradable, soluble organic impurities are consumed by EM, bacteria, and protozoa in each of these processes, and a sizable portion of the barely soluble components are bound into floc particles.

8.4 PARAMETERS USED TO ACCESS EM TECHNOLOGY

In the Armidale-Dumaresq region, some septic tanks were improved and sampled to ascertain the impact of the EM on the septic systems (Szymanski and Patterson, 2003). Due to their similarity in size and type of waste generation as well as their proximity for simpler monitoring, these septic

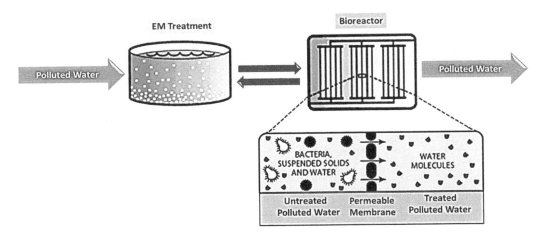

FIGURE 8.3 Schematic representation of a general secondary treatment strategy utilizing EM technology.

tanks were selected. At the inlet inspection port, the activated EM was first delivered to the septic tank. On a regular weekday, activated EM was made available. The following parameters were measured in samples taken from each device's outlet pipe septic tank before the application of EM: pH, total alkalinity, total dissolved solids (TDS), electrical conductivity (EC), suspended solids (TSS), oxygen demand, and total solids (TS).

8.4.1 pH

pH is a numerical indicator of the acidity or basicity of aqueous or other liquid solutions. When the pH levels of the different septic tanks are compared, it was evident that all but one have improved, even though there was no discernible trend between them before the application of EM. Findings also indicate that the pH environment tended to decrease to a lower pH level after the application.

8.4.2 BOD

BOD is the acronym for biological oxygen demand or biochemical oxygen demand. It is the amount of dissolved oxygen utilized by microbes to break down organic material present in a given water source. BOD is an extensively used parameter for accessing water quality.

8.4.3 EC

The ability of water to transmit electricity can be assessed using its electrical conductivity. The ability of a solution to transmit current is measured by its ionic activity, and water has a high electrical conductivity. However, additional monitoring of one septic tank indicated that it was a special instance. Additionally, it shows a common pattern among septic tanks that might indicate a higher EC when EM was used. Following the cessation of EM treatments, an overall decline in EC was also observed.

8.4.4 TA

The ability of water to neutralize acids is measured by its total alkalinity (TA). Water contains alkaline substances like hydroxides and carbonates that remove H^+ ions from the solution, reducing the water's acidity and raising pH. The amount of acid necessary to raise a specific sample's pH to 4.2

Effective Microorganisms 167

is used to determine the total alkalinity. All alkaline compounds were completely utilized at this level. The measurement of alkalinity is necessary to ascertain the capacity of water to counteract the corrosive and acidic effects of wastewater and other sources, such as rainfall.

8.4.5 TS

All of a sample's dissolved, colloidal, and suspended solids make up the total solids. Total solids, a measurement that is frequently used in the water treatment industry, includes both dissolved and suspended solids in a liquid. A high total solids level indicates that the liquid sample contains a lot of solid material. A high level of total solids could make the sample be considered contaminated depending on the evaluation criteria.

8.4.6 TSS

Plankton, algae, fine organic debris, and other particulate matter, as well as silt and clay flakes, are all examples of total suspended solids. The TSS analysis was unable to find any interesting trends, like the persistence of similar conditions after the EM application had ended.

8.4.7 TDS

Total dissolved solids refers to the amount of organic and inorganic materials, including metals, minerals, salts, and ions, that are dissolved in a given volume of water. Anything dissolved in water that isn't an H_2O molecule is essentially measured by TDS. When water comes into contact with soluble material, the substance's particles are absorbed into the water because it is a solvent, resulting in total dissolved solids. TDS can be found in water from a wide variety of sources, including natural water springs, chemicals used to treat municipal water supplies, runoff from yards and roads, and even your home plumbing system. As a result of the addition of EM, the TDS concentration decreased, exhibiting a clear pattern. Additionally, it emphasizes the occurrence of similar circumstances in some septic tanks.

8.4.8 COD

Chemical oxygen demand, also known as chemical oxygen depletion, is referred to as COD. By performing this test, you can determine how much oxygen is necessary for the organic matter in water to chemically break down or oxidize. Calculating how much of an oxidizing chemical was consumed during the test allows us to determine how much oxygen or organic matter was present. COD is a crucial factor in determining the quality of water and aids in reducing environmental and human health risks.

8.5 TYPICAL TESTING OF EFFECTIVE MICROORGANISMS

The most common types of water impurities come from synthetic, animal, or botanical sources and can get into the water in several ways, including through the industrial sector, detergents, and waste. Microorganisms are crucial in the various steps of treating contaminated water, whether in residential or commercial settings. Numerous microorganisms found in the water themselves specifically support the degradation of organic pollutants (Aracic *et al.*, 2015). Microorganisms were used in an experiment involving a septic tank. Even though microorganisms are required for the breakdown of organic waste, some people may experience health problems from them (Freitag, 2000). Among them could be bacteria and viruses found in the wastes that are produced. Because of the presence of lactic acid bacteria, which secrete organic acids, enzymes, antioxidants, and metallic chelates, the EM species of microorganisms are used because they contain a variety of organic acids. The improvement of solid-liquid separation, which is the basis for cleaning water, is facilitated by the creation of an antioxidant environment by EM (Figure 8.2) (Higa and Chinen, 1998). The volume of sludge is reduced, which is one of the main advantages of using EM. According to theory, the helpful microorganisms in EM should decompose the organic matter by turning it into carbon dioxide (CO_2) and methane (CH_4) or use it for growth and reproduction. This is allegedly the case for both septic tanks and wastewater treatment facilities, according to studies. This is emphasized by Freitag (2000), who contends that adding EM to anaerobic treatment facilities assisted in lowering the unpleasant by-products of this decomposition as well as the production of residual sludge. These elements tend to indicate that EM should help treat wastewater by enhancing the quality of water discharged and lowering the amount of sewage sludge produced.

EM significantly lowers the amount of sewage sludge produced and gets rid of the smells that go along with it (Higa and Chinen, 1998). This generally suggests that sludge and other solids in water are now more easily digestible. The primary assumption is that using EM will lead to a general trend toward a reduction in the amount of solids and, consequently, a reduction in sewage sludge. The changes in total suspended solids, total dissolved solids, and total solids over a short period are the best indicators of how many solids are being removed from water. The main goal of wastewater improvement is to find out if using efficient microorganism technology can reduce the amount of sewage sludge. Monitoring the corresponding changes in pH, electrical conductivity, and other physicochemical indicators of changing characteristics in the contaminated water after the application of EM is a secondary goal.

Except for TSS, all parameters showed the same conditions following the submission of the EM, according to the primary data, which also revealed an interesting pattern. This phenomenon happened after highly different introductory circumstances. The implication was that conditions could be created after an EM application that would help the EM survive. Sewage quality would have affected the highly variable conditions inside the tanks during application and monitoring. Another pattern that emerged following the end of EM application was a decline in total alkalinity, EC, and pH, which may have been triggered by a reduction in EM volume. Research findings suggested that applying EM could lead to a reduction in sludge volumes, although this effect will not materialize until the application is finished. Higher entrained solids levels could have been caused by organic matter breaking down in the septic tank, which would account for the initial increases in solids. According to another study, the idea that additives could chemically stir up the sludge in the tank and, as a result, explain why the results of DO, TDS, and COD are different could be supported (Elmoteleb, 2017).

According to an alternative perspective, perturbations in the microbial inhabitants could influence the entire community at a single trophic level, thereby influencing the efficacy of EM application (Boruszko, 2015). The similarity of the pattern may also have resulted from specific environmental conditions created by dominant microorganisms. Another study finding suggested that alterations in the microbial community could affect the pH of a system, other intermediates, and the system's overall effectiveness (Shalaby, 2011). The idea that being "notoriously fragile" is justified by pathogens being driven out of microbial ecosystems through competition and the introduction of beneficial EM that enhances the preexisting beneficial microorganisms is refuted by research (Namsivayam, 2011). However, another study that looked at EM and backed up this assertion with data from the Gushikawa City Library indicates that there is no longer a need for solid handling (Higa and Parr, 1994). While claims of decreased solids are supported by EM treatment, this trend did not begin until nearly 14 days following the last application of EM. As a result, it is not conclusively believed that EM tanks have had any impact on the septic system's solids content. This statement is supported by research done by Egyptian scientists on the effects of bacterial supplements on septic systems (El-Zanfaly *et al.*, 2010). It is crucial to conduct additional research after the finishing touches, including more frequent sampling (such as weekly samples to identify variations in EM applications) and extended observation (5 weeks of continuous observation), as the literature from the various studies has not been able to explain why similar conditions keep

Effective Microorganisms

occurring. This would allow for monitoring and observations of similarities in the system's conditions. When EM is used in a septic system, the amount of suspended solids present has no bearing on the tanks' pH, electrical conductivity, or alkalinity. However, research suggests that the EM may affect some system variables for on-site wastewater improvement, which allows septic tanks to display EM-relevant conditions. The findings do not support realistic solids reductions over time, which leans toward the conclusion that EM affected septic systems during the eight-week sampling period. Although more research is necessary before using EM as a supplement to lower the solids content of wastewater, suggestions have been made in this direction.

8.6 APPLICATIONS

Originally intended to address issues with productivity, mainly in organic crop systems, EM was applied to crops. As a result, technology gradually spread across all continents. Many systems related to agriculture and environmental management use EM today. These include livestock, aquaculture, and systems for the production of crops and animals, as listed in Table 8.2. Environmental management utilizes EM extensively for waste recycling of both solid and liquid wastes, as well as

TABLE 8.2

Approach and Applications for Various EM Formulations

EM Formulation	Approach to Preparation	Application
EM-1	It is primarily composed of three types of microbes. These microorganisms are lactic acid bacteria, photosynthetic bacteria, and yeast. Prof. T. Higa discovered EM-1, the original genuine effective microorganisms.	By bringing diversity and harmony to the microbial community in the soil, EM-1 improves soil health. Better root development, nutrient uptake, higher yields, and improvements to the size, quality, and flavor of your products are all facilitated by it.
EM-A	Make it by combining 5% of EM1 with an equivalent amount of sugarcane molasses, then storing the mixture for one to two weeks with the lid tightly closed at a constant temperature of about 30°C. A sweet and sour aroma and pH levels lower than 3.5 are required.	It is the most widely used method for reducing offensive smells in sewage treatment facilities. It also helps to decrease mud volume, increase sedimentation activity, and hasten sedimentation due to material decomposition.
EM- Bokashi	It is made by combining EM-A with fresh and superior organic material, such as fish meal, wheat bran, and rice bran. After that, it is permitted to ferment for up to two weeks in a tightly sealed container.	Increase in the rate of anaerobic digestion and fermentation of biodegradable waste during composting.
EM- Compost	When combined with EM-A and kept covered, animal waste, organic food scraps, and garden trimmings like leaves and other debris can promote anaerobic fermentation.	It is frequently used in agricultural processes like composting.
EM-5	In a sealed container, the mixture of EM1, vinegar, strong distillation alcohol, and water ferments for more than 30 days until it stops producing fermentation gas.	Since EM-5 works with all plant systems, it prevents destructive insects in addition to enhancing the inherent health-protective immune system.
EMX	It is an upgraded form of EM liquid.	It is safe to use on people, significantly reduces free radicals in the body, strengthens the immune system, and reduces the chance of cancerous cells forming.

for decomposition. The initial research in agriculture cleared the path for case studies and broad EM applications across a range of contexts. It is used in many different ways throughout the world, including in the USA, Europe, Africa, Asia, Central and South America, and Oceania. Agriculture and environmental management are the two primary areas into which the practical uses of EM can be broadly divided. Research projects and case studies highlighting EM's benefits in these two crucial areas have been published almost everywhere in the world.

8.6.1 Agricultural Applications

Agriculture was EM's initial application. Therefore, boosting the output of organic or natural farming systems was the initial application of EM. The amount of time needed to prepare biofertilizers was decreased because EM was directly applied to compost or organic matter added to crop fields. Additionally, EM is added in the form of bokashi (compost), which is created by mixing nitrogen-rich materials like rice, corn, wheat bran, fish meal, or oil cakes with waste products like rice husk and sawdust as a carrier. EM's efficiency in crop production has been the subject of several studies. Research on vegetables in New Zealand, papaya in Brazil, and apples in Japan provides very clear examples of this phenomenon (Daly and Stewart, 1999; Chagas et al., 1999; Fujita, 2001). These studies are just a few of many, and they all demonstrate how using EM or EM-based products, like bokashi, can eventually increase yields from conventional organic systems. Numerous nations have reported on how EM affects plant growth by preventing or suppressing pests and diseases. Sclerotinia in turf grass was controlled with EM, according to Kremer et al. (2000). In China and Australia, Phytophthora has been controlled with EM derivatives, according to Guest and Wang. In Costa Rica, EM has been successful in controlling black Sigatoka. These are just a handful of the many studies that show how successful EM is at producing crops. Each of them highlights how EM works well in a range of settings, which is crucial to its effectiveness and versatility. Numerous factors have been implicated in the causal phenomenon that led to these findings. Better photosynthesis, higher nutrient release, and increased protein activity are some of the advantages of using EM to compost organic matter (Konoplya and Higa, 2000; Xu et al., 2008; Amanullah et al., 2007). Additionally, the application of EM has been connected to improved root penetration, increased carbon mineralization, increased resistance to water stress, and enhanced soil properties (Daly and Stewart, 1999; Iriti et al., 2019; Joshi et al., 2019). It is also widely accepted that EM is used in animal husbandry in many parts of the world. Research from Belarus, Asia, where EM was first applied and is still widely used, as well as studies on the impact of EM on egg quality and chicken laying performance, attest to the advantages of EM application in swine and poultry units (Chantsavang and Watcharangkul, 1999; Konoplya and Higa, 2000; Atsbeha and Hailu, 2021). In these units, EM is added to the feed and sprayed for sanitation. In South Africa, integrated animal units and poultry farms use EM to boost productivity (Hanekom et al., 1999).

8.6.2 Environmental Management

In modern agriculture, environmental management is a crucial and contentious issue. Humanity faces issues with the disposal of farm wastes, the release of contaminated waters, and the reduction of dioxin produced by waste incineration or waste disintegration. To protect the current environment and possibly even improve it, legislation is being introduced in many nations. The contribution of EM to environmental management is crucial. This microbial solution, which was initially created for natural or organic farming systems, was further developed to address environmental problems, making it possible to reuse the majority of waste. Composting was the first instance of EM being used in environmental management. Animal and crop wastes were successfully composted to create biofertilizers. Studies from the Netherlands and Malaysia show that composting animal or plant wastes can increase crop yields compared to conventional organic systems (van Bruchem et al., 1999; Jusoh et al., 2013). Since the middle of the 1990s, very successful projects have been carried

Effective Microorganisms

out in Asia using EM to compost waste. Hanoi, Vietnam, which falls under the jurisdiction of that nation's Ministry of Science, Technology, and Environment, is a good example. In a project run by the Red Cross, the city of Pusan in Korea uses EM in more than 500 apartments to compost kitchen wastes that are then recycled into home gardens. Similar work is being done in the New Zealand city of Christchurch, which will serve as a field site for the International Conference on EM in January 2002. Water purification for reuse is another area where EM is successfully used. With great success, the Gushikawa City Library uses EM to treat recycled sewage water for the garden and restrooms. When water is treated with EM, the COD and BOD of the water are significantly reduced (Okuda and Higa, 1999; Safwat and Matta, 2021). Recently, a project on the application of EM for water treatment was started in the Australian city of McKay, on the Gold Coast. Before discharge, the water quality is improved after EM and oxygen treatment of the city sewage system. A vacation island uses EM to treat its water, and this water is recycled into odorless gardens. The water quality is in line with Australia's strict environmental regulations, and this study will be presented in New Zealand the following year. According to research conducted in South Africa, pig manure can be improved by EM before being fed to fish (Hanekom et al., 1999). The pigs' growth was stimulated by adding EM to their diet. Fish that were fed this manure were able to harvest more produce following EM treatment, which decreased the number of bacteria in their feces. EM is being used for waste management in many different countries, though not all of the projects are recorded. Concerning EM applications for cropping, livestock, and waste management, the Nature farm in Sara Buri, Thailand, provides the best example of any practitioner. Unfortunately, these findings have not been documented because the farm is a practical setting where EM technology is extensively taught to Thai and international students free of charge. Results from the most recent EM studies on environmental management were very intriguing. These would significantly contribute to the improvement of environmental quality if they could be replicated. First, a study from Belarus shows how EM can lessen radioactive contamination in impacted soils (Konoplya and Higa, 2000). The use of EM improved the uptake of Cs137 from soils contaminated by Chernobyl. Destroying these crops would reduce the level of contamination in the soils. Additionally, it was discovered that EMX, an EM derivative with antioxidant properties, acted as a radioprotective agent. A recent study conducted in Nigeria discovered that utilizing parts of efficient microbes could reduce dioxin production (Saibu et al., 2020). More importantly, a study conducted in Okinawa by Miyajima revealed that the use of EM in a commercial incinerator decreased the amount of dioxin produced (Miyajima et al., 2001). These offer worthwhile directions for future study and acceptance.

8.6.2.1 Water Quality Restoration

Numerous studies are focused on EM application to restore water quality in rivers and lakes, as listed in Table 8.3. In a study, Konin Lake in Poland was improved by EM to restore the water quality, as severe cyanobacterial water blooms were noticed (Dondajewska et al., 2019). This study was started in 2011 and completed in 2015, while a prior action to eliminate the external nutrient loads was taken before EM treatment. Some parameters, such as phytoplankton structure, water chemistry, and macrophyte distribution, were noticed during this study. The application of EM resulted in favorable changes in the ecosystem when organic matter was abundant. However, cyanobacteria blooms were also present because of the high nutrient levels. In another study, EM was applied with a loess ball to remove the contamination of harbor sediments (Ekpeghere et al., 2012). It was discovered that malodor was quickly reduced when 4% of EM loess balls were used in the EM treatment. However, the malodor was rapidly eliminated from the treatment when 0.05% molasses and 0.1% concentration of EM loess ball were added. Molasses is a carbon source for EM, and by fermentation, it provides organic acids such as acetic acid, propionic acid, and valeric acid. The combined action of EM and organic acids, molasses, and other nutrients with EM loess balls remediates the polluted marine sediments and removes associated malodor rapidly. EM with mudballs was also applied to an artificial river water treatment to examine the temperature effect on COD and TSS (Nugroho, 2017). COD reduction efficiency at 30°C was observed 59.4%, while at 25°C, it was

TABLE 8.3

Applications of EM in Wastewater Treatment

Type of Wastewater	Pollutants Studied	References
Domestic sewage and wastewater	Alkalinity, COD, BOD, TSS, TDS, NO_3	(SKR Namsivayam, 2011; El Shafei and Abd Elmoteleb, 2018)
Artificial lake	COD, TN, TP, TSS, Turbidity	(Zhao *et al.*, 2013)
Municipal wastewater	COD, TN, BOD, TSS, NH_3, TP	(Lee and Cho, 2010; Ahmad, 2017; Safwat, 2018)
Synthetic wastewater	Heavy metals, Basic dyes	(Ting *et al.*, 2013; Bhagavathi Pushpa *et al.*, 2016)
Industry wastewater	COD, TSS	(Priya *et al.*, 2015)
Artificial river water	COD, TSS	(Nugroho, 2017)
Raw sewage	BOD, COD, TDS, TSS, TP	(Monica *et al.*, 2011; Rashed and Massoud, 2015)
Beet sugar wastewater	COD, BOD, TDS	(AA Embaby, 2010)
Synthetic polluted water	NH_3-N, TP, COD	(M Ji, 2014)
Stream water	TP, TN, Xenobiotic	(Park *et al.*, 2016)

66.7%, while mudballs containing EM treatment have TSS reduction efficiency almost 99.7% and 100% at 30°C and 25°C, respectively. Efficacy of EM is increased with dissolved oxygen (DO), as also noticed in a river water treatment (M Ji, 2014). Reduction of ammonia nitrogen, COD, and TP was observed in polluted water after EM treatment increase with the abundance of DO at 34.34%, 4.59%, and 18.18%, respectively. These observations suggest that EM treatment works better in aerobic conditions. A eutrophic artificial lake was also treated with EM, aeration system, fingerlings, and aquatic vegetation near the lakes and consisted of an integrated restoration technique (Zhao *et al.*, 2013). Some features such as COD, TP, TSS, turbidity, and TN were observed, and by treatment, these were removed at 70.12%, 86.87%, 82.36%, 79.28%, and 57.36%, respectively. This integrated restoration technique has an advantage over other conventional techniques of water treatment as there was no secondary pollution observed, and it also maintains good water quality for a long time. Moreover, improvements in stream water quality were noted as a result of EM (Park *et al.*, 2016). In this study, soil balls that contained EM were used, and improvement was performed by hardener soil balls with a reduction of total nitrogen, dissolved oxygen, and total phosphorus.

8.6.2.2 Bioremediation of Wastewater

A study that used EM to improve domestic sewage found that parameters like COD, alkalinity, TDS, and BOD significantly decreased. The study's results promised that EM might be applied to clean up household waste (Monica *et al.*, 2011). In a different sewage treatment study, critical parameters like TSS, TDS, COD, and BOD were found to have decreased by 91%, 55%, 82%, and 85%, respectively, after 3.0 mL/L of EM solution was applied to sewage in an aerobic environment and left for three days. The findings of this study suggested that EM solution can aid in sewage treatments and help in maintaining a good environment. In another work, EM was immobilized using alginate and used to treat wastewater (Ting *et al.*, 2013). In this investigation, EM-free cells and alginate-immobilized cells were utilized, and a reduction of Cr^{2+}, Cu^{3+}, and Pb^{2+} was observed. It was discovered that EM immobilized in alginate was more effective than EM in free cells. The reductions in Cr^{2+}, Cu^{3+}, and Pb^{2+} were 0.859, 0.160, and 0.755 mg/mL for free cells of EM and 2.695, 0.940, and 4.011 mg/g for alginate-immobilized EM. In a different study, the impact of EM on the improvement was examined using primary settled wastewater (S Ahmed, 2013). Parameters including COD, total phosphorous, ammonia, BOD, TSS, and soluble COD were measured with removal rates of 72.01%, 62.5%, 50.4%, 75.7%, 80.9%, and 61.5%, respectively, in an anaerobic batch reactor.

The treatment was done at 35°C for 24 h, and a reduction in Salmonella, total coliform, ammonia, and sulfate-reducing bacteria were also observed. One unit of EM was applied to 1000 units of wastewater in another treatment that used an anaerobic reactor. Reductions of 60%, 84%, 62%, and 62% were seen in parameters like COD, TSS, BOD, and turbidity, respectively (Anwar *et al.*, 2013). Such investigations suggested that EM treatment in wastewater is also beneficial to improving water quality under anaerobic conditions.

8.7 INDIA'S APPROACH TO POLLUTED WATER TREATMENT

The government of India (GOI) approved the Namami Gange Program, an integrated conservation mission, as a flagship program in June 2014 with a budget outlay of Rs. 20,000 crores. It is an integrated umbrella program to ensure the effective reduction of pollution and preservation of the Ganges River and its tributaries, including the Yamuna River. By offering financial and technical support, the GOI is assisting the state governments in addressing the problems caused by pollution of the Ganga and its tributaries. Twenty-four sewerage infrastructure projects totaling Rs. 4773 crores have been approved under the Namami Gange Program to build or renovate sewage treatment plants (STPs) with a capacity of 1940 million liters per day (MLD) in the Yamuna river basin, including the Hindon and Chambal tributaries. A total of 24 projects are being undertaken, with one each in Himachal Pradesh, Haryana, Delhi, Uttar Pradesh, and Rajasthan. Five of these projects have already been finished. Fourteen common effluent treatment plants (CETPs) with a combined capacity of 161.5 MLD are located in Haryana, and the Yamuna basin in Delhi is home to 13 CETPs with a capacity of 212.3 MLD. Under the Namami Gange Programme, industrial clusters are also given financial support to upgrade or expand their CETPs. One CETP project in Mathura has been upgraded for the textile industrial cluster. The Uttar Pradesh Pollution Control Board reports that 26 polluted industries in Aligarh are known to release their treated effluent into the Aligarh drain and that 11 polluting industries are located in the Hathras district. The Yamuna River eventually flows into this drain at Morewali Dargah in Agra. With adequate effluent treatment facilities that are governed by the provisions of current environmental laws, all industries are permitted to operate. According to Aligarh Nagar Nigam, a pilot program using phytoremediation technology is being used to treat the polluted water from the Jafri and Aligarh drains. Additionally, Uttar Pradesh Jal Nigam has commenced work on a 45 MLD–capacity STP in the town.

8.8 NEW DEVELOPMENTS IN EM TECHNOLOGY

There are issues in the natural world as a result of human overconsumption, which pollutes the environment and triggers Mother Nature's retaliation. It is possible to successfully and diligently halt this catastrophe by utilizing EM's advantages. With EM, waste can be efficiently removed and converted into reasonably priced, premium organic fertilizer, giving humanity a more wholesome and healthy environment. To increase yield and the production of essential oils, research on the effects of drought stress on crops is highly sought after, particularly on medicinal and aromatic plants in dry areas. A recent study examined the effects of EM on salvia's capacity for photosynthetic and nutrient absorption, taking into consideration the role that biological fertilizers and medicinal herbs play in producing the product's yield, both quantitatively and qualitatively (Elnahal *et al.*, 2022). Many crops' growth and productivity are adversely impacted by the severe environmental factor of salinity. Combining EM with nanomaterials like Mg^{2+} to improve sweet potato performance in salinity-prone environments could be an interesting or novel application (Abd El-Mageed *et al.*, 2022). Explosive pollutants that are still present in soils around the world pose serious risks to ecological safety and human health. To test the effectiveness of combined remediation in modifying the microbial community and microenvironment of explosive-contaminated soils, trinitrotoluene (TNT)- and cyclotrimethylenetrinitramine (RDX)-contaminated soils are simulated and remedied using vetiver grass and EM (Yang *et al.*, 2022). Considering all of its advantages, EM can be utilized to enhance the capabilities of self-curing concrete. In order to address problems with early strength features and improve mechanical and microstructural properties at both early and late ages, integrated EM was utilized in concrete due to its advantageous chemical and physical properties (Memon *et al.*, 2022; Onaizi *et al.*, 2022). *Xenorhabdus nematophila*, a biocontrol alternative, is also used in a variety of applications (Yadav and Rathore, 2018, 2022b, 2022a; Yadav and Rathore, 2020). The potential use of EM as an electrolyte and biosurfactant for electrokinetic remediation applications was examined in a novel study (Tsui *et al.*, 2022). A study was conducted to compare how well pig wastewater was treated using an anaerobic baffled reactor with and without the addition of an effective microorganism (EM4) as a bioactivator during the startup process (Suryawan *et al.*, 2021). Changes in the water quality and the structure of the phytoplankton were discovered in a case study of a lake that was restored using barley straw and EM (Dondajewska *et al.*, 2019). An enhanced technique for creating EM soil balls was developed in a study (Park *et al.*, 2016). The pH and physical hardness of the soil ball was increased by changing its formulation. Using a semiconductor sequencer and next-generation DNA sequencing technology, the microbial community of the soil ball was discovered. The metabolic pathways responsible for the breakdown of xenobiotics were also assessed through the analysis and annotation of metagenome data (Park *et al.*, 2016). The Turawa Reservoir's water's physicochemical and microbiological characteristics are also described through the application of EM (Dobrzyński *et al.*, 2021). The water quality in aquaculture ponds may be continuously harmed by the endogenous release of nitrogen and phosphorus from sediment. In a study, this contaminated aquaculture sediment was restored using an integrated bioremediation method that combined EM and aeration techniques (Wang *et al.*, 2019). Most wastewater must be reduced in phosphorus before being discharged to receiving waters because too much phosphorus can lead to eutrophication in water bodies. Fermentation could be improved by adding activated EM to the anaerobic zone because it would boost the activity of phosphorus-accumulating organisms (PAOs), which would increase phosphorus release and decrease the average phosphorus concentration in the effluent (Rashed and Massoud, 2015).

8.9 CONCLUSIONS

The treatment of polluted water is found to benefit greatly from the use of efficient microorganisms. They are known to reduce sludge and produce high-quality compost. Thus, the overall cost of the treatment is decreased. They also have the benefit of making the bioremediation process odorless and non-pathogenic. Utilizing effective microorganisms in technology has good potential to solve many environmental issues. To determine its susceptibility and to create effective microorganisms using different bacterial groups, more research should be done. One should exercise caution when using EM, as demonstrated in this chapter. Although it plays a significant role in agriculture and environmental management, it is not a solution to every issue. Like all methods, EM must be used under the rules with care and diligence. Failure to do so would result in unfavorable outcomes, as in some cases, which have also received attention. However, the adoption of EM technology will make sure that the goal that everyone in this world strives to realize—a cleaner environment for the current and following generations of humans—is realized.

REFERENCES

Abd El-Mageed, T. A. *et al.* 2022. "Coapplication of Effective Microorganisms and Nanomagnesium Boosts the Agronomic, Physio-Biochemical, Osmolytes, and Antioxidants Defenses Against Salt Stress in Ipomoea batatas." *Frontiers in Plant Science* 13: 1446. https://doi.org/10.3389/FPLS.2022.883274/BIBTEX.

Ahmad, J. 2017. "Bioremediation of Petroleum Sludge Using Effective Microorganism (EM) Technology." *Petroleum Science & Technology* 35, no. 14: 1515–22. https://doi.org/10.1080/10916466.2017.1356850.

Ahmed, S., and H. A. E. R. 2013. "Treatment of Primary Settled Wastewater Using Anaerobic Sequencing Batch Reactor Seeded with Activated EM." *Civil & Environmental Research* 3: 130–7.

Amanullah, M. M., E. Somasundaram, K. Vaiyapuri, and K. Sathyamoorthi. 2007. "Poultry Manure to Crops – A Review." *Agricultural Reviews* 28: 216–22.

Anwar, Z. R., M. Ariffin, A. Hassan, I. Mahmood, and A. K. Khamis. 2013. "Treatment of Rubber Processing Wastewater by Effective Microorganisms Using Anaerobic Sequencing Batch Reactor." *Journal of Agrobiotechnology* 4: 1–15.

Aracic, Sanja, Sam Manna, Steve Petrovski, Jennifer L. Wiltshire, Gülay Mann, and Ashley E. Franks. 2015. "Innovative Biological Approaches for Monitoring and Improving Water Quality." *Frontiers in Microbiology* 6(JUL): 826. https://doi.org/10.3389/FMICB.2015.00826/BIBTEX.

Atsbeha, Alem Tadesse, and Teweldemedhn Gebretinsae Hailu. 2021. "The Impact of Effective Microorganisms (EM) on Egg Quality and Laying Performance of Chickens." *International Journal of Food Science* 2021: 8895717. https://doi.org/10.1155/2021/8895717.

Bhagavathi Pushpa, T., J. Vijayaraghavan, K. Vijayaraghavan, and J. Jegan. 2016. "Utilization of Effective Microorganisms Based Water Hyacinth Compost as Biosorbent for the Removal of Basic Dyes." *Desalination & Water Treatment* 57, no. 51: 24368–77. https://doi.org/10.1080/19443994.2016.1143405.

Boruszko, D., and A. Butarewicz. 2015. "Impact of Effective Microorganisms Bacteria on Low-Input Sewage Sludge Treatment." *Environment Protection Engineering* 41, no. 4: 83–96. https://doi.org/10.37190/epe150407.

Carpenter, S. R., E. H. Stanley, and M. J. Vander Zanden. 2011. "State of the World's Freshwater Ecosystems: Physical, Chemical, and Biological Changes." *Annual Review of Environment & Resources* 36, no. 1: 75–99. https://doi.org/10.1146/annurev-environ-021810–094524.

Chagas, P. R. R., H. Tokeshi, and M. C. Alves. 1999. "Effect of Calcium on Yield of Papaya Fruits on Conventional and Organic (Bokashi EM) Systems." In *Proceedings of the 6th International Conference on Kyusei Nature Farming*. Pretoria, South Africa. https://www.infrc.or.jp/knf/PDF%20KNF%20Conf%20Data/C6-OAP-260.pdf.

Chantsavang, S., and P. Watcharangkul. 1999. "Influence of EM on the Quality of Poultry Products." In *the Proceedings of the 5th International Conference on Kyusei Nature Farming:* 133–50. Thailand. https://www.infrc.or.jp/knf/PDF%20KNF%20Conf%20Data/C5-5-177.pdf.

Chaurasia, A., B. R. Meena, A. N. Tripathi, K. K. Pandey, A. B. Rai, and B. Singh. 2018. "Actinomycetes: An Unexplored Microorganisms for Plant Growth Promotion and Biocontrol in Vegetable Crops." *World Journal of Microbiology & Biotechnology* 34, no. 9: 132. https://doi.org/10.1007/s11274-018-2517-5.

Comesaña, D. A. *et al.* 2018. "Municipal Sewage Sludge Variability: Biodegradation Through Composting with Bulking Agent." In *Sewage*, edited by Sewage Ivan X. Zhu. IntechOpen. https://doi.org/10.5772/INTECHOPEN.75130.

Daly, M. J., and D. P. C. Stewart. 1999. "Influence of Effective Microorganisms (EM) on Vegetable Production and Carbon Mineralization, a Preliminary Investigation." *Journal of Sustainable Agriculture* 14, no. 2–3: 15–25. https://doi.org/10.1300/J064v14n02_04.

de Souza, R., Adriana Ambrosini, and Luciane M P M. P. Passaglia. 2015. "Plant Growth-Promoting Bacteria as Inoculants in Agricultural Soils." *Genetics & Molecular Biology* 38, no. 4: 401–19. https://doi.org/10.1590/S1415–475738420150053.

Dobrzyński, J., P. S. Wierzchowski, W. Stępień, and E. B. Górska. 2021. "The Reaction of Cellulolytic and Potentially Cellulolytic Spore-Forming Bacteria to Various Types of Crop Management and Farmyard Manure Fertilization in Bulk Soil." *Agronomy* 11, no. 4: 772. https://doi.org/10.3390/agronomy11040772.

Dondajewska, Renata, Anna Kozak, Joanna Rosińska, and Ryszard Gołdyn. 2019. "Water Quality and Phytoplankton Structure Changes Under the Influence of Effective Microorganisms (EM) and Barley Straw—Lake Restoration Case Study." *Science of the Total Environment* 660: 1355–66. https://doi.org/10.1016/j.scitotenv.2019.01.071.

Ekpeghere, Kalu I., Byung-Hyuk Kim, Hee-Seong Son, Kyung-Sook Whang, Hee-Sik Kim, and Sung-Cheol Koh. 2012. "Functions of Effective Microorganisms in Bioremediation of the Contaminated Harbor Sediments." *Journal of Environmental Science & Health. Part A, Toxic/Hazardous Substances & Environmental Engineering* 47, no. 1: 44–53. https://doi.org/10.1080/10934529.2012.629578.

El-Gendy, Mervat Morsy Abbas Ahmed, Salha Hassan Mastour Al-Zahrani, and Ahmed Mohamed Ahmed El-Bondkly. 2017. "Construction of Potent Recombinant Strain Through Intergeneric Protoplast Fusion in Endophytic Fungi for Anticancerous Enzymes Production Using Rice Straw." *Applied Biochemistry & Biotechnology* 183, no. 1: 30–50. https://doi.org/10.1007/s12010–017–2429–0.

Elmoteleb, M. M. E. S. 2017. "Investigate the Effect of Effective Microorganism (Em) on Improving the Quality of Sewage Water from Al-Gabal Al-Asfar Area in Egypt." *The 1st International Conference on Towards a Better Quality of Life:* 1–12. http://dx.doi.org/10.2139/ssrn.3164096

Elnahal, A. S. M., M. T. El-Saadony, A. M. Saad, E.-S. M. Desoky, A. M. El-Tahan, M. M. Rady, S. F. AbuQamar, and K. A. El-Tarabily. 2022. "The Use of Microbial Inoculants for Biological Control, Plant Growth Promotion, and Sustainable Agriculture: A Review." *European Journal of Plant Pathology* 162, no. 4: 759–92. https://doi.org/10.1007/s10658-021-02393-7.

El Shafei, M., and E. Abd Elmoteleb. 2018. "Investigate the Effect of Effective Microorganism (EM) on Improving the Quality of Sewage Water from Al-Gabal Al-Asfar Area in Egypt." *SSRN Electronic Journal*. https://doi.org/10.2139/SSRN.3164096.

El-Zanfaly, H. T., A. Mostafa, M. Mostafa, and I. Fahim. 2010. "Effect of Bacterial Additives on The Performance of Septic Tanks for Wastewater Treatment in the Upper Egypt Rural Area." *WIT Transactions on Ecology & the Environment* 142: 389–400. https://doi.org/10.2495/SW100361.

Embaby, A. A. *et al.* 2010. "Application of Effective Microorganisms in Treatment of Wastewater of Beet Sugar Factory at Bilqas, Dakahlia Governorate, Egypt." *Journal of Environmental Sciences* 39: 151–8.

Farrag, H. M., and A. A. Bakr. 2021. "Biological Reclamation of a Calcareous Sandy Soil with Improving Wheat Growth Using Farmyard Manure, Acid Producing Bacteria and Yeast.". *SVU-International Journal of Agricultural Sciences* 3, no. 1: 53–71. https://doi.org/10.21608/svuijas.2021.57919.1070.

Fayemi, O. E., and A. O. Ojokoh. 2014. "The Effect of Different Fermentation Techniques on the Nutritional Quality of the Cassava Product (fufu)." *Journal of Food Processing & Preservation* 38, no. 1: 183–92. https://doi.org/10.1111/j.1745–4549.2012.00763.x.

Freitag, D. G. 2000. *The Use of Effective Microorganisms (EM) in Organic Waste Management the Use of EM at Skywalker Ranch: A Case Study of Sustainable Farming and Composting*. Accessed October 29, 2022. http://www.emtrading.com.

Fujita, M. 2001. "Nature Farming Practices for Apple Production in Japan." *Journal of Crop Production* 3, no. 1: 119–25. https://doi.org/10.1300/J144v03n01_11.

Hanekom, D., J. F. Prinsloo, and H. J. Schoonbee. 1999. "University of the North; Sovenga, South Africa." In *Proceedings of the 6th International Conference on Kyusei Nature Farming*. Pretoria, South Africa. Aquaculture Research Unit. A comparison of the effect of Anolyte and EM on the faecal bacterial loads in the water and on fish produced in pig cum fish integrated production units. https://www.infrc.or.jp/knf/PDF%20KNF%20Conf%20Data/C6-3-225.pdf.

Hidalgo, D., F. Corona, and J. M. Martín-Marroquín. 2022. "Manure Biostabilization by Effective Microorganisms as a Way to Improve Its Agronomic Value." *Biomass Conversion & Biorefinery* 12, no. 10: 4649–64. https://doi.org/10.1007/s13399–022–02428-x.

Higa, T., and J. F. Parr. 1994. *Beneficial and Effective Microorganisms for a Sustainable Agriculture and Environment*. Atami: International Nature Farming Research Center.

Higa, T., and M. N. Chinen. 1998. *EM Treatments of Odor, Waste Water, and Environment Problems*. Atami: International Nature Farming Research Center.

Horwath, W. R. 2017. "The Role of the Soil Microbial Biomass in Cycling Nutrients." *Microbial Biomass*: 41–66. https://doi.org/10.1142/9781786341310_0002.

Iriti, Marcello, Alessio Scarafoni, Simon Pierce, Giulia Castorina, and Sara Vitalini. 2019. "Soil Application of Effective Microorganisms (EM) Maintains Leaf Photosynthetic Efficiency, Increases Seed Yield and Quality Traits of Bean (*Phaseolus vulgaris* L.) Plants Grown on Different Substrates." *International Journal of Molecular Sciences* 20, no. 9: 2327. https://doi.org/10.3390/ijms20092327.

Ji, M., and J. Z. R. Z. J. C. 2014. "Experiments on the Effects of Dissolved Oxygen on the Free and Immobilization Effective Microorganisms (EM) in Treating Polluted River Water. C.W." *Proceedings of International Conference on Material and Environmental Engineering (ICMAEE 2014)* 57: 5–7.

Joshi, H., S. Duttand, P. Choudhary, and S. L. Mundra. 2019. "Role of Effective Microorganisms (EM) in Sustainable Agriculture" *International Journal of Current Microbiology & Applied Sciences* 8, no. 3: 172–81. https://doi.org/10.20546/ijcmas.2019.803.024.

Jusoh, Mohd Lokman Che L. C., Latifah Abd Manaf, and Puziah Abdul Latiff. 2013. "Composting of Rice Straw with Effective Microorganisms (EM) and Its Influence on Compost Quality." *Iranian Journal of Environmental Health Science & Engineering* 10, no. 1: 17. https://doi.org/10.1186/1735-2746-10-17.

Konoplya, E. F., and T. Higa. 2000 Sept. 20–22. "EM Application in Animal Husbandry–Poultry Farming and Its Action Mechanisms." In *Proceedings of the International Conference on EM Technology and Nature Farming*. Pyongyang, Korea. https://www.eminfo.nl/wp-content/uploads/2013/08/EM-1-Application-In-Animal-Husbandry-Poultry-Farming-and-its-Action-Mechanisms.pdf.

Kremer, Robert, E. Ervin, M. Wood, and D. Abuchar. 2000. "Control of Sclerotinia Homoeocarpa in Turfgrass Using Effective Microorganisms." *Accesses*. https://www.researchgate.net/publication/242224354_Control_of_Sclerotinia_homoeocarpa_in_Turfgrass_Using_Effective_Microorganisms.

Lee, J., and M. H. Cho. 2010. "Removal of Nitrogen in Wastewater by Polyvinyl Alcohol (PVA)-Immobilization of Effective Microorganisms." *Korean Journal of Chemical Engineering* 27, no. 1: 193–7. https://doi.org/10.1007/s11814–009–0330–4.

Londoño, N. A. *et al.* 2015. "Bacteriocinas Producidas por Bacterias Ácido Lácticas y Su Aplicación en la Industria de Alimentos." *Revista Alimentos Hoy* 23, no. 36: 186–205.

Effective Microorganisms

López, T. R., and D. V. M. Ambiental. 2017. Uso de microorganismos eficientes para tratar aguas contaminadas. *Ingeniería Hidráulica y Ambiental* XXXVIII, no. 3: 88–100.

Luna, M. A., and J. M. 2016. "Microorganismos Eficientes y Sus Beneficios para los Agricultores." *Revista Cientifíca Agroecosistmas* 4, no. 2: 31–40.

Meena, S. K., and V. S. Meena. 2017. "Importance of Soil Microbes in Nutrient Use Efficiency and Sustainable Food Production." *Agriculturally Important Microbes for Sustainable Agriculture* 2: 3–23. https://doi.org/10.1007/978-981-10-5343-6_1.

Memon, R. P., G. F. Huseien, A. T. Saleh, Ali Taha Saleh, U. Memon, M. Alwetaishi, O. Benjeddou, and A. R. M. Sam. 2022. "Microstructure and Strength Properties of Sustainable Concrete Using Effective Microorganisms as a Self-Curing Agent." *Sustainability* 14, no. 16: 10443. https://doi.org/10.3390/su141610443.

Miyajima, T., H. Ogawa, and I. Koike. 2001. "Alkali-Extractable Polysaccharides in Marine Sediments: Abundance, Molecular Size Distribution, and Monosaccharide Composition." *Geochimica & Cosmochimica Acta* 65, no. 9: 1455–66. https://doi.org/10.1016/S0016-7037(00)00612-8.

Monica, S. *et al.* 2011. "Formulation of Effective Microbial Consortia and Its Application for Sewage Treatment." *Journal of Biochemical & Microbial Technology* 3, no. 3: 51–5. https://doi.org/10.4172/1948-5948.1000051.

Morocho, M. T., and M. L.-M. Agrícola. 2019. "Microorganismos Eficientes, Propiedades Funcionales y Aplicaciones Agrícolas." *Revista Centro Agricola* 46, no. 2: 93.

Namsivayam, S. K. R., and G. N. J. K. 2011. "Evaluation of Effective Microorganism (EM) for Treatment of Domestic Sewage." *Journal of Experimental Science* 2: 30–2.

Nugroho, L. F., F. L. Nugroho, D. Rusmaya, Y. M. Yustiani, F. I. Hafiz and R. B. T. Putri. 2017. Effect of Temperature on Removal of COD and TSS from Artificial River Water by Mudballs Made from EM4, Rice Bran and Clay. "F.H.R.P." *International Journal of GEOMATE* 12: 91–5.

Okuda, A., and T. Higa. 1999. "Purification of Waste Water with Effective Microorganisms and Its Utilization in Agriculture." In *Proceedings of the 5th International Conference on Kyusei Nature Farming*, edited by Y. D. A. Senanayake, and U. R. Sangakkara: 246–53. Bangkok, Thailand: APNAN. http://www.infrc.or.jp/knf/5th_Conf_S_8_2.html.

Onaizi, Ali M., Ghasan Fahim Huseien, Nor Hasanah A. Shukor Lim, W. C. Tang, Mohammad Alhassan, and Mostafa Samadi. 2022. "Effective Microorganisms and Glass Nanopowders from Waste Bottle Inclusion on Early Strength and Microstructure Properties of High-Volume Fly-Ash-Based Concrete." *Biomimetics* 7, no. 4: 190. https://doi.org/10.3390/biomimetics7040190.

Park, Gun-Seok, Abdur Rahim Khan, Yunyoung Kwak, Sung-Jun Hong, ByungKwon Jung, Ihsan Ullah, Jong-Guk Kim, and Jae-Ho Shin. 2016. "An Improved Effective Microorganism (EM) Soil Ball-Making Method for Water Quality Restoration." *Environmental Science & Pollution Research International* 23, no. 2: 1100–7. https://doi.org/10.1007/s11356-015-5617-x.

Priya, M., T. Meenambal, N. Balasubramanian, and B. Perumal. 2015. "Comparative Study of Treatment of Sago Wastewater Using HUASB Reactor in the Presence and Absence of Effective Microorganisms." *Procedia Earth & Planetary Science* 11: 483–90. https://doi.org/10.1016/j.proeps.2015.06.048.

Ramirez Martinez, M. A. 2006. *Tecnologia de los Microorganismos (EM), Aplicada a la Agricultura y Medio Ambiente Sostenible*: 42. Thesis submitted to Universidad Industrial De Santander Escuela De Ingenieria Quimica Especializacion Ingenieria Ambiental Bucaramanga.

Rashed, E. M., and M. Massoud. 2015. "The Effect of Effective Microorganisms (EM) on EBPR in Modified Contact Stabilization System." *HBRC Journal* 11, no. 3: 384–92. https://doi.org/10.1016/j.hbrcj.2014.06.011.

Safwat, S. M. 2018. "Performance of Moving Bed Biofilm Reactor Using Effective Microorganisms." *Journal of Cleaner Production* 185: 723–31. https://doi.org/10.1016/j.jclepro.2018.03.041.

Safwat, S. M., and M. E. Matta. 2021. "Environmental Applications of Effective Microorganisms: A Review of Current Knowledge and Recommendations for Future Directions." *Journal of Engineering & Applied Sciences* 68, no. 1: 1–12. https://doi.org/10.1186/S44147-021-00049-1/TABLES/2.

Saibu, Salametu, Sunday A. Adebusoye, Ganiyu O. Oyetibo, and Debora F. Rodrigues. 2020. "Aerobic Degradation of Dichlorinated Dibenzo-p-Dioxin and Dichlorinated Dibenzofuran by Bacteria Strains Obtained from Tropical Contaminated Soil." *Biodegradation* 31, no. 1–2: 123–37. https://doi.org/10.1007/s10532-020-09898-8.

Servin, Jacqueline A., Craig W. Herbold, Ryan G. Skophammer, and James A. Lake. 2008. "Evidence Excluding the Root of the Tree of Life from the Actinobacteria." *Molecular Biology Evolution* 25, no. 1: 1–4. https://doi.org/10.1093/molbev/msm249.

Shalaby, Emad A. 2011. "Prospects of Effective Microorganisms Technology in Wastes Treatment in Egypt." *Asian Pacific Journal of Tropical Biomedicine* 1, no. 3: 243–8. https://doi.org/10.1016/S2221-1691(11)60035-X.

Soto, J. A. *et al.* 2017. "Inoculation of Substrate with Lactic Acid Bacteria for the Development of Moringa oleifera Lam. Plantlets." *Cuban Journal of Agricultural Science* 51, no. 2.

Su, Pin, Xinqiu Tan, Chenggang Li, Deyong Zhang, J. Cheng, Songbai Zhang, Xuguo Zhou, Qingpin Yan, Jing Peng, Zhuo Zhang, Yong Liu, and Xiangyang Lu. 2017. "Photosynthetic Bacterium R *Hodopseudomonas palustris* GJ-22 Induces Systemic Resistance Against Viruses." *Microbial Biotechnology* 10, no. 3: 612–24. https://doi.org/10.1111/1751–7915.12704.

Suryawan, R. F., P. C. Susanto, N. H. Parmenas, and D. Setiadi. 2021. "Strategy to Increase Bank Satisfaction in the New Normal Era of Covid-19." *Jurnal Mantik* 5, no. 3: 1977–81.

Szymanski, N., and R. Patterson. 2003. "Effective Microorganisms (EM) and Wastewater Systems." In *Future Directions for On-Site Systems: Best Management Practice, Proceedings of On Site 2003 Conference:* 384. Armidale: Lanfax Laboratories Armidale, University of New England.

Timmusk, Salme, Eviatar Nevo, Fantaye Ayele, Steffen Noe, and Ylo Niinemets. 2020. "Fighting Fusarium Pathogens in the Era of Climate Change: A Conceptual Approach." *Pathogens* 9, no. 6: 419. https://doi.org/10.3390/pathogens9060419.

Ting, Adeline Su Yien, Nurul Hidayah Abdul Rahman, Mohamed Ikmal Hafiz Mahamad Isa, and Wei Shang Tan. 2013. "Investigating Metal Removal Potential by Effective Microorganisms (EM) in Alginate-Immobilized and Free-Cell Forms." *Bioresource Technology* 147: 636–9. https://doi.org/10.1016/j.biortech.2013.08.064.

Tsui, To-Hung, L. e. Zhang, Jingxin Zhang, Yanjun Dai, and Yen Wah Tong. 2022. "Methodological Framework for Wastewater Treatment Plants Delivering Expanded Service: Economic Tradeoffs and Technological Decisions." *Science of the Total Environment* 823: 153616. https://doi.org/10.1016/j.scitotenv.2022.153616.

van Bruchem, J., H. Schiere, and H. van Keulen. 1999. "Dairy Farming in the Netherlands in Transition Towards More Efficient Nutrient Use." *Livestock Production Science* 61, no. 2–3: 145–53. https://doi.org/10.1016/S0301-6226(99)00064-0.

Vurukonda, Sai Shiva Krishna Prasad, Davide Giovanardi, and Emilio Stefani. 2018. "Plant Growth Promoting and Biocontrol Activity of Streptomyces spp. as Endophytes." *International Journal of Molecular Sciences* 19, no. 4: 952. https://doi.org/10.3390/ijms19040952.

Wang, Y., P. Xu, Z. Nie, Q. Li, N. Shao, and G. Xu. 2019. "Growth, Digestive Enzymes Activities, Serum Biochemical Parameters and Antioxidant Status of Juvenile Genetically Improved Farmed Tilapia (*Oreochromis niloticus*) Reared at Different Stocking Densities in in-Pond Raceway Recirculating Culture System." *Aquaculture Research* 50, no. 4: 1338–47. https://doi.org/10.1111/are.14010.

Xu, H., R. Xu, F. Qin, MA, G., Y. Yu, and S. K. Shah. 2008. "Biological Pest and Disease Control in Greenhouse Vegetable Production." *Acta Horticulturae* 767: 229–38.

Xu, Rong, Kai Zhang, P. Liu, Huawen Han, Shuai Zhao, Apurva Kakade, Aman Khan, Daolin Du, and Xiangkai Li. 2018. "Lignin Depolymerization and Utilization by Bacteria." *Bioresource Technology* 269: 557–66. https://doi.org/10.1016/J.BIORTECH.2018.08.118.

Yadav, M., and Rathore, J. S. 2018. "TAome Analysis of Type-II Toxin-Antitoxin System from Xenorhabdus nematophila." *Computational Biology & Chemistry* 76: 293–301. https://doi.org/10.1016/j.compbiolchem.2018.07.010.

Yadav, M., and Rathore, J. S. 2020. "The hipBAXn Operon from Xenorhabdus nematophila Functions as a Bonafide Toxin-Antitoxin Module." *Applied Microbiology & Biotechnology* 104, no. 7: 3081–95. https://doi.org/10.1007/s00253-020-10441-1.

Yadav, M., and Rathore, J. S. 2022a. "Functional and Transcriptional Analysis of Chromosomal Encoded hip-BAXn2 Type II Toxin-Antitoxin (TA) Module from Xenorhabdus nematophila." *Microbial Pathogenesis* 162: 105309. https://doi.org/10.1016/J.MICPATH.2021.105309.

Yadav, M., and J. S. Rathore. 2022b. "In-Silico Analysis of Genomic Distribution and Functional Association of hipBA Toxin-Antitoxin (TA) Homologs in Entomopathogen Xenorhabdus nematophila." *Journal of Asia-Pacific Entomology* 25, no. 3: 101949. https://doi.org/10.1016/J.ASPEN.2022.101949.

Yang, X., J. L. Lai, Y. Zhang, and X. G. Luo. 2022. "Reshaping the microenvironment and bacterial community of TNT- and RDX-contaminated soil by combined remediation with vetiver grass (Vetiveria ziznioides) and effective microorganism (EM) flora". *The Science of the total environment*, 815: 152856.

Yang, Z., Z. Jiang, C. Y. Hse, and R. Liu. 2017. "Assessing the Impact of Wood Decay Fungi on the Modulus of Elasticity of Slash Pine (*Pinus elliottii*) by Stress Wave Non-destructive Testing." *International Biodeterioration & Biodegradation* 117: 123–7. https://doi.org/10.1016/j.ibiod.2016.12.003.

Zhao, W. Y., G. W. Yang, X. X. Chen, J. Y. Yu, and B. K. Ma. 2013. "Study on Integrated Restoration Technique for Eutrophic Artificial Lakes." *Advanced Materials Research* 807–809: 1304–10. https://doi.org/10.4028/www.scientific.net/AMR.807–809.1304.

Zhou, Qunlan, Kangmin Li, Xie Jun, and Liu Bo. 2009. "Role and Functions of Beneficial Microorganisms in Sustainable Aquaculture." *Bioresource Technology* 100, no. 16: 3780–6. https://doi.org/10.1016/j.biortech.2008.12.037.

9 Microbe-Assisted Remediation of Pesticide Residues from Soil and Water

Odangowei Inetiminebi Ogidi and Nkechinyere Richard-Nwachukwu

9.1 INTRODUCTION

Pesticides have been widely used in the management of insect-borne diseases like malaria and typhus as well as crop protection in agriculture and gardening throughout the world (Pascal-Lorber and Laurent, 2011). Due to this broad usage, persistent organic pesticides have contaminated several agricultural soils, natural water reservoirs, and rural regions (Chaudhry *et al.*, 2005). For a long time, crop protection against pests was the main objective of farming in order to maximize crop production. Meanwhile, it's likely that the toxicity of the substances used was underestimated and wasn't always the top priority when applying pesticides, with implications for both farmers and crop consumers as well as the environment.

Many of the pollutants were used on a global scale for many years before it became clear that they posed an unacceptable risk to human health (Berdowski *et al.*, 1997). These substances were frequently also enduring in uncontrolled natural settings. The analytes persist in soils and sediments for a very long time after their initial use, and from there they can enter food chains, surface water, and ground water (Li *et al.*, 2000). The hydrophobicity of these persistent organic pollutants (POPs) can cause accumulation in animal adipose tissues, which can result in biomagnification at higher trophic levels (Gavrilescu, 2004).

Over time, these elevated levels of toxic substances in the body could lead to health issues (Moerner *et al.*, 2002). Traditional POP remediation technologies include physicochemical techniques like burning, incineration, land filling, composting, and chemical amendments (Kempa, 1997; Wehtje *et al.*, 2000). Due to the fact that these techniques are mostly ex situ, excavation and shipping are expensive. These techniques are also intrusive and harmful to the ecosystem as a whole because a sizable portion of the soil is removed. As a result, in situ remediation technologies have gained popularity in recent years because they are less intrusive, inexpensive, low-maintenance, and frequently powered by solar energy (Chaudhry *et al.*, 2005).

Bioremediation, in which plant-associated microorganisms are used, and phytoremediation, in which they are not, are two "natural" remediation strategies that have proven effective in numerous situations. Both strategies rely on the inherent capacity of soil microbes and plants to take up pollutants as readily as they do nutrients and then metabolize, store, or co-metabolize them (McGuinness and Dowling, 2009; Ogidi and Njoku, 2017). Soil pollution is a significant barrier to the widespread use of these technologies. Each polluted soil has its own set of physical and chemical features and pollutant profile; therefore, each remediation strategy must be tailored to it (Glick, 2003).

Depending on the scenario, one must decide whether to use phytoremediation or bioremediation to clean up polluted soil and water. Phytoremediation uses plants instead of only soil microorganisms like bacteria and fungus (Weyens *et al.*, 2009). The classification, significance, environmental fate, effects, and biodegradation of pesticides in soil and water, as well as the mechanisms of microbial pesticide degradation, factors affecting microbial pesticide degradation, potential of

DOI: 10.1201/9781003423393-9

microorganisms in pesticide residue remediation, and application of microbial remediation, are discussed in this chapter.

9.2 CLASSIFICATION OF PESTICIDES

Different classification criteria, including chemical classes, functional groups, modes of action, and toxicity, may be used to categorize pesticides.

9.2.1 CLASSIFICATION BY ORIGIN

Chemical and biological pesticides are two types of pesticides that may be categorized according to their source. Chemical pesticides are typically organic substances that may be synthesized chemically or derived from natural sources (Kim *et al.*, 2017). Biopesticides are organic compounds that use non-toxic processes to control pests (Kumar and Singh, 2015).

9.2.1.1 Chemical Pesticides

Four groups—organophosphates, organochlorines, carbamates, and pyrethrins and pyrethroids—can be distinguished using this classification (Figure 9.1). A thorough understanding of the chemical and physical characteristics of pesticides is crucial for making informed decisions about how to apply them, what precautions should be taken during application, and how much to use (Yadav and Devi, 2017).

Organochlorines: Organic substances known as organochlorine pesticides (OCs) are hydrocarbon chains bound with at least one covalently bound chlorine atom. These substances are often used in agriculture, particularly as insecticides to manage a variety of insects. Dichlorodiphenyldichloroethane (DDD), dichlorodiphenyltrichloroethane (DDT), dieldrin, dicofol, aldrin, lindane, endosulfan, chlordane, isobenzan, isodrin, chloropropylate, and toxaphene are the most prevalent organochlorines.

These substances tend to bioaccumulate in tissues and persist in the environment because they are lipophilic and difficult to break down. OCs fall under the category of persistent organic pollutants due to their significant persistence in the environment. They may harm living things by having mutagenic, histopathological, enzyme-inducing, or enzyme-inhibiting effects, as well as carcinogenic and teratogenic effects. Exposure to organochlorines may raise the risk of breast cancer for people.

Organophosphates: Synthetic insecticides known as organophosphates (OPs) include phosphoric acid esters and thiophosphoric acid esters. The first one to be made and employed as an

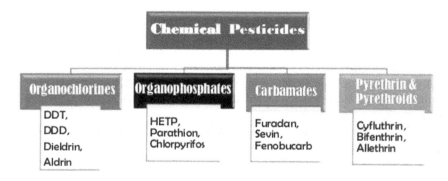

FIGURE 9.1 Classification of the chemical pesticides. HETP: Hexaethyl tetraphosphate; DDD: dichlorodiphenyldichloroethane; DDT: dichlorodiphenyltrichloroethane

agricultural pesticide was hexaethyltetraphosphate (HETP). For insects and other creatures, such as birds, amphibians, and mammals, OPs are extremely poisonous. Their toxicity is brought on via the central and peripheral nervous systems' suppression of acetylcholinesterase (AChE) (Roberts and Reigart, 2013; Rathnayake and Northrup, 2016).

This enzyme's suppression has muscarinic and nicotinic effects. Muscarinic symptoms include coughing, chest tightness, expectorating foamy discharges, pulmonary edema, and wheezing, for inhalation; hypersalivation, cramps, vomiting, nausea, tenesmus stomach, and diarrhea, for ingestion; and impaired vision, miosis, and eye discomfort for eyes. Nicotininc symptoms includes excessive perspiration, increasing flaccidity, fasciculation, and weakening of proximal muscle groups (Eddleston, 2016).

Carbamates: Compounds produced from carbamic acid are known as carbamates. An ester group and an amino group are joined in their chemical structure. R1 and R2 are often organic radicals or hydrogen. It is feasible to understand the objective by taking into account the functional group R1 if R2 is hydrogen (Dhouib *et al.*, 2016). Insecticides, herbicides, and fungicides are all conceivable when R1 is a methyl group, an aromatic group, or a benzimidazole moiety. Carbamates are also used as biocides in commercial and residential goods to control pests in homes (Struger *et al.*, 2016). Aldicarb, carbofuran, fenoxycarb, carbaryl, ethienocarb, and fenobucarb are examples of common carbamate insecticides (Arif *et al.*, 2012).

Because carbamates are inhibitors of acetylcholinesterase function like organophosphates, their toxicity affects the nervous system (Wang *et al.*, 2015). Since human natural killer (NK) cells and cytotoxic T lymphocytes, which offer defense against malignancies, are inhibited and caused to undergo apoptosis by carbamate insecticides, exposure to these substances raises the risk of non-lymphoma Hodgkin's in humans (Li *et al.*, 2015).

Pyrethrins and Pyrethroids: According to Ensley (2007), they may be separated into aromatic rings and the alcohol moiety, which has part of the terminal side chain moieties changed. In comparison to pyrethrins, pyrethroids are manufactured to have a stronger insecticidal effect and to be less sensitive to air and light. In general, sunlight quickly breaks down pyrethrins and many pyrethroids in the air, but they last a long time in the soil because of their strong bond with it (ATSDR, 2003). Since pyrethrins and pyrethroids are sprayed directly onto plants and crops, they may be discovered on leaves, fruits, and vegetables (Morgan *et al.*, 2018).

The chrysanthemic acid moiety of pyrethrin I is modified and the alcohols are esterified to produce pyrethroids, which are synthetic chemicals. These categories include (Ensley, 2007):

- Esters containing chrysanthemic acid and one alcohol, with a furan ring and terminal side chain moieties, are first-generation pyrethroids.
- In contrast to pyrethrins, second-generation pyrethroids have higher insecticidal potency and lower sensitivity to air and light because they contain 3-phenoxybenzyl alcohol derivatives in their moiety and have some of the terminal side chain moieties replaced with dichlorovinyl or dibromovinyl substitutes. The pyrethrins and many pyrethroids breakdown quickly in sunshine, but due to their high soil binding, they persist there for a long period (ATSDR, 2003). Since pyrethrins and pyrethroids are sprayed directly onto plants and crops, they may be discovered on leaves, fruits, and vegetables (Morgan *et al.*, 2018).

The neurologic system is harmed by pyrethrins and pyrethroids, which block the sodium channels in the axons (Wu *et al.*, 2017). Though less dangerous to humans, they are hazardous to insects. However, it was reported that these pesticides may cause neurological symptoms like headaches and dizziness, respiratory effects like coughing or upper respiratory irritation from breathing dust or aerosol droplets, gastrointestinal effects like nausea and vomiting, and irritation and/or cutaneous effects (Saillenfait *et al.*, 2015). Cardiovascular issues might result from pyrethroid usage (Han *et al.*, 2017).

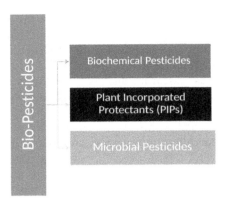

FIGURE 9.2 Classification of biopesticides.

9.2.1.2 Biopesticides

Insecticides derived from natural sources are known as "biopesticides" (animals, plants, microorganisms, and minerals). They fall into one of three broad categories according on the kind of active substance they use (Figure 9.2), namely microbial pesticides, plant-incorporated protectants (PIPs), and biochemical.

Biochemical Pesticides: They are organic substances that use non-toxic methods to manage pests. They may be created synthetically or taken from natural sources to have the same structure and functionality as natural ones (Mishra *et al.*, 2015). Chemical substances released by plants or animals are known as semiochemicals. These substances include pheromones, allomones, kairomones, and attractants. They are linked to the essential pest behaviors, including eating, mating, and oviposition (laying eggs) (Sarwar, 2015). Therefore, it is possible to alter the pest life cycle by taking advantage of their concentration.

Plant-Incorporated Protectants: When the pest feeds on plants, the plants may create plant-incorporated protectants. By adding the gene that controls a particular pesticidal protein into the genetic code of the plant, it is possible to genetically modify the plants to compel their production. The plant may create poisonous substances for certain pests in this manner (Parker and Sander, 2017).

Microbial Pesticides: Living organisms, including bacteria, fungus, algae, and viruses, are included in microbial pesticides, which kill pests. They hinder the growth of other microbes or produce toxic compounds that degrade the environment and cause illnesses in pests (Montesinos, 2003).

9.2.2 CLASSIFICATION BY TARGET

The functions they perform and the kinds of pests they target allow for the classification of pesticides. Insecticides, herbicides, rodenticides, bactericides, and fungicides are the primary categories. They may interact with pests differently and have varied toxicity levels depending on their chemical makeup. Chemical and biological substances known as insecticides target and eliminate insects. Larvicides are particular insecticides that only affect an insect's larval stage of development. In addition to being used in farming, horticulture, forestry, and gardening, these substances are also used to control insects like mosquitoes and ticks that transfer illnesses like dengue and malaria to humans and other animals (Haddi *et al.*, 2017). (Kleinschmidt *et al.*, 2018). The groups of carbamates, pyrethroids, and organophosphates include the most often used insecticides. They affect the victims' neurological systems, resulting in spasms, breathing problems, and/or death.

9.3 SIGNIFICANCE OF PESTICIDE RESIDUES IN SOIL AND WATER

Agricultural pesticides have the potential to harm wildlife because they may accumulate in the soil and water supply. Pesticides that left behind significant levels of residue in the soil for more than one growing season or until the sowing or planting of a subsequent crop were considered persistent in the past. The idea of persistence has been replaced with the idea of the amount of time needed for 50% of the active chemical to degrade (in the lab) or dissipate (in the field). Therefore, the DT50 number (the amount of time needed for 50% of the original concentration to dissipate) is comparable to the more often-used phrase "half-life" (Craven, 2000). A quantifiable indication of pesticide persistence in soil is provided by DT50 values (Beulke and Brown, 2001).

Estimates of pesticide concentrations in soil, surface water, and groundwater are included in the data package submitted for pesticide registration. The rate at which a pesticide breaks down in soil is one of the most important elements in determining its environmental impact. First, the soil's inherent properties (pH, organic matter content, microbial activity); second, the current temperature and seasonal shifts; third, the soil's moisture content and availability; fourth, pesticide persistence is influenced by a number of factors, including whether the soil is aerobic or anaerobic (Craven, 2000).

Consumers who consume crops cultivated in these soils might be at risk from translocation of soil residues into those crops. Pesticide residues in the soil may directly poison soil-dwelling animals and/or plants. Pesticides can have the following negative effects on organisms that are not their intended targets: 1) a decline in species diversity; 2) alteration of habitat due to a decline in species; 3) behavioral changes; 4) growth and reproduction changes; and 5) food quality and quantity shifts, resistance and susceptibility increases, illness susceptibility increases, and biological magnification (Ware, 2000).

Populations with resistance may develop as a consequence of selection pressures brought on by ongoing pesticide usage. Additionally, pesticides may build up in soil flora and/or fauna and be transferred to other species, which may have an impact on the populations of soil arthropods. In general, root crops tend to absorb more residues, and peel contains more residues than pulp. In some crops, residue levels are greater in the plant's lower and root regions than in its aerial section. It is possible to link chronic organohalogenated pesticide contamination in soil and sediments to point sources (waste plant effluents and industrial discharges), although diffuse or non-point sources (precipitation, agricultural runoff, particle movement) are more usually responsible (Falandysz *et al.*, 2001).

One of the most important factors that affects a pesticide's persistence in water is how well it dissolves in that medium. DDT, dieldrin, and endrin are examples of insoluble pesticides that are often associated with sediments and may thus persist in surface water for extremely long periods of time. Water chemistry, pH, temperature, and aquatic life are all important factors, as is the total amount of organic and inorganic particles floating in the water which pose a threat to human health because of their potential usage (Barbash and Resek, 1996).

9.4 ENVIRONMENTAL FATE OF PESTICIDES IN SOIL AND WATER

9.4.1 SOURCES OF PESTICIDES FOUND IN SOIL AND WATER

High pesticide concentrations can enter the soil by being applied directly to the soil surface, being incorporated into the top few inches (metric) of soil, or being applied to crops. Up to one-third of the total quantity of pesticides sprayed on crops is thought to have the potential to pollute the soil. The crop canopy affects how much is deposited into the soil, which varies. Additionally, significant quantities of pesticide may enter the soil by drift during pesticide application, air fallout, and wind-borne transport to unintended regions.

Low-volatility toxic compounds, as well as those with adequate atmospheric lifetimes (on the order of a few days or more), may be dispersed over the whole global troposphere. When pesticides are applied as granules and/or soil injections, the impact of drift is close to zero, as opposed to the 50% or more that may be lost when using spraying techniques. Variables including boom height, droplet size in the spray stream, relative humidity, temperature, and the size of the area being treated affect how much drift occurs to non-target locations. The way a pesticide reaches the atmosphere and the size of the pesticide-containing droplets have an impact on the pesticide's mobility and deposition. Negative impacts on nontarget species might result from transport or drift over many kilometers.

Numerous investigations conducted in the early 1960s showed that pesticides were present in snow and/or rainwater. Low rainfall during the time period was suggested as a possible explanation for the high levels of residues found in soil in March 1966, which could not be explained by pesticide use in the field at the time of year. The quantities of pesticides that might be present in rainfall, however, are not high enough to represent a significant cause of soil contamination. Pesticides with a range of polar chemical properties have been discovered in precipitation all over the globe (Grynkiewicz *et al.*, 2001; Majewski and Capel, 1995).

Additionally, pesticides may directly deposit on soil via air dust. Pesticides used directly on water's surface to kill weeds; algae blooms; and water-breeding insects like mosquitoes, black flies, and biting midges before they're restocked might contaminate the water. In addition, water may be tainted by drift during application and by air fallout on rain and dust. When water-soluble pesticides are applied to soil, surface runoff may transport them to neighboring water sources. Contamination might also originate from a variety of different sources, such as industrial effluents, sewage, or spills (Gerecke *et al.*, 2002). It's possible for pesticides to be leached down into the soil and away from their intended targets in the top layers (Larson *et al.*, 1997).

9.4.2 FATE OF PESTICIDES IN SOIL AND WATER

To be certain, with a high degree of certainty, that a chemical won't have a negative impact on the environment, it is crucial to predict the pesticide's environmental destiny in pre-market experiments. Pesticides migrate constantly across various environmental compartments, which makes the movement of pesticides in the environment exceedingly complicated. These exchanges may sometimes include the conveyance of pesticides across great distances in addition to nearby locations. The destiny of pesticides in the environment cannot be predicted by a single criterion. This knowledge can only be obtained by understanding how various components interact. After interaction with soil, a variety of variables affect the behavior and destiny of pesticides (Racke, 2003).

Intake by soil creatures or plants, leaching with water that is percolating downhill, adsorption to clay and organic matter, microbial degradation, movement with runoff water or eroded soil, photolysis, and chemical degradation are some of these variables. As opposed to temperate locations, the fate of pesticides in tropical areas is less well known. According to field studies of the destiny of tropical pesticides, dissipation often happens more quickly than it does for pesticides used in temperate climates. Increased volatility and rapid chemical and microbiological degradation are two mechanisms relating to the impact of tropical temperatures that contribute to this accelerated dissipation (Racke, 2003).

9.5 IMPACTS OF PESTICIDES ON ECOSYSTEM

Pesticides and other toxic compounds depend on the natural environmental reactions to operate. When a pesticide particle comes into contact with a soil particle, a water molecule, or a living organism, a complex series of chemical, biological, and physical events takes place (Gavrilescu, 2005). The transportation of pesticide from one area to another is a primary process that contributes to its alteration in the soil environment. The term "transfer process" is used to describe the movement of a pesticide through several components of an ecosystem, including biota, sediments, water, and the atmosphere. During the process of transportation, the pesticide particles are transferred from their

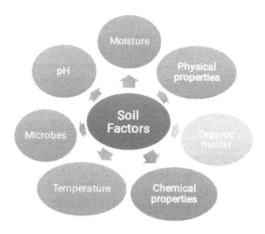

FIGURE 9.3 Soil factors that affect how contaminants are taken up and broken down by the soil.

initial location to the surrounding environment, including watercourses and groundwater bodies (Gavrilescu, 2005).

Pesticides undergo several transformations during the process, resulting in diverse fractions and forms depending on their binding capacity, release mechanism, and degradation efficiency in soil conditions. These processes depend on ambient variables, pollutant properties, and soil features. The rates of soil pollutant absorption and degradation are influenced by a multitude of variables (Figure 9.3). These factors encompass the physical and chemical attributes of the soil, such as its pH level, temperature, organic and mineral composition, and moisture level. The decomposition process is greatly affected by the existence of soil microorganisms. Conversely, the growth and pesticide degradation abilities of these soil bacteria are dependent on the specific environmental conditions of the soil (Odukkathil and Vasudevan, 2013). Tang *et al.* (2005) describe bioremediation as a complex and demanding process that considers the characteristics and constituents of the soil. For example, a decrease in soil pH enhances the absorption of ionizing herbicides, such as picloram and atrazine (Andreu and Picó, 2004).

9.5.1 Important Mechanisms Involved in Pesticides Accumulation in Soil Ecosystem

- *Pesticide transportation*: Transportation refers to the process of transferring pesticides from their original location to different parts of the environment. Most pesticides are applied through spraying, after which they might partially evaporate and be released into the atmosphere. Contaminants, like soil particles, have the ability to undergo evaporation from the surface of plants and soil particles. Following the application of pesticides and their interaction with colloidal particles, certain pollutants remain unattached to soil particles and are able to migrate towards adjacent water sources and underground water sources, resulting in contamination of these bodies of water.
- *Pesticide diffusion*: The diffusion mechanism of pesticides is significant in terms of their ability to move through soil, water, biota, and the atmosphere (Gavrilescu, 2005). Pesticides are commonly applied in both solid and liquid states; however, there is a possibility that they may undergo volatilization. Pesticides, whether in solid or liquid form, undergo a transformation into a gaseous state and are subsequently emitted into the atmosphere (Frazar, 2000; Whitford *et al.*, 1999). The dispersion of these toxins through the air can result in extensive environmental damage, extending beyond their original application site. Pesticide residues have the ability to dissolve in water, migrate from the soil to

bodies of water, and accumulate as contaminants in waterways. Moreover, residues exhibit a propensity to dissolve in precipitation and seep into dissolved compositions. The leaching process is primarily influenced by the chemical properties of sprayed pesticides and the specific geological conditions in the area. Humans and other organisms can easily consume pesticide residues present in the air, soil, and water.

- *Pesticide conversion*: Conversion is an important process that can have a substantial impact on the quantity of pesticides present in the environment. The process of converting pesticides through oxidation-reduction reactions, which can reduce their toxicity, is commonly known as pesticide breakdown. Environmental insecticides can easily undergo several conversion processes, such as degradation. Upon the application of a pesticide, the existing enzymes and microbes facilitate the degradation of the pesticide molecules' structure through the assistance of certain chemicals and enzymatic activity. In addition to the role of living organisms in the decomposition of hazardous substances, abiotic forces can also contribute to the degradation of pesticides. These contaminants can also undergo photo-degradation when exposed to sunlight. Microbial conversion is a notable remediation technique for the degradation of pesticide chemicals (Gavrilescu, 2005).

- *Pesticide transportation*: Transportation refers to the process of transferring pesticides from their original location to different parts of the environment. Most pesticides are applied through spraying, after which they might partially evaporate and enter the atmosphere. Contaminants, like soil particles, have the ability to evaporate from the surface of plants and soil particles. Following the application of pesticides and their interaction with colloidal particles, certain pollutants do not adhere to soil particles and instead become mobile, spreading to nearby watercourses and underground water sources, resulting in contamination of these bodies of water.

9.6 BIODEGRADATION OF PESTICIDES

Insecticides undergo degradation in the environment through the combined action of sunshine and microbes. Pesticides employed for crop spraying ultimately infiltrated sediments, water bodies, and soil. Various microorganisms present in soil and sediment require these chemical compounds for their growth and metabolism, leading to their degradation into simpler forms. Microbes such as bacteria, fungi, and algae have the ability to digest pesticides and convert them into non-hazardous chemicals. For instance, *Actinomycetes* and *Cyanobacteria* are very efficient degraders that reside in the soil.

Various authors have undertaken biodegradation investigations and discovered that certain species of bacteria and fungi, such as *Pseudomonas, Bacillus, Alcaligenes, Brevibacterium, Klebsiella*, and *Enterobacter*, have the ability to break down pesticide residues. Microbes are excellent bioremediators due to their high catalytic activity, rapid reproduction rate, and large surface-to-volume area, which enables them to chemically modify pesticides with ease. Certain microorganisms require a period of adjustment before they can generate enzymes that break down pesticides, whilst others do this through random genetic alterations.

9.7 MECHANISMS OF MICROBIAL DEGRADATION OF PESTICIDES

During biodegradation, microorganisms consume the pollutant chemicals as resources for their metabolic activities, transforming the pesticides into breakdown products or mineralizing them entirely. Biotransformation systems rely heavily on enzymes like hydrolases, peroxidases, and oxygenases, which modulate and catalyze the corresponding biochemical reactions. There are three distinct phases of pesticide breakdown:

Phase 1: In the first stage, pesticides are broken down by chemical processes such oxidation, reduction, or hydrolysis into less harmful and more water-soluble byproducts.

Phase 2: In the second stage, the poisonous, water-insoluble compounds of the first are converted into sugars and amino acids.

Phase 3: In the third stage, the less harmful secondary conjugates of the metabolites created in the second stage are produced. Degradation is facilitated by bacteria and fungi, which may secrete enzymes either outside or inside of their cells. Degradation time is an important factor to think about while organizing a bioremediation project. Typically, the first-order model is employed to interpret it (Khajezadeh *et al.*, 2020), and this interpretation is sensitive to the initial and final pollutant concentrations. Several variables, including as soil microbes, temperature, moisture, availability, and pesticide leaching, restrict how well this technique works (Soulas and Lagacherie, 2001).

9.7.1 BACTERIAL DEGRADATION

Over time, a number of bacterial strains that may break down pesticides in soil and water have been discovered. Each bacterium has a unique characteristic that makes it especially well suited for a degradative process. The adaptability, development, and function of a bacterial strain are influenced by the operational circumstances, including pH, temperature, water content, and the kinds of contaminants. The formation of metabolites during the degradation process might also result in extra environmental issues since they could be harder to get rid of than the original component. This must be taken into account as a disadvantage.

As an example, a principal important breakdown product of the organophosphate chlorpyrifos, which is utilized as an insecticide, is 3,5,6-trichloro-2-pyridinol (TCP). TCP produces extensive pollution in soils and aquatic ecosystems and is more water soluble than chlorpyrifos. Only a few bacteria have the ability to hydrolyze both the pesticide and its metabolite, and one of them is the bacterium *Ochabactrum* sp. JAS2 (Abraham and Silambarasan, 2016). When a bacterial consortium is utilized instead of an isolated pure culture, the degradation is often simpler (Doolotkeldieva *et al.*, 2021; Jariyal *et al.*, 2018). In the natural world, bacteria cohabit and rely on one another to survive. Each bacterium may produce metabolites that can be utilized as a substrate by other bacteria in the metabolic pathways of pesticide breakdown.

9.7.2 FUNGAL DEGRADATION

There are several fungi in agricultural soils and water that may be used to biodegrade pesticides. Molds, yeast, and filamentous fungus are all members of this group of microorganisms. The ability of fungi to synthesize a large number of enzymes engaged in degradative processes facilitates fungal breakdown. These bacteria are also able to adjust soil permeability and the capacity of soil and water to exchange ions.

Due to traits including particular bioactivity, growth morphology, and great tolerance even at high concentrations of contaminants, fungi may be superior degraders to bacteria. Since fungi may convert pesticides into a more palatable form for bacteria, using both bacteria and fungus to accelerate breakdown is a typical strategy.

9.7.3 ENZYMATIC DEGRADATION

The enzymes generated by microorganisms or plants during their metabolic activities are what cause enzymatic biodegradation. Enzymes are biological macromolecules that may speed up biochemical processes like the breakdown of pesticides. By reducing the process's own activation energy, these molecules influence the rate of the reaction. They mostly participate in the metabolic processes of oxidation, hydrolysis, reduction, and conjugation.

- The first stage of pesticide breakdown, oxidation, involves the transfer of an electron from reductants to oxidants. Enzymes such as laccase and oxygenase could be involved in this process. While laccases split the ring found in aromatic compounds, convert oxygen to water, and create free radicals, oxygenases accelerate the oxidation process by integrating one or two molecules of oxygen. Microorganisms use the heat or energy produced during the reaction for their metabolic processes.

- By introducing H or HO- groups from H_2O molecules, hydrolysis enables the breakage of substrate bonds. Thus, the original pesticide molecules are split into smaller chain compounds. Typically, cellulases, esterases, and lipases are involved in the hydrolysis routes. Reduction enables the transformation via reductive enzymes (Luo *et al.*, 2018).
- The conjugation process, which is characteristic of fungal biodegradation, is carried out utilizing currently available enzymes. In order to speed up the mineralization of pesticides, exogenous or endogenous natural chemicals are added. This procedure involves xyloxylation, alkylation, acylation, and nitrosylation, among other processes.

9.7.4 MINERALIZATION

Pesticides may degrade into inorganic substances such as carbon dioxide, salts, minerals, and water via the mineralization process. The pesticide chemicals serve as a source of nourishment for the microorganisms. The deterioration in this instance is also impacted by a number of variables, including the types of contaminants, the microbial species present, and the soil's properties. The concentration of the microbial community determines the pace of mineralization; hence, a decline in the microbial population does not accelerate the deterioration. For instance, when the soil microbial population is diminished, numerous metabolites might emerge that are more poisonous, persistent, and mobile than the original chlorothalonil (CTN), an organochlorine fungicide that is destroyed in CO_2. This is brought on by the lack of groups that are actively destroying the environment or by a decline in soil biodiversity that results in low microbial activity. Soil conditions have a role in the glyphosate mineralization process. Nguyen *et al.* analyzed a range of agricultural soils that differed in texture, organic matter concentration, pH, and exchangeable ions (2018). Using univariate and multivariate regression analyses, they determined what factors influence glyphosate mineralization.

9.7.5 CO-METABOLISM

This is the biological transformation of an organic molecule which is not required to assist microbial development through a series of processes. Due to this synergistic impact, the pesticides are eventually destroyed after being converted by microbes and enzymes into beneficial molecules for various physical, chemical, and biological transformations. Transferases (glutathione S-transferases and glucosyl transferases), hydrolytic enzymes (amidases, nitrilases, and esterases), oxidases (peroxidases and cytochrome P-450s), and reductases are just a few of the enzymes that may be engaged in the co-metabolic process. In an investigation by Ma *et al.* (2014), the co-metabolic transformation of the pesticide imidacloprid (IMI) was examined. Both organic acids and carbohydrates were tested as potential energy sources. The examined bacterium is *P. indica* CGMCC 6648, which can break down IMI through the hydroxylation route and produces the metabolites 5-hydroxy IMI and one olefin.

9.7.6 ALGAL–BACTERIAL–FUNGAL (MULTITAXON) CONSORTIA

Investigating pesticide breakdown using microorganisms in freshwater and soil habitats is vital in the field of ecological restoration research. Bacterial bioremediation, mycoremediation, and phycoremediation are all parts of microbioremediation, a powerful technique for pollution reduction and sustainable development. These include, respectively, cleansing polluted environments using bacteria, fungus, and microalgae. This process depends on bacteria's capacity to transform contaminants into safe or less harmful chemicals.

Microbe-Assisted Remediation of Pesticide Residues 189

This section describes possible microalgal-bacterial-fungal consortia for the recovery of pesticide-polluted freshwater and soil habitats and sheds light on microorganisms with the ability to degrade pesticides, either individually or in consortia. Algae, bacteria, and fungi are related to one another, offering a rare opportunity to use this mutualism as a polymicrobial association for pesticide bioremediation. The provision of alternative carbon sources by the microalgae enhances the remediation capacity of bacteria and fungi by increasing their quantity (number) and biomass. This benefit is obtained by bacteria and fungi via an algal mutualism.

In the past, bacteria, fungi, and microalgae were often utilized separately for cleanup; however, consortia have since been shown to be quite effective in a number of bioaugmentation research projects. Microalgae may be adept at collecting and eliminating one kind of contamination in consortia, whilst bacteria and fungi may concentrate on different types of contaminants in soil and aquatic habitats, increasing the overall effectiveness of bioremediation. A consortial strategy that encompasses a wider biological range improves the consortia's capacity for survival and overall competitiveness. As a result, the combination of several microbes' catabolic pathways may promote their harmonic and synergistic development in challenging settings and can prove more successful in cleaning up contaminated locations (Pino and Peñuela, 2011).

The utilization of algae-bacteria or algae-fungi consortia for pesticide bioremediation has since been the subject of several investigations. The technique for pesticide remediation offers a lot of promise for combining microalgae, bacteria, and fungi. In contrast to conventional engineering techniques, which have a number of drawbacks, including the creation of secondary contaminants, being uneconomical, and being practically inapplicable in some situations, biodegradation using multitaxon (algal-bacterial-fungal) consortia is a perfect self-sustaining approach.

9.8 FACTORS AFFECTING MICROBIAL DEGRADATION OF PESTICIDES

The rate of pesticide breakdown in the soil by microorganisms is influenced by a number of variables, including the pH of the soil, organic matter, moisture content, nitrogen and carbon concentration, and temperature. The chemical, biological, and physical features of soil, as well as microbial diversity and activity, are also impacted by practices like tillage and manuring (Somasundaram *et al.*, 1989).

9.8.1 SOIL pH AND SALINITY

The sensitivity of the soil to acid or alkaline hydrolysis has a significant impact on the influence of soil pH on pesticide degradation. Pesticide mobility, adsorption on soil surfaces, chemical speciation, and bioavailability are all impacted by soil pH. Singh *et al.* (2003) investigated how soil pH affected the breakdown of the organophosphate pesticide Chlorpyrifos in a research (Singh *et al.*, 2003). The findings showed that when the soil pH was below 6.7, the deterioration process accelerated. The rate of pesticide breakdown is significantly influenced by soil salinity. Numerous studies have shown that a high salt content slows the mechanism by which insecticides degrade.

Siddique *et al.* (2002) found that soil slurry cultures work best at pH 9, whereas liquid cultures of 1,2,3,4,5,6-hexachlorocyclohexane (HCH) degrade best at an initial pH of 8. (Siddique *et al.*, 2002). Kah *et al* . (2007) also found that pH influences the rate of degradation of four basic pesticides, Metribuzin, Terbutryn, Pirimicarb, and Fenpropimorph, as well as six acidic pesticides, including 2,4-D, Dicamba, Fluroxypyr, Fluazifop-P, and Flupyrsulfuron-methyl (Kah *et al.*, 2007). Similar research by Fang *et al.* (2010) demonstrated the need for neutral pH for the efficient breakdown of DDT and the inhibition of this process by acidic or alkaline pH by *Sphingobacterium* sp. (Fang *et al.*, 2010).

9.8.2 SOIL MOISTURE

The soil's moisture content has a big impact on deterioration since it's necessary for microbial growth and activity. Pesticide breakdown speeds up with soil moisture and slows down in dry soils.

9.8.3 Pesticide Structure

The physical and chemical characteristics of a pesticide are determined by its structure, which also influences how quickly it degrades. Polar groups such as OH, NH_2, and COOH are added to provide the microorganisms an attacking site, while substituents on the benzene ring are added to speed up the rate of breakdown.

9.8.4 Pesticide Concentration and Solubility

An essential factor in the process of deterioration is the pesticide content in the soil. The amount of attack sites in the soil will be impacted by the pesticide's high initial concentration, which will also have damaging effects on bacteria. In a research, Fang *et al.* (2010) discovered that a greater concentration of DDT reduced the bacteria *Sphingobacterium* sp.'s ability to degrade substances (Fang *et al.*, 2010). Because bacteria primarily target the dissolved portion of pesticides in soil, those with a high water solubility tend to break down more quickly than those with a low solubility (Odukkathil and Vasudevan, 2013).

9.8.5 Temperature

By altering the pesticide's solubility and hydrolysis, temperature has an impact on its adsorption in the soil. They looked at whether breakdown of these isomers was most successful at a temperature of 30°C for both liquid culture and soil slurry. Fang *et al.* (2010) tested the rate of DDT breakdown by *Sphingobacterium* sp. at various temperatures, including 20, 30, and 40°C. They noticed that the bacteria functioned best at a temperature of 30°C (Fang *et al.*, 2010).

9.8.6 Soil Organic Matter

By speeding the rate of co-metabolic breakdown, soil organic matter either increases or lowers microbial activity by promoting the adsorption process (Perucci *et al.*, 2000). Additionally, the presence of organic matter affects the local microbial flora, boosting species diversity and, eventually, the number of enzyme systems available to destroy pesticide compounds (Neumann *et al.*, 2014).

9.9 POTENTIAL OF MICROORGANISMS IN PESTICIDE RESIDUE REMEDIATION

Natural attenuation is the term used to describe the soil's capacity to naturally lower the concentration of any pollutants. The soil's microorganisms are crucial in facilitating chemical processes that reduce the molecular structures of pollutants to less hazardous compounds. Microorganisms are a promising source for bioremediation of pesticide residues owing to their quick rate of reproduction, high surface area volume ratio, and high catalytic power. Some microbe species have even been shown to spontaneously or by random mutation produce pesticide-degrading enzymes, while in other species, the presence of a specific pesticide induces the production of such enzymes (Tewari, 2012).

9.10 APPLICATION OF MICROBIAL REMEDIATION

The bioremediation processes may be used in situ, ex situ, or directly on the site. In the in situ method, treatment is done in the polluted area, and the procedure is usually aerobic. In order to do this, oxygen must be added to the soil or water. The primary in situ approaches are:

- Natural attenuation, which makes use of the soil's native microorganisms.
- Biostimulation, which involves maximizing the types and quantities of nutrients used to stimulate and encourage the development of native microorganisms.

Microbe-Assisted Remediation of Pesticide Residues

- Bioaugmentation, which involves adding microbial strains or enzymes to contaminated soils.
- Bioventing, which involves supplying oxygen via unsaturated soil zones to encourage the development of native microbes that can break down the toxins.
- Biosparging, which works by forcing air under pressure into a zone of saturated soil to raise the oxygen level and encourage microorganisms to break down the contaminant.

These techniques are inexpensive and very efficient. The fact that the contaminated soil isn't moved is their main benefit. Ex situ methods, on the other hand, involve transporting contaminated soil from polluted sites to the location where cleanup will take place. The primary methods include:

- Bioreactors, which treat polluted soil with wastewater to create slurry and encourage microbial processes that can remove pollutants.
- Composting, which involves combining contaminated soil with amendments to encourage the pesticides' aerobic degradation. This method includes biopiles and land farming.

When using on-site techniques, the soil is taken from and processed nearby the polluted site. For instance, the treatment for land farming can also be carried out on site, which lowers the operation cost in comparison to the ex situ approach.

9.10.1 Biostimulation

Nitrogen, phosphorous, carbon, and oxygen are added as part of the biostimulation process to encourage the development of the local microorganisms. The addition of nitrogen and phosphorus is typical since these elements promote biodegradation and broaden the variety of microbial species. Throughout the process, the amount of nutrients provided must be maintained under control since too few or too many stimulants may cause microbial activity and variety to decline.

Tebuconazole breakdown in soil was investigated by Bácmaga *et al.* (2019) using the biostimulation method. Tebuconazole interferes with enzymatic activity and microbial growth; as a result, its content in the soil has to be decreased. The microbial population declines in areas with high pesticide concentrations.

9.10.2 Bioaugmentation

Microorganisms with particular metabolic capacities in this manner support the biodegradation processes. Since excessive concentrations of pesticides hinder the essential processes of soil microorganisms, the concentration of pesticides in the soils is a parameter that conditions the process. In dumping grounds, Doolotkeldieva *et al.* (2018) investigated the bacterial decomposition of pesticide-contaminated soils. Doolotkeldieva *et al.* (2018) discovered many bacterial strains in the investigated soils in preliminary research. The degradation of aldrin, a dispersed chlorinated hydrocarbon insecticide, was next examined. The findings showed that *Pseudomonas fluorescens* and *Bacillus polymyxa*, two bacteria strains with particular genes (cytochrome P450), could break down aldrin in a rather quick manner. The next set of experimental tests were developed using the choice of a particular bacteria and the optimization of soil parameters including temperature, pH, and the nutrients present in the soil. Mesocosms were specifically built up with soil tainted with various chemicals and injected with the microbial community (Bhardwaj *et al.*, 2020).

Pesticide concentrations in contaminated soil may range at various depths because the pesticides leak into the soil's subsurface and adsorb on the soil particles, which reduces their bioavailability. Odukkathil and Vasudevan (2016) conducted an experimental test in a 4500-cm^3 glass column to assess the bioaugmentation procedure. The findings indicated that pesticide concentrations were high in the bottom soil as a consequence of pesticides drifting downhill during water seepage, while concentrations were low in the center soil, perhaps as a result of more microbial activity that encouraged breakdown.

9.10.3 APPLICATION OF NATURAL ATTENUATION, BIOSTIMULATION, AND BIOAUGMENTATION

The biodegradation of pesticides using natural attenuation, biostimulation, and bioaugmentation techniques has been evaluated and compared in a number of research studies. For instance, Bhardwaj *et al.* (2020) used three distinct methods to examine the biodegradation of atrazine. Each mesocosm contained 100 kg of soil that was poisoned with 300 mg of atrazine per kilogram of soil. They discovered that whereas natural attenuation suggested the soil microbiome had a built-in capacity for atrazine biodegradation, the natural process was sluggish.

On the other hand, after 35 days of biostimulation and bioaugmentation therapy, the atrazine was entirely gone. Additionally, the pollutant was degraded within 21 days, making the bioaugmentation technique faster than biostimulation. The authors advise using this approach since it is quick and inexpensive. Combining bioaugmentation and biostimulation therapies may increase the effectiveness of bioremediation of polluted soils (Villaverde *et al.*, 2018). Tests were conducted on 1-kilogram mesocosms that were lindane-polluted at a level equivalent to 2 mgkg^{-1} of soil by Raimondo *et al.* (2020). They showed that utilizing both bioaugmentation and biostimulation at the same time may boost lindane elimination and decrease pesticide half-life.

9.10.4 BIOVENTING

A method of in situ bioremediation called bioventing encourages the breakdown of organic contaminants adsorbed to the soil. By introducing nutrients and oxygen into the unsaturated zone of polluted soil via specially designed wells, the microbial activity is increased. The only oxygen required to support microbial activity and prevent the volatilization of pollutants must be provided despite the ventilation being minimal. Regarding aeration, bioventing may be carried out in an active or passive manner. In the former, air is blown into the soil using a blower, whilst in the latter, the impact of atmospheric pressure alone drives gas exchange via the vent wells. Depending on the kind and concentration of the contamination, the pace of biodegradation, and the features of the soil, such as permeability and water content, bioventing remediation may last anywhere between six months and five years.

9.10.5 BIOSPARGING

In the biosparging method, the biodegradation process is triggered by air injection into groundwater to raise the oxygen content, which in turn activates the local microorganisms. The process is similar to bioventing, except with biosparging, air is delivered directly into the saturated zone. Volatile organic chemicals may rise higher as a result to the unsaturated zone, facilitating biodegradation.

9.10.6 COMPOSTING

One method for the bioremediation of pesticides is composting. It entails combining non-hazardous organic amendments with polluted soil to encourage the growth of bacterial and/or fungal populations that may break down pesticides through a co-metabolic pathway. This strategy is especially recommended when the pesticide content is low. The pollutant's microbial bioaccessibility is essential for composting. Controlling the water content, soil composition, and characteristics of the additional amendment are crucial for this reason.

Biochar may be added to polluted soils to speed up the decomposition processes. Black carbon known as biochar is created when biomass is thermally converted under oxygen-restricted circumstances (gasification) or without oxygen (pyrolysis). High porosity and a broad surface area are two

Microbe-Assisted Remediation of Pesticide Residues

characteristics that make it conducive to pesticide adsorption. In addition, biochar is a source of carbon that encourages microbial activity, which aids in biodegradation. It has been reported that applying biochar enhances soil aeration and increases the soil's ability to store water (Varjani *et al.*, 2019).

Sun *et al.* (2020) isolated tebuconazole degrading bacterial strain *Alcaligenes faecalis* WZ-2 and used biochar for immobilization of the bacterial strain. Results showed that, in addition to hastening the breakdown of tebuconazole, the biochar-immobilized WZ-2 also returned the native soil microbial enzyme activity and microbiome community composition. Landfarming and biopiles are two methods that may be used to compost.

9.10.7 LANDFARMING

An aerobic bioremediation method called landfarming has been around for a while. In order to aerate the combination, contaminated soils are regularly tilled into the soil surface across large regions in a landfarming zone. The degrading process has sluggish kinetics and might take years. Toxic substances (both the parent chemical and its metabolites) must not leach or volatilize throughout the procedure. Before the treatment begins, the soil must have a waterproof cover in place to prevent any chance of infiltration. Particularly for soils polluted with a variety of contaminants, this treatment is used.

9.10.8 BIOPILES

Biopiles are elevated heaps of polluted dirt that are piped to the surface for ventilation. For optimal microbial activity, the top layer of soil is treated with a nutrient-rich solution and a light misting of air or oxygen. Water content, pH, temperature, and the concentration of nutrients and oxygen must all be controlled in order to maximize biodegradation. These factors all have an impact on the process.

9.10.9 SLURRY BIOREACTORS

Wastewater and soils that have been polluted are combined to create a slurry to form suspensions that are aqueous and range from 10% to 30% wv-1 in a bioreactor. Both aerobic and anaerobic operation conditions are possible for the bioreactor. Organochlorine insecticides' anaerobic biodegradation was investigated by Baczynski *et al.* in 2010 and 2012.

9.11 CONCLUSION

The use of pesticides is essential for keeping insects and other pests under control in an agricultural setting. Pesticide use has a strong negative impact on human health, since it may pollute the soil and water habitats to dangerous levels. Bioremediation technologies are effective in cleaning up these soils and water, but in many parts of the globe, physicochemical methods are also used often in extremely wide areas. We have covered a variety of methods for degrading these dangerous synthetic compounds in the soil and water bodies in this chapter. Pesticides may be bioremediated in soil and water by using either specialized or native microorganisms (bacteria and fungus), or by enzymatic breakdown. Microorganisms may be widely researched as a cost- and eco-friendly technique to solve the problem.

REFERENCES

Abraham, Jayanthi, and Sivagnanam Silambarasan. 2016. "Biodegradation of Chlorpyrifos and Its Hydrolysis Product 3,5,6-Trichloro-2-Pyridinol Using a Novel Bacterium *Ochrobactrum* sp." *Pesticide Biochemistry & Physiology* 126: 13–21. https://doi.org/10.1016/j.pestbp.2015.07.001.

Andreu, V., and Y. Picó. 2004. "Determination of Pesticides and Their Degradation Products in Soil: Critical Review and Comparison of Methods." *TrAC Trends in Analytical Chemistry* 23, no. 10–11: 772–89. https://doi.org/10.1016/j.trac.2004.07.008.

Arif, I. A., M. A. Bakir, and H. A. Khan. 2012. "Microbial Remediation of Pesticides." In *Pesticides: Evaluation of Environmental Pollution*. 1st ed., edited by H. S. Rathore, and L. M. L. Nollet: 131–44. Boca Raton, FL: Taylor & Francis Group.

Bácmaga, M., J. Wyszkowska, and J. Kucharski. 2019. "Biostimulation as a Process Aiding Tebuconazole Degradation in Soil." *Journal of Soils & Sediments* 19, no. 11: 3728–41. https://doi.org/10.1007/s11368-019-02325-3.

Baczynski, Tomasz P., Daniel Pleissner, and Tim Grotenhuis. 2010. "Anaerobic Biodegradation of Organochlorine Pesticides in Contaminated Soil—Significance of Temperature and Availability." *Chemosphere* 78, no. 1: 22–8. https://doi.org/10.1016/j.chemosphere.2009.09.058.

Baczynski, T. P., D. Pleissner, and M. Krylow. 2012. "Bioremediation of Chlorinated Pesticides in Field-Contaminated Soils and Suitability of Tenax Solid-Phase Extraction as a Predictor of Its Effectiveness." *CLEAN—Soil, Air, Water* 40, no. 8: 864–9. https://doi.org/10.1002/clen.201100024.

Barbash, J. E., and E. A. Resek. 1996. *Pesticides in Ground Water: Distribution, Trends, and Governing Factors*. Chelsea, MI: Ann Arbor Press, Inc.

Berdowski, J. J. M., J. Baas, J. P. J. Bloos, A. J. H. Visschedijk, and P. Y. J. Zandveld. 1997. "The European Emission Inventory of Heavy Metals and Persistent Organic Pollutants for 1990." In *UBA-FB*. Apeldoorn, The Netherlands: Institute of Environmental Science.

Beulke, S., and C. Brown. 2001. "Evaluation of Methods to Derive Pesticide Degradation Parameters for Regulatory Modelling." *Biology & Fertility of Soils* 33, no. 6: 558–64. https://doi.org/10.1007/s003740100364.

Bhardwaj, Pooja, Kunvar Ravendra Singh, Niti B. Jadeja, Prashant S. Phale, and Atya Kapley. 2020. "Atrazine Bioremediation and Its Influence on Soil Microbial Diversity by Metagenomics Analysis." *Indian Journal of Microbiology* 60, no. 3: 388–91. https://doi.org/10.1007/s12088-020-00877-4.

Chaudhry, Qasim, Margaretha Blom-Zandstra, Satish Gupta, and Erik J. Joner. 2005. "Utilising the Synergy Between Plants and Rhizosphere Microooganisms to Enhance Breakdown of Organic Pollutants in the Environment." *Environmental Science & Pollution Research International* 12, no. 1: 34–48. https://doi.org/10.1065/espr2004.08.213.

Craven, A. 2000. "Bound Residues of Organic Compounds in the Soil: The Significance of Pesticide Persistence in Soil and Water: A European Regulatory View." *Environmental Pollution* 108, no. 1: 15–8. https://doi.org/10.1016/s0269-7491(99)00198-0.

Dhouib, Ines, Manel Jallouli, Alya Annabi, Soumaya Marzouki, Najoua Gharbi, Saloua Elfazaa, and Mohamed Montassar Lasram. 2016. "From Immunotoxicity to Carcinogenicity: The Effects of Carbamate Pesticides on the Immune System." *Environmental Science & Pollution Research International* 23, no. 10: 9448–58. https://doi.org/10.1007/s11356-016-6418-6.

Doolotkeldieva, T., S. Bobusheva, and M. Konurbaeva. 2021. "The Improving Conditions for the Aerobic Bacteria Performing the Degradation of Obsolete Pesticides in Polluted Soils." *Air, Soil and Water Research* 14.

Doolotkeldieva, Tinatin, Maxabat Konurbaeva, and Saykal Bobusheva. 2018. "Microbial Communities in Pesticide-Contaminated Soils in Kyrgyzstan and Bioremediation Possibilities." *Environmental Science & Pollution Research International* 25, no. 32: 31848–62. https://doi.org/10.1007/s11356-017-0048-5.

Eddleston, M. 2016. "Pesticides." *Medicine* 44, no. 3: 193–6. https://doi.org/10.1016/j.mpmed.2015.12.005.

Ensley, S. 2007. "Pyrethrins and Pyrethroids." In *Veterinary Toxicology*. 2nd ed. Vols. 494–8, edited by R. C. Gupta. Cambridge, MA: Academic Press, ISBN 9780123704672.

Falandysz, J., B. Brudnowska, M. Kawano, and T. Wakimoto. 2001. "Polychlorinated Biphenyl and Organochlorines Pesticides in Soils from the Southern Part of Poland." *Archives of Environmental Contamination & Toxicology* 40, no. 2: 173–8. https://doi.org/10.1007/s002440010160.

Fang, Hua, Bin Dong, H. Yan, Feifan Tang, and Yunlong Yu. 2010. "Characterization of a Bacterial Strain Capable of Degrading DDT Congeners and Its Use in Bioremediation of Contaminated Soil." *Journal of Hazardous Materials* 184, no. 1–3: 281–9. https://doi.org/10.1016/j.jhazmat.2010.08.034.

Fishel, F. 1997. *Pesticides and the Environment. Insects and Diseases*. Missouri, CO: University Extension, University Of Missouri-Columbia.

Frazar, C. 2000. *The Bioremediation and Phytoremediation of Pesticide-Contaminated Sites*. Washington, DC: United States Environmental Protection Agency.

Gavrilescu, M. 2004. "Removal of Heavy Metals from the Environment by Biosorption." *Engineering in Life Sciences* 4, no. 3: 219–32. https://doi.org/10.1002/elsc.200420026.

Microbe-Assisted Remediation of Pesticide Residues

Gavrilescu, M. 2005. Fate of pesticides in the environment and its bioremediation. *Engineering in Life Sciences* 5, no. 6: 497–526. https://doi.org/10.1002/elsc.200520098.

Gerecke, Andreas C., Michael Schärer, Heinz P. Singer, Stephan R. Müller, René P. Schwarzenbach, Martin Sägesser, Ueli Ochsenbein, and Gabriel Popow. 2002. "Sources of Pesticides in Surface Waters in Switzerland: Pesticide Load Through Waste Water Treatment Plants-Current Situation and Reduction Potential." *Chemosphere* 48, no. 3: 307–15. https://doi.org/10.1016/s0045-6535(02)00080-2.

Glick, Bernard R. 2003. "Phytoremediation: Synergistic Use of Plants and Bacteria to Clean up the Environment." *Biotechnology Advances* 21, no. 5: 383–93. https://doi.org/10.1016/s0734-9750(03)00055-7.

Grynkiewicz, M., Z. Polkowska, T. Górecki, and J. Namieśnik. 2001. "Pesticides in Precipitation in the Gdansk Region (Poland)." *Chemosphere* 43, no. 3: 303–12. https://doi.org/10.1016/s0045-6535(00)00130-2.

Haddi, Khalid, Hudson V. V. Tomé, Yuzhe Du, Wilson R. Valbon, Yoshiko Nomura, Gustavo F. Martins, K. Dong, and Eugênio E. Oliveira. 2017. "Detection of a New Pyrethroid Resistance Mutation (V410L) in the Sodium Channel of *Aedes aegypti*: A Potential Challenge for Mosquito Control." *Scientific Reports* 7: 46549. https://doi.org/10.1038/srep46549.

Han, Jiajun, Liqin Zhou, Mai Luo, Yiran Liang, Wenting Zhao, Peng Wang, Zhiqiang Zhou, and Donghui Liu. 2017. "Nonoccupational Exposure to Pyrethroids and Risk of Coronary Heart Disease in the Chinese Population." *Environmental Science & Technology* 51, no. 1: 664–70. https://doi.org/10.1021/acs.est.6b05639.

Jariyal, Monu, Vikas Jindal, Kousik Mandal, Virash Kamal Gupta, and Balwinder Singh. 2018. "Bioremediation of Organophosphorus Pesticide Phorate in Soil Bymicrobial Consortia." *Ecotoxicology & Environmental Safety* 159: 310–6. https://doi.org/10.1016/j.ecoenv.2018.04.063.

Kah, Melanie, Sabine Beulke, and Colin D. Brown. 2007. "Factors Influencing Degradation of pesticides in Soil." *Journal of Agricultural & Food Chemistry* 55, no. 11: 4487–92. https://doi.org/10.1021/jf0635356.

Kempa, E. S. 1997. "Hazardous Wastes and Economic Risk Reduction: Case Study, Poland." *International Journal of Environmental Pollution* 7: 221–48.

Khajezadeh, Masoud, Kazem Abbaszadeh-Goudarzi, Hossein Pourghadamyari, and Farshid Kafilzadeh. 2020. "A Newly Isolated *Streptomyces rimosus* Strain Capable of Degrading Deltamethrin as a Pesticide in Agricultural Soil." *Journal of Basic Microbiology* 60, no. 5: 435–43. https://doi.org/10.1002/jobm.201900263.

Kim, Ki-Hyun, Ehsanul Kabir, and Shamin Ara Jahan. 2017. "Exposure to Pesticides and the Associated Human Health Effects." *Science of the Total Environment* 575: 525–35. https://doi.org/10.1016/j.scitotenv.2016.09.009.

Kleinschmidt, I., J. Bradley, T. B. Knox, A. P. Mnzava, H. T. Kafy, C. Mbogo, B. A. Ismail, J. D. Bigoga, A. Adechoubou, K. Raghavendra. 2018. Implications of insecticide resistance for malaria vector control with long-lasting insecticidal nets: A WHO-coordinated, prospective, international, observational cohort study. *Lancet Infectious Diseases* 18, no. 6: 640–9. https://doi.org/10.1016/S1473-3099(18)30172-5

Kumar, S., and A. Singh. 2015. "Biopesticides: Present Status and the Future Prospects." *Journal of Fertilizers & Pesticides* 6, no. 2: 2–4. https://doi.org/10.4172/2471-2728.1000e129.

Larson, S. J., P. D. Capel, and M. S. Majewski. 1997. *Pesticides in Surface Waters*. Chelsea, MI: Ann Arbor Press, Inc.

Li, Qing, Maiko Kobayashi, and Tomoyuki Kawada. 2015. "Carbamate Pesticide-Induced Apoptosis in Human T Lymphocytes." *International Journal of Environmental Research & Public Health* 12, no. 4: 3633–45. https://doi.org/10.3390/ijerph120403633.

Li, Y. F., M. T. Scholtz, and B. J. van Heyst. 2000. "Globar Gridded Emission Inventory of Alfa-hexachlorocyclohexane." *Journal of Geophysical Research: Atmospheres* 105, no. D5: 6621–32. https://doi.org/10.1029/1999JD901081.

Luo, Xiangwen, Deyong Zhang, Xuguo Zhou, Jiao Du, Songbai Zhang, and Yong Liu. 2018. "Cloning and Characterization of a Pyrethroid Pesticide Decomposing esterase Gene, Est3385, from *Rhodopseudomonas palustris* PSB-S." *Scientific Reports* 8, no. 1: 7384. https://doi.org/10.1038/s41598-018-25734-9.

Ma, Yuan, Shan Zhai, Shi Yun Mao, Shi Lei Sun, Ying Wang, Zhong Hua Liu, Yi Jun Dai, and Sheng Yuan. 2014. "Co-metabolic Transformation of the Neonicotinoid Insecticide Imidacloprid by the New Soil Isolate *Pseudoxanthomonas indica* CGMCC 6648." *Journal of Environmental Science & Health. Part. B, Pesticides, Food Contaminants, & Agricultural Wastes* 49, no. 9: 661–70. https://doi.org/10.1080/03601234.2014.922766.

Majewski, M. S., and P. D. Capel. 1995. *Pesticides in the Atmosphere. Distribution, Trends and Governing Factors*. Chelsea, MI: Ann Arbor Press, Inc.

McGuinness, Martina, and David Dowling. 2009. "Plant-Associated Bacterial Degradation of Toxic Organic Compounds in Soil." *International Journal of Environmental Research & Public Health* 6, no. 8: 2226–47. https://doi.org/10.3390/ijerph6082226.

Mishra, J., S. Tewari, S. Singh, and N. K. Arora. 2015. "Biopesticides: Where We Stand?" In *Plant Microbes Symbiosis: Applied Facets*, edited by N. K. Arora: 37–75. New Delhi, India: Springer.

Moerner, J., R. Bos, and M. Fredrix. 2002. "Reducing and Eliminating the Use of Persistent Organic Pollutants." In *Guidance on Alternative Strategies for Sustainable Pest & Vector Management*. Geneva: United Nations Environment Programme.

Montesinos, Emilio. 2003. "Development, Registration and Commercialization of Microbial Pesticides for Plant Protection." *International Microbiology* 6, no. 4: 245–52. https://doi.org/10.1007/s10123-003-0144-x.

Morgan, Marsha K., Denise K. MacMillan, Dan Zehr, and Jon R. Sobus. 2018. "Pyrethroid Insecticides and Their Environmental Degradates in Repeated duplicate-Diet Solid Food Samples of 50 Adults." *Journal of Exposure Science & Environmental Epidemiology* 28, no. 1: 40–5. https://doi.org/10.1038/jes.2016.69.

Neumann, Dominik, Anke Heuer, Michael Hemkemeyer, Rainer Martens, and Christoph C. Tebbe. 2014. "Importance of Soil Organic Matter for the Diversity of Microorganisms involved in the Degradation of Organic Pollutants." *ISME Journal* 8, no. 6: 1289–300. https://doi.org/10.1038/ismej.2013.233.

Nguyen, Nghia Khoi, Ulrike Dörfler, Gerhard Welzl, Jean Charles Munch, Reiner Schroll, and Marjetka Suhadolc. 2018. "Large Variation in Glyphosate Mineralization in 21 Different Agricultural Soils Explained by Soil Properties." *Science of the Total Environment* 627: 544–52. https://doi.org/10.1016/j.scitotenv.2018.01.204.

Odukkathil, G., and N. Vasudevan. 2013. Toxicity and Bioremediation of Pesticides in Agricultural Soil. *Reviews in Environmental Science & Bio/Technology* 12, no. 4: 421–44. https://doi.org/10.1007/s11157-013-9320-4.

Odukkathil, G., and N. Vasudevan. 2016. Residues of endosulfan in surface and subsurface agricultural soil and its bioremediation. *Journal of Environmental Management* 165: 72–80. https://doi.org/10.1016/j.jenvman.2015.09.020.

Ogidi, O. I., and O. C. Njoku. 2017. "A Review on the Possibilities of the Application of Bioremediation Methods in the Oil Spill Clean-Up of Ogoni Land." *International Journal of Biological Sciences & Technology* 9, no. 6: 48–59.

Parker, Kimberly M., and Michael Sander. 2017. "Environmental Fate of Insecticidal Plant-Incorporated Protectants from Genetically Modified Crops: Knowledge Gaps and Research Opportunities." *Environmental Science & Technology* 51, no. 21: 12049–57. https://doi.org/10.1021/acs.est.7b03456.

Pascal-Lorber, S., and F. Laurent. 2011. "Phytoremediation Techniques for Pesticide Contaminations." In *Alternative Farming Systems, Biotechnology, Drought Stress and Ecological Fertilisation*, edited by E. Lichtfouse. Cham: Springer.

Perucci, P., S. Dumontet, S. A. Bufo, A. Mazzatura, and C. Casucci. 2000. "Effects of Organic Amendment and Herbicide Treatment on Soil Microbial Biomass." *Biology & Fertility of Soils* 32, no. 1: 17–23. https://doi.org/10.1007/s003740000207.

Pino, N., and G. Peñuela. 2011. "Simultaneous Degradation of the Pesticides Methyl Parathion and Chlorpyrifos by an Isolated Bacterial Consortium from a Contaminated Site." *International Biodeterioration & Biodegradation* 65, no. 6: 827–31. https://doi.org/10.1016/j.ibiod.2011.06.001.

Racke, K. D. 2003. *Environmental Fate and Effects of Pesticides: What Do We Know About the Fate of Pesticides in Tropical Ecosystems*, edited by J. R. Coats and H. Yamamoto. Washington, DC: American Chemical Society.

Raimondo, Enzo E., Juliana M. Saez, Juan D. Aparicio, María S. Fuentes, and Claudia S. Benimeli. 2020. "Bioremediation of Lindane-Contaminated Soils by Combining of Bioaugmentation and Biostimulation: Effective Scaling-Up from Microcosms to Mesocosms." *Journal of Environmental Management* 276: 111309. https://doi.org/10.1016/j.jenvman.2020.111309.

Rathnayake, L. K., and S. H. Northrup. 2016. "Structure and Mode of Action of Organophosphate Pesticides: A Computational Study." *Computational & Theoretical Chemistry* 1088: 9–23. https://doi.org/10.1016/j.comptc.2016.04.024.

Roberts, J. R., and J. R. Reigart. 2013. "Organophosphate Insecticides." In *Recognition and Management of Pesticide Poisonings*. 6th ed. vol. 2013: 199–204. Washington, DC: United States Environmental Protection Agency.

Saillenfait, Anne-Marie, Dieynaba Ndiaye Ndiaye, and Jean-Philippe Sabaté. 2015. "Pyrethroids: Exposure and Health Effects—An Update." *International Journal of Hygiene & Environmental Health* 218, no. 3: 281–92. https://doi.org/10.1016/j.ijheh.2015.01.002.

Sarwar, M. 2015. "Information on Activities Regarding Biochemical Pesticides: An Ecological Friendly Plant Protection Against Insects." *International Journal of Engineering and Advanced Technology* 1: 27–31.

Microbe-Assisted Remediation of Pesticide Residues

Siddique, Tariq, Benedict C. Okeke, Muhammad Arshad, and William T. Frankenberger. 2002. "Temperature and pH Effects on Biodegradation of Hexachlorocyclohexane Isomers in Water and a Soil Slurry." *Journal of Agricultural & Food Chemistry* 50, no. 18: 5070–6. https://doi.org/10.1021/jf0204304.

Singh, Brajesh K., Allan Walker, J. Alun W. Morgan, and Denis J. Wright. 2003. "Effects of Soil pH on the Biodegradation of Chlorpyrifos and Isolation of a Chlorpyrifos-Degrading Bacterium." *Applied & Environmental Microbiology* 69, no. 9: 5198–206. https://doi.org/10.1128/AEM.69.9.5198-5206.2003.

Somasundaram, L., J. R. Coats, and K. D. Racke. 1989. "Degradation of Pesticides in Soil as Influenced by the Presence of Hydrolysis Metabolites." *Journal of Environmental Science & Health, Part B* 24, no. 5: 457–78. https://doi.org/10.1080/03601238909372661.

Soulas, G., and B. Lagacherie. 2001. "Modelling of Microbial Degradation of Pesticides in Soils." *Biology & Fertility of Soils* 33: 551–7.

Struger, John, Josey Grabuski, Steve Cagampan, E. Sverko, and Chris Marvin. 2016. "Occurrence and Distribution of Carbamate Pesticides and Metalaxylin Southern Ontario Surface Waters 2007–2010." *Bulletin of Environmental Contamination & Toxicology* 96, no. 4: 423–31. https://doi.org/10.1007/s00128-015-1719-x.

Sun, Tong, Jingbo Miao, Muhammad Saleem, Haonan Zhang, Yong Yang, and Qingming Zhang. 2020. "Bacterial Compatibility and Immobilization with Biochar Improved Tebuconazole Degradation, Soil Microbiome Composition and Functioning." *Journal of Hazardous Materials* 398: 122941. https://doi.org/10.1016/j.jhazmat.2020.122941.

Tang, Yinjie J., Lihong Qi, and Barbara Krieger-Brockett. 2005. "Evaluating Factors That Influence Microbial Phenanthrene Biodegradation Rates by Regression with Categorical Variables." *Chemosphere* 59, no. 5: 729–41. https://doi.org/10.1016/j.chemosphere.2004.10.037.

Tewari, L., J. K. Saini, and R. Arti. 2012. "Bioremediation of Pesticides by Microorganisms: General Aspects and Recent Advances." In *Bioremediation of Pollutants*: 25–49. New Delhi: IK International Publishing House Pvt. Ltd.

Varjani, Sunita, Gopalakrishnan Kumar, and Eldon R. Rene. 2019. "Developments in Biochar Application for Pesticide Remediation: Current Knowledge and Future Research Directions." *Journal of Environmental Management* 232: 505–13. https://doi.org/10.1016/j.jenvman.2018.11.043.

Villaverde, J., M. Rubio-Bellido, A. Lara-Moreno, F. Merchan, and E. Morillo. 2018. "Combined Use of Microbial Consortia Isolated from Different Agricultural Soils and Cyclodextrin as a Bioremediation Technique for Herbicide Contaminated Soils." *Chemosphere* 193: 118–25. https://doi.org/10.1016/j.chemosphere.2017.10.172.

Wang, Yanhua, Chen Chen, Xueping Zhao, Qiang Wang, and Yongzhong Qian. 2015. "Assessing Joint Toxicity of Four Organophosphate and Carbamate Insecticides in Common Carp (Cyprinus carpio) Using Acetylcholinesterase Activity as an Endpoint." *Pesticide Biochemistry & Physiology* 122: 81–5. https://doi.org/10.1016/j.pestbp.2014.12.017.

Ware, G. 2000. *The Pesticide Book*. 5th ed. Fresno, CA: Thomson Publications.

Wehtje, G., R. H. Walker, and J. N. Shaw. 2000. "Pesticide Retention by Inorganic Soil Amendments." *Weed Science* 48, no. 2: 248–54. https://doi.org/10.1614/0043-1745(2000)048[0248:PRBISA]2.0.CO;2.

Weyens, Nele, Daniel Van Der Lelie, Safiyh Taghavi, Lee Newman, and Jaco Vangronsveld. 2009. "Exploiting Plant-Microbe Partnerships to Improve Biomass Production and Remediation." *Trends in Biotechnology* 27, no. 10: 591–8. https://doi.org/10.1016/j.tibtech.2009.07.006.

Whitford, F., J. Nelson, H. Barrett, and M. Brichford. 1999. *Pesticides and Water Quality: Principles, Policies, and Programs*: 47907. West Lafayette, IN: Purdue University Cooperative Extension Service.

Wu, Shaoying, Yoshiko Nomura, Yuzhe Du, Boris S. Zhorov, and K. Dong. 2017. "Molecular Basis of Selective Resistance of the Bumblebee BiNav1 Sodium Channelto Tau-Fluvalinate." *Proceedings of the National Academy of Sciences of the United States of America* 114, no. 49: 12922–7. https://doi.org/10.1073/pnas.1711699114.

Yadav, I. C., and N. L. Devi. 2017. "Pesticides Classification and Its Impact on Human and Environment." *Environmental Sciences & Engineering* 6: 140–57.

10 Rejuvenation of Ponds through Phytoremediation
A Sustainable Approach for Water Quality Enhancement

*Ritambhara K. Upadhyay, Naval Kishore, Mukta Sharma,
Kenate Worku, Chandra Shekhar Dwivedi, and
Gaurav Tripathi*

10.1 INTRODUCTION

10.1.1 BACKGROUND AND SIGNIFICANCE

Ponds are essential components of freshwater ecosystems, providing habitats for a diverse range of plants, animals, and microorganisms. They play a crucial role in water storage, flood control, groundwater recharge, and nutrient cycling. However, ponds are susceptible to pollution from various sources, including industrial discharge, agricultural runoff, and urban development, leading to the deterioration of water quality and the disruption of the delicate balance within the ecosystem.

Traditional methods of pond remediation, such as chemical treatment and mechanical dredging, often come with drawbacks such as high costs, potential ecological harm, and short-term effectiveness (Pandey *et al.*, 2019). In recent years, there has been growing interest in developing sustainable and environmentally friendly approaches to restore and rejuvenate polluted ponds. Phytoremediation, an emerging green technology, has gained attention as a promising method for mitigating pond pollution and improving water quality (Licht and Isebrands, 2014, Vishnoi and Srivastava, 2007).

Phytoremediation utilizes plants and their associated microorganisms to remove, degrade, or immobilize contaminants present in soil, sediments, and water bodies. Plants possess natural abilities to accumulate, transform, and detoxify pollutants through various mechanisms, including phytoextraction, phytostabilization, rhizofiltration, and rhizodegradation (Greipsson,2015; Lajayer *et al.*, 2019). The use of plants in remediation processes offers several advantages, including cost-effectiveness, low energy requirements, aesthetic benefits, and the potential for long-term sustainable management.

The significance of phytoremediation in pond restoration lies in its ability to address the root causes of pollution, improve water quality, and promote ecological balance. By harnessing the natural processes of plants, phytoremediation can target specific pollutants, such as heavy metals, organic contaminants, nutrients, and pesticides, and facilitate their removal or transformation. Furthermore, phytoremediation can help enhance the overall health of the pond ecosystem by stimulating microbial activity, promoting biodiversity, and improving habitat conditions for aquatic organisms (Marmiroli and Maestri, 2008).

Another significant aspect of phytoremediation is its potential to engage local communities and stakeholders in environmental conservation efforts. The implementation of phytoremediation projects in ponds can raise awareness about the importance of water quality, ecological restoration, and

Rejuvenation of Ponds through Phytoremediation
199

sustainable land use practices. Additionally, phytoremediation projects can provide educational and recreational opportunities, fostering a sense of ownership and stewardship among the community members.

However, despite its promising potential, phytoremediation also faces certain challenges and limitations. Factors such as plant selection, site conditions, pollutant availability, and project scalability need to be carefully considered for successful implementation. Furthermore, long-term monitoring and management are essential to ensure the effectiveness and sustainability of phytoremediation projects.

In conclusion, the application of phytoremediation in pond restoration offers a sustainable and eco-friendly approach to address water quality issues and promote the overall health of aquatic ecosystems (Pandey *et al.*, 2019; Marmiroli and Maestri, 2008). By harnessing the natural capabilities of plants, phytoremediation can contribute to the preservation of biodiversity, the protection of water resources, and the enhancement of ecosystem services. Continued research; technological advancements; and collaborations among scientists, policymakers, and local communities are necessary to unlock the full potential of phytoremediation and foster its wider adoption in pond rejuvenation efforts.

10.1.2 OBJECTIVES OF THE CHAPTER

The objectives of this chapter on phytoremediation of ponds are as follows:

- **Evaluate the effectiveness of phytoremediation techniques:**
  The chapter aims to assess the efficiency of different phytoremediation techniques, such as phytoextraction, phytostabilization, rhizofiltration, and rhizodegradation, in removing pollutants from ponds (Lajayer *et al.*, 2019). This involves investigating the ability of various plant species to uptake, sequester, degrade, or transform contaminants present in pond water, sediments, or soils.
- **Assess the suitability of plant species for phytoremediation:**
  The chapter aims to identify and evaluate plant species that are suitable for phytoremediation in ponds. This includes examining the tolerance, adaptability, and pollutant-removal capabilities of different plant species under varying environmental conditions and pollutant types (Thakur *et al.*, 2016).
- **Investigate the mechanisms of pollutant removal by plants:**
  The chapter seeks to understand the mechanisms by which plants uptake, translocate, and detoxify pollutants in pond environments. This involves investigating the physiological, biochemical, and genetic processes involved in pollutant absorption, accumulation, and transformation within plants and their associated rhizospheric microorganisms.
- **Assess the ecological impacts of phytoremediation:**
  The chapter aims to evaluate the ecological effects of phytoremediation on pond ecosystems. This includes assessing changes in water quality parameters, such as nutrient levels, heavy metal concentrations, and organic pollutant content, as well as evaluating the impacts on biodiversity, microbial communities, and overall ecosystem functioning.
- **Identify challenges and limitations of phytoremediation in pond restoration:**
  The chapter aims to identify the potential challenges and limitations associated with the application of phytoremediation in pond restoration projects (Licht and Isebrands, 2014). This includes investigating factors that may influence the effectiveness of phytoremediation, such as plant growth limitations, pollutant availability, site-specific conditions, and long-term sustainability.
- **Explore synergistic approaches and integration with other techniques:**
  The chapter aims to explore the potential synergistic effects of integrating phytoremediation with other remediation techniques, such as bioremediation or constructed wetlands.

This involves investigating the combined use of phytoremediation with complementary technologies to enhance pollutant removal efficiency and optimize pond restoration outcomes.

- **Provide recommendations for the implementation of phytoremediation projects:** Based on the findings of the chapter, the research aims to provide practical recommendations and guidelines for the successful implementation of phytoremediation projects in pond restoration efforts. This includes considerations related to plant selection, site preparation, monitoring and maintenance, and community engagement.

By addressing these objectives, the chapter aims to contribute to the understanding of phytoremediation as a viable and sustainable approach for rejuvenating ponds and improving their water quality and to provide insights for the development of effective management strategies in pond restoration projects.

10.2 PHYTOREMEDIATION: AN OVERVIEW

Phytoremediation of ponds refers to the use of plants and associated microbial communities to remove, degrade, or immobilize pollutants present in pond water, sediments, or soils (Greipsson, 2015; Singh, 2012). It is an eco-friendly and sustainable approach that harnesses the natural abilities of plants to remediate polluted aquatic environments and improve water quality.

10.2.1 PRINCIPLES OF PHYTOREMEDIATION OF PONDS

- **Plant Uptake and Accumulation:** Phytoremediation relies on the ability of plants to absorb pollutants from the water or sediments through their root systems. Plants can accumulate pollutants in their tissues, facilitating their removal from the pond environment.
- **Transformation and Degradation:** Certain plants and their associated microbial communities possess the capability to transform or degrade pollutants. They can metabolize or break down contaminants into less harmful forms or convert them into gaseous compounds that can be released into the atmosphere.
- **Stabilization and Immobilization:** Some plant species have the ability to stabilize pollutants in the sediments, preventing their release into the water column. This process, known as phytostabilization, reduces the mobility and bioavailability of contaminants, minimizing their impact on the aquatic ecosystem.
- **Rhizofiltration:** Phytoremediation in ponds can involve the use of plants with dense root systems to filter and remove pollutants directly from the water column. These plants act as living filters, trapping contaminants in their roots or adsorbing them onto root surfaces.
- **Enhanced Microbial Activity:** The rhizosphere, the region of soil influenced by plant roots, harbors a diverse array of microorganisms (Rathika *et al.*, 2021). Phytoremediation promotes the growth of beneficial microbial communities that can enhance pollutant degradation and nutrient cycling in the pond ecosystem.
- **Plant Selection and Adaptation:** The success of phytoremediation in ponds depends on selecting plant species that are well adapted to the specific environmental conditions and capable of tolerating or accumulating the target pollutants. Native plant species are often preferred due to their adaptability and ecological compatibility.
- **Monitoring and Long-Term Management:** Phytoremediation projects require continuous monitoring of water quality parameters, pollutant concentrations, and plant health to ensure their effectiveness (Marmiroli and Maestri, 2008). Long-term management strategies, including regular maintenance, periodic harvesting, and replanting, are crucial to sustain the benefits of phytoremediation in ponds.

Rejuvenation of Ponds through Phytoremediation

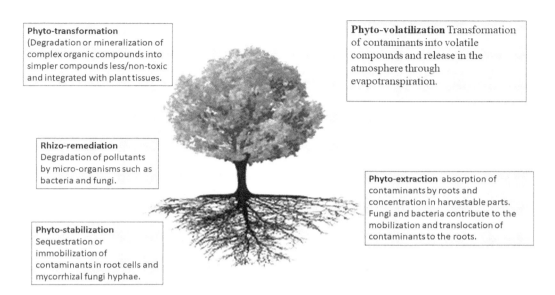

FIGURE 10.1 Mechanism of phytoremediation

These principles guide the implementation of phytoremediation techniques in ponds and contribute to the overall goal of restoring water quality, enhancing the ecological balance, and promoting the long-term sustainability of aquatic ecosystems.

10.2.2 Phytoremediation Mechanisms

Phytoremediation mechanisms play a crucial role in the rejuvenation of polluted environments. They involve various processes by which plants help remove, degrade, or immobilize contaminants, thus improving the overall environmental quality. Here are some key mechanisms utilized in phytoremediation:

- Phytoextraction
- Phytostabilization
- Rhizofiltration
- Phytovolatilization
- Phytodegradation
- Phytostimulation/rhizodegradation

These phytoremediation mechanisms collectively contribute to the rejuvenation of polluted environments by facilitating the removal, degradation, or immobilization of contaminants. The specific mechanism(s) employed in a given phytoremediation project depend on the type of pollutants, plant species selected, environmental conditions, and the desired remediation goals.

10.2.3 Advantages and Limitations of Phytoremediation

Advantages of Phytoremediation of Ponds:

- **Sustainability and Environmental Friendliness:** Phytoremediation is a sustainable and environmentally friendly approach to pond remediation. It relies on natural processes and utilizes living plants to remove or transform contaminants, reducing the need for invasive or chemical-intensive methods.

- **Cost Effectiveness:** Phytoremediation can be a cost-effective remediation technique compared to traditional methods such as dredging or soil excavation. It requires lower initial capital investment and has lower operational and maintenance costs in the long run.
- **Aesthetically Pleasing and Ecological Benefits:** Phytoremediation projects can enhance the aesthetic value of ponds by introducing and promoting the growth of vegetation. This vegetation provides habitat and food sources for various organisms, increases biodiversity, and improves the overall ecological health of the pond ecosystem.
- **Versatility and Applicability:** Phytoremediation can be applied to a wide range of pollutants, including heavy metals, organic compounds, nutrients, and pesticides (Gaur *et al.*, 2014). Different plant species can be selected based on the specific contaminants and site conditions, making phytoremediation a versatile approach for addressing multiple types of pollution.
- **Long-Term Effectiveness:** Phytoremediation offers long-term effectiveness in improving water quality and restoring the ecological balance of ponds. Once established, plants can continue to uptake, sequester, or transform contaminants, providing sustained remediation benefits over time.
- **Community Engagement and Education:** Phytoremediation projects in ponds can engage local communities and stakeholders in environmental conservation efforts. They provide opportunities for education, public participation, and awareness-building about water quality issues and sustainable management practices.

Limitations of Phytoremediation of Ponds:

- **Time-Intensive Process:** Phytoremediation is a relatively slow process compared to some other remediation techniques. It may take several years for plants to effectively remediate polluted ponds, depending on the pollutant concentrations, site conditions, and plant species selected (Licht and Isebrands, 2014).
- **Limitations in Contaminant Types:** While phytoremediation can address a wide range of pollutants, certain contaminants, such as highly persistent organic compounds, may be less amenable to degradation or uptake by plants. In such cases, complementary techniques may be required.
- **Dependence on Site-Specific Factors:** The success of phytoremediation in ponds is influenced by various site-specific factors, including soil composition, hydrology, pH, nutrient levels, and climatic conditions. Some sites may not be suitable for phytoremediation due to extreme environmental conditions or unsuitable soil characteristics.
- **Limited Depth and Mobility:** Phytoremediation is most effective in shallow or surface water environments where plant roots can access the contaminants. It may be less effective in deeper ponds or ponds with significant sediment pollution, where contaminant mobility and availability may pose challenges for plant uptake.
- **Monitoring and Maintenance Requirements:** Phytoremediation projects require ongoing monitoring of plant health, pollutant concentrations, and water quality parameters (Marmiroli and Maestri, 2008). Regular maintenance activities, such as harvesting, replanting, and controlling invasive species, may be necessary to ensure the effectiveness and longevity of the remediation efforts.
- **Regulatory and Approval Challenges:** The implementation of phytoremediation projects in ponds may require regulatory approvals and permits, especially if the ponds are in protected or sensitive areas. Obtaining the necessary permissions and navigating the regulatory processes can present challenges and delays in project implementation.

Rejuvenation of Ponds through Phytoremediation

It is important to consider these advantages and limitations when planning and implementing phytoremediation projects in ponds. Site-specific assessments, careful plant selection, and regular monitoring are crucial for maximizing the effectiveness and success of phytoremediation in rejuvenating polluted ponds.

10.3 TYPES OF POND CONTAMINANTS

10.3.1 NUTRIENT ENRICHMENT (EUTROPHICATION)

One of the primary contaminants that affect ponds is nutrient enrichment, leading to a phenomenon called eutrophication. Eutrophication occurs when excessive amounts of nutrients, particularly nitrogen and phosphorus, enter the pond ecosystem. These nutrients can originate from various sources, including agricultural runoff, sewage discharge, and the use of fertilizers in residential areas. The main causes of nutrient enrichment in ponds include:

- **Agricultural Activities:** Runoff from agricultural fields can carry excess fertilizers and manure, containing high levels of nitrogen and phosphorus, into ponds. These nutrients promote the growth of algae and aquatic plants, leading to eutrophication.
- **Urban Runoff:** Urban areas contribute to nutrient enrichment through stormwater runoff that carries fertilizers, pet waste, and other pollutants into ponds. Impermeable surfaces, such as roads and parking lots, prevent natural infiltration, increasing the volume and speed of runoff.
- **Wastewater Discharge:** Wastewater treatment plants, septic systems, and industrial facilities can discharge nutrient-rich effluents into water bodies, including ponds. Improperly treated or untreated wastewater can significantly contribute to nutrient enrichment and eutrophication (Barakat, 2011).
- **Land Development:** Construction activities associated with land development can disturb natural landscapes, leading to increased soil erosion. Sediments rich in nutrients can enter ponds through runoff, further exacerbating eutrophication.

The consequences of nutrient enrichment and eutrophication in ponds include:

- **Excessive Algal Growth:** Increased nutrient levels promote the growth of algae in ponds, resulting in algal blooms. These blooms can cause water discoloration, reduce water clarity, and deplete dissolved oxygen levels during decomposition, leading to detrimental effects on aquatic life.
- **Oxygen Depletion:** Algal blooms can lead to oxygen depletion in the water due to increased decomposition of dead algae by bacteria. This process, known as oxygen demand, can create low-oxygen or anoxic conditions, negatively impacting fish and other aquatic organisms.
- **Loss of Biodiversity:** Eutrophication can lead to shifts in species composition and a decline in biodiversity. Some species may thrive in nutrient-rich environments, while others, particularly those adapted to low-nutrient conditions, may suffer or be outcompeted.
- **Harmful Algal Blooms:** Certain types of algae, such as cyanobacteria (blue-green algae), can produce toxins during blooms. These toxins can be harmful to humans, pets, and aquatic organisms, posing risks to public health and ecosystem integrity.

Addressing nutrient enrichment and eutrophication in ponds requires implementing management strategies such as:

- **Nutrient Source Control:** Implementing best management practices in agriculture and urban areas to reduce nutrient runoff, such as proper fertilizer use, soil conservation measures, and green infrastructure for stormwater management (Rathika *et al.*, 2021).

- **Vegetation Management:** Planting and maintaining a buffer zone of native vegetation around ponds can help filter and absorb nutrients from runoff, preventing their entry into the water.
- **Wetland Construction:** Constructing or restoring wetlands adjacent to ponds can act as nutrient sinks and provide natural filtration, reducing nutrient loads before they enter the pond ecosystem.
- **Biological Controls:** Introducing or promoting the growth of aquatic plants, such as submerged or floating species, can help absorb excess nutrients and reduce algal blooms. These plants compete with algae for nutrients and provide shading that limits algal growth.

By addressing nutrient enrichment and eutrophication, the water quality and ecological health of ponds can be improved, fostering a balanced and sustainable aquatic ecosystem.

10.3.2 HEAVY METALS

Another significant group of contaminants that can affect ponds is heavy metals. Heavy metals are naturally occurring elements with high atomic weights and include substances such as lead, mercury, cadmium, arsenic, chromium, and copper (Greipsson,2015; Saxena *et al.*, 2020). These metals can enter pond ecosystems through various sources, including industrial discharge, mining activities, atmospheric deposition, and agricultural runoff. Some of the main causes and sources of heavy metal contamination in ponds are:

- **Industrial Activities**: Industrial processes, such as mining, metal smelting, manufacturing, and waste disposal, can release heavy metals into the environment. Effluents from industrial facilities may contain elevated levels of heavy metals, which can enter ponds through runoff or direct discharges.
- **Atmospheric Deposition:** Heavy metals can be present in atmospheric emissions from industrial activities, vehicle exhaust, and incineration. These metals can settle onto the pond surface through atmospheric deposition, especially in areas near industrial or heavily trafficked regions.
- **Agricultural Runoff:** The use of certain fertilizers, pesticides, and soil amendments in agriculture can contribute to heavy metal contamination in ponds (Rathika *et al.*, 2021). These contaminants may originate from the application of sludge, biosolids, or manure, which can contain elevated levels of heavy metals.
- **Urban Runoff:** Urban areas can be a source of heavy metal contamination due to runoff from roads, roofs, and other impervious surfaces. These metals can be derived from vehicle emissions, industrial activities, construction materials, and household products.

The consequences of heavy metal contamination in ponds include:

- **Toxicity to Aquatic Organisms:** Heavy metals can be highly toxic to aquatic organisms, including fish, invertebrates, and algae (Rice *et al.*, 2014). They can accumulate in tissues and disrupt various physiological processes, leading to impaired growth, reproduction, and overall health.
- **Bioaccumulation and Biomagnification:** Some heavy metals have the ability to accumulate in organisms over time, leading to bioaccumulation within the food chain. Predatory species at higher trophic levels may exhibit higher concentrations of heavy metals due to biomagnification, posing risks to organisms at the top of the food web, including humans (Bebianno *et al.*, 2004).
- **Water Quality Degradation:** Heavy metals can alter water quality parameters, such as pH and dissolved oxygen levels, affecting the overall ecological balance of the pond ecosystem

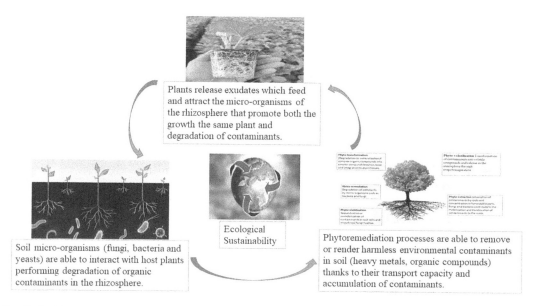

FIGURE 10.2 Phytoremediation of heavy metals.

(Gaur *et al.*, 2014). They can also cause aesthetic issues, such as water discoloration or turbidity.
- **Ecological Imbalance:** Heavy metal contamination can disrupt the natural balance of the pond ecosystem by affecting the abundance and diversity of species. Sensitive species may decline, while metal-tolerant species may become dominant, leading to shifts in community composition.

Addressing heavy metal contamination in ponds requires implementing appropriate management strategies, including:

- **Source Control and Pollution Prevention:** Implementing pollution prevention measures in industries, such as proper waste management and treatment, can minimize the release of heavy metals into the environment. Similarly, promoting sustainable agricultural practices and reducing the use of heavy metal-containing products can help prevent contamination.
- **Sediment Management:** Heavy metals can accumulate in pond sediments over time. Techniques such as sediment dredging, capping, or stabilization can be employed to reduce the bioavailability and mobility of heavy metals in sediments, minimizing their impact on the water column.
- **Phytoremediation:** Phytoremediation, as discussed earlier, can be a valuable approach for addressing heavy metal contamination. Certain plant species have the ability to accumulate heavy metals in their tissues, reducing the metal concentration in the pond water or sediments. These plants can be harvested and properly disposed of to remove the accumulated heavy metals from the ecosystem.
- **Monitoring and Assessment:** Regular monitoring of heavy metal concentrations in water, sediments, and biota is essential to assess the effectiveness of remediation efforts and determine any potential risks to aquatic life and human health.

By implementing these strategies, the contamination of ponds with heavy metals can be mitigated, helping to restore and protect the ecological health and water quality of these important aquatic ecosystems.

10.3.3 Organic Pollutants

Organic pollutants are another group of contaminants that can impact ponds. These pollutants are carbon-based compounds derived from various sources, including industrial activities, agricultural practices, domestic waste, and urban runoff (Greipsson,2015). They can enter pond ecosystems through surface runoff, atmospheric deposition, or direct discharge. Some common types of organic pollutants found in ponds include:

- **Pesticides and Herbicides:** Agricultural activities often involve the use of pesticides and herbicides to control pests and weeds. These chemical compounds can enter ponds through runoff or drift, leading to contamination of the water and sediments.
- **Industrial Chemicals:** Industrial processes can release a wide range of organic pollutants into the environment. Examples include solvents, petroleum hydrocarbons, polychlorinated biphenyls (PCBs), and dioxins. Industrial discharges or accidental spills can introduce these pollutants into ponds.
- **Petroleum Products:** Oil spills or leakage from storage tanks and pipelines can introduce petroleum-based pollutants, such as crude oil and refined products, into ponds. These pollutants can have significant adverse effects on aquatic organisms and the overall ecosystem.
- **Polycyclic Aromatic Hydrocarbons (PAHs):** PAHs are a group of organic pollutants that are formed during incomplete combustion processes. They can be present in coal tar, gasoline, vehicle exhaust, and certain industrial emissions. PAHs can enter ponds through atmospheric deposition or runoff from contaminated areas.
- **Personal Care Products and Pharmaceuticals:** Personal care products, including cosmetics, soaps, and lotions, as well as pharmaceuticals, can be sources of organic pollutants in ponds. These compounds can enter the water through wastewater discharges or septic system leachate (Barakat, 2011).

The consequences of organic pollutant contamination in ponds include:

- **Toxicity to Aquatic Organisms:** Many organic pollutants can be toxic to aquatic organisms, including fish, invertebrates, and algae. They can cause acute or chronic effects, such as reduced growth, reproductive abnormalities, impaired immune function, and even mortality.
- **Bioaccumulation and Biomagnification:** Some organic pollutants have the potential to bioaccumulate in organisms, particularly those that are lipophilic (soluble in fats)(Bebianno *et al.*, 2004). Through biomagnification, these compounds can increase in concentration as they move up the food chain, posing risks to higher trophic levels, including predators and humans.
- **Water Quality Degradation:** Organic pollutants can lead to water quality degradation in ponds. They can contribute to increased turbidity, reduced water clarity, and changes in pH levels. Some pollutants can also affect oxygen levels and result in hypoxic or anoxic conditions, negatively impacting aquatic life.
- **Ecological Imbalance:** Organic pollutants can disrupt the balance of the pond ecosystem by affecting the composition and abundance of species. Sensitive species may decline, while pollutant-tolerant species may become dominant, leading to shifts in community structure and potential loss of biodiversity.

Addressing organic pollutant contamination in ponds requires implementing appropriate management strategies, including:

- **Source Control and Pollution Prevention:** Implementing best management practices in agriculture, industry, and urban areas can help minimize the release of organic pollutants

Rejuvenation of Ponds through Phytoremediation

(Lamichhane *et al.*, 2018, Vinodhini and Narayanan, 2008). This includes proper storage, handling, and disposal of chemicals, promoting sustainable agricultural practices, and raising awareness about the proper use and disposal of personal care products.

- **Wastewater Treatment:** Ensuring adequate wastewater treatment for domestic, industrial, and agricultural effluents can help remove organic pollutants before they enter ponds. This can involve the use of advanced treatment technologies, such as activated carbon filtration or advanced oxidation processes (Barakat, 2011).
- **Wetland Construction and Restoration:** Constructing or restoring wetlands in the vicinity of ponds can act as natural filters, facilitating the removal and degradation of organic pollutants through plant uptake, microbial processes, and sedimentation.
- **Monitoring and Assessment:** Regular monitoring of organic pollutant concentrations in water, sediments, and biota is crucial to assess the effectiveness of remediation efforts and determine any potential risks to the aquatic ecosystem and human health.

By implementing these strategies, the contamination of ponds with organic pollutants can be mitigated, helping to restore and protect the ecological health and water quality of these important aquatic ecosystems (Marmiroli and Maestri, 2008).

10.4 PHYTOREMEDIATION TECHNIQUES FOR POND REJUVENATION

10.4.1 PHYTOEXTRACTION

It is the process by which plants uptake contaminants from the soil, sediments, or water and accumulate them in their tissues (Lajayer *et al.*, 2019). Plants absorb pollutants through their roots and transport them to the above-ground parts, such as leaves, stems, and roots. Contaminants can be stored in these plant tissues or undergo further transformations. It is particularly effective for heavy metals and certain organic pollutants.

10.4.2 PHYTOSTABILIZATION

It involves reducing the mobility and bioavailability of pollutants by plants, thus minimizing their potential to migrate and cause harm. In this mechanism, plants immobilize contaminants in the soil or sediments through root uptake and subsequent precipitation or complexation. It is especially useful for pollutants such as heavy metals and metalloids, preventing their leaching into groundwater or their uptake by organisms.

10.4.3 RHIZOFILTRATION

It refers to the filtration of pollutants from water using plant root systems. Plants with dense root structures, such as wetland plants, act as living filters. As water passes through the root zone, pollutants are trapped by the roots or adsorbed onto their surfaces. This mechanism is effective for removing a wide range of contaminants, including heavy metals, organic compounds, and nutrients, from water bodies.

10.4.4 PHYTODEGRADATION

It involves the direct breakdown or transformation of organic contaminants within the plant tissues. Plants possess enzymes and metabolic pathways that can metabolize or break down organic pollutants into less toxic or more easily degradable forms. Phytodegradation is particularly relevant for organic compounds such as hydrocarbons, pesticides, and industrial chemicals.

10.4.5 Phytovolatilization

It is the process by which plants uptake contaminants from the soil or water and release them into the atmosphere in the form of volatile compounds. This mechanism is particularly applicable to volatile organic compounds (VOCs) and certain heavy metals. Plants absorb the pollutants through their roots or leaves and subsequently transform them into volatile forms that can be released through transpiration or metabolic processes.

10.4.6 Phytostimulation/Rhizodegradation

It involves the breakdown or transformation of contaminants in the rhizosphere, the soil region influenced by plant roots. Plant roots release various exudates, including enzymes and organic acids, which can stimulate microbial activity and enhance the degradation of organic pollutants. The synergistic interactions between plants and rhizospheric microorganisms play a vital role in the rhizodegradation process.

10.5 SELECTION OF SUITABLE PLANT SPECIES

10.5.1 Characteristics of Phytoremediation Plants

Phytoremediation plants, also known as hyperaccumulators or remediation plants, possess certain characteristics that make them suitable for the process of phytoremediation (Malayeri *et al.*, 2008, Peer *et al.*, 2005). These characteristics include:

- **Metal Tolerance:** Phytoremediation plants have the ability to tolerate high concentrations of heavy metals without showing significant negative effects. They have developed mechanisms to prevent the toxic effects of heavy metals, such as metal sequestration in cell vacuoles, metal chelation, and regulation of metal transport (Rice *et al.*, 2014; Thakur *et al.*, 2016).
- **Accumulation Capacity:** Phytoremediation plants are capable of accumulating heavy metals from the environment into their tissues. They can take up metals from the soil or water through their roots and translocate them to the above-ground parts of the plant. Some plants have a high capacity for metal accumulation, which is essential for effective phytoremediation.
- **Root System:** Phytoremediation plants often have a well-developed and extensive root system. This allows them to explore a large volume of soil or water and enhance metal uptake from the contaminated environment. Additionally, some plants have specialized root structures, such as root hairs or mycorrhizal associations, which can further enhance metal uptake.
- **Fast Growth and High Biomass Production:** Phytoremediation plants are typically fast-growing species that can quickly establish themselves in contaminated areas. Their rapid growth and high biomass production allow them to extract large amounts of metals from the environment and facilitate the remediation process.
- **Adaptability:** Phytoremediation plants exhibit adaptability to different environmental conditions, including variations in soil pH, moisture levels, and nutrient availability (Rathika *et al.*, 2021)). They can grow in a wide range of soil types and climates, making them versatile for phytoremediation applications in diverse regions.
- **Wide Range of Plant Families:** Phytoremediation plants belong to various plant families, including Brassicaceae, Poaceae, Fabaceae, and Asteraceae, among others. Different plant families may have specific mechanisms for metal uptake and tolerance, allowing for a diverse range of plant options for phytoremediation (Thakur *et al.*, 2016).

Rejuvenation of Ponds through Phytoremediation

- **Large Leaf Surface Area:** Plants with large leaf surface areas are desirable for phytoremediation as they can maximize metal uptake through foliar absorption. Leaves provide a significant surface area for the exchange of gases and the absorption of metals from the atmosphere or water.
- **Longevity and Persistence:** Some phytoremediation plants exhibit longevity and persistence in contaminated environments. They can survive and continue to remove metals over multiple growing seasons, enhancing the overall effectiveness of the phytoremediation process.

It is important to note that not all phytoremediation plants possess all of these characteristics. Different plants may have specific adaptations and efficiencies in metal uptake and accumulation. Therefore, the selection of appropriate phytoremediation plants depends on the specific contaminant, site conditions, and remediation goals.

10.5.2 Native and Non-Native Plant Species

When considering plant species for phytoremediation, there are two categories to consider: native and non-native plant species. Let's explore the characteristics and considerations associated with each category:

Native Plant Species:

Native plants are indigenous to a particular region or ecosystem and have naturally evolved and adapted to the local environmental conditions.

Benefits: Using native plant species for phytoremediation offers several advantages, such as:

- **Ecological Fit:** Native plants are well adapted to the local climate, soil conditions, and ecological interactions (Rathika *et al.*, 2021). They are more likely to establish successfully and persist in the remediation site, enhancing ecosystem stability and biodiversity.
- **Ecosystem Services:** Native plants often provide additional ecosystem services, such as supporting pollinators, providing habitat for wildlife, and preventing erosion. They contribute to the overall ecological functioning and health of the site.
- **Genetic Diversity:** Native plant populations typically exhibit higher genetic diversity, which can increase their resilience to changing environmental conditions and enhance their ability to adapt to site-specific contaminants.
- **Site-Specific Considerations:** When selecting native plant species for phytoremediation, consider the following:
- **Metal Tolerance:** Native plants that naturally occur in metal-contaminated environments, known as metallophytes, may possess innate metal tolerance and accumulation capabilities (Thakur *et al.*, 2016).
  - **Local Availability:** Native plants are often readily available from local nurseries or seed banks, making them accessible and cost-effective for restoration projects.
  - **Regulatory Considerations:** In some cases, using native plants may align with regulatory requirements or conservation plans that promote the restoration of native plant communities.

Non-Native Plant Species:

Non-native, also referred to as exotic or introduced, plant species originate from outside the region or ecosystem where they are being utilized.

Benefits: Non-native plant species can offer certain advantages for phytoremediation, including:

- **Specific Metal Accumulation:** Some non-native plants, particularly hyperaccumulators from other regions, may exhibit exceptional metal accumulation capabilities that can aid in the remediation of specific contaminants (Malayeri *et al.*, 2008).
- **Rapid Growth and Biomass Production:** Certain non-native species may have faster growth rates and higher biomass production than native species, allowing for more rapid phytoremediation outcomes (Bebianno *et al.*, 2004).
- **Experimental Approach:** Non-native species can be utilized in research and experimental studies to test their effectiveness in remediating specific contaminants or to explore new phytoremediation strategies.
- **Site-Specific Considerations:** When considering non-native plant species for phytoremediation, it is important to address the following considerations:
  - **Invasiveness:** Non-native species may possess invasive traits that can negatively impact native plant communities, disrupt ecological interactions, and reduce biodiversity. Careful assessment and monitoring are required to prevent unintended ecological consequences.
  - **Genetic Contamination:** Non-native plants may hybridize with native species, potentially leading to genetic contamination and alteration of local gene pools.
  - **Regulatory Considerations:** Introducing non-native species may require approval or adherence to regulations to ensure they do not pose a risk to native ecosystems or biodiversity.

Ultimately, the selection of native or non-native plant species for phytoremediation depends on site-specific factors, project goals, ecological considerations, and regulatory requirements. A balanced approach that integrates both native and non-native species, when appropriate, can maximize the effectiveness of phytoremediation while minimizing potential ecological risks.

10.5.3 PLANT TOLERANCE AND ACCUMULATION

Plant tolerance and accumulation are two key characteristics related to the ability of plants to withstand and remediate contaminants, particularly heavy metals (Thakur *et al.*, 2016). Let's explore these concepts further:

Plant Tolerance:

Plant tolerance refers to the ability of plants to survive and grow in environments with high levels of contaminants, such as heavy metals.
- **Tolerance mechanisms:** Plants have developed various mechanisms to tolerate and cope with high concentrations of contaminants (Thakur *et al.*, 2016). These mechanisms include:
  - **Metal Exclusion:** Some plants prevent the entry of heavy metals into their tissues by restricting metal uptake through roots or reducing metal translocation to the shoots.
  - **Metal Sequestration:** Plants can sequester heavy metals within their cell vacuoles or by binding them to complex molecules, such as phytochelatins or metallothioneins. This sequestration reduces the toxicity of metals within plant cells.
  - **Metal Detoxification:** Plants can activate detoxification mechanisms that involve enzymatic processes, such as metal chelation and complexation, to render heavy metals less harmful.

Rejuvenation of Ponds through Phytoremediation

- **Enhanced Antioxidant Defense:** Contaminant-stressed plants often produce higher levels of antioxidants, such as glutathione and ascorbic acid, to combat the oxidative stress caused by heavy metals.

Tolerant Plant Species: Some plant species have naturally evolved or adapted to contaminated environments and exhibit inherent tolerance to specific heavy metals. These species are known as metallophytes.

Tolerance Range: Different plant species have varying degrees of tolerance to heavy metals. Some can tolerate high metal concentrations, while others may be sensitive even at low levels. The specific tolerance range depends on the plant species and the specific metal being considered.

Plant Accumulation:

Plant accumulation, also known as metal uptake or phytoextraction, refers to the ability of plants to uptake and accumulate heavy metals from the soil or water.

Accumulation Mechanisms: Plants employ various mechanisms to accumulate heavy metals, including:

- **Active Uptake:** Plants can actively take up heavy metals from the soil or water through their root systems. This uptake is facilitated by metal transporters present in the root cells.
- **Metal Translocation:** After uptake, plants can translocate heavy metals from the roots to the shoots through the xylem or phloem, allowing for metal accumulation in above-ground plant parts.
- **Metal Sequestration:** Once accumulated, plants may store heavy metals in specialized tissues or structures, such as vacuoles or trichomes, to prevent metal toxicity in essential organs.

Hyperaccumulators: Certain plant species, known as hyperaccumulators, have an exceptional ability to accumulate high concentrations of specific heavy metals in their tissues (Malayeri *et al.*, 2008; Rathika *et al.*, 2021). These plants can accumulate levels of metals far exceeding those found in non-accumulator species.

Phytostabilization: Some plants may not accumulate high levels of metals but can facilitate their immobilization and stabilization in the soil through root exudates or binding to soil particles, reducing their bioavailability and potential for environmental impact.

The selection of plant species for phytoremediation should consider both tolerance and accumulation characteristics. Tolerant plant species can thrive in contaminated environments, while accumulators can effectively remove and concentrate heavy metals from the soil or water. It is important to match the specific contaminants and site conditions with plant species that possess the appropriate tolerance and accumulation capabilities for effective phytoremediation.

10.6 MANAGEMENT STRATEGIES FOR PHYTOREMEDIATION IN PONDS

10.6.1 PLANTING DESIGN AND ARRANGEMENT

Planting design and arrangement play crucial roles in the success of phytoremediation projects. The proper arrangement of plants can optimize their remediation potential, maximize contaminant uptake, and enhance overall efficiency. Here are some key considerations for planting design and arrangement in phytoremediation:

- **Plant Species Selection:** Choose plant species that are known to have high tolerance and accumulation capabilities for the specific contaminants present in the site. Consider native species whenever possible, as they are likely to be well adapted to local conditions.

- **Plant Density and Spacing:** The density and spacing of plants should be determined based on factors such as the contaminant concentration, soil characteristics, and the growth habits of the selected plant species. Dense planting can maximize contaminant uptake and provide efficient coverage of the remediation area.
- **Plant Distribution Patterns:** Different distribution patterns can be employed based on site characteristics and project goals. Some common patterns include:

  - **Uniform Distribution:** Plants are evenly spaced throughout the remediation area, providing consistent coverage and uptake.
  - **Clustered Distribution:** Plants are grouped in clusters, which can be effective in targeting specific contaminated hotspots.
  - **Gradient Distribution:** Plants are arranged in a gradient pattern, with higher densities near the contamination source and decreasing densities as the distance from the source increases.

Interplanting and Polyculture: Consider incorporating a variety of plant species with complementary traits and accumulation capabilities. This approach, known as interplanting or polyculture, can enhance the overall remediation potential by targeting multiple contaminants and improving ecosystem resilience.

Succession Planning: In cases where phytoremediation is a long-term process, plan for successive plantings or the introduction of new plant species to ensure continuous remediation. This can involve planting fast-growing pioneer species initially and later introducing slower-growing, long-lived species.

Root Depth and Distribution: Consider the depth and distribution of plant roots, as they play a vital role in contaminant uptake. Select plants with root systems that can effectively explore the soil profile and reach the contaminant source. Deep-rooted species are suitable for remediating contaminants that are located deeper in the soil.

Site Preparation: Adequate site preparation, such as soil amendment, removal of competing vegetation, and soil aeration, may be necessary before planting to optimize plant growth and enhance contaminant uptake.

Monitoring and Maintenance: Regular monitoring of plant health, growth, and contaminant uptake is essential. Implement appropriate maintenance practices, such as irrigation, weed control, and nutrient management, to ensure optimal plant performance.

Considerations for Aquatic Systems: In the case of phytoremediation in ponds or other aquatic systems, choose plant species that can tolerate waterlogged conditions and have the ability to uptake contaminants from the water column or sediments. Plant placement in specific areas of the pond, such as along the shoreline or in floating rafts, can enhance remediation efficiency.

It is important to note that the specific planting design and arrangement may vary depending on site-specific factors, project goals, and the characteristics of the contaminants being remediated. Consulting with experts in phytoremediation or landscape design can help in developing an effective planting plan for the specific site.

10.6.2 Nutrient Management

Nutrient management is an essential aspect of phytoremediation projects, as it can influence the growth and performance of plants and their ability to remediate contaminants effectively. Here are some key considerations for nutrient management in phytoremediation:

- **Soil Nutrient Analysis:** Conduct a thorough analysis of the soil to determine its nutrient content and any potential nutrient deficiencies or imbalances. This analysis helps in developing an appropriate nutrient management plan.

Rejuvenation of Ponds through Phytoremediation

- **Nutrient Amendments:** Based on the soil analysis, apply appropriate nutrient amendments to address any deficiencies or imbalances. Common amendments include organic matter (compost, manure), inorganic fertilizers, and mineral amendments. The choice of amendments should consider the specific nutrient requirements of the selected plant species.
- **Nutrient Release Rate:** Consider the release rate of nutrients from amendments to ensure a steady and balanced supply over time. Slow-release fertilizers or organic amendments can help provide nutrients gradually and reduce the risk of nutrient leaching or runoff.
- **Nutrient Application Timing:** Timing of nutrient application is important to support plant growth and establishment. Apply nutrients prior to planting or during specific growth stages to meet the nutritional needs of plants during critical periods.
- **Nutrient Uptake Monitoring:** Regularly monitor plant nutrient uptake and assess any potential nutrient deficiencies or excesses. Leaf tissue analysis can provide insights into the nutrient status of plants and help adjust nutrient management practices accordingly.
- **Phytotoxicity Prevention:** Excessive nutrient application can lead to phytotoxicity, negatively impacting plant health and performance. Avoid over-fertilization and closely monitor nutrient levels to prevent toxicity issues.
- **Nutrient Cycling and Recycling:** Promote nutrient cycling and recycling within the phytoremediation system by incorporating organic matter, such as plant residues or mulch, into the soil. This helps enhance soil fertility, improve nutrient availability, and support the long-term sustainability of the remediation site.
- **Irrigation Management:** Proper irrigation practices are crucial for nutrient management. Ensure that plants receive adequate water without excessive leaching of nutrients. Irrigation scheduling should consider plant water needs, soil moisture levels, and nutrient availability.
- **Nutrient Runoff and Leaching Mitigation:** Implement strategies to minimize nutrient runoff and leaching from the remediation site. These can include implementing vegetative buffers, erosion control measures, or using appropriate application techniques to prevent nutrient loss.
- **Environmental Considerations:** Consider the potential impacts of nutrient management practices on the surrounding environment, such as water bodies. Avoid excessive nutrient inputs that can contribute to eutrophication or water pollution.

It is important to tailor nutrient management practices to the specific needs of the phytoremediation project, including the chosen plant species, soil conditions, and contaminant types. Consulting with soil and plant nutrition experts can provide valuable guidance in developing an effective and environmentally sustainable nutrient management plan for phytoremediation projects.

10.6.3 Water Level and Flow Management

Water level and flow management are crucial aspects of phytoremediation projects, particularly in aquatic or wetland environments. Proper management of water levels and flow can impact the growth and performance of plants, as well as the transport and distribution of contaminants. Here are some key considerations for water level and flow management in phytoremediation:

- **Hydrological Assessment:** Conduct a thorough hydrological assessment of the site to understand the natural water regime, including rainfall patterns, groundwater dynamics, and surface water flows. This assessment helps determine the baseline conditions and guides water management decisions.

- **Water Level Monitoring:** Establish a system for monitoring water levels within the phytoremediation system. Regular monitoring allows for adjustments in water management strategies based on the specific needs of the plant species and the contaminants being targeted.
- **Water Source:** Determine the water source for the phytoremediation system, such as surface water, groundwater, or a combination of both. Consider the quality of the water source and potential impacts on plant health and contaminant transport.
- **Water Addition or Removal:** Depending on site conditions and project goals, water may need to be added or removed from the phytoremediation system. This can be achieved through methods such as irrigation, drainage systems, pumps, or diversion channels.
- **Contaminant Transport:** Consider the movement and transport of contaminants within the water system. Proper water flow management can facilitate the transport of contaminants to the plant roots for uptake and removal from the system.
- **Flow Control Structures:** Install flow control structures, such as weirs, gates, or sluice gates, to manage water levels and flow rates. These structures allow for manipulation of water depths and control over the movement of water through the system.
- **Water Retention Time:** Optimize the water retention time in the phytoremediation system. Longer retention times can enhance contact between contaminants and plant roots, allowing for greater contaminant uptake.
- **Seasonal Variations:** Consider seasonal variations in water availability and flow rates. Adjust water management strategies accordingly to ensure optimal plant growth and contaminant removal throughout different seasons.
- **Erosion and Sediment Control:** Implement erosion control measures, such as vegetative buffers or erosion control blankets, to prevent soil erosion and sedimentation that could affect water flow patterns and plant performance.
- **Environmental Considerations:** Take into account the potential environmental impacts of water management practices. Ensure that water discharges from the site meet regulatory requirements and do not contribute to water pollution or negative ecological effects downstream.

Water level and flow management should be tailored to the specific needs of the phytoremediation project, including the plant species, contaminant types, and site conditions. Consulting with hydrological experts and considering local regulations and guidelines can provide valuable insights and guidance for effective water management in phytoremediation projects.

10.6.4 Harvesting and Disposal of Plant Biomass

In phytoremediation projects, the harvesting and disposal of plant biomass are important steps to consider once the plants have completed their remediation cycle or reached maturity. Proper handling of plant biomass ensures the containment and management of any accumulated contaminants and promotes the overall success and sustainability of the remediation process (Bebianno *et al.*, 2004). Here are some considerations for harvesting and disposing of plant biomass in phytoremediation:

- **Contaminant Concentration Analysis:** Before harvesting, conduct a thorough analysis of the plant biomass to determine the concentration of contaminants present. This analysis helps assess the effectiveness of the remediation process and informs decisions regarding biomass handling and disposal.

Rejuvenation of Ponds through Phytoremediation

- **Biomass Harvesting Techniques:** Select appropriate harvesting techniques based on the plant species, growth habit, and biomass quantity. Common methods include manual cutting, mowing, or using specialized harvesting equipment. Take care to minimize soil disturbance and prevent recontamination during harvesting.
- **Biomass Segregation:** Separate harvested biomass into different categories based on contaminant levels or plant parts. For example, separate roots, stems, leaves, and seeds if applicable. This segregation allows for targeted disposal and management of different biomass components.
- **Contaminated Biomass Handling:** Contaminated biomass should be handled with caution to prevent any further contamination. Implement measures such as wearing personal protective equipment (PPE) and using appropriate containment and storage practices to minimize the risk of contaminant release.
- **Contaminated Biomass Disposal Options:** Depending on the contaminant levels and local regulations, several disposal options may be available:

  - **Secure Landfill:** Contaminated biomass can be disposed of in a secure landfill facility designed to handle hazardous or contaminated waste. Ensure compliance with waste disposal regulations and coordinate with the appropriate authorities.
  - **Incineration:** Incineration of contaminated biomass can be considered, especially for biomass with high contaminant concentrations. However, careful consideration should be given to air emissions and the potential release of contaminants during the incineration process.
  - **Phytomining:** In some cases, the harvested biomass may contain valuable metals or contaminants that can be extracted and recovered using specialized techniques. Phytomining can be an option to recover these valuable resources while reducing waste volumes.
  - **Bioenergy Generation:** Biomass with lower contaminant levels can be used as a feedstock for bioenergy generation through processes such as anaerobic digestion or combustion. This allows for energy recovery while reducing biomass volume.
  - **Composting:** If the contaminant levels are low and do not pose a significant risk, the biomass can be composted under controlled conditions. This process can help degrade organic matter and reduce the overall volume of the biomass.
  - **Non-Contaminated Biomass Reuse:** If the harvested biomass is found to have low contaminant levels or is free from contaminants, it may be suitable for beneficial reuse. Consider options such as composting, mulching, or using biomass as a soil amendment in non-sensitive areas.
  - **Regulatory Compliance:** Ensure compliance with local, regional, and national regulations regarding the handling, transportation, and disposal of contaminated biomass. Coordinate with relevant authorities and waste management agencies to adhere to applicable guidelines and permits.
  - **Monitoring and Documentation:** Maintain proper documentation of the harvested biomass, its disposal methods, and any associated testing or analysis results. Regularly monitor the disposal site to ensure proper containment and management of the disposed biomass.

It is important to note that the specific harvesting and disposal methods may vary depending on the site-specific conditions, plant species, and contaminant types. Consult with environmental experts, waste management professionals, and regulatory authorities to develop appropriate strategies for the harvesting and disposal of plant biomass in phytoremediation projects.

10.7 CASE STUDIES AND SUCCESS STORIES

10.7.1 PHYTOREMEDIATION IN NUTRIENT-ENRICHED PONDS

Here are a few case studies and success stories that highlight the application of phytoremediation in nutrient-enriched ponds:

CASE STUDY: PHYTOREMEDIATION OF NUTRIENT-ENRICHED PONDS USING FLOATING WETLAND PLANTS

In a study conducted in the USA, floating wetland plants were employed to remediate nutrient-enriched ponds. The researchers used a combination of native plant species, including various grasses, sedges, and rushes, which were grown on floating mats. These floating wetland mats helped to absorb excess nutrients, such as nitrogen and phosphorus, from the water column through the plant roots. Over time, the plants effectively reduced nutrient levels, resulting in improved water quality and reduced algal blooms.

SUCCESS STORY: PHYTOREMEDIATION OF NUTRIENT-ENRICHED PONDS IN CHINA

In China, phytoremediation techniques have been successfully applied to address nutrient enrichment in ponds and lakes. One notable success story involves the use of water hyacinth (*Eichhornia crassipes*), an invasive plant species with high nutrient uptake capacity. Water hyacinth was introduced into nutrient-enriched ponds, where it rapidly absorbed excess nutrients, including nitrogen and phosphorus, from the water. As a result, water quality significantly improved, and algal blooms were mitigated. This success story demonstrated the effectiveness of phytoremediation in restoring the ecological balance of nutrient-enriched aquatic systems.

CASE STUDY: PHYTOREMEDIATION OF NUTRIENT-ENRICHED PONDS USING MACROPHYTES

A study conducted in Brazil focused on the phytoremediation of nutrient-enriched ponds using macrophytes. The researchers selected native plant species, including water lettuce (*Pistia stratiotes*) and water hyacinth (*Eichhornia crassipes*), due to their high nutrient uptake capacity. These plants were introduced into nutrient-enriched ponds, and over time, they effectively removed excess nutrients from the water, resulting in a significant reduction in nutrient concentrations. This case study demonstrated the potential of using native macrophytes for phytoremediation in ponds with nutrient pollution.

SUCCESS STORY: PHYTOREMEDIATION OF NUTRIENT-ENRICHED PONDS IN EUROPE

In Europe, phytoremediation has been successfully implemented to address nutrient enrichment in ponds and small lakes. One notable success story comes from the Netherlands, where constructed floating wetlands were utilized to remediate nutrient-enriched ponds. Native plant species, including cattails (*Typha* spp.) and reeds (*Phragmites* spp.), were grown on floating platforms. These plants effectively absorbed excess nutrients, such as nitrogen and phosphorus, from the water, resulting in improved water quality and reduced nutrient concentrations. The success of this project led to the wider adoption of phytoremediation techniques for nutrient management in aquatic systems across Europe.

These case studies and success stories highlight the effectiveness of phytoremediation in addressing nutrient enrichment in ponds and lakes. By harnessing the natural capabilities of plants to absorb and assimilate nutrients, phytoremediation offers a sustainable and eco-friendly approach to improve water quality and restore the ecological balance of nutrient-enriched aquatic systems.

10.7.2 Phytoremediation of Heavy Metal–Contaminated Ponds

Here are a few case studies and success stories that demonstrate the application of phytoremediation in heavy metal–contaminated ponds:

CASE STUDY: PHYTOREMEDIATION OF HEAVY METAL–CONTAMINATED PONDS USING AQUATIC PLANTS

A case study conducted in Thailand focused on the phytoremediation of heavy metal–contaminated ponds using aquatic plants. Native plant species, including water hyacinth (*Eichhornia crassipes*) and water lettuce (*Pistia stratiotes*), were cultivated in the contaminated ponds. These plants effectively accumulated heavy metals such as cadmium, lead, and copper in their tissues. Over time, the concentrations of heavy metals in the water decreased significantly, indicating successful remediation. This study highlighted the potential of aquatic plants for phytoremediation of heavy metal–contaminated ponds.

SUCCESS STORY: PHYTOREMEDIATION OF HEAVY METAL–CONTAMINATED PONDS IN SPAIN

In Spain, a success story involves the phytoremediation of heavy metal–contaminated ponds using the common reed (*Phragmites australis*). The reeds were planted in ponds contaminated with metals such as copper, zinc, and cadmium. The roots of the reeds absorbed and accumulated the heavy metals, effectively reducing their concentrations in the water. The success of this project demonstrated the ability of native plant species to remediate heavy metal pollution in ponds.

CASE STUDY: PHYTOREMEDIATION OF HEAVY METAL–CONTAMINATED PONDS USING FLOATING PLANTS

A case study in Poland investigated the phytoremediation of heavy metal–contaminated ponds using floating plants, specifically water fern (*Azolla filiculoides*) and duckweed (*Lemna minor*). These floating plants were able to absorb heavy metals, including zinc, copper, and lead, from the water. The study found that the plants effectively reduced the concentrations of heavy metals, providing a sustainable and cost-effective approach to remediate the ponds.

SUCCESS STORY: PHYTOREMEDIATION OF HEAVY METAL–CONTAMINATED PONDS IN INDIA

In India, a success story involves the phytoremediation of heavy metal–contaminated ponds using vetiver grass (*Chrysopogon zizanioides*). Vetiver grass was planted in ponds contaminated with metals such as arsenic, cadmium, and chromium. The grass effectively absorbed and accumulated the heavy metals in its tissues, thereby reducing their concentrations in the water. The success of this project highlighted the potential of vetiver grass for phytoremediation of heavy metal contamination in ponds.

These case studies and success stories illustrate the effectiveness of phytoremediation in addressing heavy metal contamination in ponds. By harnessing the natural capabilities of plants to accumulate and sequester heavy metals, phytoremediation offers a sustainable and environmentally friendly approach to remediate heavy metal–contaminated aquatic systems.

10.7.3 Phytoremediation of Organic Pollutants in Ponds

Here are a few case studies and success stories that highlight the application of phytoremediation in ponds contaminated with organic pollutants:

CASE STUDY: PHYTOREMEDIATION OF ORGANIC POLLUTANTS IN POND SEDIMENTS

A case study conducted in the United States focused on the phytoremediation of organic pollutants in pond sediments. Native wetland plants, such as cattails (*Typha* spp.) and bulrushes (*Scirpus* spp.), were planted in the pond to facilitate the degradation and removal of organic pollutants, including polycyclic aromatic hydrocarbons and polychlorinated biphenyls. Over time, the plants enhanced microbial activity in the sediments, promoting the breakdown of organic pollutants. This case study demonstrated the effectiveness of phytoremediation in reducing the levels of organic pollutants in pond sediments.

SUCCESS STORY: PHYTOREMEDIATION OF ORGANIC POLLUTANTS IN CONTAMINATED PONDS IN INDIA

In India, a success story involved the phytoremediation of ponds contaminated with organic pollutants, particularly dyes and textile effluents. Various plant species, including water hyacinth (*Eichhornia crassipes*) and water lettuce (*Pistia stratiotes*), were utilized in the phytoremediation process. These plants effectively absorbed and metabolized the organic pollutants, resulting in improved water quality and reduced pollutant concentrations. The success of this project demonstrated the potential of phytoremediation to remediate organic pollutants in ponds, particularly in the textile industry.

CASE STUDY: PHYTOREMEDIATION OF ORGANIC POLLUTANTS IN INDUSTRIAL WASTEWATER PONDS

A case study conducted in Brazil focused on the phytoremediation of industrial wastewater ponds contaminated with organic pollutants. Native plant species, such as duckweed (*Lemna* spp.) and water fern (*Salvinia* spp.), were employed in the phytoremediation process. These

Rejuvenation of Ponds through Phytoremediation

plants effectively absorbed and degraded the organic pollutants present in the wastewater, including hydrocarbons and phenolic compounds. The study found significant reductions in pollutant concentrations, indicating successful remediation. This case study highlighted the potential of phytoremediation in treating organic pollutants in industrial wastewater ponds.

SUCCESS STORY: PHYTOREMEDIATION OF ORGANIC POLLUTANTS IN AGRICULTURAL RUNOFF PONDS

In a success story from the Netherlands, phytoremediation was used to address organic pollutants in agricultural runoff ponds. Constructed wetlands were established using a combination of native plant species, including reeds (*Phragmites* spp.) and bulrushes (*Scirpus* spp.). These plants effectively absorbed and transformed organic pollutants, such as pesticides and herbicides, present in the agricultural runoff. The phytoremediation system improved water quality by reducing the levels of organic pollutants and preventing their downstream transport. This success story demonstrated the potential of phytoremediation in mitigating organic pollutant contamination in agricultural runoff ponds.

These case studies and success stories illustrate the effectiveness of phytoremediation in addressing organic pollutant contamination in ponds. By utilizing appropriate plant species and their associated microbial communities, phytoremediation offers a sustainable and nature-based approach to remediate ponds contaminated with organic pollutants.

10.8 CHALLENGES AND LIMITATIONS

10.8.1 Site-Specific Considerations

Phytoremediation of ponds, while an effective and sustainable approach, is not without its challenges and limitations. Site-specific considerations play a crucial role in determining the success and feasibility of phytoremediation projects. Here are some of the challenges and limitations associated with phytoremediation in ponds:

Contaminant Composition and Concentration: The type and concentration of contaminants present in the pond can significantly impact the effectiveness of phytoremediation. Some contaminants may be more challenging for plants to uptake, metabolize, or detoxify, making remediation more difficult. High concentrations of contaminants can also overwhelm the plants' capacity to accumulate or degrade them, limiting the efficacy of phytoremediation.

Plant Selection and Adaptability: The selection of suitable plant species is crucial for successful phytoremediation. However, not all plant species are equally effective in remediating specific contaminants. Site-specific factors such as soil type, nutrient availability, water quality, and climatic conditions must be considered when selecting plants that can thrive and efficiently remediate contaminants in the specific pond environment.

Growth and Establishment: Establishing plants in ponds can be challenging due to factors such as fluctuating water levels, wave action, and competition from existing vegetation. Ensuring proper plant establishment and growth is essential for the long-term success of phytoremediation projects. Adequate plant density, rooting depth, and suitable planting techniques should be considered to overcome these challenges.

Remediation Timeframe: Phytoremediation is generally a slower process compared to other remediation techniques. It may take several years for plants to effectively reduce contaminant levels in ponds, depending on various factors such as contaminant concentration, plant species, and environmental conditions. Patience and long-term commitment are required to achieve desired remediation outcomes.

Maintenance and Monitoring: Phytoremediation projects require ongoing maintenance and monitoring to ensure the health and effectiveness of the plants. Regular monitoring of contaminant levels, plant growth, and overall system performance is necessary to assess progress and make any necessary adjustments. Adequate resources and expertise are essential for the proper maintenance and monitoring of phytoremediation systems.

Regulatory and Permitting Requirements: Phytoremediation projects may be subject to regulatory and permitting requirements, depending on the jurisdiction and the nature of contaminants involved. Compliance with environmental regulations, obtaining necessary permits, and coordinating with regulatory authorities can add complexity and time to project implementation.

Limitations of Contaminant Mobility: Phytoremediation is most effective for contaminants that are present in the water column or readily available for plant uptake. Contaminants bound to sediments or with low mobility may pose challenges for phytoremediation. In such cases, additional techniques such as sediment remediation or enhanced plant–microbe interactions may be required to address the contaminants effectively.

Scale and Size Limitations: Phytoremediation may have limitations in treating large-scale or deep ponds due to challenges in establishing and maintaining vegetation throughout the entire area. The scale of the pond, as well as logistical and practical constraints, should be considered during project planning to ensure realistic expectations and successful implementation.

Addressing these challenges and limitations requires a comprehensive understanding of site-specific conditions, careful planning, and regular evaluation of project performance. Phytoremediation should be approached as part of an integrated remediation strategy, considering other remediation techniques if necessary, to achieve the desired remediation goals in ponds contaminated with various pollutants.

10.8.2 Plant Stress and Mortality

Plant stress and mortality are significant challenges and limitations in phytoremediation projects. Several factors can contribute to plant stress and mortality, which can impact the effectiveness of phytoremediation. Here are some of the challenges and limitations related to plant stress and mortality in phytoremediation:

Contaminant Toxicity: High concentrations of contaminants in the pond can be toxic to plants, leading to stress and mortality. Some contaminants, such as heavy metals and certain organic pollutants, can have direct toxic effects on plant tissues, inhibiting their growth and survival. Plants may show symptoms of leaf chlorosis, wilting, stunted growth, or complete plant death in severe cases.

Nutrient Imbalance: Imbalances in nutrient availability can cause stress and affect plant health in phytoremediation projects. Excessive or insufficient nutrient levels can impact plant growth, nutrient uptake, and overall physiological functions. Nutrient deficiencies can make plants more vulnerable to stressors, while excessive nutrients can disrupt plant metabolism and cause toxicity.

Water Quality and Availability: Inadequate water quality and availability can lead to plant stress and mortality. Factors such as high salinity, waterlogged conditions, drought, or excessive water flow can negatively affect plant health. Plants may struggle to establish roots, uptake nutrients, and maintain proper water balance, leading to stress and eventual mortality.

Rejuvenation of Ponds through Phytoremediation

Climate and Environmental Factors: Climate and environmental conditions play a significant role in plant stress and mortality. Extreme temperatures, frost, high winds, and prolonged periods of drought or heavy rainfall can stress plants and increase mortality rates. Unsuitable climatic conditions can affect plant growth, nutrient uptake, and physiological processes, compromising their ability to effectively remediate contaminants.

Pathogens and Diseases: Plants in phytoremediation projects are susceptible to various pathogens, pests, and diseases. Pathogenic attacks can weaken plant defenses and make them more vulnerable to stressors. Fungal, bacterial, or viral infections can lead to plant decline, reduced biomass, and even plant death, limiting their capacity to remediate contaminants effectively (Keiblinger *et al.*, 2018).

Invasive Species Competition: Invasive plant species can outcompete and negatively impact the growth and survival of selected phytoremediation plants. Invasive species can quickly colonize the area, depriving target plants of resources and space. This competition can lead to stress, reduced growth, and potential mortality of the desired phytoremediation plants.

Genetic Variability: Genetic variability within plant populations can influence their tolerance to contaminants and overall stress resilience. Some plant individuals within a species may exhibit higher tolerance and better remediation capacity than others. However, limited genetic diversity and lack of suitable genotypes can reduce the overall success of phytoremediation projects.

Addressing plant stress and mortality requires careful selection of appropriate plant species and genotypes that are well-adapted to the site-specific conditions and contaminants. Adequate plant nutrition, irrigation, and maintenance practices are crucial to minimize stress and mortality risks. Monitoring plant health, early detection of stress symptoms, and prompt mitigation measures can help optimize plant survival and enhance phytoremediation performance. Additionally, diversifying plant species and incorporating native plant communities can enhance ecological resilience and reduce vulnerability to stressors.

10.8.3 Long-Term Effectiveness

Ensuring long-term effectiveness is a crucial challenge in phytoremediation projects. While phytoremediation offers sustainable and cost-effective remediation solutions, several factors can affect its long-term effectiveness. Here are some challenges and limitations associated with the long-term effectiveness of phytoremediation:

Contaminant Rebound: In some cases, contaminants can re-enter the system over time due to factors such as sediment re-suspension, leaching from surrounding areas, or continued pollution sources. This can lead to a gradual increase in contaminant concentrations, compromising the long-term effectiveness of phytoremediation. Regular monitoring and addressing potential sources of contamination are necessary to sustain remediation outcomes.

Plant Senescence and Decline: Over time, phytoremediation plants may experience senescence and decline, leading to reduced biomass, lower remediation capacity, and increased mortality. Factors such as aging, nutrient depletion, disease, or changes in environmental conditions can contribute to plant senescence. Incorporating strategies for plant rejuvenation, periodic replacement, or diversifying plant species can help maintain long-term effectiveness.

Slow Remediation Rates: Phytoremediation is often a slower process compared to other remediation techniques. It may take several years or even decades to achieve desired remediation goals, depending on contaminant concentrations, plant species, and environmental

222   Eco-Restoration of the Polluted Environment

conditions. Patience and ongoing commitment are required to ensure long-term effectiveness and sustained contaminant reduction.

Plant Succession and Ecological Dynamics: Phytoremediation systems may undergo natural plant succession and ecological changes over time. Early successional plant species used initially may give way to other plant species as the environment evolves. Changes in plant community composition can affect the overall remediation capacity and effectiveness. Understanding and managing ecological dynamics within the system is essential for long-term success.

Maintenance and Monitoring: Regular maintenance and monitoring are essential to sustain long-term effectiveness in phytoremediation projects. Monitoring contaminant levels, plant health, and overall system performance allows for timely adjustments and intervention, ensuring remediation goals are met over the long term. Adequate resources and commitment to maintenance and monitoring are critical for sustained effectiveness.

Financial and Institutional Support: Long-term effectiveness relies on consistent financial and institutional support. Adequate funding is necessary for ongoing monitoring, maintenance, and potential plant replacement. Institutional commitment, including stakeholder engagement and regulatory compliance, is crucial to ensure sustained efforts and the continued success of phytoremediation projects.

Addressing these challenges and limitations requires a holistic and adaptive approach. Incorporating monitoring and adaptive management strategies, regular assessment of remediation goals, and implementing necessary modifications are essential for sustaining long-term effectiveness. Additionally, integrating phytoremediation with other complementary techniques, such as physical or chemical remediation methods, can provide synergistic effects and enhance long-term remediation outcomes.

10.9 FUTURE DIRECTIONS AND RESEARCH NEEDS

10.9.1 INTEGRATING PHYTOREMEDIATION WITH OTHER RESTORATION TECHNIQUES

Integrating phytoremediation with other restoration techniques holds great potential for enhancing the overall effectiveness of remediation and ecosystem restoration efforts. Here are some future directions and research needs in this area:

Comprehensive Assessment of Combined Approaches: Future research should focus on systematically evaluating and comparing the performance of integrated approaches that combine phytoremediation with other restoration techniques, such as bioremediation, physical remediation, or engineered solutions. This includes assessing their synergistic effects, cost-effectiveness, and long-term sustainability in addressing complex contamination scenarios.

Ecological Interactions and Ecosystem Services: Understanding the ecological interactions and ecosystem services associated with the integration of phytoremediation and restoration techniques is essential. Research should explore the impact of combined approaches on biodiversity, soil quality, hydrological functions, carbon sequestration, and other ecosystem services. Assessing the resilience and stability of restored ecosystems under various stressors is also important.

Optimization of Combined Approaches: Future research should aim to optimize the integration of phytoremediation with other restoration techniques to achieve the best results. This includes investigating the optimal timing, sequencing, and spatial arrangement of different remediation and restoration activities. Research should also explore the most suitable plant species and combinations for specific contaminants and restoration goals.

Rejuvenation of Ponds through Phytoremediation

Monitoring and Long-Term Evaluation: Long-term monitoring and evaluation of integrated approaches are crucial to assess their performance over time. Research should focus on developing standardized monitoring protocols and indicators that capture both remediation and restoration outcomes. This will enable better comparisons between different projects and facilitate the identification of best practices.

Stakeholder Engagement and Decision-Making: Integrating phytoremediation with other restoration techniques often involves multiple stakeholders and decision-makers. Future research should explore effective stakeholder engagement strategies and decision-making processes to ensure successful project implementation and long-term support. Social and economic aspects, including cost-benefit analyses and community acceptance, should be integrated into the evaluation of integrated approaches.

Knowledge Exchange and Case Studies: Sharing knowledge and experiences through case studies and collaborative platforms is crucial for advancing the integration of phytoremediation with other restoration techniques. Future research should promote the dissemination of successful case studies, lessons learned, and best practices. Collaborative networks and platforms can facilitate knowledge exchange and foster innovation in the field.

Emerging Technologies and Techniques: Research should explore the integration of emerging technologies and techniques, such as molecular tools, nanoremediation, or smart monitoring systems, with phytoremediation and other restoration approaches. Investigating the potential synergies and challenges associated with these innovative approaches can contribute to the development of more efficient and effective integrated strategies.

Overall, future research should focus on understanding the ecological, technical, and socioeconomic aspects of integrating phytoremediation with other restoration techniques. This will help unlock the full potential of combined approaches for the remediation and restoration of contaminated environments, leading to more sustainable and resilient ecosystems.

10.9.2 Genetic Engineering and Plant Breeding for Enhanced Phytoremediation

Integrating phytoremediation with other restoration techniques holds great potential for enhancing the overall effectiveness of remediation and ecosystem restoration efforts. Here are some future directions and research needs in this area:

Comprehensive Assessment of Combined Approaches: Future research should focus on systematically evaluating and comparing the performance of integrated approaches that combine phytoremediation with other restoration techniques, such as bioremediation, physical remediation, or engineered solutions. This includes assessing their synergistic effects, cost-effectiveness, and long-term sustainability in addressing complex contamination scenarios.

Ecological Interactions and Ecosystem Services: Understanding the ecological interactions and ecosystem services associated with the integration of phytoremediation and restoration techniques is essential. Research should explore the impact of combined approaches on biodiversity, soil quality, hydrological functions, carbon sequestration, and other ecosystem services. Assessing the resilience and stability of restored ecosystems under various stressors is also important.

Optimization of Combined Approaches: Future research should aim to optimize the integration of phytoremediation with other restoration techniques to achieve the best results. This includes investigating the optimal timing, sequencing, and spatial arrangement of different remediation and restoration activities. Research should also explore the most suitable plant species and combinations for specific contaminants and restoration goals.

Monitoring and Long-Term Evaluation: Long-term monitoring and evaluation of integrated approaches are crucial to assess their performance over time. Research should focus on developing standardized monitoring protocols and indicators that capture both remediation and restoration outcomes. This will enable better comparisons between different projects and facilitate the identification of best practices.

Stakeholder Engagement and Decision-Making: Integrating phytoremediation with other restoration techniques often involves multiple stakeholders and decision-makers. Future research should explore effective stakeholder engagement strategies and decision-making processes to ensure successful project implementation and long-term support. Social and economic aspects, including cost-benefit analyses and community acceptance, should be integrated into the evaluation of integrated approaches.

Knowledge Exchange and Case Studies: Sharing knowledge and experiences through case studies and collaborative platforms is crucial for advancing the integration of phytoremediation with other restoration techniques. Future research should promote the dissemination of successful case studies, lessons learned, and best practices. Collaborative networks and platforms can facilitate knowledge exchange and foster innovation in the field.

Emerging Technologies and Techniques: Research should explore the integration of emerging technologies and techniques, such as molecular tools, nanoremediation, or smart monitoring systems, with phytoremediation and other restoration approaches. Investigating the potential synergies and challenges associated with these innovative approaches can contribute to the development of more efficient and effective integrated strategies.

Overall, future research should focus on understanding the ecological, technical, and socioeconomic aspects of integrating phytoremediation with other restoration techniques. This will help unlock the full potential of combined approaches for the remediation and restoration of contaminated environments, leading to more sustainable and resilient ecosystems.

10.9.3 Ecological and Socioeconomic Impacts of Phytoremediation

Understanding the ecological and socioeconomic impacts of phytoremediation is crucial for its successful implementation and wider adoption. Here are some future directions and research needs in this area:

Ecological Impact Assessment: Future research should focus on conducting comprehensive ecological impact assessments of phytoremediation projects. This includes evaluating the effects of phytoremediation on soil quality, water quality, biodiversity, and overall ecosystem functioning. Long-term monitoring studies are needed to assess the resilience and stability of ecosystems after phytoremediation and to identify any unintended ecological consequences.

Restoration of Ecosystem Services: Phytoremediation can have positive impacts on ecosystem services such as water purification, carbon sequestration, and habitat provision. Research should quantify and evaluate the restoration of these ecosystem services following phytoremediation efforts. Understanding the long-term benefits and trade-offs of phytoremediation for ecosystem services will help inform decision-making and promote the integration of phytoremediation into broader ecosystem management strategies.

Ecological Interactions and Succession: Investigating the ecological interactions and succession dynamics during and after phytoremediation is essential. Research should explore how phytoremediation plants interact with native plant communities, soil microorganisms, and other organisms within the ecosystem. Understanding the ecological processes and successional trajectories following phytoremediation will contribute to the development of more resilient and sustainable remediation strategies.

Rejuvenation of Ponds through Phytoremediation

Socioeconomic Assessment: Research should assess the socioeconomic impacts of phytoremediation, including its effects on local communities, economy, and human well-being. This involves evaluating the social acceptance, perceptions, and attitudes towards phytoremediation, as well as its economic feasibility and cost-effectiveness compared to other remediation methods. Incorporating socioeconomic factors into the assessment of phytoremediation projects will provide a more holistic understanding of their overall impact.

Stakeholder Engagement and Participation: Future research should explore effective strategies for stakeholder engagement and participation in phytoremediation projects. Engaging local communities, landowners, regulators, and other stakeholders from the early stages of project planning to implementation and monitoring can help address concerns, build trust, and ensure the integration of local knowledge and values. Participatory approaches can enhance the social acceptance and long-term success of phytoremediation initiatives.

Knowledge Transfer and Capacity Building: Promoting knowledge transfer and capacity building is crucial for the widespread adoption of phytoremediation. Future research should focus on developing educational resources, training programs, and outreach initiatives to enhance awareness and understanding of phytoremediation among scientists, practitioners, policymakers, and the general public. Building capacity in phytoremediation techniques and best practices will facilitate the effective implementation of projects and foster innovation in the field.

Policy and Regulatory Frameworks: Research should contribute to the development of appropriate policy and regulatory frameworks for phytoremediation. This includes evaluating existing regulations, identifying gaps and barriers, and providing evidence-based recommendations for the integration of phytoremediation into environmental policies and guidelines. Effective policy frameworks will support the adoption and mainstreaming of phytoremediation as a sustainable and cost-effective remediation approach.

By addressing these research needs, we can gain a deeper understanding of the ecological and socioeconomic impacts of phytoremediation. This knowledge will help optimize the implementation of phytoremediation projects, enhance ecosystem restoration, and ensure the sustainable and responsible use of this technology.

10.10 CONCLUSION

This chapter provided a comprehensive understanding of phytoremediation as a promising approach for pond rejuvenation and as an eco-friendly and sustainable technique for rejuvenating ponds and improving their water quality. By presenting case studies, discussing challenges, and highlighting the potential benefits, this chapter seeks to contribute to the growing body of knowledge on pond restoration and promote the adoption of phytoremediation as a viable option for environmental conservation. The case studies highlight the effectiveness of various phytoremediation techniques and the importance of plant selection and management strategies. Additionally, the chapter identifies the challenges and limitations of phytoremediation in pond ecosystems and suggests future research directions to improve the efficiency and sustainability of this approach. Overall, this contributes to the knowledge base on pond restoration and promotes the use of phytoremediation as a valuable tool in environmental management and conservation.

REFERENCES

Barakat, M. A. 2011. "New Trends in Removing Heavy Metals from Industrial Wastewater." *Arabian Journal of Chemistry* 4, no. 4: 361–77. https://doi.org/10.1016/j.arabjc.2010.07.019.

Bebianno, M. J., F. Géret, P. Hoarau, M. A. Serafim, M. R. Coelho, M. Gnassia-Barelli, and M. Roméo. 2004. "Biomarkers in Ruditapes decussatus: A Potential Bioindicator Species." In *Biomarkers* 9, no. 4–5: 305–30. https://doi.org/10.1080/13547500400017820.

Gaur, Nisha, Gagan Flora, Mahavir Yadav, and Archana Tiwari. 2014. "A Review with Recent Advancements on Bioremediation-Based Abolition of Heavy Metals." *Environmental Science: Processes & Impacts* 16, no. 2: 180–93. https://doi.org/10.1039/c3em00491k.

Greipsson, S. 2015. 2011. "Phytoremediation." *Nature Education Knowledge* 3, no. 10: 7. http://www.nature.com/scitable/knowledge/library/phytoremediation-. . . 1 of 5 1/1/2015: 11: 41 PM. Figure: 2.

Keiblinger, Katharina M., Martin Schneider, Markus Gorfer, Melanie Paumann, Evi Deltedesco, Harald Berger, Lisa Jöchlinger, Axel Mentler, Sophie Zechmeister-Boltenstern, Gerhard Soja, and Franz Zehetner. 2018. "Assessment of Cu Applications in Two Contrasting Soils—Effects on Soil Microbial Activity and the Fungal Community Structure." *Ecotoxicology* 27, no. 2: 217–33. https://doi.org/10.1007/s10646-017-1888-y.

Lajayer, Asgari, Behnam, Nader Khadem Moghadam, Mohammad Reza Maghsoodi, Mansour Ghorbanpour, and Khalil Kariman. 2019. "Phytoextraction of Heavy Metals from Contaminated Soil, Water and Atmosphere Using Ornamental Plants: Mechanisms and Efficiency Improvement Strategies." *Environmental Science & Pollution Research International* 26, no. 9: 8468–84. https://doi.org/10.1007/s11356-019-04241-y.

Lamichhane, J. R., E. Osdaghi, F. Behlau, J. Köhl, J. B. Jones, and J. N. Aubertot. 2018. "Thirteen Decades of Antimicrobial Copper Compounds Applied in Agriculture. A Review." *Agronomy for Sustainable Development* 38, no. 3. https://doi.org/10.1007/s13593-018-0503-9.

Licht, L. A., and J. G. Isebrands. 2014. "Linking Phytoremediated Pollutant Removal to Biomass Economic Opportunities. Biomass and for Phytoremediation of Contaminants." In *Emerging Technologies & Management of Crop Stress Tolerance* vol. 2: 449–70. Elsevier ISBN 9780128010877.

Malayeri, Behrouz E., Abdolkarim Chehregani, Nafiseh Yousefi, and Bahareh Lorestani. 2008. "Identification of the Hyper Accumulator Plants in Copper and Iron Mine in Iran." *Pakistan Journal of Biological Sciences* 11, no. 3: 490–2. https://doi.org/10.3923/pjbs.2008.490.492.

Marmiroli, N., and E. Maestri. 2008. "Trace Elements Contamination and Availability: Human Health Implications of Food Chain and Biofortification." In *Trace Elements as Contaminants and Nutrients: Consequences in Ecosystems and Human Health*: 23–53. New York: John Wiley & Sons Inc.

Pandey, Sandeep K., Ritambhara K. Upadhyay, Vineet Kumar Gupta, Kenate Worku, and Dheeraj Lamba. 2019. "Phytoremediation Potential of Macrophytes of Urban Waterbodies in Central India." *Journal of Health & Pollution* 9, no. 24: 191206. https://doi.org/10.5696/2156-9614-9.24.191206.

Peer, W. A., I. R. Baxter, E. L. Richards, J. L. Freeman, and A. S. Murphy. 2005. "Phytoremediation and Hyperaccumulator Plants." In *Topics in Current Genetics* 14: 299–340 ISBN 3540221751. https://doi.org/10.1007/4735_100.

Rathika, R., P. Srinivasan, Jawaher Alkahtani, L. A. Al-Humaid, Mona S. Alwahibi, R. Mythili, and T. Selvankumar. 2021. "Influence of Biochar and EDTA on Enhanced Phytoremediation of Lead Contaminated Soil by Brassica juncea." *Chemosphere* 271: 129513. https://doi.org/10.1016/j.chemosphere.2020.129513.

Rice, Kevin M., Ernest M. Walker, Miaozong Wu, Chris Gillette, and Eric R. Blough. 2014. "Environmental Mercury and Its Toxic Effects." *Journal of Preventive Medicine & Public Health* 47, no. 2: 74–83. https://doi.org/10.3961/jpmph.2014.47.2.74.

Saxena, Gaurav, Diane Purchase, Sikandar I. Mulla, Ganesh Dattatraya Saratale, and Ram Naresh Bharagava. 2020. "Phytoremediation of Heavy Metalcontaminated Sites: Eco-environmental Concerns, Field Studies, Sustainability Issues, and Future Prospects." In *Reviews of Environmental Contamination & Toxicology* 249: 71–131. https://doi.org/10.1007/398_2019_24.

Singh, S. 2012. "Phytoremediation: A Sustainable Alternative for Environmental Challenges." *International Journal of Green and Herbal Chemistry* 1: 133–9.

Thakur, Sveta, Lakhveer Singh, Zularisam Ab Wahid, Muhammad Faisal Siddiqui, Samson Mekbib Atnaw, and Mohd Fadhil Md. F. M. Din. 2016. "Plant-Driven Removal of Heavy Metals from Soil: Uptake, Translocation, Tolerance Mechanism, Challenges, and Future Perspectives." *Environmental Monitoring & Assessment* 188, no. 4: 206. https://doi.org/10.1007/s10661-016-5211-9.

Vinodhini, R., and M. Narayanan. 2008. "Bioaccumulation of Heavy Metals in Organs of Fresh Water Fish Cyprinus carpio (Common Carp)." *International Journal of Environmental Science & Technology* 5, no. 2: 179–82. https://doi.org/10.1007/BF03326011.

Vishnoi, S. R., and P. N. Srivastava. 2007. *Phytoremediation—Green for Environmental Clean*: Vol. 1016, 1021. Proceedings of the Proceedings of Taa l2007: The 12th World Lake Conference. Jaipur, Rajasthan, India.

11 Simple Techniques for Isolation and Characterisation of Bacteria with Potential for Degradation of DDT from Contaminated Soil

Murtala Ya'u

It is not a nuclear war that will destroy our planet. . . . No; it is a silent and ever-increasing accumulation of anthropogenic contaminants that will destroy the planet completely.

Murtala Ya'u

About 24% of all global deaths are linked to environmental degradation, which is roughly 13.7 million deaths a year.

World Health Organization

Biological systems are undoubtedly promising tools for combating the carnage induced by forces of environmental contamination.

Murtala Ya'u

Why is a whole chapter dedicated to just a set of protocols for isolation of bacteria with pesticide (DDT) degradation potential despite so much pre-existing literature on that? Is there a new approach that makes this chapter different from others? Of course, yes: it is the precision, simplicity and coherence of the information that make this chapter unique and captivating, especially for beginners who are intimidated by the sophistication in the narration of the vast majority of techniques presented in other works.

11.1 INTRODUCTION

Advances in modern agricultural practices, industrialisation and other anthropogenic activities have cumulatively heightened the accumulation of hazardous chemicals in the ecosystem and the subsequent biomagnification of these toxic contaminants in biological systems (Carvalho, 2006; Mansouri *et al.*, 2017). Different varieties of these toxic chemicals are commonly found in the environment (UNEP, 2002). The majority of them are the well-known pesticides, inorganic chemicals and other organic compounds used in automobiles and other industrial processes. Although pesticide production, for instance, is strictly regulated to reasonably minimise the negative impact on the environment and human health, excessive application may still raise serious health concerns in

DOI: 10.1201/9781003423393-11

227

humans and other biotic elements of the ecosystem. Substantial proportions of these pesticides are organochlorines which have been classified among the persistent organic pollutants (POPs) (UNEP, 2002). These environmental pollutants are of high concern based on their recalcitrance, toxicity and tendency to bioaccumulate. Some of them, such as dichlorodiphenyltrichloroethane (DDT), are semi-volatile, capable of undergoing long-range atmospheric transport, enabling their global distribution (UNEP, 2002).

DDT is among the most notorious persistent organic pollutants that are still drawing the interest of many researchers due to their degrading impact on ecology and human health. A population becomes exposed to DDT and its metabolites, dichlorodiphenyldichloroethylene (DDE) and dichlorodiphenyldichloroethane (DDD), via consumption of contaminated foods such fish, meat, dairy products and poultry and plant-based foods like fruits, vegetables, legumes and cereals (Abdul Kader, 2019; Hu et al., 2019; Mendes et al., 2019). Its toxic impact is more serious for the environment when there is co-contamination with metals. Despite the ban on the use of DDT by the Stockholm Convention (which is an international treaty signed in 2001 that became effective in 2004, aiming at a restriction of production and elimination of persistent organic pollutants) for its toxicity, DDT is, however, still in use in developing countries in Africa, Asia and the South Pacific (UNEP, 2002; WHO, 2011; UNEP, 2019). This has largely increased the global burden of DDT and its metabolites. Exposure to DDT is primarily implicated in mammalian endocrine disruption, muscular dysfunction, cognitive decline and probable involvement in the impairment of reproduction and triggering of carcinogenicity (Arrebola et al., 2015; Harada et al., 2016; Singh et al., 2016; Cohn et al., 2019; Truong et al., 2019). The recalcitrant nature of DDT (with a half-life of 20–30 years, depending on the soil topology) and its negative implications for the ecosystem necessitated an urgent need to clear this contaminant. Following from this, a large number of studies have tried to solve this problem. However, not much clearance of persistent DDT residues has been achieved (Beunink and Rehm, 1988; Aislabie et al., 1997; Bidlan et al., 2002; Bao et al., 2012).

It has been noted that a significant number of techniques applied for environmental clean-up of DDT and its metabolites are either partially effective or too expensive to manage (Sudharshan et al., 2012). This, however, necessitated the need to develop a subtle approach that involves effective and economically feasible waste remediation technologies using available microbial consortia that could be rapidly deployed in a range of physical settings (Yu et al., 2011; Xie et al., 2018).

One of the major advancements in environmental management in recent times is the deliberate employment of microorganisms for the clean-up of contaminated environments (Barragán-Huerta et al., 2007; Fang et al., 2010). Bacteria have gained much popularity as a convenient biotool for decontamination of the environment, especially pesticide-contaminated sites (Foght et al., 2001; Ahuja et al., 2003; Cao et al., 2012). Their involvement in cleaning up a wide range of toxic contaminants has proved more cost effective and reliable than most of the conventional methods of physicochemical treatments, such as volatilisation, adsorption and photolysis (Gao et al., 2011). Bioremediation of DDT and its metabolites needs an effective mineralisation process by efficient microbial consortia. The major requirements for effective bacterial DDT degradation include the presence of an adequate microbial population greater than 10^4 CFU/g of active degraders. Also, appropriate pH, temperature, nutrients, moisture content and a terminal electron acceptor (TEA) are essential for successful biodegradation of the DDT contaminants. Some bacterial species possess a pool of enzymes that degrade DDT and other organochlorine pesticides via several biological reactions, rendering them at least less toxic (Nadeau et al., 1998; Kantachote et al., 2003; Mwangi et al., 2010; Pant et al., 2013; Pan et al., 2016; Murtala et al., 2020b). This chapter aims to present a step-by-step approach for the isolation of DDT degraders from soil using some of the simpler techniques that can be successfully used even in developing countries with limited availability of equipment, especially for beginners in the area of bioremediation technologies.

Techniques for Isolation and Characterisation of Bacteria

11.2 CHOICE OF SOIL HORIZON FOR ISOLATION OF DDT-DEGRADING BACTERIA

In the isolation of a bacterial entity that can be used for the degradation of pesticides, particularly DDT from contaminated soil, a good choice of soil sampling procedure, particularly soil sample collection, is pertinent to the overall success of the isolation process. Some notable ingredients are essential to the success of soil sample collection. These ingredients include considering the spatio-vertical distribution of DDT in the soil horizons and the soil horizon that favours the highest microbial activity. Following are highlights of these ingredients.

11.2.1 SPATIO-VERTICAL DISTRIBUTION OF DDT IN THE SOIL HORIZONS

Regarding the spatial distribution of DDT in a vertical direction, the contaminated soil can be classified into three layers of horizons. These horizons markedly differ in their depth (usually in cm) and the DDT load.

11.2.1.1 Surface Horizon

The surface horizon mostly ranges between 0–15 cm vertically down the soil surface where the soil sample is to be collected. According to some literature, this horizon contains the least amount of the total DDT concentration relative to the other horizons. However, the surface horizon is known to favour microbial activity more than most of the other horizons due to its high contents of organic matter (humus) that support microbial growth.

11.2.1.2 Subsurface Horizon

This horizon occupies the second layer of the horizons, which ranges 15–30 cm downwards. It contains a higher sum of the total DDT concentration compared to the surface horizon. But its humus content decreases vertically, resulting in lower microbial activity relative to the surface horizon. The vertical increase in the DDT concentration in this horizon results from its ability to percolate downwards.

11.2.1.3 Deep Horizon

The deep horizon ranges 30–45 cm downwards, with most of the DDT percolating vertically into the horizon, making it the horizon with the highest DDT concentration. However, bacterial activity is very low in this horizon due to the decrease in substrate utilisation, resulting from decreased organic matter in the horizon.

11.2.1.4 Soil Horizon with Favourable Microbial Activity

A horizon with significant microbial activity is essential to the successful isolation of microbial entities. Favourable microbial activity is generally known to occur on the surface horizon. Therefore, soil sampling for the isolation of DDT degradation bacteria is recommended on this horizon, despite the higher concentrations of DDT in the deep horizons.

11.3 SOIL SAMPLE COLLECTION AND PROCESSING

As earlier recommended, a soil sample is normally collected at the surface horizon (0–15 cm) of the contaminated soil using a sterile spatula. The soil sample is mixed evenly, and 20 g is carefully put into a cold sterile container, preferably an icepack, and transported to the laboratory at 4°C for bacterial isolation. In cases where there is high contamination of DDT in a given area, the soil sample can be taken at a subsurface horizon, mainly at the depth of 20 cm.

There are two stages in the growth and isolation of bacteria with DDT degradation potential from a soil sample. The first stage involves the growth of different bacterial colonies from the soil sample using appropriate culture media (Murtala *et al.*, 2021). The second stage consists of direct isolation of a bacterial isolate with a DDT degradation capacity using a minimal sail medium (MSM) supplemented with a small quantity of DDT (Wang *et al.*, 2010, 2011; Pant *et al.*, 2013).

11.4 GROWTH OF BACTERIAL ISOLATES

11.4.1 Culture Media Choice and Conditions for the Growth of Bacteria from a Soil Sample

After a successful field sampling, a researcher has to prepare for the growth of the target bacteria using standard isolation protocols. The first thing that will come to the mind of the researcher is the culture medium that is good enough to grow the bacteria from the soil sample. Here, the researcher has two commonly used culture media at his or her disposal: soil extract medium (SEM) or Luria-Bertani (LB) medium.

11.4.2 Preparation of SEM for Bacterial Growth

A researcher can prepare SEM by mixing agar and an aqueous sampled soil fraction, that is, soil:distilled water by a ratio of 1:1 p/v. The researcher can simply obtain this aqueous soil fraction by dissolving 20 g of sampled soil into 20 mL of distilled water and mixing thoroughly, agitating for 1 hour and then decanting it overnight. The soluble fraction obtained is used to hydrate 1.5% agar, and the mixture is autoclaved and plated. However for convenience, LB medium is preferable since it allows for faster growth and precipitation of good growth yields. Additionally, LB medium permits extreme reproducibility of the physiological state of bacteria during steady-state growth, which is advantageous to the researcher. Therefore, in this chapter, elaboration on bacterial isolation using LB medium will be made in an easy-to-follow manner.

11.4.3 Preparation of LB Medium for Bacterial Growth

The researcher can easily prepare a litre of LB medium by dissolving 10.0 g Bacto Tryptone, 5.0 g yeast extract and 10.0 g NaCl in a litre of distilled water. The pH of the culture can be adjusted to 7.0 by adding 1 N NaOH and then autoclaving. This medium allows for the rapid growth of the isolates. Additionally, for growth on a solid surface, the media is simply supplemented with 15.0 g Bacto Agar. Table 11.1 shows a recipe for the preparation of 1 litre of LB medium.

11.4.4 Isolation of Bacteria from Soil Using LB Medium

Recall that LB medium was recommended earlier as the preferred medium for the growth of isolates from soil samples. Now the researcher can go ahead and isolate the bacteria following a simple protocol as follows: an air-dried soil sample (0.5 g) is suspended in 25 mL of the prepared LB

TABLE 11.1
One-Litre Recipe of LB Medium for the Growth of Soil Bacteria

Component	Amount (g)
Bacto Trypone	10.0
Yeast extract	5.0
NaCl	10.0
Distilled water	1000 (mL)

Techniques for Isolation and Characterisation of Bacteria

medium. The suspension is kept for 48 hours at 30°C on a shaker. After 48-hour incubation, turbidity is observed as an indication of bacterial growth. The cultured medium can be stored at a lower temperature before the preparation of a minimal salt medium. The next stage will be the isolation of bacteria with DDT degradation potential. As earlier stated, to isolate a bacterium that degrades DDT, the researcher must use a minimal salt enrichment medium in which a small quantity of DDT is supplemented.

11.5 ISOLATION OF DDT-DEGRADING BACTERIA

11.5.1 Preparation of DDT–Minimal Salt Enrichment Medium (DDT-MSM)

A minimal salt medium (MSM) is a selective medium that contains only salts, nitrogen and a preferentially supplemented carbon source. In the case of DDT-MSM, DDT is supplemented as the only carbon source, ensuring that only bacterial species that can degrade and utilise DDT as energy and carbon source can grow in the medium. In the laboratory, a researcher can prepare a litre of DDT minimal salt enrichment medium by dissolving into 1 litre of distilled water the following components: 0.1 g $CaCl_2.2H_2O$, 0.08 g $Ca(NO_3)_2.4H_2O$, 0.5 g $MgCl_2.6H_2O$, 1.0 g Na_2SO_4, 1.0 g KH_2PO_4 and 0.05 mg mL^{-1} DDT. Additionally, vitamins can be supplemented as may be required. Table 11.2 contains a summary of the components required for the preparation of DDT minimal salt enrichment medium.

11.5.2 Isolation of DDT-Degrading Isolates Using MSM

This is the final step in the isolation protocol. It is also the most important step that separates the target bacterium from the other bacterial species grown in the LB medium. Here the idea of survival based on fitness comes into play; only a bacterium that can utilise DDT can survive and grow. After preparing the minimal salt enrichment medium, the researcher is ready to go to the isolation of the desired bacterium. An aliquot (120 µL) from the cultured LB supernatant is pipetted into a tube containing 5 mL of DDT-MSM enrichment medium. The mixture is incubated for 1 week at 30°C on a rotary shaker at 100 rpm. After incubation, 100 µL of the bacterial suspension is transferred into 5 mL of fresh DDT-enriched MSM, and the incubation step is repeated. After five sequential sub-cultivations, the isolate is then inoculated onto MSM agar plates enriched with 0.05 mg mL^{-1} of DDT and incubated for 72 hours at 30°C, and the isolate formed is preserved. This ensures adequate exposure of the isolate to the DDT pesticide. At this stage, the isolate that adapts and grows in the DDT-MSM agar is ready to be harvested from the medium and identified, preferably at the genetic level. Therefore, in the next segment of this chapter, the focus will be on simple procedures for the genetic identification of a bacterial isolate.

TABLE 11.2
Summary of the Components of DDT Minimal Salt Enrichment Medium

Component	Amount (g)
$CaCl_2.2H_2O$	0.1
$Ca(NO_3)_2 4H_2O$	0.08
$MgCl_2.6H_2O$	0.5
Na_2SO_4	1.0
KH_2PO_4	1.0
DDT	0.05

11.6 GENETIC IDENTIFICATION OF THE BACTERIAL ISOLATE

There are many simpler identification techniques for bacterial identification such as morphological and biochemical techniques. But with the advancement of biotechnological methods, these techniques are considered to have low precision and less reliability (Stackebrandt and Goebel, 1994; Hong and farrence, 2015). That is why this chapter will only focus on genetic identification, which is more accurate and reliable. After successful isolation of the DDT-degrading bacterium, as mentioned earlier, this chapter will focus on some simple procedures for genetic identification of the isolate. This can be done in any molecular biology laboratory. Genetic identification of bacteria involves minimal but careful lab work in addition to bioinformatics. The series of work to be done includes extraction of bacterial genomic DNA, 16S ribosomal RNA gene (16S rRNA) amplification, Agarose gel purification and sequencing of amplified 16S rRNA gene, DNA alignment using BLAST search program and phylogenetic analysis. So, let's continue the journey.

11.6.1 SIMPLE PROCEDURE FOR THE EXTRACTION OF BACTERIAL DNA

To get enough bacterial DNA for the subsequent analyses, the isolate must be grown in a freshly prepared and autoclaved LB medium. The protocol involves inoculating the isolate into 8 mL LB medium and incubating overnight at 37°C and 200 rpm for 24 hours. The resulting bacterial suspension ($OD_{600} = 0.6$) is pelleted at 10,000 rpm for 5 min, and the genomic DNA is extracted as follows: the cell pellets are thawed on ice in 4.5 mL of 40 mM EDTA, 0.75 M sucrose and 50 mM Tris-HCl (pH 8.3). Lysozyme is then added to a final concentration of 1 mg mL⁻¹; the suspension is incubated at 37°C for 30 min, followed by the addition of 800 µg of proteinase K and sodium dodecyl sulphate (SDS) to a final concentration of 0.5% (w/v); and the mixture is incubated for an additional 2 hours at 37°C. The protein and polysaccharide complexes are removed by extraction with an equal volume of chloroform-isoamyl alcohol (24:1), followed by extraction with phenol-chloroform-isoamyl alcohol (50:49:1). The genomic DNA is recovered by the addition of 0.6 volumes of isopropanol and centrifugation for 10 min at 10,000 rpm, and the pellet is suspended to a concentration of between 50 and 100 µg mL⁻¹ in a Tris-EDTA buffer. The recovered DNA can be kept in a clean Eppendorf tube, ready for 16S rRNA amplification. It is also recommended to use a commercial DNA extraction kit from a reliable company.

11.6.2 PROTOCOL FOR 16S RIBOSOMAL RNA GENE AMPLIFICATION

This technique allows amplification of the desired gene into millions of copies from a smaller DNA copy. The technique involves a controlled polymerase chain reaction in which a template DNA is polymerised into many copies in a binary fashion. To amplify about 1.5 Kb gene from the bacterial DNA, 16S ribosomal RNA (16S rRNA) gene primers (BAC27F and BAC1492R) [16SRNA BAC27F: 5'-AGA GTT TGA TCC TGG CTC AAG-3' and 16SRNA BAC1492R: 5'- GGT TAC CTT GTT ACG ACT T-3'] are used. These universal primers can be purchased from any reliable chemical or biological company. The polymerase chain reaction (PCR) can be done using any reliable thermocycler. The total reaction volume is 15 µL, in which the reaction mix consists of 1 µL of the genomic DNA, 1.5 µL of 10X TaqA buffer, 0.5 µL of each of 10 µM forward and reverse primers, 0.75 µL of 1.25 mM of $MgCl_2$, 0.15 µL of 0.25 mM of dNTP and 0.12 µL of Taq DNA polymerase in ddH₂O. PCR conditions are set as follows: the initial melting temperature is 95°C for 5 min, 35 cycles each at a melting temperature of 94°C for 0.5 min. The annealing temperature is 52°C for 0.5 min and the extension at 72°C for 1 min. The final elongation is set for 10 min at 72°C (Martinez-Murcia et al., 1992; Goto et al., 2000).

After the final elongation cycle, the size of the DNA fragment is compared with a marker such as Hyper Ladder-1K. The 3 µL PCR product is mixed with 5XDNA loading buffer blue (1.5 µL) and loaded onto 1.5% agarose gel electrophoresis that has been stained with ethidium bromide.

Techniques for Isolation and Characterisation of Bacteria   233

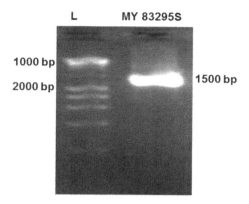

FIGURE 11.1 Gel electrophoretic image of the 16S rRNA gene from *Bacillus velezensis* strain MY 83295S. L represents Hyper Ladder-1K marker Bioline (Lot No: H4-q111B). This image is adapted from Murtala *et al.* (2021).

The electrophoresis is run for 35 min under 120 V and 300 mA current. The product is then visualised with any reliable gel documentation system. The presence of a product of the expected size is regarded as a positive result. For instance, Figure 11.1 shows a clear image of an electrophoretic gel of 16S rRNA gene from *Bacillus velezensis* strain MY 83295. Now the next step is a protocol for recovery of the desired DNA fragment from agarose gel. Then automated sequencing of the amplified 16S rRNA gene is next to be done, though the detailed sequencing protocol will not be discussed in this chapter.

11.6.3 Protocol for Agarose Gel Purification of Amplified 16S rRNA Gene

This technique ensures a recovery or isolation of the desired fragment of intact DNA from an agarose gel following agarose gel electrophoresis. It is recommended to use a prepared kit for gel purification that is readily available. However, the kit must be from a reliable biochemical company to ensure a good result. Whenever the kit is purchased, it is mostly accompanied by the manufacturer's instructions and protocol. The protocol is easy to follow, and the result is often remarkable. Here, there is a description of a protocol for using a PrepEase gel purification kit by following the manufacturer's protocol. About 100 mg of the fragment is excised from the gel using a clean, nuclease-free scalpel, and excess agarose is cut off to minimise the gel volume. It is weighed and transferred into a clean microcentrifuge tube. A 200 µL of Neutralize Tagment (NT) buffer is then added to solubilise the gel, and the mixture is incubated at 50°C until the gel is completely dissolved. To bind the DNA to the column, PrepEase clean-up column is then placed into a 2 mL collecting tube, and the sample is directly added to the centre of the clean-up column and then centrifuged at 11,000g for 1 minute and the flow-through is discarded. A 600 µl of NT3 buffer is then added directly to the clean-up column and then centrifuged at 11,000g for 1 min to wash the column. To dry the column and remove excess NT3 buffer, the column is further centrifuged at 11,000g for 2 minutes. Then incubation at 70°C for 3 minutes follows to completely remove the residual ethanol from NT3 buffer. The PrepEase clean-up column is then placed into a clean 1.5-mL microcentrifuge tube, and 25 µL of NE buffer pre-heated at 70°C is added. Then it is incubated at room temperature for 1 minute, and the DNA is eluted by centrifugation at 11,000 g for 1 minute. After successful elution of the DNA from agarose gel, the gel-purified DNA product undergoes nucleotide sequencing, normally using Sanger's sequencing method that, as mentioned earlier, will not be discussed in this chapter.

After a successful sequencing, DNA alignment is next in line. DNA alignment is a tool in bioinformatics that allows for the arrangement of sequences of DNA, RNA or protein to identify regions

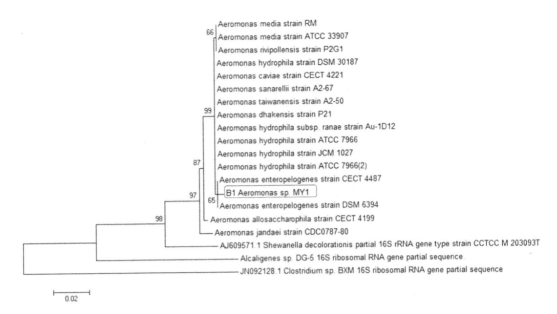

FIGURE 11.2 Evolutionary relationships of taxa of *Aeromonas* sp. strain MY1 isolate. The strain was shown in a box coded as B1 *Aeromonas* sp. MY1. Adapted from Murtala et al. (2020a).

of similarity that give a sense in terms of structural, functional or evolutionary relationships between the sequences. It is done using ClustalW version 2.0.12, available at http://www.clustal.org/, and the sequence is compared to sequences in the public databases with the BLAST search program on the National Center for Biotechnology Information (NCBI) website (http://www.ncbi.nlm.nih.gov/) to identify the isolate. The programs used for sequence alignment and BLAST search are user friendly. They provide a user with a set of instructions on how they can be managed. After the successful running of these programs, the genus of the isolate can be known to the researcher. However, more information on the evolutionary descent of the isolate is required. This information can be obtained from phylogenetic analysis, which is the next bioinformatics tool to be discussed in this chapter.

11.6.4 Phylogenetic Analysis of the Isolate

The root of phylogenetic information resides in the formation of a phylogenetic tree. A phylogenetic tree reveals evolutionary relationships among various biological species about their similarities and differences in their genetic characteristics. Phylogenetic or evolutionary history of the isolate is constructed using the neighbour-joining method. The evolutionary distances are computed using the maximum composite likelihood method. The phylogenetic tree is constructed using MEGA version 6 software program (Takahashi et al., 1999; Tamura et al., 2013). It is also user friendly, containing operational instructions for the researcher. Figure 11.2 illustrates a typical phylogenetic tree showing evolutionary relationships among a set of bacterial species that belong to the genus *Aeromonas* (Murtala et al., 2021).

11.7 GROWTH CHARACTERISATION OF AN ISOLATE IN DDT ENRICHMENT MEDIUM USING ONE-FACTOR-AT-A-TIME METHOD

Before closing this chapter, the researcher will find it beneficial to understand an easy method for optimisation of the isolate towards DDT degradation and subsequent growth in a DDT enrichment medium. The best and most simple experimental design of optimisation for beginners is monothetic

Techniques for Isolation and Characterisation of Bacteria

analysis, also called the one-factor-at-a-time (OFAT) method. This method involves testing factors or variables one at a time instead of multiple variables simultaneously. This method greatly minimises experimental errors compared to factor effects, which must be additive and independent of each other.

To characterise the growth of the isolate in DDT enrichment medium, optimum DDT concentration (as a sole carbon source), pH, temperature and incubation time are determined using one factor at a time. Turbidity of the medium is an index of growth of the isolate in DDT enrichment medium, which is determined spectrophotometrically as optical density (OD) at 600_{nm}.

Before inoculation, the isolate is pre-cultured in LB broth for 48 h and then centrifuged. The supernatant is discarded, and the cell pellets are washed with fresh MSM medium and used as an inoculum source. The growth of the isolate in DDT enrichment medium is determined *in vitro* after adjusting the optical density to 600_{nm} of cell density of the inoculum source to 0.6; the cells (100 µL) are then inoculated into MSM media containing varying concentrations DDT (10, 20, 30. 40 50, 60 and 70 mg L^{-1}) at various pH values (5.5, 6.0, 6.5, 7.0, 7.5, 8.0 and 8.5) and incubated at different temperatures (20, 25, 30, 35, 40 and 45°C) under shaking (150 rpm) at different incubation periods of 24, 48, 72, 96, 120, 144 and 168 hours. For the screening of optimum inoculum size, varying inoculum concentrations of 50, 100, 150, 200 and 250 µL are used. The experiments are conducted one factor at a time in triplicate. The best value for each factor determined is used for the subsequent determination of the next factor. For instance, the optimum temperature determined is used for the determination of an optimum pH, and so on. The OD is measured spectrophotometrically at 600_{nm} after blanking with freshly prepared MSM medium containing DDT for the determination of cell growth as an index of degradation capacity of DDT by the isolate (Murtala *et al.*, 2020b).

11.8 CONCLUSION

The researcher is now familiar with techniques and methodologies for isolation and basic characterisation of a bacterial isolate that can degrade DDT pesticide. He or she is equipped with techniques in soil sampling for bacterial isolation, preparation of media for isolation of bacteria from a soil sample, genetic identification of bacterial isolate and characterisation of bacteria with DDT degradation potential. These techniques can also be used to isolate and characterise soil bacteria capable of degrading the majority of pesticides and other chemical contaminants.

REFERENCES

Abdul Kader, M. 2019. "Domination of Pollutant Residues Among Food Products of South-East Asian Countries." *Acta Scientific Pharmaceutical Sciences* 3, no. 9: 75–9.

Ahuja, Rajiv, and Ashwani Kumar. 2003. "Metabolism of DDT [1,1,1-Trichloro-2,2-Bis(4-Chlorophenyl) Ethane] by *Alcaligenes denitrificans* ITRC-4 Under Aerobic and Anaerobic Conditions." *Current Microbiology* 46, no. 1: 65–9. https://doi.org/10.1007/s00284-002-3819-1.

Aislabie, J. M., N. K. Richards, and H. L. Boul. 1997. "Microbial Degradation of DDT and Its Residues-A Review." *New Zealand Journal of Agricultural Research* 40, no. 2: 269–82. https://doi.org/10.1080/002 88233.1997.9513247.

Arrebola, Juan P., Hidaya Belhassen, Francisco Artacho-Cordón, Ridha Ghali, Hayet Ghorbel, Hamouda Boussen, Francisco M. Perez-Carrascosa, José Expósito, Abderrazek Hedhili, and Nicolás Olea. 2015. "Risk of Female Breast Cancer and Serum Concentrations of Organochlorine Pesticides and Polychlorinated Biphenyls: A Case-Control Study in Tunisia." *Science of the Total Environment* 520: 106–13. https://doi.org/10.1016/j.scitotenv.2015.03.045.

Bao, Peng, Zheng-Yi Hu, Xin-Jun Wang, Jian Chen, Yu-Xin Ba, Jing Hua, Chun-You Zhu, Min Zhong, and Chun-Yan Wu. 2012. "Dechlorination of p,p'-DDTs Coupled with Sulfate Reduction by Novel Sulfate- Reducing Bacterium Clostridium sp. BXM." *Environmental Pollution* 162: 303–10. https://doi. org/10.1016/j.envpol.2011.11.037.

Barragán-Huerta, B. E., C. Costa-Pérez, J. Peralta-Cruz, J. Barrera-Cortés, F. Esparza-García, and R. Rodríguez-Vázquez. 2007. "Biodegradation of Organochlorine Pesticides by Bacteria Grown in Microniches of the

Porous Structure of Green Bean Coffee." *International Biodeterioration & Biodegradation* 59, no. 3: 239–44. https://doi.org/10.1016/j.ibiod.2006.11.001.

Beunink, J., and H. J. Rehm. 1988. "Synchronous Anaerobic and Aerobic Degradation of DDT by an Immobilized Mixed Culture System." *Applied Microbiology & Biotechnology* 29, no. 1: 72–80. https://doi.org/10.1007/BF00258354.

Bidlan, R., and H. K. Manonmani. 2002. "Aerobic Degradation of Dichlrodiphenyltrichloroethane (DDT) by *Serratia marcescens* DT-1P." *Process Biochemistry* 38, no. 1: 49–56. https://doi.org/10.1016/S0032-9592(02)00066-3.

Cao, Fang, Tong Xu Liu, Chun Yuan Wu, Fang Bai Li, Xiao Min Li, Huan Yun Yu, Hui Tong, and Man Jia Chen. 2012. "Enhanced Biotransformation of DDTs by an Iron- and Humic-Reducing Bacteria Aeromonas hydrophila HS01 upon Addition of Goethite and Anthraquinone-2,6-Disulphonic Disodium Salt (AQDS)." *Journal of Agricultural & Food Chemistry* 60, no. 45: 11238–44. https://doi.org/10.1021/jf303610w.

Carvalho, P. F. 2006. "Agriculture, Pesticides, Food Security and Food Safety. A Review." *Environmental Science & Policy* 9: 985–692.

Cohn, Barbara A., Piera M. Cirillo, and Mary Beth Terry. 2019. "DDT and Breast Cancer: Prospective Study of Induction Time and Susceptibility Windows." *Journal of the National Cancer Institute* 111, no. 8: 803–10. https://doi.org/10.1093/jnci/djy198.

Fang, Hua, Bin Dong, H. Yan, Feifan Tang, and Yunlong Yu. 2010. "Characterization of a Bacterial Strain Capable of Degrading DDT Congeners and Its Use in Bioremediation of Contaminated Soil." *Journal of Hazardous Materials* 184, no. 1–3: 281–9. https://doi.org/10.1016/j.jhazmat.2010.08.034.

Foght, J., T. April, K. Biggar, and J. Aislabie. 2001. "Bioremediation of DDT-Contaminated Soils: A Review." *Bioremediation Journal* 5, no. 3: 225–46. https://doi.org/10.1080/20018891079302.

Gao, B., W. Leu, W. B. Jia, L. J. Jia, L. Xu, and J. Xie. 2011. "Isolation and Characterization of an *Alcaligenes sp.* strain DG-5 Capable of Degrading DDTs Under Aerobic Conditions." *Journal of Environmental Science & Health. Part B* 46: 57–263.

Goto, Keiichi, Tomoko Omura, Yukihiko Hara, and Yoshito Sadaie. 2000. "Application of the Partial 16S rDNA Sequence as an Index for Rapid Identification of Species in the Genus Bacillus." *Journal of General & Applied Microbiology* 46, no. 1: 1–8. https://doi.org/10.2323/jgam.46.1.

Harada, Takanori, Makio Takeda, Sayuri Kojima, and Naruto Tomiyama. 2016. "Toxicity and Carcinogenicity of Dichlorodiphenyltrichloroethane (DDT)." *Toxicological Research* 32, no. 1: 21–33. https://doi.org/10.5487/TR.2016.32.1.021.

Hong, S., and C. E. Farrence. 2015. "Is It Essential to Sequence the Entire 16S RRNA Gene for Bacterial Identification?." *American Pharmaceutical Review* 18, no. 7: 1–7.

Hu, X., S. Li, P. Cirillo, N. Krigbaum, V. Tran, T. Ishikawa, M. A. La Merill, D. P. Jones, and B. Cohn. 2019. "Metabolome Wide Association Study of Serum DDT and DDE in Pregnancy and Early Postpartum." *Reproductive Toxicology*, pii: S0890–6238(18): 30588–4.

Kantachote, D., I. Singleton, N. McClure, R. Naidu, M. Megharaj, and B. D. Harch. 2003. "DDT Resistance and Transformation by Different Microbial Strains Isolated from DDT-Contaminated Soils and Compost Materials." *Compost Science & Utilization* 11, no. 4: 300–10. https://doi.org/10.1080/1065657X.2003.10702139.

Mansouri, Ahlem, Mickael Cregut, Chiraz Abbes, Marie-Jose Durand, Ahmed Landoulsi, and Gerald Thouand. 2017. "The Environmental Issues of DDT Pollution and Bioremediation: A Multidisciplinary Review." *Applied Biochemistry & Biotechnology* 181, no. 1: 309–39. https://doi.org/10.1007/s12010-016-2214-5.

Martinez-Murcia, A. J., S. Benlloch, and M. D. Collins. 1992. "Phylogenetic Interrelationships of Members of the Genera Aeromonas and Plesiomonas as Determined by 16S Ribosomal DNA Sequencing: Lack of Congruence with Results of DNA-DNA Hybridizations." *International Journal of Systematic Bacteriology* 42, no. 3: 412–21. https://doi.org/10.1099/00207713-42-3-412.

Mendes, R. A., M. O. Lima, R. J. A. de Deus, A. C. Medeiros, K. C. F. Faial, I. M. Jesus, K. R. F. Faial, and L. S. Santos. 2019. "Assessment of DDT and Mercury Levels in Fish and Sediments in the Iriri River, Brazil: Distribution and Ecological Risk." *Journal of Environmental Science & Health. Part B* 9: 1–10.

Murtala, Y., B. C. Nwanguma, and L. U. S. Ezeanyika. 2020a. "Characterization of a Novel *p,p'*-DDT Degrading Bacterium: *Aeromonas* sp. strain MY1." *Asian Journal of Biotechnology & Bioresource Technology* 6, no. 4: 12–22.

Murtala, Y., B. C. Nwanguma, and L. U. S. Ezeanyika. 2020b. "*Staphylococcus* sp. strain MY 83295F: A Potential *p,p'*-DDT-Degrading Bacterium Isolated from Pesticide Contaminated Soil." *Acta Biologica Marisiensis* 3, no. 2: 22–35. https://doi.org/10.2478/abmj-2020-0008.

Murtala, Y., B. C. Nwanguma, I. Bala, and L. U. S. Ezeanyika. 2021. "*Bacillus velezensis* strain MY 83295S: A *p,p*'-DDT-Degrader Isolated from a Tropical Irrigation Site." *Acta Biologica Turcica* 34, no. 2: 76–85.

Mwangi, K., H. I. Boga, A. W. Muigai, C. Kiiyukia, and M. K. Tsanuo. 2010. "Degradation of Dichlorodiphenyltrichloroethane (DDT) by Bacterial Isolates from Cultivated and Uncultivated Soil." *African Journal of Microbiology Research* 4, no. 3: 185–96.

Nadeau, L. J., G. S. Sayler, and J. C. Spain. 1998. "Oxidation of 1,1,1-Trichloro-2,2-Bis(4-Chlorophenyl) Ethane (DDT) by *Alcaligenes eutrophus* A5." *Archives of Microbiology* 171, no. 1: 44–9. https://doi.org/10.1007/s002030050676.

Pan, Xiong, Dunli Lin, Yuan Zheng, Qian Zhang, Yuanming Yin, Lin Cai, Hua Fang, and Yunlong Yu. 2016. "Biodegradation of DDT by *Stenotrophomonas* sp. DDT-1: Characterization and Genome Functional Analysis." *Scientific Reports* 6: 21332. https://doi.org/10.1038/srep21332.

Pant, G., S. K. Mistry, and G. Sibi. 2013. "Isolation, Identification and Characterization of p, p-DDT Degrading Bacteria from Soil." *Journal of Environmental Science & Technology* 6, no. 3: 130–7. https://doi.org/10.3923/jest.2013.130.137.

Singh, Z., J. Kaur, R. Kaur, and S. S. Hundal. 2016. "Toxic Effects of Organochlorine Pesticides: A Review." *American Journal of Bioscience* [Special Issue]: Recent Trends in Experimental Toxicology 4 3: 11–8. https://doi.org/10.11648/j.ajbio.s.2016040301.13.

Stackebrandt, E., and B. M. Goebel. 1994. "Taxonomic Note: A Place for DNA-DNA Reassociation and 16S RRNA Sequence Analysis in the Present Species Definition in Bacteriology." *International Journal of Systematic & Evolutionary Microbiology* 44, no. 4: 846–9. https://doi.org/10.1099/00207713-44-4-846.

Sudharshan, Simi, Ravi Naidu, Megharaj Mallavarapu, and Nanthi Bolan. 2012. "DDT Remediation in Contaminated Soils: A Review of Recent Studies." *Biodegradation* 23, no. 6: 851–63. https://doi.org/10.1007/s10532-012-9575-4.

Takahashi, T., I. Satoh, and N. Kikuchi. 1999. "Phylogenetic Relationships of 38 Taxa of the Genus *Staphylococcus* Based on 16S RRNA Gene Sequence Analysis." *International Journal of Systematic Bacteriology* 49, no. 2: 725–8. https://doi.org/10.1099/00207713-49-2-725.

Truong, Kim M., Gennady Cherednichenko, and Isaac N. Pessah. 2019. "Interactions of Dichlorodiphenyltrichloroethane (DDT) and Dichlorodiphenyldichloroethylene (DDE) with Skeletal Muscle Ryanodine Receptor Type 1." *Toxicological Sciences* 170, no. 2: 509–24. https://doi.org/10.1093/toxsci/kfz120.

Tamura, Koichiro, Glen Stecher, Daniel Peterson, Alan Filipski, and Sudhir Kumar. 2013. "MEGA6: Molecular Evolutionary Genetics Analysis Version 6.0." *Molecular Biology Evolution* 30, no. 12: 2725–9. https://doi.org/10.1093/molbev/mst197.

UNEP. 2002. *"UNEP/Chemicals/2002/9", Stockholm Convention on Persistent Organic Pollutants (POPs)*. Geneva: United Nations Environment Programme.

UNEP. 2019. *DDT Expert Group and Its Report on the Assessment of Scientific, Technical, Environmental and Economic Information on the Production and Use of DDT and Its Alternatives for Disease Vector Control*. Conference of the Parties to the Stockholm Convention on Persistent Organic Pollutants Ninth meeting, April 29, 2010, Geneva.

Wang, G. L., Meng Bi, Bin Liang, J. D. Jiang, and S. P. Li. 2011. "*Pseudoxanthomonas jiangsuensis* sp. nov. A DDT degrading Bacterium Isolated from a Long Term DDT Polluted Site." *Current Microbiology* 62, no. 6: 1760–6. https://doi.org/10.1007/s00284-011-9925-1.

Wang, Guangli, J. Zhang, L. Wang, Bin Liang, Kai Chen, Shunpeng Li, and Jiandong Jiang. 2010. "Co-metabolism of DDT by the Newly Isolated Bacterium, *Pseudoxanthomonas sp.* Wax." *Brazilian Journal of Microbiology* 41, no. 2: 431–8. https://doi.org/10.1590/S1517-83822010000200025.

WHO. 2011. *The Use of DDT in Malaria Vector Control: WHO Position Statement*. Geneva: World Health Organization. Accessed January 20, 2020. http://whqlibdoc.who.int/hq/2011/WHO_HTM_GMP_2011_eng.pdf.

Xie, Hui, Lusheng Zhu, and Jun Wang. 2018. "Combined Treatment of Contaminated Soil with a Bacterial *Stenotrophomonas* strain DXZ9 and Ryegrass (*Lolium perenne*) Enhances DDT and DDE Remediation." *Environmental Science & Pollution Research International* 25, no. 32: 31895–905. https://doi.org/10.1007/s11356-018-1236-7.

Yu, H. Y., L. S. Bao, Y. Liang, and E. J. Zeng. 2011. "Field Validation of Anaerobic Degradation Pathways for Dichlorodiphenyltrichloroethylene (DDT) and 13 Metabolites in Marine Sediment Cores from China." *Environmental Science & Technology* 45: 5245–52.

12 Bacterial Reduction of Molybdenum as a Tool for Its Bioremediation

Mohd Yunus Abd Shukor

12.1 INTRODUCTION

Right now, human actions are the greatest threat to our environment. Intensive industrialization, coupled with agriculture and urbanization, has caused various damage to the surrounding environment as the human population has grown. The difficulties are exacerbated by the overexploitation of natural resources and by humans' disregard for the rules of nature. Hydrocarbon and metal ion pollution have both increased over time all over the world (Kristanti *et al.*, 2012; Guo *et al.*, 2014; Abd Gafar and Shukor, 2018; Hanapiah *et al.*, 2018; Roy *et al.*, 2020; Shuhaimi *et al.*, 2021). There have been documented instances of metal and chemical toxicity, both acute and chronic, in high-exposure occupational and environmental contexts. Heavy metals are typically found almost everywhere. Anthropogenic activities have led to a dramatic rise in heavy metal levels since the pre-industrial era (Liu *et al.*, 2020; Allamin, 2021; Cheng *et al.*, 2021; Al-Saidi *et al.*, 2022; dos Reis Ferreira *et al.*, 2023) When heavy metal concentrations rise above a certain threshold, damage may be done to humans and wildlife. Metals such as cadmium, cobalt, chromium, lead, mercury, silver, copper, molybdenum, arsenic, nickel, and zinc are hazardous in both their fundamental and compound forms (Francisco *et al.*, 2002; Retamal-Morales *et al.*, 2018; Hofmann *et al.*, 2021). Due to their inert nature, metals accrue in the food chain and pose health and environmental risks. Heavy metal contamination is a chief cause of worry for public health around the world, making the need for its clean-up and removal all the more pressing.

Molybdenum is an essential co-factor for more than 50 enzymes; it is thus one of the most crucial trace elements (Zha *et al.*, 2013; Wu *et al.*, 2014). The metal is essential in animal and plant physiology; for instance, it acts as a catalyst for numerous redox transfer and hydroxylation reactions, all of which contribute to improved cellular function (Pandey and Singh, 2002). Molybdenum's widespread use in industry means that people are increasingly at risk of being exposed to its toxicity through things like ceramics, contact lens solution, glass, metallurgical processes, pigments, lubricants, electronic devices, color additives, catalysts, and cosmetics (Zha *et al.*, 2013; Pandey and Singh, 2002). The World Health Organization (WHO) recommends no more than 0.07 mg/L of molybdenum in drinking water; however, reports show that levels in the groundwater around mining regions have risen to as high as 0.5 mg/L (Geng *et al.*, 2014). Hypocuprosis and molybdenosis are conditions that can develop, especially in ruminants, after prolonged exposure to molybdenum in drinking water or in feed (Zha *et al.*, 2013).

The enigma of how microorganisms reduce molybdate to molybdenum blue, or Mo-blue, has persisted in the scientific community for almost a century. Enzymatic rather than abiotic reduction processes (Yong *et al.*, 1997) have been only recently demonstrated (Rahman *et al.*, 2013). Prior research on bacterial molybdate reduction centered on identifying strains with enhanced Mo-blue synthesis or molybdenum resistance. However, most contaminated sites have both organic and inorganic contaminants, making successful cleanup a difficult challenge. For the past five years, researchers have been trying to identify microbes that have the capacity to reduce and/or degrade

238 DOI: 10.1201/9781003423393-12

Bacterial Reduction of Molybdenum for Its Bioremediation

many contaminants at once, which would make them useful for cleaning up co-contaminated sites. Multiple molybdenum-reducing bacteria capable of degrading a wide range of toxic pollutants have been discovered in recent years. Solving the phenomenon of molybdenum (sodium molybdate) reduction to Mo-blue may require a deeper understanding of the inhibitory kinetics of reduction and the mechanism of reduction of molybdenum to Mo-blue. This is a major stage in the implementation of research findings.

12.1.1 MOLYBDENUM (MO)

As a metal, molybdenum is classified in the transition series group VI, period V. It has a melting point of 2623°C and a boiling point of 4639°C; its atomic number is 42, and its relative atomic mass is 95.94 g/mol. Due to its scarcity in nature, molybdenum is more commonly found in its sulfide (MoS_2), molybdate ($PbMoO_4$), and molybdate ($CaMoO_4$) forms. Molybdenite can be acquired through direct mining, which is a common practice. Additionally, the metal can be salvaged as a waste product during the copper mining process (Kosaka and Wakita, 1978; Nasernejad *et al.*, 1999; Battogtokh *et al.*, 2014). Molybdenum is put to extensive use in the industrial production of a wide variety of products, such as non-ferrous alloys, specific steels, contacts in electrical applications, spark plugs, X-ray tubes, glass-to-metal seals, and pigments, to name a few. Other products that benefit from the use of molybdenum include extrusion tools, spray coatings, glass melting furnace electrodes, metalworking tools, and spacecraft materials. As a lubricant, molybdenum disulfide is an appropriate component for lubricants because it possesses a number of desirable properties and is also a major source of pollution (Barceloux *et al.*, 1999; WHO, 2011).

A molybdate is an oxyanion molecule containing molybdenum at its sixth-highest oxidation state in inorganic chemistry. Molybdenum is capable of forming a wide variety of oxyanions, which can take the shape of either single- or multi-stranded polymeric structures. The bigger oxyanions belong to a class of compounds known as polyoxometalates, with isopolymetalates a special class because the latter include just one kind of metal atom, while heteropolymetalates involve the participation of anions that include silicate or tungstate. Isolated molybdenum oxyanions can be as small as MoO_4^{2-}, which is molybdate, or as large as 154 Mo atoms, as seen, for example, in isopoly-Mo-blues. Molybdenum deviates from the other group 6 elements in its behavior. Chromium, a relatively more toxic element, forms CrO^{2-}, Cr_2O7^{2-}, $Cr_3O_{10}^{2-}$, and $Cr_4O_{13}^{2-}$ ions. Tungsten, like molybdenum, produces several tungstates with 6-coordinate tungsten, making it chemically identical to molybdenum.

In terms of solubility, molybdenum trioxide (MoO_3) is only somewhat soluble in water, while metallic molybdenum (Mo), molybdenum chloride ($MoCl_5$), molybdenite (MoS_2), and calcium molybdate ($CaMoO_4$) are all totally insoluble (Barceloux *et al.*, 1999; Sebenik *et al.*, 2013). Molybdenum is a crucial trace element required mainly as a cofactor (Dixit *et al.*, 2015). There are over 50 enzymes in the nitrogen, sulfur, and carbon cycles that require molybdenum as a cofactor at the active site (Zha *et al.*, 2013; Wu *et al.*, 2014).

12.1.2 ANIMALS' MOLYBDENUM ENTRY PATHWAYS

Molybdenum and molybdenum compounds can enter the body through inhalation or through the digestive system. While ingestion results in practically instantaneous absorption, information about inhalation absorption efficiency and rate is currently lacking (Sebenik *et al.*, 2013). The rate of absorption after oral consumption is affected by factors like the compound's solubility and the nutritional makeup of the meal (Sebenik *et al.*, 2013). The bioavailability of molybdenum is conditional on the animal species and is also significantly impacted by the chemical form in which it exists (WHO, 2011; Wuana and Okieimen, 2011; Olaniran *et al.*, 2013; Stafford *et al.*, 2016; Kaplan *et al.*, 2013). Hexavalent molybdenum (Mo^{6+}) is easily absorbed after oral administration, and non-ruminant animals have a higher absorption capacity for it than ruminants do. Molybdenum

240   Eco-Restoration of the Polluted Environment

tetraoxide (MoO_4), on the other hand, is not easily absorbed (Miller *et al.*, 1972). Between 30 and 70% of molybdenum from food is absorbed in the human digestive system (Robinson *et al.*, 1973). There is a strong probability that molybdenum will get over the placental barrier because it enters the blood circulation and organs so rapidly. The buildup of molybdenum in human tissues has not been demonstrated with sufficient evidence; however, it is absorbed mostly by the liver, bones, and kidneys (Schroeder *et al.*, 1970; WHO, 2011). In terms of animal excretion, molybdenum compounds are primarily eliminated by urine in rats and, to a lesser extent, via feces (Pitt, 1976; WHO, 2011). However, because of poor absorption, molybdenum is typically excreted in both feces and urine in sheep, cattle, and horses (WHO, 2011; Pitt, 1976; Miller *et al.*, 1972), while molybdenum is largely balanced between ingestion and excretion in most non-ruminants, including humans (Schroeder *et al.*, 1970). Due to molybdenum's complex interaction with copper and sulfate, it has been suggested that animals experiencing a copper deficit are more vulnerable to molybdenum poisoning. It has been suggested that copper-deficient animals be provided with dietary sulfate to alleviate the negative effects of molybdenum poisoning, though the opposite is true for non-ruminants (Pitt, 1976; WHO, 2011).

12.1.3 MOLYBDENUM ACUTE AND CHRONIC TOXICITIES

The evaluation of the potentially hazardous consequences that molybdenum compounds have can be either acute (short-term) or chronic (long-term) exposure. It is still difficult to have a good understanding of the acute toxicity of molybdenum compounds, and there is only a limited quantity of evidence available on long-term toxicity in humans. Because the majority of chemicals are developed for a specific purpose and can have a range of effects depending on the organism they come into contact with, it is essential to test the toxicity of these chemicals on a wide variety of living things. Molybdenum compounds can be found in a broad diversity of different chemical arrangements, all of which need to be taken into consideration. A wide variety of animal models have been utilized in several studies that have reported on the harmful properties of molybdenum compounds. A serious effect of molybdenum toxicity is towards oogenesis and spermatogenesis in general, as well as toxicity towards ruminants in particular.

12.1.3.1 Toxicity of Molybdenum to Oogenesis and Spermatogenesis

In general, exposure to molybdenum shows adverse effects on spermatogenesis and oogenesis in various animal models. In one study, Long-Evans rats were fed a meal containing various doses of 5 to 20 mg/kg copper and molybdenum (0.1 to 14 mg/kg/day) for 13 weeks, causing the seminiferous tubules of the male rats' reproductive organs to degenerate, leading to pronounced sterility. Rats that received the two highest dosages also had more pronounced results. When comparing the weight gain of breastfeeding mothers and their puppies in the groups receiving lower doses to those groups receiving higher doses, those in the higher-dose groups showed less weight gain. Furthermore, 75% of males fed a diet containing 80 and 140 mg/Kg of molybdenum from weaning were sterile. Testicular degeneration was detected by the limited histology analysis. While molybdenum intake during pregnancy and fetal development was unaffected, there was some impairment with regular lactation (Jeter and Davis, 1954).

Degeneration of testicular shape and function has been described at oral gavage doses of sodium molybdate more than 30 mg/kg body weight. The relative weights of the epididymides, testes, ventral prostate, and seminal vesicles, as well as their absolute weights, were significantly reduced at higher treatment levels. Reduced sperm count and motility are symptoms of sperm abnormalities, which were also detected. Marker testicular enzyme activities, including lactate dehydrogenase (increases), sorbitol dehydrogenase (decreases), and gamma-glutamyl transpeptidase (increases), were significantly altered alongside histological changes in the testes. Molybdenum was also shown to accumulate in the scrotum, epididymis, and seminal vesicles. Molybdenum's potential to alter testicular histoarchitecture and sperm morphology was revealed by this study. Male-mediated developmental

Bacterial Reduction of Molybdenum for Its Bioremediation

toxic effects may be the result of alterations in the testicles and the spermatotoxic consequences (Pandey and Singh, 2002). Epididymal weight, morphologic abnormalities, count, sperm motility, and histological modifications were all significantly reduced in rats treated with 12 mg of tetrathiomolybdate per kg/day for up to 60 days (Lyubimov *et al.*, 2004). Testes from rabbits fed with 39 mg Mo/kg dry weight of carrots showed the degeneration of cells and seminiferous tubules, including several syncytial giant cells in addition to the regular spermatogenic cells (Bersényi *et al.*, 2008). Toxic effects on mouse testes (impotence and infertility) were observed at molybdenum concentrations above 100 mg/L, along with changes in malondialdehyde (MDA) and the enzymes superoxide dismutase and glutathione peroxidase levels (Zhai *et al.*, 2013). Molybdenum's ability to regulate the intricate processes of oxidative stress in the testes explains how it influences sperm quality.

12.1.3.2 Molybdenum Toxicity in Ruminants

Molybdenum concentrations of 10 mg/kg body weight promote copper deficiency in the tissue of ruminants, and sulfate in the diet makes the problem worse (WHO, 2011; Yamaguchi *et al.*, 2007). The intricate connection between copper, molybdenum, and sulfate metabolisms in regard to kinetics, bioavailability, and one of the most crucial biological connections between animal diet and disease is illustrated by copper's widespread use. However, this interdependence is still poorly understood. As a result of reactions with sulfides, various molybdenum compounds in ruminants generate thiomolybdate compounds, which sequester copper to make an insoluble compound that the body has a hard time absorbing. Because of this, a number of copper-dependent proteins cannot be synthesized or used effectively (Barceloux *et al.*, 1999; Stafford *et al.*, 2016; Pitt, 1976; Rajagopalan, 1988). As a result, the diminished copper bioavailability will lead to a secondary copper deficit. Excessive molybdenum consumption causes an increase in urinary copper loss. The use of tetrathiomolybdates may be useful in the treatment of chronic toxicity of copper. Due to insufficient formation of the copper chelant thiomolybdates by monogastric intestinal flora, nonruminants can develop a tolerance to molybdenum toxicity (Miller *et al.*, 1972; Pitt, 1976).

Impaired copper consumption and metabolism due to secondary copper insufficiency is a main symptom of molybdenum poisoning (molybdenosis) (Barceloux *et al.*, 1999; Haywood *et al.*, 2004; Majak *et al.*, 2004). A herd's difficulty with morbidity of up to 80% in cattle is a common indicator of this. Most of them have symptoms of scours, which include chronic diarrhea and green, liquid feces. In addition, black animals tend to have depigmentation of the hair coat due to poor tyrosinase activity and decreased melanin production (Silver and Phung, 1996). Joint discomfort, microcytic hypochromic anemia, concomitant osteoporosis and lameness have all been linked to a weakening in bone mineralization due to molybdenum and phosphorous competition that is very similar in size. Sheep, especially lambs younger than 30 days old, can exhibit swayback or enzootic ataxia manifested as the stiffness of the back and legs, causing difficulty in rising. In heifers, the same symptoms manifest as impaired milk output, fertility, and weight loss throughout puberty (Ward, 1978). Animals with aberrant connective tissue and growth typically begin showing symptoms within one to two weeks after chronic molybdenum exposure. When cattle or sheep experience acute poisoning, they may show signs of anorexia and lethargy for up to three days, with mortality beginning in the first week (WHO, 2011; Majak *et al.*, 2004).

12.1.4 Bioremediation

Bacterial-based environmental remediation has garnered a lot of consideration as a potentially game-changing technology in recent years, with several commercial bacterial-based metal remediation products such as AMT-Bioclaim and Bio-Fix in the market (Brierley *et al.*, 1986; Dayana *et al.*, 2013). The unique catabolic property of microbes is utilized for the elimination, reduction, or transformation of contaminants into less dangerous products such as water, carbon dioxide, microbial biomass, and inorganics (Elekwachi *et al.*, 2014). Microorganisms are not capable of degrading elemental toxicants such as heavy metals, but they can convert one toxic oxidation state to another less toxic state. Most microbes require heavy metals for vital metabolic functions, but in excess, these metal ions become toxic to the microorganisms. For adsorption capacity, the total microbial biomass and the geochemistry of the system need to be optimal. Several metal oxyanions do not participate in microbial adsorption, necessitating the focus of bioremoval on the conversion of the oxidation states from soluble into precipitable forms (Dixit *et al.*, 2015).

12.1.5 MOLYBDENUM POLLUTION

Its versatility has made molybdenum valuable in numerous products, including components of nickel-base alloys, lubricants in the form of molybdenum disulfide, components of additives, glass, paint composition, pigment composition, and electrical and electronic components. It is the effluents from these industries that form a major source of molybdenum pollution (Davis, 1991). Most naturally occurring waters have Mo contents below 10 µg/L (Smedley and Kinniburgh, 2017). Heavy metal contamination, particularly molybdenum, has reached alarming levels in the Black Sea and Tokyo Bay as a result of decades of intensive manufacturing, which has resulted in levels of molybdenum far higher than the background level mentioned previously (Davis, 1991). Contamination from periods of industrial discharge has led to molybdenum levels in soils feeding ruminants in Tyrol, Austria, that surpass as much as 200 parts per million, causing fatalities and scouring in cattle that eventually were remediated using microorganisms from sewage (Neunhäuserer *et al.*, 2001). Soil molybdenum pollution was the outcome of a light-bulb facility constructed in the early 1970s in China with molybdenum levels in the soils ranging from 0.25 to 252 mg/kg (Geng *et al.*, 2014). The air in Islamabad, Pakistan, has been discovered to have abnormally high levels of molybdenum and niobium. It was hypothesized that human activities were to blame for the increased concentrations of these metals (Qadir *et al.*, 2012). Molybdenum levels in soil can reach 11.7 mg/kg in parts of Jordan where oil shale is mined and farmed extensively. Exposure to molybdenum by ruminants grazing on these soils may cause undocumented diseases (Al Kuisi *et al.*, 2015). To better reflect the occurrence of molybdenum pollution globally, a Scopus keywords search focused on the topics of "molybdenum" AND "pollution" yielded 235 journal articles. VOSViewer software was utilized to generate the keyword co-occurrence of the documents, which is shown in Figure 12.1. The study's goal is to conduct a complete bibliometric evaluation of the study landscape of molybdenum pollution in soils and water bodies and its sources of pollution. Seven clusters were discovered, with China reporting many cases of molybdenum pollution, with mining and smelting as the main sources of pollution, and groundwater and river water pollution of molybdenum are reported more often (Figure 12.2). In recent years, molybdenum pollution is more reported in the countries of China, Saudi Arabia, and the United States than anywhere else, and studies are oriented towards the effect of molybdenum in agricultural soils (Figure 12.3).

Pollution from mining operations is a significant contributor to molybdenum pollution, along with that from manufacturing facilities. Molybdenum is the most abundant metal found in the mine effluent, which is likely because of the high pH values used in many molybdenum, copper, and uranium flotation mill processing plants. At these high pH values, molybdenum metal is highly soluble as molybdate ions. Many of the Red River's flora and fauna perished after being exposed to molybdenum and other harmful elements emitted from the Molycorp molybdenum mine, which exceeded the New Mexico Environmental Standard (Jacobs and Testa, 2014). Molybdenum was found in a tailings dam as a byproduct of copper mining at Iran's Miduk Copper Complex at elevated levels (Kargar *et al.*, 2011). Heavy metals including lead, copper, and molybdenum have poisoned approximately 300 square kilometers of land in Armenia, originating from the Alaverdi copper-molybdenum mine. The level of contamination is 20–40 times the maximum permissible levels in the soil close to the active mining area (Simeonov *et al.*, 2011). In Mo mining in northeast China, the effluent has polluted the Nver River. Several geochemical methods were employed to determine the molybdenum species and amounts of Mo in addition to copper, zinc, and iron concentrations. The amount of Mo exceeds about 1130 times the concentration of Mo normally found in noncontaminated soil, as determined by X-ray diffraction and X-ray fluorescence. This study demonstrates a

Bacterial Reduction of Molybdenum for Its Bioremediation

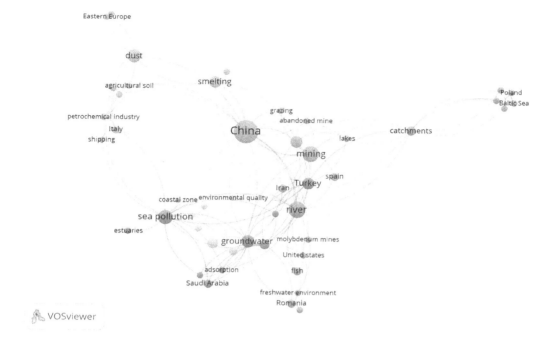

FIGURE 12.1 Density visualization of bibliometric keyword search results with the topics of "molybdenum" and "pollution".

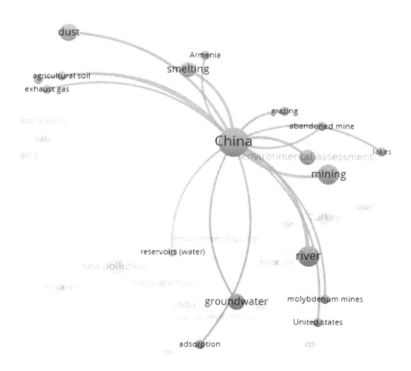

FIGURE 12.2 Density and network visualization of bibliometric keyword search results with the topics of "molybdenum" and "pollution" focusing on China.

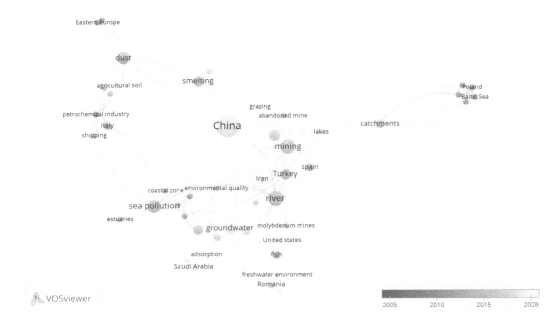

FIGURE 12.3 Overlay visualization of bibliometric keyword search results with the topics of "molybdenum" and "pollution".

possible link between Mo tailings ponds and the presence of large sedimental levels of these metals, suggesting that sedimentary molybdenum may constitute a significant threat to the local ecosystem (Simeonov et al., 2011).

Estimates of dissolved Mo concentrations of 900 ppm in the ground have been reported in an American uranium mill in southern Colorado, while concentrations of dissolved Mo as high as 25 ppm have been found in the effluent output of a molybdenum mine in Colorado (LeGendre and Runnells, 1975). Molybdenum concentrations of 20 mM (or about 2000 ppm) has been found in contaminated soils and water (Runnells et al., 1976). During the calving seasons of 2007 and 2008, 11 cow farmers in the dark brown soil zone of Saskatchewan experienced a total of 46 cow deaths. One of the leading causes of death was found to be molybdenum poisoning (Furber, 2009). Molybdenite is a byproduct of the gold and copper mining industries and can be used to showcase molybdenum. Leakage from mining sites has polluted thousands of acres of paddy land intermittently during the past few years (Yong, 2000), while in Malaysia, lubricating oil wastes with 1–5% Mo is below-the-radar pollution, with the highest molybdenum concentration found at 17.86 mg/Kg soil at one spot, and is a dominant and mostly ignored source of molybdenum contamination (Yakasai, 2017). In a more recent study, 25 surface sediments and 8 core sediments are sampled from the Taipu River of the Yangtze River delta in China to determine the total Mo and Mo in reducible, oxidizable, acid-extractable, and residual fractions. The typical Mo content in surface sediments was 0.58 mg/kg, with values varying from 0.22 to 3.47 mg/kg. The most contaminated samples had Mo in oxidizable concentrations of over 74%, while deep samples had a concentration of around 22%. Textile production, dyeing, and metal casting are suggested as possible polluting sources (Li et al., 2021).

12.1.6 MOLYBDENUM BIOREMEDIATION

Bioreduction, sequestration or bioaccumulation, bioprecipitation, biosorption, and efflux pumping are some of the techniques microorganisms use to immobilize and detoxify metal ions like Mo^{6+}, Cr^{6+}, Cu^{2+}, and Hg^{2+} (Francisco et al., 2002; Retamal-Morales et al., 2018; Lloyd, 2003; Shukor et al.,

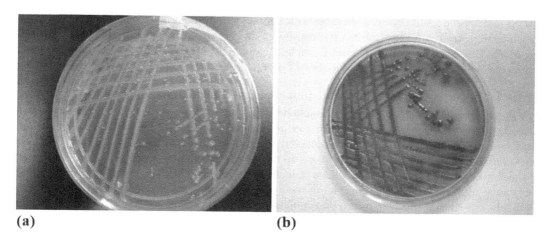

FIGURE 12.4 Molybdenum-reducing bacterium on low phosphate molybdate agar after 20 h (a) and 36 h (b).

2009; Halmi *et al.*, 2013; Othman *et al.*, 2013). Molybdate (Mo^{6+}) reduction by microorganisms to Mo-blue (Figure 12.4) was originally described in the common bacterium *E. coli* by Capaldi and Proskauer in the last century (Levine, 1925). However, it has only been during the last three decades that the specifics of the reduction phenomena have been revealed, thanks to the work of Campbell *et al.* in *E. coli* K12 (1985). The groundbreaking work of Sugio *et al.* (1988) on *Acidithiobacillus ferrooxidans* (previously known as *Thiobacillus ferrooxidans*) was then followed by work on the heterotrophic bacterium *Enterobacter cloacae* strain 48 (EC 48) (Ghani *et al.*, 1993) before numerous other Mo-reducing bacteria were characterized. Sulfide, a product of sulphate-reducing bacteria like *Desulfovibrio vulgaris* and *D. desulphuricans*, reduces molybdenum to MoS_2 (Mo^{4+}), a lower oxidation state that is highly insoluble. This is another potential route for molybdenum bioremediation where the insoluble molybdenum disulfide forms precipitates (Tucker *et al.*, 1997; Biswas *et al.*, 2009). Biosorption is another route. The plant *Posidonia oceanica* was among the first biosorbents utilized to remove molybdenum (Pennesi *et al.*, 2013). The fast-growing *Chlorella sorokiniana TU5* and *Picochlorum oklahomensis* show good efficiency in adsorbing molybdenum from an aqueous solution (Tambat *et al.*, 2023).

In a recent study, molybdenum from a spent hydrodesulfurization catalyst is adsorbed onto phosphazene chitosan derivatives as a remediation tool (Hala *et al.*, 2022). In another study, amendments of molybdenum-contaminated agricultural soils with biochar immobilized with nanoscale zero-valent iron (BC-nZVI), ferrous minerals (magnetite and ferrihydrite), and drinking water treatment residues (WTR) reduced the bioavailability of molybdenum significantly (Wang *et al.*, 2021). The first instance of molybdenum bioremediation was carried out in Tyrol, Austria. Large molybdenum-contaminated pastures were cleaned up with the help of a plant-microbe partnership. The contamination had resulted from industrial discharges. Sewage bacteria were used to convert the soluble hazardous molybdenum into insoluble molecules, thus detoxifying the molybdenum (Neunhäuserer *et al.*, 2001). The application of molybdenum-reducing bacteria in this scenario, and potential future applications of co-remediation of other toxicants that have recently been discovered (AbdEl-Mongy *et al.*, 2018; AbdEl-Mongy *et al.*, 2021; Huang *et al.*, 2019; Idris *et al.*, 2019; Kabir *et al.*, 2019; Mohammed *et al.*, 2019; Rusnam *et al.*, 2019; Rusnam and Gusmanizar, 2020, 2022; Alhassan *et al.*, 2020; Saeed *et al.*, 2020; Rusnam *et al.*, 2021, Rusnam and Gusmanizar, 2022; Abu Zeid *et al.*, 2021; Yakasai *et al.*, 2022; Xing *et al.*, 2023), represent a long-term solution to molybdenum contamination. More recently, phytoremediation has also been attempted, where *Vallisneria natans* (Lour.) Hara assisted by the bacterium *Serratia marcescens* A2 was able to reduce molybdenum concentration in an aqueous environment by 8.42% (Xing *et al.*, 2023). Aerobic granular sludge

(AGS) is a byproduct of sewage treatment with the potential to be used to adsorb molybdate from groundwater. At an initial pH of 2.0, the maximum adsorption capacity for molybdenum is 190.4 mg Mo/g. Molybdenum of up to 0.5 mg/L can be remedied using a granule dosage of 5 g/L to levels below the WHO-recommended threshold of 70 µg/L (Yuan et al., 2023).

12.1.6.1 Bacteria-Catalyzed Molybdenum (Sodium Molybdate) Reduction to Mo-Blue

The proposed mechanism of bacterial reduction of molybdenum to Mo-blue was first studied in *E. cloacae* strain 48 (Ghani *et al.*, 1993; Ariff *et al.*, 1997). In this proposed method, molybdenum's oxidation state was assumed to be first reduced from Mo^{6+} to Mo^{5+} via enzyme-catalyzed redox, and then phosphate came in to generate Mo-blue. In the work of Campbell *et al.* (1985), an observation of the similarity of the bacterial molybdate reduction spectrum and the ascorbate-reduced phosphomolybdate in the phosphate determination spectrum led the researchers to suggest the identity of the bacterial-reduced Mo-blue as a reduced phosophomolybdate (Shukor *et al.*, 2003). Since most heterotrophs undertake fermentation-producing organic acids under semi-aerobic circumstances—this is possible even in shaking mode at high cell densities and under static growth conditions—the acidic conditions promote phoshophomolybdate species formation at low phosphate concentrations (<5 mM). The medium's pH will drop, and molybdate ions will be transformed into phosphomolybdate as a result, which is the hallmark of molybdenum chemistry (Lee, 1977; Shukor *et al.*, 2007). Molybdenum exists as molybdate ions (MoO_2^{4+}) in an aqueous solution with a pH of 7. However, molybdate ions in an acidic solution will create a special complex structure known as polymolybdate anions. This characteristic is unique to molybdenum chemistry. After combining with additional anions such as SO_4^{2-}, SiO_4^{2-}, or PO_4^{2-}, the polymolybdate ions are transformed into a large complex (Figure 12.5) referred to as heteropolymolybdate (Lee, 1977; Sims, 1961). Molybdenum (sodium molybdate) reduction to Mo-blue is where this phenomenon is said to have initially appeared. The generated heteropolymolybdate ($PMo_{12}O^{3+}$, 12-molybdophosphate) can be

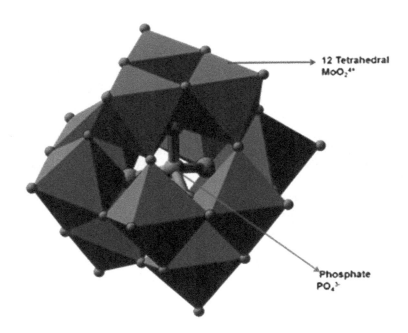

FIGURE 12.5 Structure of a typical 12-phosphomolybdate, derived from that of phosphotungstate. The image is under a Creative Commons license. The phosphotungstate structure, "Phosphotungstate-3D-polyhedra", image was created by Ben Mills (benjah-bmm27). This image is in the public domain. Source: https://en.wikipedia.org/wiki/File:Phosphotungstate-3D-polyhedra.png.

converted to molybdenum-blue by donating two electrons to it from reducing agents such as Fe^{2+} or Sn^{2+} ions and also dithionite. By using a thermally induced hopping mechanism, electron spin resonance demonstrates that the two contributed electrons are dispersed across the entire polymetallate sphere (ESR). According to nuclear magnetic spectroscopy, the valence of the 12 molybdenum atoms is determined by the steady motion of the two-electron reduction. The Keggin complex is responsible for the transition between 6+ and 5+ oxidation levels of molybdenum (Kazansky and Fedotov, 1980; Chae *et al.*, 1993).

Interestingly, the ascorbic acid-reduced phosphate colorimetric detection method is Mo-blue, which is very similar to the Mo-blue generated by numerous molybdenum-reducing bacteria in terms of their absorption spectra. Typically, the bacterial Mo-blue spectrum will have a peak at 865 nm and a characteristic shoulder at 700 nm, while the spectrum for Mo-blue resulting from phosphate measurement will give a typical peak at 890 nm and a lower absorbance or shoulder at 710 nm (Shukor *et al.*, 2007). According to a study, molybdate must come into touch with cells in order to be reduced to Mo-blue (Campbell *et al.*, 1985). Mo-blue is being speculated to be reduced phosphomolybdate. It is difficult to distinguish between the Mo-blue species formed during molybdenum (sodium molybdate) reduction to Mo-blue by bacteria, as there are several lacunary species of phosphomolybdate, and these species can vary in tandem with changes in pH. It has been proposed that spectral absorbance scanning is an effective way to separate out the various heteropolymolybdates. These include phosphomolybdate, sulphomolybdate, and silicomolybdate (Sims, 1961). Since the pH drops during the fermentation of most Mo-reducing bacteria and the ascorbic acid technique exhibits an absorption spectrum similar to Mo-blue, Shukor *et al.* (2007) suggest that phosphomolybdate is generated chemically as an intermediary step in the bacterial reduction of molybdenum (Figure 12.6). Apart from that, trapping *Enterobacter cloacae* 48 in a semipermeable membrane such as dialysis tubing (Shukor *et al.*, 2009) gives supporting evidence for Campbell's claim that cell-to-cell contact is necessary for Mo-blue formation in microbes (Campbell *et al.*, 1985; Shukor *et al.*, 2008).

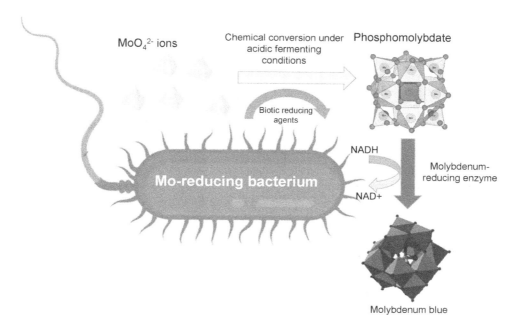

FIGURE 12.6 This simplified diagram shows the most likely pathway by which molybdate is converted to Mo-blue.

12.1.6.2 Molybdenum-Reducing Bacteria Characteristics

To date, at least 12 different bacterial genera (*Acinetobacter, Raoultella, Klebsiella, Burkholderia, Morganella, Bacillus, Escherichia, Enterobacter, Pseudomonas, Clostridium, Pantoea,* and *Serratia*) have been isolated from various soils and contaminated aquatic bodies globally that are capable of reducing molybdenum. Nearly 50% of the molybdenum-reducing bacteria are isolated from Malaysian soil, while the remainders are isolated from China, Nigeria, Egypt, Indonesia, Sudan, Pakistan, Iraq, the United States, and Antarctica (Table 12.1). *Bacillus subtilis* LM 4–2, a molybdenum-resistant bacterium from China, was discovered in a molybdenum mine (You *et al.*, 2015). However, Mo-blue reduction was not reported. It was reported that the genome of this molybdenum-resistant bacterium had been sequenced in its entirety; however, the characterization study was left out. As a result, the sequencing of this isolate presents a new challenge in the field of genome annotation of molybdenum-reducing bacteria. *Staphylococcus* spp. and *Bacillus* spp. are both Gram-positive rod and cocci and are two of the 12 genera presently acknowledged for molybdenum reduction. Most of the remaining genera are rod-shaped bacteria, and they are all Gram negative (Table 12.1). Eighty percent of the isolated reducers exhibited an optimum temperature of between 30 and 37°C, while 10% needed temperatures between 25 and 30°C (AbdEl-Mongy *et al.*, 2021; Sabo *et al.*, 2021; Yakasai *et al.*, 2021; Abu Zeid *et al.*, 2021), 5% between 15 and 20°C (Ahmad *et al.*, 2013), and 5% 40°C (Shukor *et al.*, 2010a, 2010b). Two of them are thermophilic bacteria identified in Egypt, while one is an Antarctic psychrotolerant bacterium (Table 12.1). Molybdate reduction is most favorable at pHs between 5 and 6 (Campbell *et al.*, 1985), 5.5 and 6.8 (Shukor *et al.*, 2008a, 2009, 2010; AbdEl-Mongy *et al.*, 2015; Masdor *et al.*, 2015; Khayat *et al.*, 2016; Sabullah *et al.*, 2016), 7 and 8 (Halmi *et al.*, 2013; Ghani *et al.*, 1993; Shukor *et al.*, 2008a; Rahman *et al.*, 2009; Yunus *et al.*, 2009; Lim *et al.*, 2012; Abo-Shakeer *et al.*, 2013; Othman *et al.*, 2015), and 6.5 and 7.5 (Ahmad *et al.*, 2013). For the most efficient synthesis of Mo-blue, 80% of Mo-reducing bacteria require simple sugar glucose as the best electron donor source, 20% require sucrose, and 5% require fructose (Table 12.1). Most Mo reducers depend on these conditions to obtain a readily assimilated carbon supply, which demonstrates that molybdenum reduction is a growth-connected endeavor. The general fact that phosphate concentrations above 5 mM impede molybdenum (sodium molybdate) reduction to Mo-blue suggests that the phosphate's physical contact with the phosphomolybdate substrate is responsible for this inhibition rather than an enzymatic one (Shukor *et al.*, 2008a). The ideal concentration of phosphate for Mo reducers to produce Mo-blue is between 2 and 7.5 mM (Table 12.1).

Optimal Mo-blue production may be sustained and tolerated at molybdenum (sodium molybdate) concentrations as high as 80 mM in a number of different molybdenum-reducing bacteria, including *Escherichia coli* K12 and *Klebsiella oxytoca* strain hkeem. The optimal reduction for most Mo-reducing bacteria occurs between 25 and 55 mM of sodium molybdate (Table 12.1). *Pseudomonas* sp., found in Nigerian soil, had the highest optimal concentration of molybdate at 100 mM (Table 12.1). In light of reports of molybdenum contamination reaching as high as 2000 ppm in the past, the fact that molybdenum reducers can now reduce and tolerate amounts exceeding 20 mM is encouraging (Runnells *et al.*, 1976). Metals such as arsenic, cadmium, silver, copper, chromium, mercury, zinc, and lead, as well as respiratory inhibitors, block practically all Mo reducers in a similar fashion (Table 12.1).

Nearly all molybdenum reducers have been described as vulnerable to the same hazardous heavy metals (Table 12.1). Toxic hexavalent chromate reduction to the less toxic trivalent state by *Bacillus* sp. (Elangovan *et al.*, 2006) and *Enterobacter cloacae* strain H01 (Rege *et al.*, 1997) are inhibited by mercury and copper. This is a problem since anionic metals like chromate, molybdate, and vanadate are difficult to clean up in the presence of these toxic cationic metals. These cationic metal ions, such as those found in metal reductases, bind to the sulfhydryl, phosphoryl, carboxyl, amine, and amide groups of numerous ubiquitous enzymes, rendering them inactive. To counteract this, bioremediation can begin with the addition of chelating or reacting substances that are safe for humans and animals, such as manganese oxide, phosphate, calcium carbonate, and magnesium

TABLE 12.1

Summary of the Characteristics of Various Molybdenum-Reducing Bacteria

Bacteria	Country of Origin	Unique Ability	Optimal pH and Temperature	Best Carbon Source	PoO$_4$ (mM)	MoO$_4$ (mM)	Heavy Metal Inhibition	Best Primary Model	Kinetics of Reduction	Optimization Method	Author
Raoultella ornithinolytica A1	Iraq	Psychrotolerant	pH 7.0 and 25°C	Glucose	n.a.	10	n.a.	n.a.		OFAT	(Xing *et al.*, 2023a)
Enterobacter aerogenes strain Amr-18	Egypt	Acrylamide can act as electron donor. Acrylamide, acetamide, and propionamide as N source for growth	6.3 and 6.8	Glucose	7.5	15–20	Ag$^+$, Cu^{2+}, Hg^{2+}	n.a.		OFAT	(Yakasai *et al.*, 2022)
Pseudomonas sp. strain Neni-4	Indonesia	Growth on phenol, benzoate, salicylic acid, and catechol	6.3 25 and 40°C	Glucose	5.0–7.5	15–20	n.a.	n.a.		OFAT	(Rushnam *et al.*, 2022)
Pseudomonas putida strain Neni-3	Indonesia	Decolorization of Congo Red	6–6.5 25 to 37°C	Glucose	2.5–7.5	10–15	n.a.	n.a.		OFAT	(Rusnam and Gusmanizar, 2022)
Serratia sp. strain Amr-4	Egypt	Growth on the pesticides carbamates, carbofuran, and carbaryl	6.0 and 6.8 and between 30 and 34°C	Glucose	2.5–7.5	20–30	Ag$^+$, Pb^{2+}, Cu^{2+}, Hg^{2+}	n.a.		OFAT	(AbdEl-Mongy *et al.*, 2021)
Escherichia coli strain Amr-13	Egypt	Growth on PEG 200, 300, and 600	5.5 and 8.0 30 and 37°C	Glucose	5	10–30	n.a.	Modified Gompertz		OFAT	(Shuhaimi *et al.*, 2021)
Bacillus sp. strain Zeid 15		Growth on acrylamide and propionamide as well as sources of electron donor for reduction	6.0 25 and 34°C	Glucose	2.5–5	15–20	Cu^{2+}, Hg^{2+}, Ag$^+$, Cr^{6+}	Modified Gompertz		OFAT	(Abu Zeid *et al.*, 2021)
Bacillus sp. strain Neni-8	Indonesia	Growth on various PEGs such as 200, 300, and 600	6.3 and 6.5, and between 30 and 37°C	Glucose	2.5–7.5	20–30	Cu^{2+}, Hg^{2+}, Ag$^+$, Cr^{6+}	Modified Gompertz		OFAT	(Rushnam *et al.*, 2021)
Bacillus amyloliquefaciens strain Neni-9	Indonesia	Growth on the pesticides carbaryl and carbofuran	pH 6.3 and 6.5, 30–37°C	Glucose	5.0–7.5	20–30	Ag$^+$, Cr^{6+}, Cu^{2+}, Hg^{2+}	1o model model using modified Gompertz		OFAT	(Rusnam and Gusmanizar, 2019)

(Continued)

TABLE 12.1 *(Continued)*

Summary of the Characteristics of Various Molybdenum-Reducing Bacteria

Bacteria	Country of Origin	Unique Ability	Optimal pH and Temperature	Best Carbon Source	PoO$_4$ (mM)	MoO$_4$ (mM)	Heavy Metal Inhibition	Best Primary Model	Kinetics of Reduction	Optimization Method	Author
Pseudomonas sp.	Nigeria		pH 6.5–7.5 37 ºC	Glucose	3.5–7.5	100	n.a.	n.a.		OFAT	(Alhassan *et al.*, 2020)
Pantoea sp. strain HMY-P4	Nigeria		pH 6.0–8.0 35–40 ˚C	Glucose	5.0	20–40	n.a.	Aiba	$q$max, Ks, and Ki of 0.89 µmol Mo-blue per h, 5.84 mM, and 32.23 mM, respectively	OFAT	(Idris *et al.*, 2019; Yakasai and Manogaran, 2020)
Enterobacter cloacae	Nigeria		pH 6.5–7.0 35–40°C	Glucose	5.0–7.5	80–100	n.a.	Monod	$q$max and Ks of 2.77 µmole Mo-blue hr^{-1} and 12.42 mM, respectively	OFAT	(Kabir *et al.*, 2019; Yakasai and Manogaran, 2020)
Morganella sp.	Nigeria		pH 6.0–7.5 35°C	Glucose	3.5	40	n.a.	Teissier-Edward	$q$max, Ks, and Ki of 7.77 mmole Mo-blue hr^{-1}, 26.63 mM, and 51.39 mM, respectively	OFAT	(Mohammed *et al.*, 2019)
Pseudomonas. strain Dr. Y Kertih	Malaysia	Growth on various xenobiotics: phenol, sodium dodecyl sulfate (SDS), acrylamide, acetamide, nicotinamide, propionamide, iodoacetamide, acetamide, and diesel	pH 6.0–6.3 25–40°C.	Glucose	5.0	20	Ag$^+$, Pb^{2+}, As^{5+}, Hg^{2+}	n.a.		OFAT	(Kesavan *et al.*, 2018)
Clostridium pasteurianum BC1 (USA)	USA	Metallic (Mo0) nanoparticles 5–20 nm in size Degradation of methyl orange	pH 6.8 n.a.	Peptone	1.74	20.67				OFAT	(Nordmeier *et al.*, 2017)

Organism	Country	Function	Conditions	Carbon source			Metals	Model	Method	Reference
Microbial electrolysis cell consortium (China)	China	Tungsten reduction and acetate biodegradation Hydrogen production	pH 3.0 22°C	Acetate	n.a.	1	n.a.	n.a.	OFAT	(Huang et al., 2019)
Raoultella ornithinolytica strain Mo1	Egypt		pH 6, 30°C	Glucose		20	n.a.	n.a.	OFAT	(Saeed et al., 2019)
Raoultella planticola strain MoI (Iraq)	Iraq		pH 6, 30°C	Glucose		20	n.a.	n.a.	OFAT	(Saeed et al., 2019)
Bacillus sonorensis strain Pharon3 (MK078035)	Egypt	Thermophilic bacterium	pH 7.07 52.2°C	Glucose	4.0	10		n.a.	RSM (CCD)	(Saeed et al., 2020)
Bacillus tequilensis strain Pharon2 (MK078034)	Egypt	Thermophilic bacterium	pH 7.02 46.1°C	Sucrose	4.0	10		n.a.	RSM (CCD)	(Saeed et al., 2020)
Bacillus sp. strain Neni-12	Indonesia	Growth on coumaphos	pH 6.3 25–37°C	Glucose	5.0	15–20	Ag^+, Cr^{6+}, Hg^{2+}	1o model, coumaphos growth model using modified Gompertz	OFAT	(Rushnam et al., 2019)
Pseudomonas sp.	Nigeria		pH 6.5–7.0 35- 40°C	Glucose	3.5	40–60	n.a.	n.a.	OFAT	(Gafasa et al., 2019)
Burkholderia vietnamiensis AQ5–12	Malaysia	Growth on glyphosate	pH 6.25–8.0 30–40°C	Glucose	5.0	40–60	n.a.	n.a.	OFAT	(Manogaran et al., 2018)
Burkholderia sp. AQ5–13	Malaysia	Growth on glyphosate	pH 6.25–8.0 35–40°C	Glucose	5.0	40–50	n.a.	n.a.	OFAT	(Manogaran et al., 2018)
Serratia marcescens strain KIK-1	Nigeria	Decolorizes various azo and triphenyl methane dyes	pH 5.8–6.5 34–37°C	Glucose	5.0	10–25	Ag^+, Cr^{6+}, Hg^{2+}, Cu^{2+}	n.a.	OFAT	(Karamba and Yakasai, 2018)
Pseudomonas putida strain Egypt-15	Egypt	Growth on PEG 4000	pH 6.5 34°C	Glucose	5.0	20	n.a.	1o model, PEG 4000 growth model using modified Gompertz	OFAT	(AbdEl-Mongy et al., 2018)
Bacillus amyloliquefaciens	Malaysia	Growth on SDS	pH 5.8–6.3 25–34°C	Glucose	5.0–7.5	30–50	Hg^{2+}, Cu^{2+}, Ag^+			(Maarof et al., 2018)

(Continued)

TABLE *12.1* (Continued)

Summary of the Characteristics of Various Molybdenum-Reducing Bacteria

Bacteria	Country of Origin	Unique Ability	Optimal pH and Temperature	Best Carbon Source	PoO$_4$ (mM)	MoO$_4$ (mM)	Heavy Metal Inhibition	Best Primary Model	Kinetics of Reduction	Optimization Method	Author
Serratia sp. strain HMY1	Nigeria	Growth on cyanide	pH 6.5–7.0 30–35°C	Sucrose	3.95	55	n.a.	1o model, model using modified Gompertz		RSM (CCD)	(Yakasai *et al.*, 2016, 2018, 2019)
Enterobacter sp. Strain Saw-2	Malaysia	Growth on phenol and catechol	pH 6.3–6.8 34–37°C	Glucose	5.0	15–30	n.a.	n.a.		OFAT	(Yakasai *et al.*, 2017b)
Serratia sp. strain HMY3	Nigeria	Growth on cyanide	pH 6.5 35°C	Sucrose	3.95	55–57.5	As^{3+}, Cr^{6+}, Hg^{2+}, Cu^2	Luong,	*qmax*, *Ks*, *Sm*, and *n* were 25.32 hr^{-1}, 113.4 mM, 55.43 mM, and 1.42, respectively	OFAT	(Mansur *et al.*, 2017)
Bacillus sp. strain Neni-10	Indonesia	Growth on dye Metanil Yellow	pH 6.3 34°C	Glucose	2.5–7.5	20	Ag$^+$, Cu^{2+}, Cr^{6+}, Hg^{2+}	1o model best model using Baranyi-Roberts		OFAT	(Mansur *et al.*, 2017; Yakasai and Manogaran, 2020)
Pseudomonas sp. strain 135	Malaysia	Growth on acrylamide, acetamide, and propionamide Acrylamide can support Mo-blue production	pH 6.0–6.3 25–40°C	Glucose	5.0–7.5	15–25	Ag$^+$, Cu^{2+}, Cd^{2+}, Hg^{2+}	n.a.		OFAT	(Yakasai *et al.*, 2017a)
Serratia marcescens strain DR.Y10	Malaysia	Growth on acrylamide, propionamide, and acetamide	pH 6.0–6.5 30–37°C	Glucose	5.0	10–30	Ag$^+$, Cu^{2+}, Cr^{6+}, Hg^{2+}	n.a.		OFAT	(Chee *et al.*, 2017)
Pseudomonas aeruginosa strain KIK-11	Malaysia	Growth on diesel and SDS	pH 5.8–6.0 25–34°C	Glucose	5.0–7.5	30–40	Ag$^+$, Cu^{2+}, Hg^{2+}	n.a.		OFAT	(Mohamad *et al.*, 2017)
Serratia sp. strain MIE2	Malaysia		pH 6.0 27 to 35oC	Sucrose	3.95	20	Hg^{2+}, Zn^{2+}, Cu^2	Teissier-Edwards	*qmax* and *Ks* and *Ki* 0.89 mmole Mo-blue hr^{-1}, 5.84 mM, and 32.23 mM, respectively	RSM (Box-Behnken and CCD)	(Halmi *et al.*, 2016; Aziz *et al.*, 2017)

Bacillus sp. strain khayat	Malaysia	Growth on SDS and diesel	pH 5.8-6.8 34°C	Glucose	5–7.5	10–20	Ag$^+$, As^{3+}, Pb^{2+}, Hg^{2+}, Cu^{2+}	n.a.	OFAT	(Khayat *et al.*, 2016)
Burkholderia sp. strain neni-11	Indonesia	Growth on acrylamide	pH 6.0–6.3 30–37°C	Glucose	5	15	Ag$^+$, Cr^{6+}, Hg^{2+}	1o model, model using modified Gompertz	OFAT	(Mansur *et al.*, 2016)
Enterobacter sp. strain Aft-3 (Pakistan)	Pakistan	Growth on azo dye	pH 5.8–6.5 37°C	Glucose	5	20–25	Ag$^+$, Cu^2, Hg^{2+}	n.a.	OFAT	(Shukor *et al.*, 2016)
Klebsiella oxytoca strain saw-5	Malaysia	Growth on glyphosate	pH 6.3–6.8 34°C	Glucose	5	20–30	Ag$^+$, Cd^{2+}, Cr^{6+}, Hg^{2+}, Cu^{2+}	n.a.	OFAT	(Sabullah *et al.*, 2016)
P. aeruginosa strainAmr-11	Egypt	Growth on phenol	pH 6.3–6.8 34°C	Glucose	2.5–7.5	20–30	Ag$^+$, As^{3+}, Pb^{2+}, Cd^{2+}, Cr^{6+}, Hg^{2+}, Cu^{2+}	n.a.	OFAT	(Ibrahim *et al.*, 2015)
Klebsiella oxytoca strain Aft-7	Pakistan	Growth on SDS	pH 5.8–6.3 25–34°C	Glucose	5–7.5	5–20	Ag$^+$, As^{3+}, Pb^{2+}, Cd^{2+}, Cr^{6+}, Hg^{2+}, Cu^{2+}	n.a.	OFAT	(Masdor *et al.*, 2015)
Enterobacter sp. strain Zeid-6 (Sudan)	Sudan	Growth on azo dye Orange G	pH 5.5–8.0 30–37°C	Glucose	5	20	Ag$^+$, Pb^{2+}, Hg^{2+}, Cu^{2+}	n.a.	OFAT	(Othman *et al.*, 2015)
Pseudomonas putida strain Amr-12	Egypt	Growth on phenol catechol	pH 6.0–7.0 20–30°C	Glucose	5.0–7.5	20–30	Ag$^+$, Cr^{6+}, Hg^{2+}	n.a.	OFAT	(AbdEl-Mongy *et al.*, 2015)
Enterobacter sp. Strain Neni-13	Indonesia	Growth on SDS	pH 6.0–6.5 37°C	Glucose	2.5–5.0	15	Ag$^+$, Cd^{2+}, Hg^{2+}, Cu^{2+}	1o model, SDS growth model using modified Gompertz	OFAT	(Rahman *et al.*, 2016)
Bacillus sp. strain Zeid 14	Sudan	Growth on amides and acetonitrile Acrylamide can support reduction	pH 6.0–6.8 25–34°C	Glucose	5.0–7.5	10–20	Ag$^+$, Cd^{2+}, Cr^{6+}, Hg^{2+}, Cu^{2+}	1o model, model using modified Gompertz	OFAT	(Adnan *et al.*, 2016)
Klebsiella oxytoca strain DRY14	Malaysia	Growth on SDS	pH 7.0 25°C	Glucose	5	25–30	Ag$^+$, Pb^{2+}, Cd^{2+}, Cr^{6+}, Hg^{2+}, Cu^{2+}	n.a.	OFAT	(Halmi *et al.*, 2013)

(Continued)

TABLE *12.1* *(Continued)*

Summary of the Characteristics of Various Molybdenum-Reducing Bacteria

Bacteria	Country of Origin	Unique Ability	Optimal pH and Temperature	Best Carbon Source	PoO$_4$ (mM)	MoO$_4$ (mM)	Heavy Metal Inhibition	Best Primary Model	Kinetics of Reduction	Optimization Method	Author
Bacillus pumilus strain lbna	Malaysia		pH 7.0–8.0 37°C	Glucose	2.5–5	40	As^{3+}, Pb^{2+}, Zn^{2+}, Cd^{2+}, Cr^{6+}, Hg^{2+}, Cu^{2+}	Luong,	$q$max, Ks, Sm, and n values of 27.3 μmol Mo-blue hr^{-1}, 115.8 mM, 57.83 mM, and 1.405, respectively	OFAT	(Abo-Shakeer *et al.*, 2017)
Bacillus sp. strain A.rzi	Malaysia		pH 7.3 28–30°C	Glucose	4	50	Cd^{2+}, Cr^{6+}, Cu^{2+}, Ag$^+$, Pb^{2+}, Hg^{2+}, Co^{2+}, Zn^{2+}	Luong	$q$max, Ks, Sm, and n values of 5.88 mole Mo-blue hr^{-1}, 70.36 mM, 108.22 mM, and 0.74, respectively	OFAT	(Othman *et al.*, 2013)
Pseudomonas sp. strain DRY1	Antarctica		pH 6.5–7.5 15–20°C	Glucose	5	30–50	Cd^{2+}, Cr^{6+}, Cu^{2+}, Ag$^+$, Pb^{2+}, Hg^{2+}	n.a.		OFAT	(Ahmad *et al.*, 2013)
Klebsiella oxytoca strain hkeem	Malaysia		pH 7.3 30°C	Fructose	4.5	80	Cu^{2+}, Ag$^+$, Hg^{2+}	n.a.			(Abo-Shakeer *et al.*, 2013)
Pseudomonas sp. strain DRY2	Malaysia		pH 6.0 40°C	Glucose	5	15–20	Cr^{6+}, Cu^{2+}, Pb^{2+}, Hg^{2+}	n.a.		OFAT	(Shukor *et al.*, 2010b)
Acinetobacter calcoaceticus strain Dr.Y12	Malaysia		pH 6.5 37°C	Glucose	5	20	Cd^{2+}, Cr^{6+}, Cu^{2+}, Pb^{2+}, Hg^{2+}	n.a.		OFAT	(Shukor *et al.*, 2010a)
Enterobacter sp. strain Dr.Y13	Malaysia		pH 6.5 37°C	Glucose	5	25–50	Cr^{6+}, Cd^{2+}, Cu^{2+}, Ag$^+$, Hg^{2+}	n.a.		OFAT	Shukor *et al.*, 2009)
S. marcescens strain Dr.Y9	Malaysia		pH 7.0 37°C	Sucrose	5	20	Cr^{6+}, Cu^{2+}, Ag$^+$, Hg^{2+}	n.a.		OFAT	(Yunus *et al.*, 2009)
Serratia sp. strain Dr.Y8	Malaysia		pH 6.0 37°C	Sucrose	5	50	Cr^{6+}, Cu^{2+}, Ag$^+$, Hg^{2+}	n.a.		OFAT	(Shukor *et al.*, 2008b)

5 *Serratia* sp. strain DrY5	Malaysia	The first purification of Mo-reducing enzyme	pH 7.0 37°C	Sucrose	5	30	Cu^{2+}	1o model, best model using Huang model	OFAT	(Rahman *et al.*, 2009; Shukor *et al.*, 2008b Shukor *et al.*, 2014; Halmi *et al.*, 2013; Syed *et al.*, 2020)
Serratia marcescens strain DRY6	Malaysia		pH 7.0 35°C	Sucrose	5	15–25	Cr^{6+}, Cu^{2+}, Hg^{2+}	n.a.	OFAT	(Shukor *et al.*, 2008b)
Enterobacter cloacae strain 48	Malaysia		pH 7.0 30°C	Sucrose	2.9	20	Cr^{6+}, Cu^{2+}	n.a.	OFAT	(Ghani *et al.*, 1993)
Escherichia coli K12	n.a.		pH 7.0 30–36°C	Glucose	5	80	Cr^{6+}	n.a.	OFAT	(Levine, 1925)

Note: OFAT= One factor at a time; CCD= Central composite design, RSM= Response surface method.

hydroxide (Hettiarachchi *et al.*, 2000; Deeb and Altalhi, 2009). Another option, tailored to molybdenum, is immobilization of the Mo-reducing bacterium in semipermeable membrane or dialysis tubing that has been found to mitigate the harmful effects of cationic metal ions' co-presence (Halmi *et al.*, 2014).

12.2 RSM VS OFAT APPROACH

The response surface method (RSM) is an example of a statistically based optimization methodology that combines statistical and mathematical methods to analytically construct a model. It's a common method used nowadays for making approximate models from real-world data (Box and Wilson, 1951; Montgomery and Runger, 1994; Sharma *et al.*, 2009). Artificial neural networks are a cutting-edge method. There are benefits and drawbacks to both RSM and ANN (Zin *et al.*, 2020). However, there are more reports of RSM, allowing for a more accurate comparison to previously reported RSM statistics. Box and Wilson in 1951 developed RSM. It employs a well-planned series of trials in the pursuit of an optimal solution (output variable). There are a number of factors that contribute to this outcome (input variables). *Serratia* sp. strain MIE2 is one example of a bacterium in which RSM using a central composite design (CCD) has been applied to maximize molybdenum reduction in *Serratia* sp. strain HMY1 and in several other optimization works, and it has been a great improvement over OFAT.

12.3 MATHEMATICAL MODELS OF THE KINETICS OF MOLYBDENUM REDUCTION

Many mechanisms associated with bacterial growth go through a distinctive stage. There is a lag time (μ) because the specific growth rate launches at zero and then speeds up to its maximum value (μmax) over some period of time. In the final stage of the growth curves, the rate decreases until it reaches zero, marking the arrival of an asymptote (A). A sigmoidal curve is typically obtained during bacterial reduction of molybdenum, with the lag phase typically occurring after $t = 0$ (Figure 12.7). After an initial phase of rapid growth, the next two stages are relatively stable, and the final stage is death (Zwietering *et al.*, 1990). It has been hypothesized that the sigmoid-shaped delay seen during storage occurs because microbial cells are readjusting their growth activity to fit with an unfamiliar

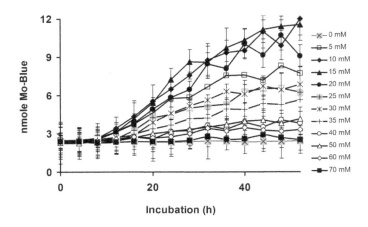

FIGURE 12.7 Production of Mo-blue at various concentrations of sodium molybdate as a function of time for *Serratia* sp. strain DRY5. The error bars are mean ± standard deviation ($n = 3$). Reprinted with permission from Syed *et al.* (2020), which is licensed under CC BY 4.0.

Bacterial Reduction of Molybdenum for Its Bioremediation

environment after a vegetative state. Throughout this time, two independent systems interact and adjust to one another; this is known as the lag period. Compared to a mechanical explanation, the inclusion of this lag time or parameter has been quite straightforward (Baranyi and Roberts, 1994). Each bacterial cell is assumed to develop at a unique pace based on its initial inoculum. The measurement of growth rates, as proposed by a number of authors, provides an example of a nonlinear distribution (Baranyi and Roberts, 1994; Buchanan *et al.*, 1997).

The value of q_m, or the maximum specific growth rate value, is sometimes used in secondary modeling exercises to represent the impacts of product, substrate, temperature, or pH on the bacterial growth rate (Zwietering *et al.*, 1990). Normally, growth values must be transformed logarithmically, and the exponential growth or expansion of the activity or organisms can be shown by the slope of the line (Fujikawa, 2010). The slope of the curve may also be determined through linear regression, which in turn can be calculated by manually calculating the parameter from the linear part of the curve. One of the most popular approaches is to translate sigmoidal data into a linear form using the natural logarithm (Othman *et al.*, 2013). Mo-blue production is associated with cellular increase or growth (Halmi *et al.*, 2013; Shukor *et al.*, 2009, 2010, 2008; Lim *et al.*, 2012; Halmi *et al.*, 2014; Mansur *et al.*, 2016; Shukor and Shukor, 2015; Gusmanizar *et al.*, 2016). Various primary modeling, for example, Gompertz (Zwietering *et al.*, 1990; Gompertz, 1825), logistic (Zwietering *et al.*, 1990;), Schnute (Zwietering *et al.*, 1990) Richards (Zwietering *et al.*, 1990; Richards, 1959), von Bertalanffy (Babák *et al.*, 2012; López *et al.*, 2004), Baranyi-Roberts (Baranyi *et al.*, 1995), Buchanan three-phase (Buchanan *et al.*, 1997), and Huang model, which is the most recent, (Huang, 2013) can be used to model Mo-blue production over time (Table 12.2). The maximum specific growth rate, normally used in bacterial growth and denoted by μ_m, needs to be converted to q_m, which stands for a maximum specific reduction rate. The best fit model governing the Mo-blue production profile is usually the modified Gompertz model in many cases (Table 12.2). An example of a typical modeling exercise to find the best model for fitting a particular Mo-blue production over time in bacteria is shown in Figure 12.8. The best model is then chosen based on error function analyses such as adjusted correlation coefficient (adj $R2$), root mean square error (RMSE), adjusted Akaike information criterion (AICc), Bayesian information criterion (BIC), and accuracy and bias factors (AF and BF) (Halmi *et al.*, 2014). This best model is then utilized to model all of the Mo-blue production curves at diverse sodium molybdate concentrations over time to obtain the valuable maximum specific reduction rate (Figure 12.9) for further secondary modelling.

The inhibitory effect of substrate (molybdate) at high concentrations on the reduction rate can be modeled once the values of the specific reduction rate at various concentrations of molybdate are acquired from the first modeling exercise. Mathematical modeling of the reduction process yields several crucial parameters, including the specific reduction rate, the theoretical maximum reduction, and the determination of the inhibition constant of reduction at high molybdenum concentrations, all of which have an impact on the lag period of reduction and substrate inhibition for the production of Mo-blue (Halmi *et al.*, 2014). In addition, secondary modeling can be applied to find the precise rate of reduction; models like Haldane's, Teissier's, Monod's, Yano's, Aiba's, and Luong's are good examples (Figure 12.10). Research into the biodegradation of organic compounds has made use of these models, but they are rarely used in the modeling of reduction kinetics concerning metal-reducing microbial activities. Researchers have reported a Haldane-type inhibition when using models to predict the reduction kinetics of mercury (Glusczak *et al.*, 2006), chromate (Sukumar, 2010), and arsenate (Soda *et al.*, 2006), whereby a Monod kinetic was reported (Truex *et al.*, 1997) for the reduction of uranium. Certain model kinetics also yielded useful information about the kinetics of bioremediation with respect to these heavy metals (Truex *et al.*, 1997). Hence, the substrate's inhibitory effect on the reduction rate (Monod) can be assessed using a secondary modeling activity (Haldane, Aiba, Yano, Teissier, and Luong). Error functions or statistical tests like corrected AICc, root-mean-square error, accuracy factor, bias factor, and

TABLE 12.2
Models to Fit the Mo-Blue Production Profile

Model	p	Equation	Best Model for Mo-Reducing Bacterium	Ref
Modified logistic	3	$y = \dfrac{A}{1 + exp\left[\dfrac{4q_m}{A}(\lambda - t) + 2\right]}$	*Bacillus* sp. strain Khayat	(Uba *et al.*, 2022)
Modified Gompertz	3	$y = A exp\left\{-exp\left[\dfrac{q_m^e}{A}(\lambda - t + 1)\right]\right\}$	*Bacillus amyloliquefaciens* strain Neni-9 *Serratia* sp. strain HMY1 *Bacillus* sp. strain Neni-12 *Bacillus* sp. strain Zeid 14 *Burkholderia* sp. strain neni-11	(Adnan *et al.*, 2016; Mansur *et al.*, 2016; Yakasai *et al.*, 2016; Rusnam and Gusmanizar, 2019 Rusnam and Gusmanizar, 2020)
Modified Richards	4	$y = A\left\{1 + v exp(1 + v) exp\left[\dfrac{q_m}{A}(1 + v)(1 + \dfrac{1}{v})(\lambda - t)\right]\right\}^{\left(\frac{-1}{v}\right)}$	nil	
Modified Schnute	4	$y = \left(q_m \dfrac{(1 - \beta)}{a}\right)\left[\dfrac{1 - \beta exp(a\lambda + 1 - \beta - at)}{1 - \beta}\right]^{\frac{1}{\beta}}$	nil	
Baranyi-Roberts	4	$y = A + q_m x + \dfrac{1}{q_m} ln(e^{-q_m x} + e^{-h_o} - e^{-q_m x - h_o})$ $-ln\left(1 + \dfrac{e^{q_m x + \frac{1}{q_m}ln(e^{-q_m x} + e^{-h_o} - e^{-q_m x - h_o})} - 1}{e(y_{max} - A)}\right)$	*Bacillus* sp. strain Neni-10	(Yakasai and Manogaran, 2020)
Von Bertalanffy	3	$y = K\left[1 - \left[1 - \left(\dfrac{A}{K}\right)^3\right] exp^{-\left(\frac{q_m x - \frac{1}{3}}{3K}\right)}\right]^3$	nil	

| Huang | 4 | $y = A + y_{max} - ln(e^A + (e^{y_{max}} - e^A)e^{-q_m B(x)})$ $$B(x) = x + \frac{1}{a} ln \frac{1 + e^{-a(x-\lambda)}}{1 + e^{a\lambda}}$$ | *Serratia* sp. strain DrY5 | (Syed *et al.*, 2020) |
| Buchanan three-phase linear model | 3 | $y = A$, if $x < $ lag $y = A + k(x-\lambda)$, if $\lambda \le x \ge x_{max}$ $y = y_{max}$, if $x \ge x_{max}$ | nil | |

Note:

q_m = maximum specific Mo-blue production rate

p = no. of parameters

A = Mo-blue lower asymptote

v = effects near which asymptote maximum Mo-blue production occurs

y_{max} = Mo-blue upper asymptote

λ = lag time

t = sampling time

e = exponent (2.718281828)

h_0 = a dimensionless parameter quantifying the initial physiological state of the reduction process

The lag time (h^{-1}) can be calculated as equal to h_0/q_{max}

α, β, k = curve-fitting parameters

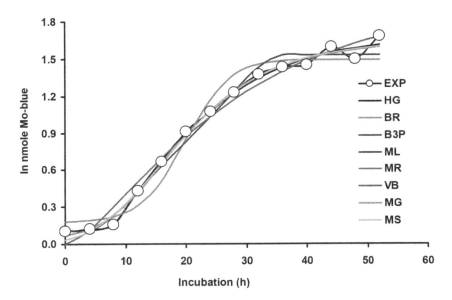

FIGURE 12.8 Mo-blue production as a function of time at 20 mM of sodium molybdate for *Serratia* sp. strain DRY5 fitted to various models. The models utilized were Huang (HG), von Bertalanffy (VB), Buchanan-three phase (B3P), modified Gompertz (MG), modified Richards (MR), Baranyi-Roberts (BR), modified logistics (ML), and modified Schnute (MS). Reprinted by permission from Hibiscus Publishers Enterprise. Reprinted with permission from Syed *et al.* (2020), which is licensed under CC BY 4.0.

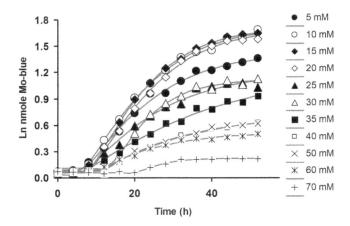

FIGURE 12.9 The profile of Mo-blue production by the bacterium *Serratia* sp. strain DRY5 at several concentrations of sodium molybdate fitted (red line) using the Huang model. Reprinted with permission from Syed *et al.* (2020), which is licensed under CC BY 4.0.

adjusted coefficient of determination (R^2) are frequently used to evaluate whether model is better (Halmi *et al.*, 2014).

It seems that more bacteria can have their molybdate inhibition on Mo reduction described by the Luong model than by any other model (Table 12.3). Unlike the more often discussed Haldane model, the concentration of substrate that is critical that completely inhibits Mo-blue formation can be calculated using the Luong and Teissier models.

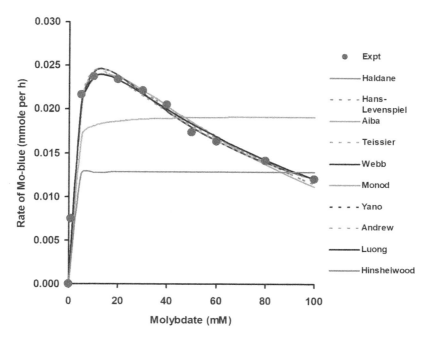

FIGURE 12.10 Effect of substrate concentration (sodium molybdate) to the rate of Mo-blue production (mmole per h) by *Serratia* sp. strain DRY5. Unpublished results.

12.4 CHARACTERIZATION OF PARTIALLY PURIFIED MOLYBDENUM-REDUCING ENZYMES FROM VARIOUS BACTERIA

E. cloacae strain 48 is the first organism studied in an effort to isolate and purify a Mo-reducing enzyme using a less efficient molybdate is utilized as a substrate (Ariff *et al.*, 1997). Partial purification using ammonium sulfate fractionation and followed by the Sephadex G-200 gel filtration is utilized, but only a mediocre outcome is achieved. It is shown that the partially purified enzyme is extremely sensitive to heat, losing 50% of its activity after only 10 minutes of incubation at 40°C and 100% after only 1 minute of incubation at 100°C. The optimal enzymatic activity occurs at 30°C and pH 8.0. Using a Lineweaver-Burk plot, it was seen that the rate of molybdate (Mo^{6+}) reduction rises with concentration and plateaus at around 100 mM. The Km (apparent) is 16.5 mM, and Vmax is 0.02 μmol/mL.h.

Using phosphomolybdate as an electron acceptor instead of molybdate, several purification steps of the enzyme from *E. cloacae* strain 48 are employed, such as ammonium sulfate (40–50%) fractionation, separation based on charge using an ion-exchange method on DEAE-cellulose, and separation based on size using a gel filtration or size exclusion method on Sephacryl S-200. A 6.5-fold enzyme purification was obtained at the ammonium sulfate fractionation step. After dialyzing the ammonium sulfate fraction and subjecting the dialyzed fraction to ion-exchange chromatography on DEAE-cellulose, a single peak was seen. The activity was significantly reduced throughout the gel filtration and ion-exchange processes, despite the excellent purification. At the final stage, the specific activity was 240 units/mg protein, but the initial crude only contributed 1.6%, indicating significant enzyme denaturation or loss during the processing steps. After undergoing partial purification using ion exchange and size exclusion or gel filtration chromatography, the final purification fold is 40. Three distinct protein bands with estimated molecular weights of 95, 100, and 105 kDa are obtained in the SDS-PAGE of the concentrated gel filtration step. This indicates that the purification of the enzyme is unsuccessful (Shukor *et al.*, 2003).

TABLE 12.3
Mathematical Models for Reduction Kinetics Involving Substrate Inhibition

Author	p	First Reported by	Reduction Rate	Best Model for Mo-Reducing Bacterium	Ref
Monod	2	(Monod, 1949)	$q_{max}\dfrac{S}{K_s+S}$	*Enterobacter cloacae*	(Yakasai and Manogaran, 2020; Syed *et al.*, 2020)
Haldane	3	(Boon and Laudelout, 1962)	$q_{max}\dfrac{S}{S+K_s+\dfrac{S^2}{K_i}}$	Nil	
Teissier-Edward	3	(Teissier, 1942)	$q_{max}\left(1-\exp\left(-\dfrac{S}{K_i}\right)-\exp\left(\dfrac{S}{K_s}\right)\right)$	*Morganella* sp.	(Mohammed *et al.*, 2019)
Aiba	4	(Aiba *et al.*, 1968)	$\mu_{max}\dfrac{S}{K_s+S}\exp\left(\dfrac{-S}{K_i}\right)$	*Pantoea* sp. strain HMY-P4	(Yakasai, 2017)
Yano and Koga	4	(Yano and Koga, 1969)	$\dfrac{q_{max}S}{S+K_s+\left(\dfrac{S^2}{K_1}\right)\left(1+\dfrac{S}{K}\right)}$	Nil	
Han and Levenspiel	5	(Han and Levenspiel, 1988)	$q\max\left[1-\left(\dfrac{S}{S_m}\right)^m\right]\dfrac{S}{S+K_s\left(1-\dfrac{S}{S_m}\right)^m}$	Nil	
Luong	4	(Luong, 1987)	$q\max\dfrac{S}{S+K_s}\left[1-\left(\dfrac{S}{S_m}\right)^n\right]$	*Bacillus pumilus* strain Ibna; *Bacillus* sp. strain A.rzi; *Serratia* sp. strain HMY3	(Othman *et al.*, 2013; Abo-Shakeer *et al.*, 2013; Yakasai, 2017)

Note:

q_{max} = maximal reduction rate (h^{-1})

S_m = maximal concentration of substrate tolerated and (mM)

K_s = half saturation constant for maximal reduction (mM)

m, n, K = curve parameters

p = product concentration (mM)

S = substrate concentration (mM)

Bacterial Reduction of Molybdenum for Its Bioremediation

Finally, the bacterium's molybdenum-reducing enzyme has been purified for the first time from *Serratia* sp. Dr. Y5 (Shukor *et al.*, 2014). The purification step involves the use of a stronger ion exchanger Mono Q, with a NaCl gradient set up from 0 to 500 mM NaCl. The protein with the activity peak was obtained at 330 mM. This was then pooled, concentrated, and further purified using gel filtration on an Agilent Zorbax GF-250 column with the aid of an HPLC (Agilent 1100 series). The isolated enzyme has been found to be monomeric, with an estimated molecular weight of 100 kDa (Shukor *et al.*, 2014).

The purified fraction of strain Dr Y5 has been characterized, its optimal pH is found to be 6, and its temperature stability is found to be between 25 and 35°C (Shukor *et al.*, 2014). The Dr.Y5 Mo-reducing enzyme was also discovered to have a catalytic efficiency (k_{cat}/K_m) of 5.47 $M^{-1}s^{-1}$, which is roughly 2×10^5 less efficient than the comparable metal reductase, chromate reductase from *Thermus scotoductus* (Opperman *et al.*, 2008).

Serratia sp. strain MIE molybdenum reductase was also successfully purified (Halmi, 2012) using gel filtration from a 40–50% ammonium sulfate fraction. The fold of purification is 20. The enzyme is monomeric, with a molecular weight of 100 kDa, and its optimal pH and temperature for activity are 5.0 and 35°C. The catalytic efficiency, which is defined as the ratio of K_{cat}/K_m is 7.89. The Mo-reducing enzyme was most recently isolated from the bacteria *Serratia* sp. strain UPM-FR1, which also degrades cyanide (Yakasai, 2017). With a final fold of 5.69, purification is lower than previous enzymes from the previous Mo-reducing bacteria. A 50–60% cut of an ammonium sulphate fraction is utilized instead, and this is followed by gel filtration on a Zorbax GF-250 column. In keeping with previous studies, this enzyme is also 100 kDa in size and has a monomeric structure. For maximum activity, it requires a pH of 5.0 and between 25 and 35°C. The catalytic efficiency (K_{cat}/K_m) is 5.35 $M^{-1}.s^{-1}$.

According to the research of Opperman *et al.*, 2008, the purified chromate reductase identity in *Thermus scotoductus* SA-01 has been discovered to be a previously known enzyme identified as the Old Yellow enzyme that incidentally exhibits novel chromate-reducing activity. Because of this, chromate reductase has not been assigned an E.C. number by the enzyme commission. Only selenate reductase (Schröder *et al.*, 1997) and mercury reductase have been shown to be true unique enzymes (Freedman *et al.*, 2012). Therefore, it is critical to sequence the purified enzyme in order to identify the Mo-reducing enzyme, which to date has not been successful due to the very low amount of enzyme obtained.

12.5 MOLYBDENUM REDUCTION LOCATION CHARACTERIZATION WITH ELECTRON TRANSPORT CHAIN INHIBITORS

The true identity of the Mo-reducing enzyme has been historically pinpointed as a component of the electron transport system or chain (Ghani *et al.*, 1993). The chain is made up of several different parts, including soluble enzymes, heme-proteins, Fe-S clusters, coenzymes, and a wide variety of metal ion cofactors. ATPase is the fifth and final complex of the chain, which is composed of complexes I, II, III, and IV (complex V) (Shukor, 2014). Amytal, antimycin A, rotenone, sodium azide, carbon monoxide, and cyanides are only a few of the many chemical agents that are known to inhibit respiratory enzymes. The enzyme NADH dehydrogenase is inhibited by rotenone, while cytochrome oxidase is inhibited by sodium azide and cyanide. Although other cytochrome b inhibitors such as antimycin A and hydroxyquinoline-N-oxide were explored, cyanide was found to be the sole inhibitor capable of stopping molybdenum reduction (Ghani *et al.*, 1993). Cyanide inhibits molybdenum reduction in *E. cloacae* 48, and this was suggested to be the location of the Mo-reducing enzyme, which is a site downstream from cytochrome (Ghani *et al.*, 1993). None of the respiratory inhibitors examined had any effect on the Mo-reducing enzyme in the newly isolated Mo-reducing bacteria (Shukor *et al.*, 2009; Othman *et al.*, 2013; Ahmad *et al.*, 2013; Shukor *et al.*, 2010a, 2010b; Yunus *et al.*, 2009; Lim *et al.*, 2012; Halmi *et al.*, 2016). As adding cyanide to a reaction mixture can raise the pH by turning the mixture into a very alkaline solution, pH adjustment

264 Eco-Restoration of the Polluted Environment

must be carried out prior to enzyme assay. After adding cyanide to a mixture, the pH is adjusted to neutral since phosphomolybdate is particularly unstable in alkaline conditions. This time, however, cyanide failed to inhibit molybdenum reduction in *E. cloacae* strain 48 (Shukor, 2014). Therefore, cyanide is unlikely to be an inhibitor to molybdenum reduction, and the location of the enzyme at the ETC needs to be revised.

12.6 BIOREMOVAL OF MOLYBDENUM FROM AQUEOUS ENVIRONMENTS

In general, immobilization of bacterial cells can make the cells more tolerant than free cells in resisting the toxic effects of xenobiotics during their biodegradation. Furthermore, immobilized cells are less susceptible to cytotoxicity caused by metal ions in their environment. This is why numerous bacteria opt to form biofilms on many supports as a means of survival. A substrate inhibition kinetics study is one of the best methods to demonstrate the reduction in toxicity of toxicants to cells when immobilized. In one study, the degradation of phenol by *Pseudomonas fluorescens* in the free or immobilized form shows a clear transition from the toxicity-indicating model of Haldane to the nontoxic-indicating model of Monod for the immobilized cells. Cells can be shielded from damage caused by other metal ions if they are immobilized. In nitrogen wastewater treatment, for instance, heavy metal inhibition of denitrification is an issue, although its solutions are rarely investigated. Sodium alginate-kaolin-immobilized *Pseudomonas brassicacearum* LZ-4 was used to shield denitrifiers against hexavalent chromium in another investigation (Yu *et al.*, 2020).

Immobilization in hollow containers by entrapment would also allow insoluble remediated products to be trapped and removed continuously. This works well for chromate and molybdate reductions, where the bacterial-reduced products are relatively insoluble. In chromate reduction by the chromate-reducing bacterium *E. cloacae*, the reduction of the more toxic and highly soluble Cr^{6+} form into the less toxic and insoluble Cr^{3+} form allows the entrapment and removal of the reduced product (Komori *et al.*, 1990). Immobilized Mo-reducing bacteria not only allow for the continuous removal of molybdenum, as the Mo-blue form is trapped in the dialysis tubing, but also confer some resistance to toxic metal ions such as mercury and copper (Rahman *et al.*, 2013; Halmi *et al.*, 2013, 2014; Rahman *et al.*, 2009; Shukor and Shukor, 2015). Immobilization of Mo-reducing bacterium in dialysis tubing was not only utilized for remediation purposes but for proving the reduction of molybdenum (sodium molybdate) by heterotrophs is enzymatically mediated and not abiotically mediated. Mo-blue is a molybdenum-reduced product that is a well-known indicator for the occurrence of reducing agents, and initial studies on the supposedly bacterial origin for the reduction of molybdate by *Thiobacillus* spp. (now *Acidithiobacillus*) to Mo-blue are complicated by the conditions of growth of this genus, where acidic conditions and ferrous ions are added into the growth environment. Acidic conditions chemically convert the molybdate ions into large polymolybdate ions complexes that can be straightforwardly reduced to Mo-blue by mild reducing agents such as ferrous ions and ascorbate (Lee, 1977; Shukor *et al.*, 2007; Greenwood and Earnshaw, 1984). This reduction process is capitalized in the colorimetric determination method for phosphate, where the Mo-blue product is very similar to the Mo-blue forms in heterotrophic bacterial reduction of molybdate. The complications resulting from the likely chemical (abiotic) reductions instead of biotic in chemolithotrophic bacteria, including *Acidothiobacillus* spp. (Yong *et al.*, 1997), cause the slow demise of molybdate reduction works in this genus. Munch and Ottow (1981) came up with a clever method employing a dialysis tube to demonstrate that the bacterium *Clostridium butyricum* catalyzed the reduction of iron (Fe^{3+} to Fe^{2+}). The experiment was performed because it was hypothesized that iron reduction could occur without enzymes in a fermenting, low-redox potential environment (Hem, 1972). By placing insoluble hematite in a dialysis tube and immersing it in media containing *C. butyricum*, the bacterium was physically extracted from the hematite (Fe^{3+}). An enzymatic reaction, rather than Fe-reducing chemicals, was shown by Munch and Ottow (1981) into be the direct source of the reduction of Fe^{3+} or Fe_2O_3 (hematite) to Fe^{2+}. Since it is not taken into account that Fe-reducing compounds are created enzymatically when using the standard approach of boiling

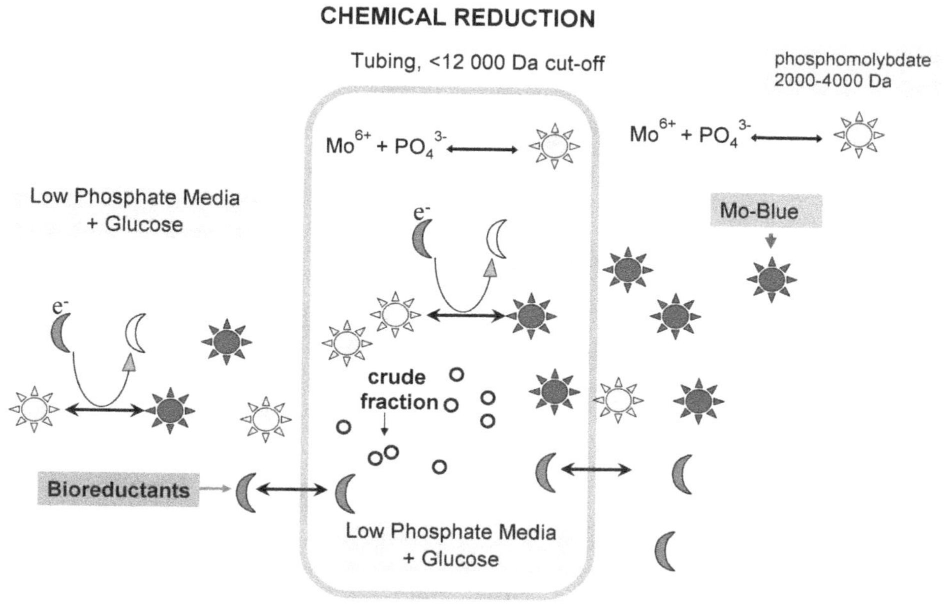

CHEMICAL REDUCTION

FIGURE 12.11 Scenario for a chemical reduction mechanism of molybdate to Mo-blue by reducing agents (bioreductants) produced by bacteria. Yellow and blue stars designate unreduced and reduced phosphomolybdate ions, respectively. Red and white crescents denote reduced and oxidized bioreductants, respectively.

cells (and cell fractions) to verify enzymatic reduction, the dialysis tubing method was adopted. When cells are heated to high temperatures, the enzymes that create bioreductants and organic acids, which may be responsible for the conversion of ferrous to ferric ions, become denatured. As a result, it will be impossible to tell if a reduction was achieved via chemical or enzymatic means.

Retention in a dialysis tube with a pore size of between 1 and 5 nm is possible because Mo-blue, the result of molybdate reduction, is a colloid with particle sizes varying from 1 to 100 nm (Sidgwick, 1951). To investigate the role of chemical Mo-reducing compounds in molybdate reduction in EC 48, the aforementioned principle can be put to use. Since Mo-reducing chemicals have small molecular weights, the dialysis tube method assumes that if molybdenum (sodium molybdate) reduction to Mo-blue in the bacterium EC 48 were solely due to Mo-reducing chemicals and not Mo-reducing enzymes or bacteria, then a similar proportion of Mo-blue would be present inside and outside the dialysis tube (Figure 12.11). Since most enzymes and Mo-blue itself are bigger than the pore size of the dialysis tube, most of the Mo-blue should be detected in the tube if the reduction were entirely enzymatic (extracellular) or bacterial bound, as illustrated in Figure 12.12. The results show that Mo-blue is discovered in the dialysis tubing after 24 h of static incubation (Figure 12.13), indicating that molybdenum (sodium molybdate) reduction to Mo-blue in this bacterium is enzymatic in nature. The experiment was replicated in several other heterotrophic Mo-reducing bacteria and showed the same results, indicating an enzymatic route for molybdenum reduction in heterotrophic bacteria.

Dialysis tubing has a high removal efficiency and can handle high molybdenum content, making it an attractive molybdenum removal option. The maximum rate of Mo-blue synthesis ($V_{MoblueMax}$) by *Serratia* sp. strain Dry5 in dialysis tubing using the Monod model is determined to be 0.2640 mmole/Mo-blue/hr, while the half-maximal rate of reduction (K_{Mo}) is determined to be 21.783 mM molybdate (Halmi *et al.*, 2014). In another study, *S. marcescens* strain Dr. Y6 entrapment in dialysis tubing showed inhibition at high concentrations of molybdenum. The Haldane model is used to generate the resulting inhibitory kinetics profile. Maximum Mo-blue output is estimated to be 0.153

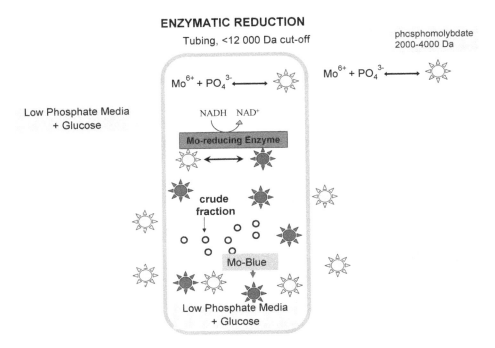

FIGURE 12.12 Scenario for an enzymatic reduction (produced by Mo-reducing bacterium) route of molybdate to Mo-blue. Yellow and blue stars signify unreduced and reduced phosphomolybdate ions, respectively.

FIGURE 12.13 Dialysis tubing experiment at time 0 (left) and after 24 h of incubation (right) of 20 mL of washed cells of Mo-reducing bacterium in minimal salts media containing glucose and 10 mM of sodium molybdate.

Bacterial Reduction of Molybdenum for Its Bioremediation

mmole Mo-blue/hr, the molybdate concentration producing a half-maximal rate of reduction (K_{Mo}) is estimated to be 0.22 mM, and the inhibition constant (K_i) is 506 mM. Based on the findings, the dialysis tubing and Mo-reducing bacteria system shows promise as a means of molybdenum bioremoval from solution (Rahman *et al.*, 2013).

12.7 CONCLUSION

This chapter is an effort to evaluate and update what is currently known about molybdenum-reducing bacteria. The capacity of several molybdenum reducers to degrade additional xenobiotics has also been upgraded and is discussed in this review. Some of these xenobiotics, like acrylamide and propionamide, can be employed as electron-donating substrates for molybdenum reduction. Future research should focus on the electron-donating potential of other xenobiotics so that more than one toxicant can be remediated simultaneously. Future remediation efforts will benefit from research into how to circumvent the blocking impact of cationic heavy metals like copper, mercury, and silver on molybdenum reduction, as well as on other anionic reductions like arsenate and chromate microbial reductions. As of now, the molybdenum-reducing enzyme's true identity has remained a mystery, but future protein sequencing of the purified enzyme may shed light on the process's underlying mechanism. Studies of inhibitory kinetics reveal that molybdate is hazardous at high concentrations, suggesting that the bacteria's potential to clean up a highly contaminated site with molybdenum, as well as molybdenum-rich effluents from mine tailings, has a finite upper limit. More research is needed to compare Mo-blue production's activation energy to that of other anionic metal reductions and xenobiotic degradation processes.

REFERENCES

Abdel-Mongy, M. A., S. A. Aqlima, M. S. Shukor, S. Hussein, A. P. K. Ling, M. Y. Shukor, and A. Peg. 2018 Sept. 30. "A PEG 4000-Degrading and Hexavalent Molybdenum-Reducing *Pseudomonas Putida* Strain Egypt-15." *Journal of the National Science Foundation of Sri Lanka* 46, no. 3: 431–42. https://doi.org/10.4038/jnsfsr.v46i3.8495.

Abdel-Mongy, M. A., M. F. Rahman, and M. Y. Shukor. 2021Dec.31. "Isolation and Characterization of a Molybdenum-Reducing and Carbamate-Degrading *Serratia* sp. Strain Amr-4 in Soils from Egypt." *Asian Journal of Plant Biology* 3, no. 2: 25–32. https://doi.org/10.54987/ajpb.v3i2.639.

Abdel-Mongy, M. A., M. S. Shukor, S. Hussein, A. P. K. Ling, N. A. Shamaan, and M. Y. Shukor. 2015. "Isolation and Characterization of a Molybdenum-Reducing, Phenol- and Catechol-Degrading *Pseudomonas putida* strain amr-12 in Soils from Egypt." *Scientific Study& Research: Chemistry & Chemical Engineering, Biotechnology, Food Industry* 16, no. 4: 353–69.

Abo-Shakeer, A., S. A. Ahmad, M. Y. Shukor, N. A. Shamaan, and M. A. Syed. 2013. "Isolation and Characterization of a Molybdenum-Reducing *Bacillus pumilus* strain Lbna." *Journal of Environmental Microbiology & Toxicology* 1, no. 1: 9–14. https://doi.org/10.54987/jemat.v1i1.18.

Abo-Shakeer, L. K. A., M. F. A. Rahman, M. H. Yakasai, N. A. Bakar, A. R. Othman, M. A. Syed, N. Abdullah, and M. Y. Shukor. 2017. "Kinetic Studies of the Partially Purified Molybdenum-Reducing Enzyme from *Bacillus pumilus* strain Lbna." *Bioremediation Science & Technology Research* 5, no. 1: 18–23, July. https://doi.org/10.54987/bstr.v5i1.354.

Abu Zeid, I. M. A., M. F. Rahman, and M. Y. Shukor. 2021 Dec. 31. "Isolation of A Molybdenum-Reducing *Bacillus* sp. Strain Zeid 15 and Modeling of Its Growth on Amides." *Bulletin of Environmental Science & Sustainable Management* 5, no. 2: 19–27. https://doi.org/10.54987/bessm.v5i2.650.

Adnan, M., I. Abu Zeid, S. A. Ahmad, M. Effendi Halmi, S. Abdullah, and M. Shukor. 2016 Jan. 1. "A Molybdenum-Reducing *Bacillus* sp. Strain Zeid 14 in Soils from Sudan That Could Grow on Amides and Acetonitrile." *Malaysian Journal of Soil Science* 20: 111–34.

Ahmad, S. A., M. Y. Shukor, N. A. Shamaan, MACC, W. P. Mac Cormack, and M. A. Syed. 2013. "Molybdate Reduction to Molybdenum Blue by an Antarctic Bacterium." *BioMed Research International* 2013: Article number 871941. https://doi.org/10.1155/2013/871941.

Aiba, S., M. Shoda, and M. Nagatani. 1968. "Kinetics of Product Inhibition in Alcohol Fermentation." *Biotechnology & Bioengineering* 10, no. 6: 845–64. https://doi.org/10.1002/bit.260100610.

Al Kuisi, Mustafa, Mohammad Al-Hwaiti, Kholoud Mashal, and Abdulkader M. Abed. 2015. "Spatial Distribution Patterns of Molybdenum (Mo) Concentrations in Potable Groundwater in Northern Jordan." *Environmental Monitoring & Assessment* 187, no. 3: 148. https://doi.org/10.1007/s10661-015-4264-5.

Alhassan, A. Y., A. Babandi, G. Uba, and H. M. Yakasai. 2020. "Isolation and Characterization of Molybdenum-Reducing *Pseudomonas* sp. from Agricultural Land in Northwest-Nigeria." *Journal of Biochemistry, Microbiology & Biotechnology* 8, no. 1: 23–8. https://doi.org/10.54987/jobimb.v8i1.505.

Allamin, I. A. 2021Dec.31. "Phytoremediation of Heavy Metals in Contaminated Soils: A Review." *Journal of Biochemistry, Microbiology & Biotechnology* 9, no. 2: 7–14. https://doi.org/10.54987/jobimb.v9i2.610.

Al-Saidi, H. M., A. A. Gahlan, and O. A. Farghaly. 2022. "Decontamination of Zinc, Lead and Nickel from Aqueous Media by Untreated and Chemically Treated Sugarcane Bagasse: A Comparative Study." *Egyptian Journal of Chemistry* 65, no. 3: 711–20.

Ariff, A. B., M. Rosfarizan, B. Ghani, T. Sugio, and M. I. A. Karim. 1997. "Molybdenum Reductase in Enterobacter cloacae." *World Journal of Microbiology & Biotechnology* 13, no. 6: 643–7. https://doi.org/10.1023/A:1018562719751.

Aziz, N. F., M. I. E. Halmi, and W. L. W. Johari. 2017Dec.31. "Statistical Optimization of Hexavalent Molybdenum Reduction by Serratia sp. strain MIE2 Using Central Composite Design (CCD)." *Journal of Biochemistry, Microbiology & Biotechnology* 5, no. 2: 8–11. https://doi.org/10.54987/jobimb.v5i2.341.

Babák, L., P. Šupinová, and R. Burdychová. 2013. "Growth Models of Thermus aquaticus and Thermus Scotoductus." *Acta Universitatis Agriculturae & Silviculturae Mendelianae Brunensis* 60, no. 5: 19–26. https://doi.org/10.11118/actaun201260050019.

Baranyi, J., and T. A. Roberts. 1994. "A Dynamic Approach to Predicting Bacterial Growth in Food." *International Journal of Food Microbiology* 23, no. 3–4: 277–94. https://doi.org/10.1016/0168-1605(94)90157-0.

Baranyi, J., and T. A. Roberts. 1995. "Mathematics of Predictive Food Microbiology." *International Journal of Food Microbiology* 26, no. 2: 199–218. https://doi.org/10.1016/0168-1605 (94)00121-l.

Barceloux, D. G., and Donald G. Barceloux. 1999. "Manganese." *Chemical Toxicology* 37, no. 2: 217–30.

Battogtokh, B., J. M. Lee, and N. Woo. 2014. "Contamination of Water and Soil by the Erdenet Copper-Molybdenum Mine in Mongolia." *Environmental Earth Sciences* 71, no. 8: 3363–74. https://doi.org/10.1007/s12665-013-2727-y.

Bersényi, András, Erzsébet Berta. István Kádár, Róbert Glávits, Mihály Szilágyi, and Sándor György Fekete. 2008. "Effects of High Dietary Molybdenum in Rabbits." *Acta Veterinaria Hungarica* 56, no. 1: 41–55. https://doi.org/10.1556/AVet.56.2008.1.5.

Biswas, Keka C., Nicole A. Woodards, Huifang Xu, and Larry L. Barton. 2009. "Reduction of Molybdate by Sulfate-Reducing Bacteria." *BioMetals* 22, no. 1: 131–9. https://doi.org/10.1007/s10534-008-9198-8.

Boon, B., and H. Laudelout. 1962. "Kinetics of Nitrite Oxidation by Nitrobacter Winogradskyi." *Biochemical Journal* 85, no. 3: 440–7. https://doi.org/10.1042/bj0850440.

Box, G. E. P., and K. B. Wilson. 1951. "On the Experimental Attainment of Optimum Conditions." *Journal of the Royal Statistical Society: Series B* 13, no. 1: 1–38. https://doi.org/10.1111/j.2517-6161.1951.tb00067.x.

Brierley, J., C. Brierley, and G. M. Goyak. 1986. "AMT-Bioclaim: A New Wastewater Treatment and Metal Recovery Technology." *No Source Inf Available* 4, January 1.

Buchanan, R. L., R. C. Whiting, and W. C. Damert. 1997. "When Is Simple Good Enough: A Comparison of the Gompertz, Baranyi, and Three-Phase Linear Models for Fitting Bacterial Growth Curves." *Food Microbiology* 14, no. 4: 313–26. https://doi.org/10.1006/fmic.1997.0125.

Campbell, A. M., A. Del Campillo-Campbell, and D. B. Villaret. 1985. "Molybdate Reduction by *Escherichia coli* K-12 and Its chl Mutants." *Proceedings of the National Academy of Sciences of the United States of America* 82, no. 1: 227–31. https://doi.org/10.1073/pnas.82.1.227.

Chae, H. K., W. G. Klemperer, and T. A. Marquart. 1993Oct.1. "High-Nuclearity oxomolybdenum (V) Complexes." *Coordination Chemistry Reviews* 128, no. 1–2: 209–24. https://doi.org/10.1016/0010-8545(93)80031-Y.

Chee, H. S., M. Manogaran, Z. Suhaili, M. H. Yakasai, M. F. A. Rahman, N. A. Shamaan, N. A. Yasid, and A. R. Othman. 2017. "Isolation and Characterisation of a Mo-Reducing Bacterium from Malaysian Soil." *Bioremediation Science & Technology Research* 5, no. 2: 17–24. https://doi.org/10.54987/bstr.v5i2.359.

Cheng, J., J. Gao, J. Zhang, W. Yuan, S. Yan, J. Zhou, J. Zhao, and S. Feng. 2021. "Optimization of Hexavalent Chromium Biosorption by *Shewanella putrefaciens* Using the Box-Behnken Design." *Water, Air, & Soil Pollution* 232, no. 3: 92. https://doi.org/10.1007/s11270-020-04947-7.

Davis, G. K. 1991. "Molybdenum." In *Metals and Their Compounds in the Environment, Occurrence, Analysis and Biological Relevance*, edited by E. Merian: 1089–100. Weinheim and New York: VCH.

Dayana, K., C. V. Sowjanya, and C. V. Ramachandramurthy. 2013. "Eco-friendly Remediation of Industrial Effluents via Biosorption Technology—An Overview." *International Journal of Engineering Research & Technology*

2, no. 11, November 13. Accessed June 15, 2023. https://www.ijert.org/research/eco-friendly-remediation-of-industrial-effluents-via-biosorption-technology-an-overview-IJERTV2IS110328.pdf.https://www.ijert.org/eco-friendly-remediation-of-industrial-effluents-via-biosorption-technology-an-overview-2.

Deeb, B. E., and A. D. Altalhi. 2009. "Degradative Plasmid and Heavy Metal Resistance Plasmid Naturally Coexist in Phenol and Cyanide Assimilating Bacteria." *American Journal of Biochemistry & Biotechnology* 5, no. 2: 84–93. https://doi.org/10.3844/ajbbsp.2009.84.93.

Dixit, R., D. Wasiullah, D. Malaviya, K. Pandiyan, U. Singh, A. Sahu, R. Shukla, B. Singh, J. Rai, P. Sharma, H. Lade, and D. Paul. 2015. "Bioremediation of Heavy Metals from Soil and Aquatic Environment: An Overview of Principles and Criteria of Fundamental Processes." *Sustainability* 7, no. 2: 2189–212. https://doi.org/10.3390/su7022189.

dos Reis Ferreira, Gustavo Magno, Josiane Ferreira Pires, Luciana Silva Ribeiro, Jorge Dias Carlier, Maria Clara Costa, Rosane Freitas Schwan, and Cristina Ferreira Silva. 2023. "Impact of Lead (Pb2+) on the Growth and Biological Activity of *Serratia Marcescens* Selected for Wastewater Treatment and Identification of Its zntR Gene—A Metal Efflux Regulator." *World Journal of Microbiology & Biotechnology* 39, no. 4: 91, February. https://doi.org/10.1007/s11274-023-03535-1.

Elangovan, R. A., S. B. Abhipsa, B. C. Rohit, P. A. Ligy, and K. B. Chandraraj. 2006. "Reduction of Cr(VI) by a Bacillus sp." *Biotechnology Letters* 28, no. 4: 247–52. https://doi.org/10.1007/s10529-005-5526-z.

Elekwachi, C. O., J. Andresen, and T. C. Hodgman. 2014. "Global Use of Bioremediation Technologies for Decontamination of Ecosystems." *Journal of Bioremediation & Biodegradation* 5, no. 4: 1–9. https://doi.org/10.4172/2155-6199.1000225.

Francisco, R., M. C. Alpoim, and P. V. Morais. 2002. "Diversity of Chromium-Resistant and -Reducing Bacteria in a Chromium-Contaminated Activated Sludge." *Journal of Applied Microbiology* 92, no. 5: 837–43. https://doi.org/10.1046/j.1365-2672.2002.01591.x.

Freedman, Zachary, Chengsheng Zhu, and Tamar Barkay. 2012 "Mercury Resistance and Mercuric Reductase Activities and Expression Among Chemotrophic Thermophilic Aquificae." *Applied & Environmental Microbiology* 78, no. 18: 6568–75, September. https://doi.org/10.1128/AEM.01060-12.

Fujikawa, Hiroshi. 2010. "Development of a New Logistic Model for Microbial Growth in Foods." *Biocontrol Science* 15, no. 3: 75–80. https://doi.org/10.4265/bio.15.75.

Furber, D. 2009. "Is Molybdenum Lurking in Your Forages?" *Canadian Cattlemen* [Internet]. Accessed March 13, 2021. https://www.canadian cattlemen.ca/features/is-molybdenum-lurking-in-your-forages/.

Gafar, A. A., and M. Y. Shukor. 2018. "Characterisation of an Acrylamide-Degrading Bacterium and Its Degradation Pathway." *Journal of Environmental Microbiology & Toxicology* 6, no. 2: 29–33, December 31. https://doi.org/10.54987/jemat.v6i2.441.

Gafasa, M. A., S. S. Ibrahim, A. Babandi, N. Abdullahi, D. Shehu, Ya *et al.* 2019. "Characterizing the Molybdenum-Reducing Properties of Pseudomonas sp. Locally Isolated from Agricultural Soil in Kano Metropolis Nigeria." *Bioremediation Science & Technology Research* 7, no. 1: 34–40, July 31. https://doi.org/10.54987/bstr.v7i1.462.

Geng, C., Y. Gao, D. Li, X. Jian, and Q. Hu. 2014. "Contamination Investigation and Risk Assessment of Molybdenum on an Industrial Site in China." *Journal of Geochemical Exploration.* PB144: 273–81. https://doi.org/10.1016/j.gexplo.2013.12.014.

Ghani, B., M. Takai, N. Z. Hisham, N. Kishimoto, A. K. M. Ismail, T. Tano *et al.* 1993. "Isolation and Characterization of a Mo6+-Reducing Bacterium." *Applied & Environment Microbiology* 59, no. 4: 1176–80.

Glusczak, Lissandra, Denisedos Santos Miron, Márcia Crestani, Milene Braga da Fonseca, Fábiode Araújo Pedron, Marta Frescura Duarte, and Vânia Lúcia Pimentel Vieira. 2006. "Effect of Glyphosate Herbicide on Acetylcholinesterase Activity and Metabolic and Hematological Parameters in Piava (Leporinus Obtusidens)." *Ecotoxicology & Environmental Safety* 65, no. 2: 237–41, October. https://doi.org/10.1016/j.ecoenv.2005.07.017.

Gompertz, B. 1825. "On the Nature of the Function Expressive of the Law of Human Mortality, and on a New Mode of Determining the Value of Life Contingencies." *Philosophical Transactions of the Royal Society of London* 115: 513–85.

Greenwood, N. N., and A. Earnshaw. 1984. *Chemistry of the Elements*. Oxford: Pergamon Press.

Guo, Qingwei, Rui Wan, and Shuguang Xie. 2014. "Simazine Degradation in Bioaugmented Soil: Urea Impact and Response of Ammonia-Oxidizing Bacteria and Other Soil Bacterial Communities." *Environmental Science & Pollution Research International* 21, no. 1: 337–43. https://doi.org/10.1007/s11356-013-1914-4.

Gusmanizar, N., M. I. E. Halmi, R. Mansur, M. F. Abd Rahman, M. S. Shukor, N. S. Azmi, and M. Y. Shukor. 2016. "Molybdenum-Reducing and Azo-Dye Decolorizing Serratia Marcescens Strain Neni-1 from

Indonesian Soil." *Journal of Urban & Environmental Engineering* 10, no. 1: 113–23. https://doi. org/10.4090/juee.2016.v10n1.113123.

Halmi, M. I. E. 2012. *Faculty of Biotechnology and Biomolecular Sciences.* A thesis submitted in partial fulfilment of the requirements for the degree of Doctor of Philosophy in the Department of Biochemistry. Universiti Putra Malaysia, Malaysia.

Halmi, M. I. E., S. R. S. Abdullah, W. L. W. Johari, M. S. M. Ali, N. A. Shaharuddin, A. Khalid *et al.* 2016. "Modelling the Kinetics of Hexavalent Molybdenum (Mo^{6+}) Reduction by the Serratia sp. Strain MIE2 in Batch Culture." *Rendiconti Lincei.* Accessed July 15, 2016. http://link.springer.com/10.1007/s12210-016-0545-3.

Halmi, M. I. E., M. S. Shukor, W. L. W. Johari, and M. Y. Shukor. 2014. "Mathematical Modeling of the Growth Kinetics of *Bacillus* sp. on Tannery Effluent Containing Chromate." *Journal of Environmental Bioremediation & Toxicology* 2, no. 1: 6–10. https://doi.org/10.54987/jebat.v2i1.139.

Halmi, M. I. E., S. W. Zuhainis, M. T. Yusof, N. A. Shaharuddin, W. Helmi, Y. Shukor, M. A. Syed, and S. A. Ahmad. 2013. "Hexavalent Molybdenum Reduction to Mo-Blue by a Sodium-Dodecyl-Sulfate-Degrading *Klebsiella oxytoca* Strain DRY14." *BioMed Research International* 2013: 384541. https://doi. org/10.1155/2013/384541.

Han, K., and O. Levenspiel. 1988. "Extended Monod Kinetics for Substrate, Product, and Cell Inhibition." *Biotechnology & Bioengineering* 32, no. 4: 430–47. https://doi.org/10.1002/bit.260320404.

Hanapiah, Munirah, Syaizwan Zahmir Zulkifli, Muskhazli Mustafa, Ferdaus Mohamat-Yusuff, and Ahmad Ismail. 2018. "Isolation, Characterization, and Identification of Potential Diuron-Degrading Bacteria from Surface Sediments of Port Klang, Malaysia." *Marine Pollution Bulletin* 127: 453–7. https://doi. org/10.1016/j.marpolbul.2017.12.015.

Haywood, S., Z. Dincer, B. Jasani, and M. J. Loughran. 2004. "Molybdenum-Associated Pituitary Endocrinopathy in Sheep Treated with Ammonium Tetrathiomolybdate." *Journal of Comparative Pathology* 130, no. 1: 21–31. https://doi.org/10.1016/s0021-9975(03)00065-3.

Hem, J. D. 1972. "Chemical Factors That Influence the Availability of Iron and Manganese in Aqueous Systems." *Geological Society of America Bulletin* 83, no. 2: 443–50. https://doi.org/10.1130/0016-7606(1972)83[443:CFTITA]2.0.CO;2.

Hettiarachchi, G. M., G. M. Pierzynski, and M. D. Ransom. 2000. "In Situ Stabilization of Soil Lead Using Phosphorus and Manganese Oxide." *Environmental Science & Technology* 34, no. 21: 4614–19. https:// doi.org/10.1021/es001228p.

Hofmann, Marika, Thomas Heine, Luise Malik, Sarah Hofmann, Kristin Joffroy, Christoph Helmut Rudi Senges, Julia Elisabeth Bandow, and Dirk Tischler. 2021. "Screening for Microbial Metal-Chelating Siderophores for the Removal of Metal Ions from Solutions." *Microorganisms* 9, no. 1: 111, January. https://doi.org/10.3390/microorganisms9010111.

Huang, Liping, Fuping Tian, Yuzhen Pan, Liyuan Shan, Yong Shi, and Bruce E. Logan. 2019. "Mutual Benefits of Acetate and Mixed Tungsten and Molybdenum for Their Efficient Removal in 40L Microbial Electrolysis Cells." *Water Research* 162: 358–68, October 1. https://doi.org/10.1016/j.watres.2019.07.003.

Huang, L. 2013. "Optimization of a New Mathematical Model for Bacterial Growth." *Food Control* 32, no. 1: 283–8. https://doi.org/10.1016/j.foodcont.2012.11.019.

Ibrahim, Y., M. Abdel-Mongy, M. S. Shukri Shukor, S. Hussein, A. P. K. Ling, and M. Y. Shukor. 2015. "Characterization of a Molybdenum-Reducing Bacterium with the Ability to Degrade Phenol, Isolated in Soils from Egypt." *BioTechnologia* 3, no. 3: 234–45. https://doi.org/10.5114/bta.2015.56573.

Hala, Ibrahium A., B. M. Atia, S. Awwad Nasser, A. A. Nayl, H. A. Radwan, and M. A. Gado. 2022. "Efficient Preparation of Phosphazene Chitosan Derivatives and Its Applications for the Adsorption of Molybdenum from Spent Hydrodesulfurization Catalyst." *Journal of Dispersion Science and Technology*: 1–16, May 4.

Idris, D., M. A. Gafasa, S. S. Ibrahim, A. Babandi, D. Shehu, Ya *et al.* 2019. "Pantoea sp. strain HMY-P4 Reduced Toxic Hexavalent Molybdenum to Insoluble Molybdenum Blue." *Journal of Biochemistry, Microbiology & Biotechnology* 7, no. 1: 31–7, July 31. https://doi.org/10.54987/jobimb.v7i1.450.

Jacobs, J. A., and S. M. Testa. 2014. *Acid Drainage and Sulfide Oxidation: Introduction: Acid Mine Drain Rock Drain Acid Sulfate Soils: Causes, Assessment, Prediction, Prevention, and Remediation*: 1–8. Hoboken, NJ: John Wiley & Sons, Inc.

Jeter, M. A., and G. K. Davis. 1954. "The Effect of Dietary Molybdenum upon Growth, Hemoglobin, Reproduction and Lactation of Rats." *Journal of Nutrition* 54, no. 2: 215–20, October 31. https://doi. org/10.1093/jn/54.2.215.

Kabir, Z. M., M. A. Gafasa, H. T. Kabara, S. S. Ibrahim, A. Babandi, M. Ya.'u *et al.* 2019. "Isolation and Characterization of Molybdate-Reducing Enterobacter Cloacae from Agricultural Soil in Gwale LGA

Kano State, Nigeria." *Journal of Environmental Microbiology & Toxicology* 7, no. 1: 1–6, July 31. https://doi.org/10.54987/jemat.v7i1.464.

Kaplan, D. 2013. "Absorption and Adsorption of Heavy Metals by Microalgae." In *Handbook of Microalgal Culture: Applied Phycology and Biotechnology*, edited by Amos Richmond and Qiang Hu: 602–11, London: Blackwell Science Ltd.

Karamba, I. K., and H. Yakasai. 2018. "Isolation and Characterization of a Molybdenum-Reducing and Methylene Blue-Decolorizing Serratia Marcescens StrainKIK-1." *In Soils from Nigeria. Bioremediation Science Journal of Technology Research* 6, no. 1: 1–8, July 31.

Kargar, M., N. Khorasani, M. Karami, G. Rafiee, and R. Naseh. 2011. "Study of Aluminum, Copper and Molybdenum Pollution in Groundwater Sources Surrounding (Miduk) Shahr-E-Babak Copper Complex Tailings Dam." *World Academy of Science, Engineering & Technology* 52, no. 4: 278–82.

Kazansky, L. P., and M. A. Fedotov. 1980. "Phosphorus-31 and Oxygen-17 N.M.R. Evidence of Trapped Electrons in Reduced 18-Molybdodiphosphate(V), P2Mo18O628-." *Journal of the Chemical Society. Chemical Communications* 14: 644–6.

Kesavan, V., A. Mansur, Z. Suhaili, M. S. R. Salihan, M. F. A. Rahman, and M. Y. Shukor. 2018. "Isolation and Characterization of a Heavy Metal-Reducing Pseudomonas sp. Strain Dr. Y. Kertih with the Ability to Assimilate Phenol and Diesel." *Bioremediation Science & Technology Research* 6, no. 1: 14–22, July 31. https://doi.org/10.54987/bstr.v6i1.394.

Khayat, M. E., M. F. A. Rahman, M. S. Shukor, S. A. Ahmad, N. A. Shamaan, and M. Y. Shukor. 2016. "Characterization of a Molybdenum-Reducing Bacillus sp. strain Khayat with the Ability to Grow on SDS and Diesel." *Rendiconti Lincei* 27, no. 3: 547–56. Accessed July 15, 2016. http://link.springer.com/10.1007/s12210-016-0519-5. https://doi.org/10.1007/s12210-016-0519-5.

Komori, K., A. Rivas, K. Toda, and H. Ohtake. 1990. "A Method for Removal of Toxic Chromium Using Dialysis-Sac Cultures of a Chromate-Reducing Strain of Enterobacter Cloacae." *Applied Microbiology & Biotechnology* 33, no. 1: 117–19. https://doi.org/10.1007/BF00170582.

Kosaka, H., and K. Wakita. 1978. "Some Geologic Features of the Mamut Porphyry Copper Deposit, Sabah, Malaysia." *Economic Geology* 73, no. 5: 618–27. https://doi.org/10.2113/gsecongeo.73.5.618.

Kristanti, Risky Ayu, Masahiro Kanbe, Tony Hadibarata, Tadashi Toyama, Yasuhiro Tanaka, and Kazuhiro Mori. 2012. "Isolation and Characterization of 3-Nitrophenol-Degrading Bacteria Associated with Rhizosphere of Spirodela Polyrrhiza." *Environmental Science & Pollution Research International* 19, no. 5: 1852–8. https://doi.org/10.1007/s11356-012-0836-x.

Lee, J. D. 1977. *Concise Inorganic Chemistry*. New York: Van Reinhold Co.

LeGendre, G. R., and D. D. Runnells. 1975. "Removal of Dissolved Molybdenum from Wastewaters by Precipitates of Ferric Iron." *Environmental Science & Technology* 9, no. 8: 744–9. https://doi.org/10.1021/es60106a010.

Levine, V. E. 1925. "The Reducing Properties of Microorganisms with Special Reference to Selenium Compounds." *Journal of Bacteriology* 10, no. 3: 217–63. https://doi.org/10.1128/jb.10.3.217-263.1925.

Li, F., Y. Wang, L. Mao, H. Tao, and M. Chen. 2021 Dec. 3. "Molybdenum Background and Pollution Levels in the Taipu China." *Environmental Chemistry Letters* 20.

Lim, H. K., M. A. Syed, and M. Y. Shukor. 2012. "Reduction of Molybdate to Molybdenum Blue by Klebsiella sp. Strain Hkeem." *Journal of Basic Microbiology* 52, no. 3: 296–305. https://doi.org/10.1002/jobm.201100121.

Liu, Cenwei, Jing Ye, Y. Lin, Jian Wu, G. W. Price, D. Burton, and Yixiang Wang. 2020. "Removal of Cadmium (II) Using Water Hyacinth (*Eichhornia crassipes*) Biochar Alginate Beads in Aqueous Solutions." *Environmental Pollution* 264: 114785, September 1. https://doi.org/10.1016/j.envpol.2020.114785.

Lloyd, Jonathan R. 2003. "Microbial Reduction of Metals and Radionuclides." *FEMS Microbiology Reviews* 27, no. 2–3: 411–25. https://doi.org/10.1016/S0168-6445(03)00044-5.

López, S., M. Prieto, J. Dijkstra, M. S. Dhanoa, and J. France. 2004. "Statistical Evaluation of Mathematical Models for Microbial Growth." *International Journal of Food Microbiology* 96, no. 3: 289–300. https://doi.org/10.1016/j.ijfoodmicro.2004.03.026.

Luong, J. H. T. 1987. "Generalization of Monod Kinetics for Analysis of Growth Data with Substrate Inhibition." *Biotechnology & Bioengineering* 29, no. 2: 242–8. https://doi.org/10.1002/bit.260290215.

Lyubimov, Alexander V., Jeffry A. Smith, Serge D. Rousselle, Michael D. Mercieca, Joseph E. Tomaszewski, Adaline C. Smith, and Barry S. Levine. 2004. "The Effects of Tetrathiomolybdate (TTM, NSC-714598) and Copper Supplementation on Fertility and Early Embryonic Development in Rats." *Reproductive Toxicology* 19, no. 2: 223–33. https://doi.org/10.1016/j.reprotox.2004.07.006.

Maarof, M., Y. Shukor, O. Mohamad, K. Karamba, H. M. Effendi, M. Rahman, and H. Yakasai. 2018 Jul. 31. "Isolation and Characterization of a Molybdenum-Reducing *Bacillus amyloliquefaciens* Strain KIK-12 in Soils from Nigeria with the Ability to Grow on SDS." *Journal of Environmental Microbiology and Toxicology* 6, no. 1: 13–20. https://doi.org/10.54987/jemat.v6i1.401.

Majak, W., D. Steinke, J. Mcgillivray, and T. Lysyk. 2004. "Clinical Signs in Cattle Grazing High Molybdenum Forage." *Rangeland Ecology & Management* 57, no. 3: 269–74. https://doi.org/10.2111/1551-5028(2004)057[0269:CSICGH]2.0.CO;2.

Manogaran, Motharasan, Siti Aqlima Ahmad, Nur Adeela Yasid, Hafeez Muhammad Yakasai, and Mohd Yunus Shukor. 2018. "Characterisation of the Simultaneous Molybdenum Reduction and Glyphosate Degradation by Burkholderia Vietnamiensis AQ5–12 and Burkholderia sp. AQ5–13." *3 Biotech* 8, no. 2: 117, February 7. https://doi.org/10.1007/s13205-018-1141-2.

Mansur, R., N. Gusmanizar, F. A. Dahalan, N. A. Masdor, S. A. Ahmad, M. S. Shukor *et al.* 2016. "Isolation and Characterization of a Molybdenum-Reducing and Amide-Degrading Burkholderia cepacia strain neni-11 in Soils from West Sumatera, Indonesia". *IIOAB Journal.* 7, no. 1: 28–40.

Mansur, Rusnam, Neni Gusmanizar, Muhamad Akhmal Hakim A. H. Roslan, Siti Aqlima Ahmad, and Mohd Yunus Shukor. 2017. "Isolation and Characterisation of a Molybdenum-Reducing and Metanil Yellow Dye-Decolourising Bacillus sp. Strain Neni-10 in Soils from West Sumatera, Indonesia." *Tropical Life Sciences Research* 28, no. 1: 69–90, January. https://doi.org/10.21315/tlsr2017.28.1.5.

Masdor, N., M. S. Abd Shukor, A. Khan, M. I. E. Bin Halmi, S. R. S. Abdullah, N. A. Shamaan *et al.* 2015. "Isolation and Characterization of a Molybdenum-Reducing and SDS-Degrading Klebsiella Oxytoca Strain Aft-7 and Its Bioremediation Application in the Environment." *Biodiversitas* 16, no. 2: 238–46.

Miller, J. K., B. R. Moss, M. C. Bell, and N. N. Sneed. 1972. Comparison of 99Mo metabolism in young cattle and swine. *Journal of Animal Science.* 34, no. 5: 846–50.

Mohamad, O., H. M. Yakasai, K. I. Karamba, M. I. E. Halmi, M. F. Rahman, and M. Y. Shukor. 2017. "Reduction of Molybdenum by Pseudomonas Aeruginosa Strain KIK-11 Isolated from a Metal-Contaminated Soil with Ability to Grow on Diesel and Sodium Dodecyl Sulphate." *Journal of Environmental Microbiology & Toxicology* 5, no. 2: 19–26, December 31. https://doi.org/10.54987/jemat.v5i2.411.

Mohammed, S., M. A. Gafasa, H. T. Kabara, A. Babandi, D. Shehu, M. Ya'u *et al.* 2019. "Soluble Molybdenum Reduction by Morganella sp., Locally Isolated from Agricultural Land in Kano." *Bioremediation Science Technology Research* 7, no. 1: 1–7, July 3.

Monod, J. 1949. "The Growth of Bacterial Cultures." *Annual Review of Microbiology* 3, no. 1: 371–94. https://doi.org/10.1146/annurev.mi.03.100149.002103.

Montgomery, D. C., and G. C. Runger. 1994. *Applied Statistics and Probability for Engineers.* Chichester and New York: John Wiley & Sons.

Munch, J. C., and J. C. G. Ottow. 1983. "Reductive Transformation Mechanism of Ferric Oxides in Hydromorphic Soils." *Environmental Biogeochemistry. International Symposium Stockholm* 1981: 383–94.

Nasernejad, B., T. Kaghazchi, M. Edrisi, and M. Sohrabi. 1999Dec. "Bioleaching of Molybdenum from Low-Grade Copper Ore." *Process Biochemistry* 35, no. 5: 437–40. https://doi.org/10.1016/S0032-9592(99)00067-9.

Neunhäuserer, C., M. Berreck, and H. Insam. 2001. "Remediation of Soils Contaminated with Molybdenum Using Soil Amendments and Phytoremediation." *Water, Air, & Soil Pollution* 128, no. 1–2: 85–96.

Nordmeier, A., J. Woolford, L. Celeste, and D. Chidambaram. 2017. "Sustainable Batch Production of Biosynthesized Nanoparticles." *Materials Letters* 191: 53–6, March 15. https://doi.org/10.1016/j.matlet.2017.01.032.

Olaniran, Ademola O., Adhika Balgobind, and Balakrishna Pillay. 2013. "Bioavailability of Heavy Metals in Soil: Impact on Microbial Biodegradation of Organic Compounds and Possible Improvement Strategies." *International Journal of Molecular Sciences* 14, no. 5: 10197–228. https://doi.org/10.3390/ijms140510197.

Opperman, Diederik Johannes, Lizelle Ann Piater, and Esta Van Heerden. 2008. "A Novel Chromate Reductase from Thermus Scotoductus SA-01 Related to Old Yellow Enzyme." *Journal of Bacteriology* 190, no. 8: 3076–82. https://doi.org/10.1128/JB.01766-07.

Othman, A. R., I. M. Abu Zeid, M. F. Rahman, F. Ariffin, and M. Y. Shukor. 2015. "Isolation and Characterization of a Molybdenum-Reducing and Orange G-Decolorizing Enterobacter sp. Strain Zeid-6 in Soils from Sudan." *Bioremediation Science Technology Research* 3, no. 2: 13–19.

Othman, A. R., N. A. Bakar, M. I. E. Halmi, W. L. W. Johari, S. A. Ahmad, H. Jirangon, M. A. Syed, and M. Y. Shukor. 2013. "Kinetics of Molybdenum Reduction to Molybdenum Blue by Bacillus sp. Strain A. Rzi." *BioMed Research International* 2013: 371058. https://doi.org/10.1155/2013/371058.

Pandey, Ratna, and S. P. Singh. 2002. "Effects of Molybdenum on Fertility of Male Rats." *Bio Metals* 15, no. 1: 65–72. https://doi.org/10.1023/a:1013193013142.

Bacterial Reduction of Molybdenum for Its Bioremediation

Pennesi, Chiara, Cecilia Totti, and Francesca Beolchini. 2013. "Removal of Vanadium(III) and Molybdenum (V)." *PLOS ONE* 8, no. 10: E76870. https://doi.org/10.1371/journal.pone.0076870.

Pitt, M. A. 1976. "Molybdenum Toxicity: Interactions Between Copper, Molybdenum and Sulphate." *Agents & Actions* 6, no. 6: 758–69. https://doi.org/10.1007/BF02026100.

Qadir, Muhammad Abdul, Jamshaid Hussain Zaidi, Shaikh Asrar Ahmad, Asad Gulzar, Muhammad Yaseen, Sadia Atta, and Asma Tufail. 2012. "Evaluation of Trace Elemental Composition of Aerosols in the Atmosphere of Rawalpindi and Islamabad Using Radio Analytical Methods." *Applied Radiation & Isotopes* 70, no. 5: 906–10. https://doi.org/10.1016/j.apradiso.2012.02.047.

Rahman, M. A., S. A. Ahmad, S. Salvam, M. I. E. Halmi, M. T. Yusof, M. Y. Shukor *et al.* 2013. "Dialysis Tubing Experiment Showed That Molybdenum Reduction in S. Marcescens Strain DrY6 Is Mediated by Enzymatic Action." *Journal of Environmental Bioremediation Toxicology* 1, no. 1: 25–7.

Rahman, M. F., M. Rusnam, N. Gusmanizar, N. A. Masdor, C. H. Lee, M. S. Shukor, M. A. H. Roslan, and M. Y. Shukor. 2016. "Molybdate-Reducing and SDS-Degrading Enterobacter sp. Strain Neni-13." *Nova Biotechnologica & Chimica* 15, no. 2: 166–81, December 1. https://doi.org/10.1515/nbec-2016-0017.

Rahman, M. F. A., M. Y. Shukor, Z. Suhaili, S. Mustafa, N. A. Shamaan, and M. A. Syed. 2009. "Reduction of Mo(VI) by the Bacterium Serratia sp. strain DRY5." *Journal of Environmental Biology* 30, no. 1: 65–72.

Rajagopalan, K. V. 1988. "Molybdenum: An Essential Trace Element in Human Nutrition." *Annual Review of Nutrition* 8: 401–27. https://doi.org/10.1146/annurev.nu.08.070188.002153.

Rege, M. A., J. N. Petersen, D. L. Johnstone, C. E. Turick, D. R. Yonge, and W. A. Apel. 1997. "Bacterial Reduction of Hexavalent Chromium by Enterobacter cloacae strain H01 Grown on Sucrose." *Biotechnology Letters* 19, no. 7: 691–4. https://doi.org/10.1023/A:1018355318821.

Retamal-Morales, Gerardo, Marika Mehnert, Ringo Schwabe, Dirk Tischler, Claudia Zapata, Renato Chávez, Michael Schlömann, and Gloria Levicán. 2018. "Detection of Arsenic-Binding Siderophores in Arsenic-Tolerating Actinobacteria by a Modified CAS Assay." *Ecotoxicology & Environmental Safety* 157: 176–81, August 15. https://doi.org/10.1016/j.ecoenv.2018.03.087.

Richards, F. J. 1959. "A Flexible Growth Function for Empirical Use." *Journal of Experimental Botany* 10, no. 2: 290–301. https://doi.org/10.1093/jxb/10.2.290.

Robinson, M. F., J. M. McKenzie, C. D. Tomson, and A. L. van Rij. 1973. "Metabolic Balance of Zinc, Copper, Cadmium, Iron, Molybdenum and Selenium in Young New Zealand Women." *British Journal of Nutrition* 30, no. 2: 195–205. https://doi.org/10.1079/bjn19730025.

Roy, Dipankar Chandra, Sudhangshu Kumar Biswas, Md Moinuddin Sheam, Md Rockybul Hasan, Ananda Kumar Saha, Apurba Kumar Roy, Md Enamul Haque, Md Mizanur Rahman, and Swee-Seong Tang. 2020. "Bioremediation of Malachite Green Dye by Two Bacterial Strains Isolated from Textile Effluents." *Current Research in Microbial Sciences* 1: 37–43. https://doi.org/10.1016/j.crmicr.2020.06.001.

Runnells, D. D., D. S. Kaback, and E. M. Thurman. 1976. "Geochemistry and Sampling of Molybdenum in Sediments, Soils, and Plants in Colorado." In *Molybdenum in the Environment*, edited by W. R. Chappel and K. K. Peterson. New York: Marcel and Dekker, Inc.

Rusnam, N., and N. Gusmanizar. 2022. "Isolation and Characterization of a Molybdenum-Reducing and the Congo Red Dye-Decolorizing *Pseudomonas putida* Strain Neni-3 in Soils from West Sumatera, Indonesia". *Journal of Biochemistry. Journal of Microbiology & Biotechnology* 10, no. 1: 17–24. https://doi.org/10.54987/jobimb.v10i1.658.

Rusnam, Gusmanizar N., N. Gusmanizar, M. F. Rahman, and N. A. Yasid. 2019. "Characterization of a Molybdenum-Reducing and Phenol-Degrading Pseudomonas sp. Strain Neni-4 from Soils in West Sumatera, Indonesia." *Bulletin of Environmental Science & Sustainable Management* 6, no. 1: 1–8. https://doi.org/10.54987/bessm.v6i1.670.

Rusnam, Gusmanizar N., and N. Gusmanizar. 2019 Dec.31. "Isolation and Characterization of a Molybdenum-Reducing and Coumaphos-Degrading Bacillus sp. Strain Neni-12 in Soils from West Sumatera, Indonesia." *Journal of Environmental Microbiology & Toxicology* 7, no. 2: 20–5. https://doi.org/10.54987/jemat.v7i2.492.

Rusnam, N., and N. Gusmanizar. 2020. "Isolation and Characterization of a Molybdenum-Reducing and Carbamate-Degrading Bacillus Amyloliquefaciens Strain Neni-9 in Soils from West Sumatera, Indonesia." *Bioremediation Science & Technology Research* 8, no. 1: 17–22, July 31. https://doi.org/10.54987/bstr.v8i1.511.

Rusnam, N., and N. Gusmanizar. 2022. "Isolation and Characterization of a Molybdenum-Reducing and the Congo Red Dye-Decolorizing Pseudomonas Putida Strain Neni-3 in Soils from West Sumatera, Indonesia." *Journal of Biochemistry, Microbiology & Biotechnology* 10, no. 1: 17–24, July 31. https://doi.org/10.54987/jobimb.v10i1.658.

Rusnam, Rahman M. F., N. Gusmanizar, H. M. Yakasai, and M. Y. Shukor. 2021 Jul. 31. "Molybdate Reduction to Molybdenum Blue and Growth on Polyethylene Glycol by Bacillus sp. Strain Neni-8." *Bulletin of the Environmental Science Sustainable Management (e-ISSN 5353)* 5, no. 1: 12–19.

Sabo, I. A., S. Yahuza, and M. Y. Shukor. 2021. "Molybdenum Blue Production from Serratia sp. Strain DRY5: Secondary Modeling." *Bioremediation Science & Technology Research* 9, no. 2: 21–4, December 31. https://doi.org/10.54987/bsr.v9i2.622.

Sabullah, M. K., M. F. Rahman, S. A. Ahmad, M. R. Sulaiman, M. S. Shukor, N. A. Shamaan, and M. Y. Shukor. 2016. "Isolation and Characterization of a Molybdenum-Reducing and Glyphosate-Degrading Klebsiella Oxytoca Strain Saw-5 in Soils from Sarawak." *Agrivita Journal of Agricultural Science* 38, no. 1: 1–13. https://doi.org/10.17503/agrivita.v38i1.654.

Saeed, A. M., E. El Shatoury, and R. Hadid. 2019. "Production of Molybdenum Blue by Two Novel Molybdate-Reducing Bacteria Belonging to the Genus Raoultella Isolated from Egypt and Iraq." *Journal of Applied Microbiology* 126, no. 6: 1722–8, June. https://doi.org/10.1111/jam.14254.

Saeed, Ali M., Hayam A. E. Sayed, and Einas H. El-Shatoury. 2020 May 1. "Optimizing the Reduction of Molybdate by Two Novel Thermophilic Bacilli Isolated from Sinai, Egypt." *Current Microbiology* 77, no. 5: 786–94. https://doi.org/10.1007/s00284-020-01874-y.

Schröder, I., S. Rech, T. Krafft, and J. M. Macy. 1997. "Purification and Characterization of the Selenate Reductase from Thauera Selenatis." *Journal of Biological Chemistry* 272, no. 38: 23765–8, September 19. https://doi.org/10.1074/jbc.272.38.23765.

Schroeder, H. A., D. V. Frost, and J. J. Balassa. 1970. "Essential Metals in Man: Selenium." *Journal of Chronic Diseases* 23, no. 4: 227–43. https://doi.org/10.1016/0021-9681(70)90003-2.

Sebenik, R. F., A. R. Burkin, and R. R. Dorfler. 2013. "Molybdenum and Molybdenum Compounds." In *Ullmann's Encyclopedia of Industrial Chemistry*. Volume 23. Weinheim: Wiley-VCH.

Sharma, Y. C., and S. N. Upadhyay. 2009. "Removal of a Cationic Dye from Wastewaters by Adsorption on Activated Carbon Developed from Coconut Coir." *Energy & Fuels* 23, no. 6: 2983–8. https://doi.org/10.1021/ef9001132.

Shuhaimi, N., M. A. Abdel-Mongy, N. A. Shamaan, C. H. Lee, M. A. Syed, and M. Y. Shukor. 2021. "Isolation and Characterization of a PEG-Degrading and Mo-Reducing Escherichia Coli Strain Amr-13 in Soils from Egypt." *Journal of Environmental Microbiology & Toxicology* 9, no. 2: 23–9, December 31. https://doi.org/10.54987/jemat.v9i2.643.

Shukor, M. S., A. Khan, N. Masdor, M. I. E. Halmi, S. R. S. Abdullah, and M. Y. Shukor. 2016. "Isolation of a Novel Molybdenum-Reducing and Azo Dye Decolorizing Enterobacter sp. Strain Aft-3 from Pakistan." *Chiang Mai University Journal of Natural Sciences* 15, no. 2: 95–114.

Shukor, M. S., and M. Y. Shukor. 2015. "Bioremoval of Toxic Molybdenum Using Dialysis Tubing." *Chemical Engineering Research Bulletin* 18, no. 1: 6–11. https://doi.org/10.3329/cerb.v18i1.26215.

Shukor, M. Y. 2014. "Revisiting the Role of the Electron Transport Chain in Molybdate Reduction by *Enterobacter cloacae* Strain 48." *Indian Journal of Biotechnology* 13, no. 3: 404–7.

Shukor, M. Y., N. Gusmanizar, N. A. Azmi, M. Hamid, J. Ramli, N. A. Shamaan, and M. A. Syed. 2009. "Isolation and Characterization of an Acrylamide-Degrading *Bacillus cereus*" *Journal of Environmental Biology* 30, no. 1: 57–64.

Shukor, M. Y., M. I. E. Halmi, M. F. A. Rahman, N. A. Shamaan, and M. A. Syed. 2014. "Molybdenum Reduction to Molybdenum Blue in Serratia sp. Strain DRY5 Is Catalyzed by a Novel Molybdenum-Reducing Enzyme." *BioMed Research International* 2014: 853084. https://doi.org/10.1155/2014/853084.

Shukor, M. Y., C. H. Lee, I. Omar, Syed M. A. Mia Karim, and N. A. Shamaan. 2003. "Isolation and Characterization of a Molybdenum-Reducing Enzyme in Enterobacter Cloacae Strain 48." *Pertanika Journal Science & Technology* 11, no. 2: 261–72.

Shukor, M. Y., M. F. A. Rahman, N. A. Shamaan, C. H. Lee, M. I. A. Karim, and M. A. Syed. 2008a. "An Improved Enzyme Assay for Molybdenum-Reducing Activity in Bacteria." *Applied Biochemistry & Biotechnology* 144, no. 3: 293–300. https://doi.org/10.1007/s12010-007-8113-z.

Shukor, M. Y., S. H. Habib, M. F. Rahman, H. Jirangon, M. P. Abdullah, N. A. Shamaan and M. A. Syed. 2008b. "Hexavalent molybdenum reduction to molybdenum blue by S. marcescens strain Dr. Y6". *Applied Biochemistry & Biotechnology*. 149, no 1: 33–43.

Shukor, M. Y., M. F. Rahman, Z. Suhaili, N. A. Shamaan, and M. A. Syed. 2010a. "Hexavalent Molybdenum Reduction to Mo-Blue by *Acinetobacter calcoaceticus*." *Folia Microbiologica (Praha)* 55, no. 2: 137–43. https://doi.org/10.1007/s12223-010-0021-x.

Shukor, M. Y., Ahmad, S. A., Nadzir, M. M. M., Abdullah, M. P., Shamaan, N. A. and Syed, M. A. 2010b. "Molybdate reduction by Pseudomonas sp. strain DRY2". *Journal of Applied Microbiology*, 108: 2050–2058.

Shukor, Y., H. A. Adam, K. I. Ithnin, I. Y. Yunus, N. A. S. Shamaan, and A. Syed. 2007. "Molybdate Reduction to Molybdenum Blue in Microbe Proceeds via a Phosphomolybdate Intermediate." *Journal of Biological Sciences* 7, no. 8: 1448–52. https://doi.org/10.3923/jbs.2007.1448.1452.

Sidgwick, N. V. 1951. *The Chemical Elements and Their Compounds*. Oxford, London: Clarendon Press.

Silver, S., and L. T. Phung. 1996. "Bacterial Heavy Metal Resistance: New Surprises." *Annual Review of Microbiology* 50: 753–89. https://doi.org/10.1146/annurev.micro.50.1.753.

Simeonov, L. I., M. V. Kochubovski, and B. G. Simeonova. 2011. *Environmental Heavy Metal Pollution and Effects on Child Mental Development* (NATO Science for Peace and Security Series C: Environmental Security; Vol. 1). Dordrecht: Springer Netherlands.

Sims, R. P. A. 1961. "Formation of Heteropoly Blue by Some Reduction Procedures Used in the Micro-Determination of Phosphorus." *Analyst* 86, no. 1026: 584–90. https://doi.org/10.1039/an9618600584.

Smedley, P. L., and D. G. Kinniburgh. 2017. "Molybdenum in Natural Waters: A Review of Occurrence, Distributions and Controls." *Applied Geochemistry* 84: 387–432, September 1. https://doi.org/10.1016/j.apgeochem.2017.05.008.

Soda, Satoshi O., Shigeki Yamamura, Hong Zhou, Michihiko Ike, and Masanori Fujita. 2006. "Reduction Kinetics of As (V) to As (III) by a Dissimilatory Arsenate-Reducing Bacterium, Bacillus sp. SF-1." *Biotechnology & Bioengineering* 93, no. 4: 812–15. https://doi.org/10.1002/bit.20646.

Stafford, Jennifer M., Charles E. Lambert, Justin A. Zyskowski, Cheryl L. Engfehr, Oscar J. Fletcher, Shanna L. Clark, Asheesh Tiwary, Cynthia M. Gulde, and Bradley E. Sample. 2016. "Dietary Toxicity of Soluble and Insoluble Molybdenum to Northern Bobwhite Quail (*Colinus virginianus*)." *Ecotoxicology* 25, no. 2: 291–301. https://doi.org/10.1007/s10646-015-1587-5.

Sugio, T., Y. Tsujita, T. Katagiri, K. Inagaki, and T. Tano. 1988. "Reduction of Mo6+ with Elemental Sulfur by Thiobacillus Ferrooxidans." *Journal of Bacteriology* 170, no. 12: 5956–9. https://doi.org/10.1128/jb.170.12.5956-5959.1988.

Sukumar, M. 2010. "Reduction of Hexavalent Chromium by Rhizopus Oryzae." *African Journal of Environmental Science & Technology* 4, no. 7: 412–18.

Syed, M. A., N. A. Shamaan, and M. Y. Shukor. 2020. "Mathematical Modeling of the Molybdenum Blue Production from Serratia sp. Strain DRY5." *Journal of Environmental Microbiology & Toxicology* 8, no. 2: 12–17, December 31. https://doi.org/10.54987/jemat.v8i2.565.

Tambat, V. S., Y. Tseng, P. Kumar, C. W. Chen, R. R. Singhania, J. S. Chang, C. Dong, and A. K. Patel. 2023. "Effective and Sustainable Bioremediation of Molybdenum Pollutants from Wastewaters by Potential Microalgae." *Environmental Technology & Innovation* 30: 103091, May 1. https://doi.org/10.1016/j.eti.2023.103091.

Teissier, G. 1942. "Growth of Bacterial Populations and the Available Substrate Concentration." *Review of Scientific Instruments* 3208: 209–14.

Truex, M. J., B. M. Peyton, N. B. Valentine, and Y. A. Gorby. 1997. "Kinetics of U(VI) Reduction by a Dissimilatory Fe(III)-Reducing Bacterium Under Non-Growth Conditions." *Biotechnology & Bioengineering* 55, no. 3: 490–6, August 5. https://doi.org/10.1002/(SICI)1097-0290(19970805)55:3<490::AID-BIT4>3.0.CO;2-7.

Tucker, M. D., L. L. Barton, and B. M. Thomson. 1997. "Reduction and Immobilization of Molybdenum by Desulfovibrio Desulfuricans." *Journal of Environmental Quality* 26, no. 4: 1146–52. https://doi.org/10.2134/jeq1997.00472425002600040029x.

Uba, G., A. Abubakar, H. M. Yakasai, and M. E. Khayat. 2022. "Mathematical Modeling of Molybdenum Blue Production from Bacillus sp. Strain Khayat." *Bulletin of Environmental Science & Sustainable Management* 6, no. 2: 8–13. https://doi.org/10.54987/bessm.v6i2.743.

Wang, Xiaoqing, Gianluca Brunetti, Wenjie Tian, Gary Owens, Yang Qu, Chaoxi Jin, and Enzo Lombi. 2021. "Effect of Soil Amendments on Molybdenum Availability in Mine Affected Agricultural Soils." *Environmental Pollution* 269: 116132, January 15. https://doi.org/10.1016/j.envpol.2020.116132.

Ward, G. M. 1978. "Molybdenum Toxicity and Hypocuprosis in Ruminants: A Review." *Journal of Animal Science* 46, no. 4: 1078–85. https://doi.org/10.2527/jas1978.4641078x.

WHO. 2011. *Molybdenum in Drinking-Water Background Document for Development of WHO Guidelines for Drinking-Water Quality*. Geneva: WHO.

Wu, Songwei, Chengxiao Hu, Qiling Tan, Zhaojun Nie, and Xuecheng Sun. 2014. "Effects of Molybdenum on Water Utilization, Antioxidative Defense System and Osmotic-Adjustment Ability in Winter Wheat (Triticum aestivum) Under Drought Stress." *Plant Physiology & Biochemistry* 83: 365–74. https://doi.org/10.1016/j.plaphy.2014.08.022.

Wuana, R. A., and F. E. Okieimen. 2011. "Heavy Metals in Contaminated Soils: A Review of Sources, Chemistry, Risks and Best Available Strategies for Remediation." *ISRN Ecology* 2011: 1–20. https://doi.org/10.5402/2011/402647.

Xing, Jie, Chunyan Li, Wanting Li, Xuemei Zhang, Zhaoquan Li, and Ang Li. 2023. "Isolation and Identification of the Molybdenum-Resistant Strain Raoultella Ornithinolytica A1 and Its Effect on MoO42– in the Environment." *Biodegradation* 34, no. 2: 169–80, April 1. https://doi.org/10.1007/s10532-022-10011-4.

Xing, Jie, Chunyan Li, Zhaoquan Li, Wanting Li, Ailun Fang, and Ang Li. 2023. "Submerged Macrophytes Mediated Remediation of Molybdenum-Contaminated Sediments." *Environmental Science & Pollution Research International* 30, no. 17: 48962–71, February 10. https://doi.org/10.1007/s11356-023-25537-0.

Yakasai, H. M., K. I. Karamba, N. A. Yasid, M. I. E. Halmi, M. F. Rahman, S. A. Ahmad, and M. Y. Shukor. 2019. "Response Surface-Based Optimization of a Novel Molybdenum-Reducing and Cyanide-Degrading *Serratia* sp. Strain HMY1." *Desalination & Water Treatment* 145: 220–31. https://doi.org/10.5004/dwt.2019.23734.

Yakasai, H. M., M. F. Rahman, M. Manogaran, N. Adeela Yasid, M. A. Syed, N. A. Shamaan, and M. Y. Shukor. 2021. "Microbiological Reduction of Molybdenum to Molybdenum Blue as a Sustainable Remediation Tool for Molybdenum: A Comprehensive Review." *International Journal of Environmental Research & Public Health* 18, no. 11: 5731, January. https://doi.org/10.3390/ijerph18115731.

Yakasai, H. M., M. F. Rahman, N. A. Yasid, S. A. Ahmad, M. I. E. Halmi, and M. Y. Shukor. 2017a. "Elevated Molybdenum Concentrations in Soils Contaminated with Spent Oil Lubricant." *Journal of Environmental Microbiology & Toxicology* 5, no. 2: 1–3. https://doi.org/10.54987/jemat.v5i2.407.

Yakasai, H. M., M. F. A. Rahman, M. A. El-Mongy, N. A. Shamaan, C. H. Lee, M. A. Syed *et al.* 2022 Dec. 31. "Isolation and Characterization of a Molybdenum-Reducing *Enterobacter aerogenes* Strain Amr-18 in Soils from Egypt That Could Grow on Amides." *Bulletin of Environmental Science and Sustainable Management (e-ISSN 2716-5353)* 6, no. 2: 40–7.

Yakasai, H. M., N. A. Yasid, and M. Y. Shukor. 2018. "Temperature Coefficient and Q10 Value Estimation for the Growth of Molybdenum-Reducing *Serratia* sp. Strain HMY1." *Bioremediation Science* and *Technology Research* 6, no. 2: 22–4.

Yakasai, M. H. 2017. *Faculty of Biotechnology and Biomolecular Sciences*. A thesis submitted in partial fulfilment of the requirements for the degree of Doctor of Philosophy in the Department of Biochemistry. Universiti Putra Malaysia, Malaysia.

Yakasai, M. H., K. K. Ibrahim, N. A. Yasid, M. I. E. Halmi, M. F. A. Rahman, and M. Y. Shukor. 2016 Dec. 31. "Mathematical Modelling of Molybdenum Reduction to Mo-Blue by a Cyanide-Degrading Bacterium." *Bioremediation Science & Technology Research* 4, no. 2: 1–5. https://doi.org/10.54987/bstr.v4i2.368.

Yakasai, M. H., M. F. A. Rahman, M. B. H. A. Rahim, M. E. Khayat, N. A. Shamaan, and M. Y. Shukor. 2017b. "Isolation and Characterization of a Metal-Reducing *Pseudomonas* sp. Strain 135 with Amide-Degrading Capability." *Bioremediation Science & Technology Research* 5, no. 2: 32–8. https://doi.org/10.54987/bstr.v5i2.361.

Yakasai, M. H., and M. Manogaran. 2020. "Kinetic Modelling of Molybdenum-Blue Production by Bacillus sp. Strain Neni-10." *Journal of Environmental Microbiology & Toxicology* 8, no. 1: 5–10. https://doi.org/10.54987/jemat.v8i1.515.

Yamaguchi, Sonoko, Chiemi Miura, Aki Ito, Tetsuro Agusa, Hisato Iwata, Shinsuke Tanabe, Bui Cach Tuyen, and Takeshi Miura. 2007. "Effects of Lead, Molybdenum, Rubidium, Arsenic and Organochlorines on Spermatogenesis in Fish: Monitoring at Mekong Delta Area and In Vitro Experiment." *Aquatic Toxicology* 83, no. 1: 43–51. https://doi.org/10.1016/j.aquatox.2007.03.010.

Yano, T. and S. Koga. 1969. "Dynamic behaviour of the chemostat subject to substrate inhibition". *Biotechnology and Bioengineering* vol 11 no. 2: 139–153.

Yong, F. S. 2000. "Mamut Copper Mine—The Untold Story." In *Minerals: Underpinning Yesterday's Needs, Today's Development and Tomorrow's Growth. Sabah* [Internet]. Kota Kinabalu, Sabah: Malaysian Chamber of Mines, Malaysia, Pacific Sutera Hotel, June 22–24. Available from Proceedings of the Minerals: Underpinning Yesterday's Needs, Today's Development and Tomorrow's Growth.

Yong, N. K., M. Oshima, R. C. Blake, II, and T. Sugio. 1997. "Isolation and Some Properties of an Iron-Oxidizing Bacterium Thiobacillus Ferrooxidans Resistant to Molybdenum Ion." *Bioscience, Biotechnology, & Biochemistry* 61, no. 9: 1523–6. https://doi.org/10.1271/bbb.61.1523.

You, X. Y., H. Wang, G. Y. Ren, J. J. Li, X. Duan, H. J. Zheng, and Z. Jiang. 2015. "Complete Genome Sequence of the Molybdenum-Resistant Bacterium Bacillus subtilis strain LM 4-2." *Standards in Genomic Sciences* 10, no. 1, December. https://doi.org/10.1186/s40793-015-0118-6.

Yu, Xuan, Juanjuan Shi, Aman Khan, Hui Yun, Pengyun Zhang, Peng Zhang, Apurva Kakade, Yanrong Tian, Yaxin Pei, Yiming Jiang, Haiying Huang, Kejia Wu, and Xiangkai Li. 2020. "Immobilized-Microbial Bioaugmentation Protects Aerobic Denitrification from Heavy Metal Shock in an Activated-Sludge Reactor." *Bioresource Technology* 307: 123185, July 1. https://doi.org/10.1016/j.biortech.2020.123185.

Yuan, T., J. Xu, Z. Wang, Z. Lei, M. Kato, K. Shimizu, and Z. Zhang. 2023Jul.15. "Efficient Removal of Molybdate from Groundwater with Visible Color Changes Using Wasted Aerobic Granular Sludge." *Separation & Purification Technology* 317: 123849. https://doi.org/10.1016/j.seppur.2023.123849.

Yunus, Shukor Mohd, Hamdan Mohd Hamim, Othaman Mohd Anas, Shamaan Nor Aripin, and Syed Mohd Arif. 2009. "Mo(VI) Reduction to Molybdenum Blue by Serratia Marcescens Strain Dr. Y 9." *Polish Journal of Microbiology* 58, no. 2: 141–7.

Zeid, I. M. A., M. F. Rahman, and M. Y. Shukor. "Isolation of A Molybdenum-Reducing Bacillus sp. strain Zeid 15 and Modeling of Its Growth on Amides." *Bulletin of Environmental Science and Sustainable Management (e-ISSN 2716-5353, no. 2021)* 5, no. 2: 19–27, December 31.

Zhai, Xiao-Wei, Yu-Ling Zhang, Qiao Qi, Y. Bai, Xiao-Li Chen, Li-Jun Jin, Xue-Gang Ma, Run-Zhe Shu, Zi-Jun Yang, and Feng-Jun Liu. 2013. "Effects of Molybdenum on Sperm Quality and Testis Oxidative Stress." *Systems Biology in Reproductive Medicine* 59, no. 5: 251–5. https://doi.org/10.3109/19396368.2013.791347.

Zin, Khairunnisa' Mohd, Mohd Izuan Effendi Halmi, Siti Salwa Abd Gani, Uswatun Hasanah Zaidan, A. Wahid Samsuri, and Mohd Yunus Abd Shukor. 2020. "Microbial Decolorization of Triazo Dye, Direct Blue 71: An Optimization Approach Using Response Surface Methodology (RSM) and Artificial Neural Network (ANN)." [Internet]. *BioMed Research International*: 2734135. Hindawi Publishing. Accessed September 10, 2020. https://www.hindawi.com/journals/bmri/2020/2734135/. https://doi.org/10.1155/2020/2734135.

Zwietering, M. H., I. Jongenburger, F. M. Rombouts, and K. van 't Riet. 1990. "Modeling of the Bacterial Growth Curve." *Applied & Environmental Microbiology* 56, no. 6: 1875–81. https://doi.org/10.1128/aem.56.6.1875-1881.1990.

13 Health Hazards and Bacterial Bioremediation of Endocrine-Disrupting Chemicals — A Concise Discussion on Phthalic Acid Esters and the Organophosphorus Pesticide Malathion

Shalini Chandel, Rishi Mahajan, and Subhankar Chatterjee

13.1 INTRODUCTION

Endocrine glands are among the vital hormone-secreting systems which play a pivotal role in human development, adaptation, and maintenance of many biological processes. They play a pivotal role in regulating many essential biological and physiological functions. Therefore, any dysfunction in the endocrine system can lead to many irreversible health complications or even death (Figure 13.1). The majority of human-made chemicals, commonly known as endocrine-disrupting chemicals (EDCs), are defined as exogenous (synthetic) chemicals, or a mixture of chemicals, which hamper the normal action of hormones. In the past 30 years, there has been burgeoning scientific evidence of EDCs' impact on human and animal health. Based on field research, epidemiological data on humans, and laboratory research with cell cultures and animal models, a detailed mechanism of how EDCs cause biological changes and how that may lead to acute health effects has been explored. Synthetic chemicals, like EDCs, that find their way into the human system can disrupt the action of hormones either directly by acting on a hormone receptor-protein complex or indirectly via action on a particular protein that controls the deliverance of hormones. EDCs can get absorbed into the body and mimic or block hormone action. They disrupt the endocrine system by changing hormone levels, affecting the synthesis or metabolism of hormones or altering the way hormones function (Hamid *et al.*, 2021).

13.2 ENVIRONMENTAL OCCURRENCE AND HEALTH EFFECTS OF ENDOCRINE-DISRUPTING CHEMICALS

In today's world, human dependence on synthetic products is unavoidable, and consequently, human exposure to EDCs is a universal and omnipresent problem (Metcalfe *et al.*, 2022). Some common examples of EDCs are DDT, various groups of pesticides, bisphenol A (BPA), and phthalates. These chemicals are frequently used in children's toys, personal care products, food containers, and flame retardants (used in furniture and floor coverings). As a result of the existence of EDCs in the environment, common exposure locations include residences, workplaces, agricultural areas, the

DOI: 10.1201/9781003423393-13

Health Hazards and Bacterial Bioremediation of EDCs 279

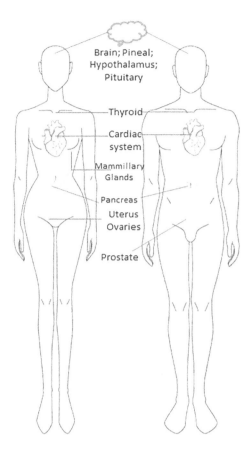

FIGURE 13.1 Site of action of endocrine-disrupting chemicals on major endocrine glands in females (left) and males (right) in the human body.

atmosphere, food, and potable water (Lauretta *et al.*, 2019). Table 13.1 provides a concise summary of the environmental origins and impacts on health associated with EDCs. On average, 1000 chemical products out of 85,000 are characterized as potential endocrine disruptors (Street *et al.*, 2018). Recent scientific research into the quantification of chemicals, notably EDCs, in bodily fluids and tissues has shown that a significant portion of the human population has been affected by exposure to EDCs. Numerous studies, on the other hand, demonstrated the potential interference of EDCs (pesticides like dichlorodiphenyltrichloroethane, methoxychlor (MXC), vinclozolin and atrazine; surfactants and detergents such as octyphenol, nonylphenol, and bisphenol-A; plasticizers such as phthalates; industrial compounds (polychlorinated biphenyls (PCBs)) with other endocrine systems (Hamid *et al.*, 2021). Some major reproductive hormones like progestins, androgens, and estrogens are among the primary targets of several EDCs (Mahajan *et al.*, 2022). Evidence of reproductive malfunction, like reduced fertility and unusual sexual development in wildlife and humans, has been observed due to exposure to EDCs (Table 13.1).

13.2.1 Phthalic Acid Esters and the Organophosphate Pesticide Malathion: Two Potent Endocrine-Disrupting Chemicals

Within this section, we will explore two crucial EDCs that have a substantial impact on the well-being of both humans and animals.

TABLE 13.1

List of Important Environmentally Persistent EDCs and Their Health Effects

EDCs: Environmental Source/Common Occurrence and Ill-Health Effects

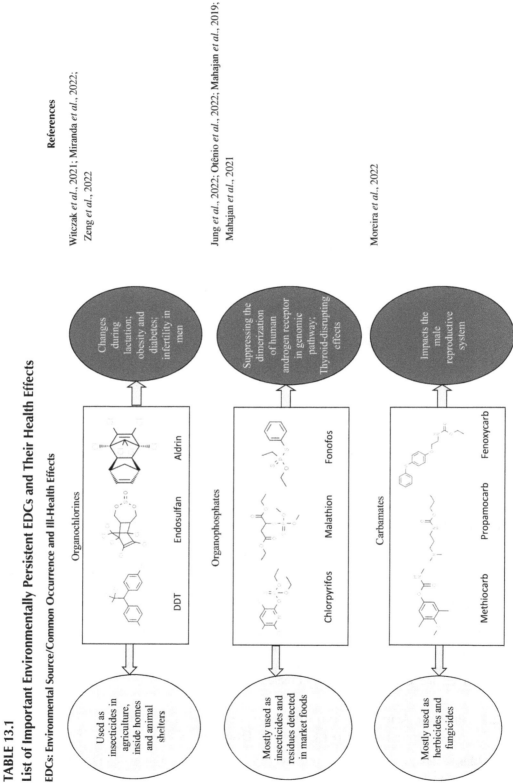

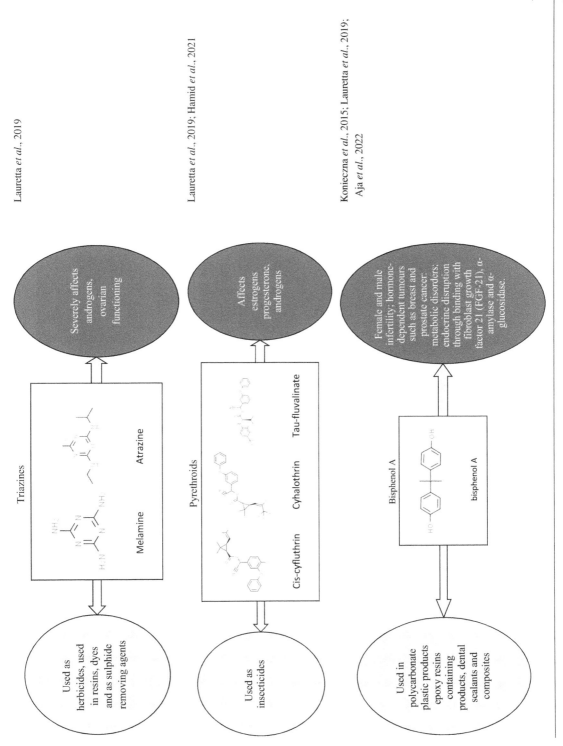

(Continued)

TABLE 13.1 *(Continued)*

List of Important Environmentally Persistent EDCs and Their Health Effects

EDCs: Environmental Source/Common Occurrence and Ill-Health Effects

Tiwari *et al.*, 2022

Results from burning processes like backyard burning, commercial/municipal waste incineration, and the use of fuels like wood, coal and oil generates dioxins

Dioxin

Dioxin

Incapacity to maintain pregnancy, decreased fertility, birth defects, endometriosis, immune system suppression, skin disorders, lowered testosterone levels, type 2 diabetes, human sex hormone-binding globulin inhibitors

Hu *et al.*, 2023

Anti-fungal agent often used in a variety of cosmetics and personal-care products, food preservative, also used as a fungicide in drosophila food media

Methylparaben

Methylparaben

Skin aging and DNA damage; decrease mRNA expression in rats, affects steroidogenesis, decreased estradiol levels in peripubertal female rats

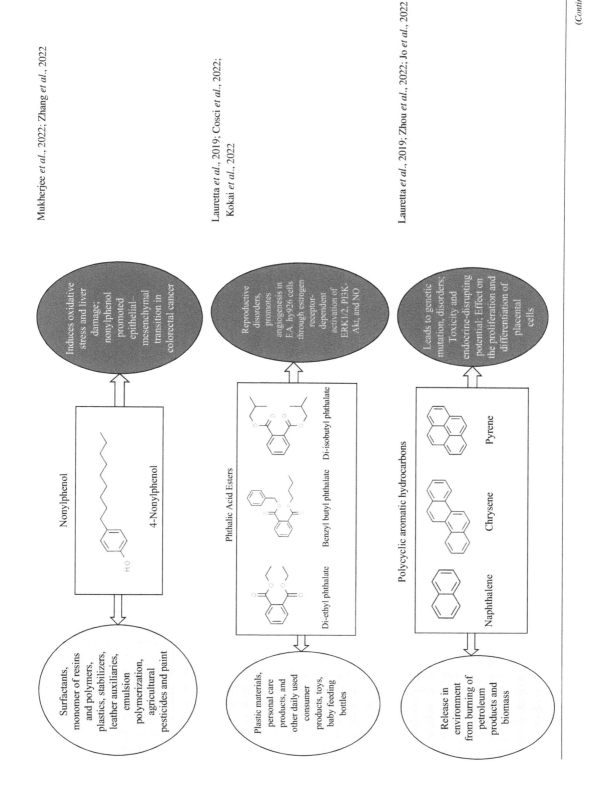
(Continued)

TABLE 13.1 (Continued)
List of Important Environmentally Persistent EDCs and Their Health Effects

EDCs: Environmental Source/Common Occurrence and Ill-Health Effects

Polybrominated diphenyl ethers (PBDEs)

Used as flame retardants

Neurobehavioral effects, especially neurodevelopment and cognitive performance; inhibit reproductive function by suppressing multiple aspects of the associated hypothalamic circuitry; induce depression like symptoms

References

Ramhøj *et al.*, 2022; Arowolo *et al.*, 2022; Xu *et al.*, 2023

13.2.1.1 Phthalic Acid Esters and Their Deleterious Health Effects

The term *phthalate* originated from phthalic acid—an isomer of ortho-isomer or phthalic acid, terephthalic acid, and meta-isomer or iso-phthalic acid. Phthalic acid esters are generally used in polyvinyl chloride production and terephthalic acid esters find their primary use in the production of carbonated beverage containers. Within the array of EDCs, a significant group of environmental contaminants with widespread industrial production and use are phthalic acid esters (PAEs), also known as 1,2-benzene dicarboxylic acid esters. These compounds have been shown to disrupt reproductive and behavioral health in both wildlife and humans.

Phthalates, commonly known as ortho-phthalates, are produced by employing a catalyst to react an excess of either straight or branched chain alcohols with phthalic anhydride. Phthalic acid esters are not chemically bonded through covalent bonds to the polymer and can migrate to the surface of the polymer matrix. PAEs can be discharged into the environment through various physical mechanisms. Phthalates represent the most widely utilized plasticizers on a global scale. (Eales *et al.*, 2022). They find application in the production of children's toys, personal hygiene items, food containers, pharmaceuticals, cleaning products, dietary supplements, lubricants, solvents, pesticides, paints, glues, and coatings. The advantageous properties of PAEs, characterized by their low melting points and high boiling points, render them highly desirable for applications as heat-transfer fluids and carriers. These PAEs have been divided into two groups, distinguishing between their chemical complexities: low-molecular-weight (LMW) phthalates and high-molecular-weight (HMW) phthalates.

As demonstrated in recent research, PAEs have become prevalent chemical pollutants in various industries due to their haphazard utilization. Over an extended period, the utilization of products containing phthalates has resulted in the release of this chemical pollutant into diverse environmental settings. One major ecological worry pertains to their persistence in the natural environment over an extended duration. Furthermore, there is evidence that PAEs can lead to serious health risks, such as reproductive abnormalities, increases angiogenesis in EA. hy926 cells by activating the ERK1/2, PI3K-Akt, and NO signaling pathways in a receptor-dependent manner (Lauretta *et al.*, 2019; Cosci *et al.*, 2022; Kokai *et al.*, 2022).

13.2.1.2 The Organophosphate Pesticide Malathion and Its Deleterious Health Effects

Organophosphate pesticides (OPPs) are used widely in agricultural practices, especially in developing countries because of their easy availability and low manufacturing cost. Among all OP pesticides, malathion ranks among the oldest and most frequently applied insecticides on a global scale. Malathion is a colorless liquid in its pure form with a garlic-like odor. The molecular mass of malathion is 330.36 grams per mole (g/mol) with molecular formulae $C_{10}H_{19}O_6PS_2$ (diethyl-2-dimethoxyphosphino-thioyl-sulfanyl-butane-dioate) (Figure 13.2).

Over the years, malathion has been widely used in agricultural fields, fruit crops, and commercial and household practices to control a broad spectrum of sucking and chewing insects, including fleas, aphids, hornets, mites, fruit flies, mosquitoes, spiders, ticks, wasps, weevils, ants, and animal parasites (ectoparasites). The most common agricultural food and feed crops malathion is used on

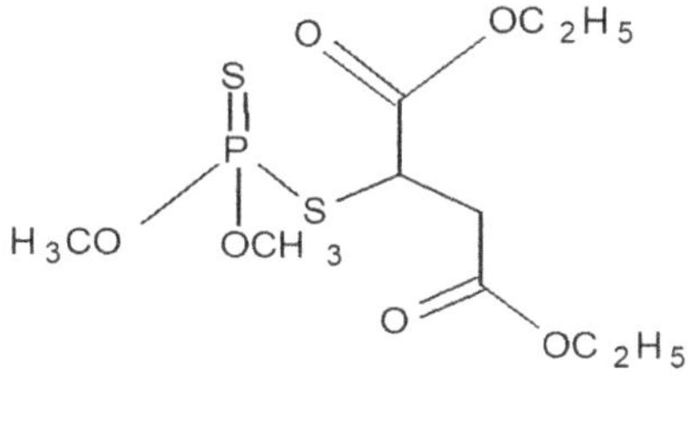

FIGURE 13.2 Chemical structure of malathion (diethyl 2-dimethoxy phosphino thioyl sulfanyl butanedioate).

include raspberries, strawberries, blueberries, cotton, garlic, limes, cherries, apple, paddy, sorghum, pea, brinjal, castor, mustard, tomato, mango, dates, and celery (Vasseghian *et al.*, 2022).

Malathion has both acute and chronic toxic effects in humans. Acute symptoms include nausea, headache, tightness in the chest, excessive sweating, salivation, vomiting, abdominal cramps, and tremors. Farm workers and their children are at the highest risk of pesticide exposure. Pesticide production facilities can be locations where occupational exposure to pesticides results in toxicity among workers. Research has revealed that individuals with prolonged exposure to pesticides in such manufacturing plants exhibited DNA damage (Morsi *et al.*, 2022).

The primary mechanism behind the toxicity of OPP against human beings and other targeted pests is the irreversible inhibition of the acetylcholinesterase enzyme. OPP pesticides mimic acetylcholine molecules and attach to the acetylcholine esterase enzyme active site, blocking it permanently. Scientific findings revealed that when malathion in combination with estrogen was administered to rats, kidney cells showed a significantly higher rate of glomerular hypertrophy, signs of tubular damage, and atypical proliferation of the hilum and cortical zone (Alfaro-Lira *et al.*, 2012). Hepatotoxicity, renal dysfunction, and reproductive dysfunction are all reported in cases of malathion toxicity. *In vivo* studies on animals showed that malathion exposure also leads to chromosomal damage. In addition to the genotoxicity, it was also reported that cumulative doses of malathion (prolonged exposure to low doses of malathion) decrease humoral (antibody-mediated immune response) immunity. A high dose of malathion is responsible for causing suppression in humoral immune responses (Morsi *et al.*, 2022; Abo *et al.*, 2021).

13.3 BACTERIAL BIOREMEDIATION

Microorganisms are one of the best bets to remove these EDCs from the environment, as they work sustainably. However, the dilemma that confronts the scientific community is the availability of novel and efficient EDC-degrading microorganisms. Within this framework, exploring unique microorganisms from sites contaminated with endocrine-disrupting compounds holds promise, as the microorganisms thriving in these polluted environments often exhibit robust tolerance to elevated EDC concentrations. They may have genetically evolved enzymatic systems to utilize EDCs as a nutrient source. Over the last few decades, several scientific studies have suggested that microbial-based remediation techniques can offer positive results in degrading EDCs.

13.3.1 REMEDIATION OF PHTHALIC ACID ESTERS BY BACTERIA

Bacteria capable of breaking down phthalic acid esters have been discovered in various environmental settings, including mangrove sediments, activated sludge, wastewater, and waste disposal sites (Chatterjee and Dutta, 2008a; Mahajan *et al.*, 2019; Kumar *et al.*, 2017) and are presented in detail in Figure 13.3.

Isolating microorganisms with the ability to degrade phthalic acid esters in a wide range of environmental conditions is significant considering the widespread distribution and presence of phthalic acid esters. *Ex-situ* biodegradation strategies vary depending on the environmental conditions, and therefore, niche-specific phthalic acid ester–degrading microorganisms have to be employed to achieve significant bioremediation. The comprehensive biodegradation process of phthalic acid esters such as DBP by bacteria is initiated by hydrolysis of the ester side-chain, which results in MnBP and n-butanol; following this, MnBP is further degraded to PA, while n-butanol is degraded into butyraldehyde. Further degradation to lower-pathway metabolites follows (Chatterjee and Dutta, 2008b).

13.3.2 BACTERIAL BIOREMEDIATION OF THE ORGANOPHOSPHATE PESTICIDE MALATHION

The remediation of particular organic contaminants through biological means employs various enzymatic mechanisms from diverse bacteria and fungi. Mineralization of pesticides in the soil is mainly due to the phosphatase activity of the soil microbes. Numerous research investigations have

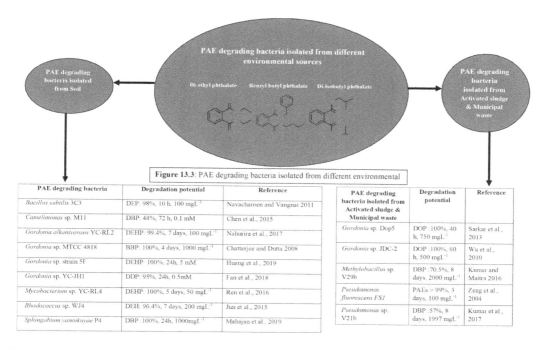

FIGURE 13.3 PAE degrading bacteria isolated from different environments.

been carried out involving the bioremediation of organophosphorus pesticides employing various fungi and bacteria. A collection of the most encouraging bacteria capable of degrading malathion are presented in Table 13.2, which highlights the prominent role of Proteobacteria followed by Firmicutes as potential malathion biodegraders.

13.4 EVALUATING THE ENZYMATIC BIODEGRADATION OF EDC (ORGANOPHOSPHATE PESTICIDE MALATHION)

The potential of enzymatic biodegradation of organophosphates is well known. This section presents a compiled and tabulated overview (Table 13.3) of the enzymes participating in the malathion degradation process. The information has been broadly divided into three sections: oxidoreductase enzyme–, esterase enzyme–, and phosphatase enzyme–mediated biodegradation.

13.4.1 ORGANO PHOSPHORUS HYDROLASES IN MALATHION BIOREMEDIATION

In the present section, we discuss the potentiality of organo phosphorus hydrolases (OPHs) as a potent candidate for *in silico* and *in vivo* protein engineering for enhanced malathion biodegradation.

Thus far, a number of organo phosphorus hydrolases have been identified, originating from various protein families. These include the organo phosphorus acid an-hydrolases (OPAAs; EC 3.4.13.9), para-oxonases (PONs; EC 3.1.8.1), phospho-tri-esterases (PTEs; EC 3.1.8.1) and phospho-tri-esterases like lactonases (PLLs; EC 3.1.1.81) from the amido hydrolase superfamily, and OPHs from the metallo-β-lactamase superfamily.

Studies on the enzyme organophosphorus hydrolase (EC 3.1.8.1), encoded by the organophosphate-degrading *(opd)* gene, have been carried out in *Pseudomonas diminuta* and *Flavobacterium* spp. (strain ATCC27551). The enzyme cleaves the phosphorus–ester bonds such as P–O, P–CN, P–F, P–S. However, in the case of malathion, for catalysis of the P-S bond, the catalytic efficiency of the

TABLE 13.2

Malathion-Degrading Bacteria Identified Based on 16S Ribosomal rRNA Gene Signature Sequences as Reported in the NCBI Database

Organism	Isolation Source	NCBI Accession Number	Reported Malathion Degradation
Pseudomonas putida strain IS82	Malathion-contaminated soil	FJ596989	Malathion (50 mg/l) was completely degraded in the inoculated cultures after 6 days
Alpha proteobacterium IS168	Malathion-contaminated soil	FJ596988	Not available
Brevibacillus sp. KB2	Malathion-contaminated field soil	HQ156457	Strain KB2 was able to degrade 72.20% of malaoxon (an analog of malathion) and 36.22% of malathion after 7 days of incubation
Bacillus cereus strain PU	Soil	FJ387129	Strain PU degraded 87.40% of malaoxon and 49.31% of malathion after 7 days of incubation
Pseudomonas sp. M4	Not available	DQ299933	Not available
Pseudomonas pseudoalcaligenes strain M31	Not available	DQ286456	Not available
Escherichia coli strain EWM	Drinking water	FJ538198	Not available
Bacillus xiamenensis strain KB1	Soil	KY848336	Not available
Bacillus sp. S4(2016b)	Soil	KX158862	Not available
Alcaligenes sp. S3 (2016)	Soil	KX158861	Not available
Bacillus aryabhattai strain S2	Soil	KX158860	Not available
Bacillus megaterium strain S1	Soil	KX158859	Not available
Serratia marcescens strain BNA	Not available	KT351729	A biodegradation efficiency of 65% was achieved within 5 days
Bacillus licheniformis strain ML-1	Contaminated soil	KM893455	Not available
Acinetobacter sp. CGEB9	Contaminated site	KC707807	*Acinetobacter* sp. CGEB9 was chosen as a model organism to show biodegradation of 90% of the initial malathion concentration (1000 ppm) within 4 days
Acinetobacter sp. CGEB8	Contaminated site	KC707806	Not available
Bacillus sp. CGEB7	Contaminated site	KC707805	Not available
Acinetobacter sp. CGEB1	Contaminated site	KC707804	Not available
Lysinibacillus sp. KB1	Cow dung	HQ156458	Not available
Acinetobacter johnsonii strain MA19	Not available	DQ864703	Nearly 100% of an initial amount of malathion (100 mg/L) could be degraded in 84 h in the presence of 2.572 g/l sodium succinate or 1.562 g/l sodium acetate

TABLE 13.3
Enzymes Participating in the Malathion Degradation Pathway

	Enzymes in the Malathion Pathway	Reactions Catalyzed
Oxidoreductase Enzymes	Malathion dicarboxylate oxidoreductase	Malathion dicarboxylate–> Dimethyldithiophosphate + Oxaloacetate
	Dimethylthiophosphate oxidoreductase	Dimethylthiophosphate–> Dimethylphosphate
	Dimethyldithiophosphate oxidoreductase	Dimethyldithiophosphate–> Dimethylthiophosphate
	Malathion oxidoreductase	Malathion–>Malaoxon
	Thiophosphate oxidoreductase	Thiophosphate–> Phosphate
	Diethyl succinate esterase	Diethyl succinate–> Ethyl succinate
	Ethylsuccinate esterase	Ethyl succinate–> Succinate
	Malathion monocarboxylate esterase	Malathion monocarboxylate–> Ethanol + Malathion dicarboxylate
Esterase Enzymes	Methylphosphate phosphomonoesterase	Methylphosphate–> Methanol + Phosphate
	Methylthiophosphate phosphomonoesterase	Methylthiophosphate–> Methanol + Thiophosphate
	S-(diethylsuccinyl)-O-methylphosphorothioate phosphodiesterase	S-(Diethylsuccinyl)-O-methylphosphorothioate–> Methanol + S-(Diethylsuccinyl)-phosphorothioate
	Malathion phosphodiesterase	Malathion–> Methanol + Desmethyl malathion
	Dimethylthiophosphate phosphodiesterase	Dimethylthiophosphate–>Methanol + Methylthiophosphate
	Dimethylphosphate phosphodiesterase	Dimethylphosphate–> Methanol + Methylphosphate
	Malaoxon phosphodiesterase	Malaoxon–> Methanol + S-(Diethylsuccinyl)-O-methylphosphorothioate
Phosphatase Enzymes	Aryldialkylphosphatase	S-(Diethylsuccinyl)-phosphorothioate–> Phosphate + Diethyl 2-mercaptosuccinate

enzymes has to be improved. Schofield and DiNovo (2010) reported mutation of organophosphorus hydrolase for improved biodegradation of the organophosphate pesticides malathion and demeton. For the improvement, the researchers made changes to the amino acid of the OPH protein using site-directed mutagenesis (the initial template utilized in the study was a genetically modified variant), with amino acid changes (a) K185R, D208G, and R319S, which alter functional expression of the OPH, and (b) A80V and I274N that improve hydrolysis.

Schofield and DiNovo (2010) introduced characteristic amino acid variations into OPH using site-directed mutagenesis and identified eight mutants. Variation of all of these mutant cell lysates was observed in their protein-specific activities when tested against malathion. Among the eight mutants, three (mutants 1, 2, and 4) were similar to the wild type and did not account for the specific activity of the protein. However, mutants 5, 6, 7, and 8 resulted in a significant increase in the specific activity. In particular, mutant 7 (mutations in amino acid, e.g. G60V, A80V, I106V, F132D, K185R, D208G, H257W, I274N, F306V, and R319S) resulted in an increase in specific activity by 77-fold against malathion. Most of these substitution positions are placed within the active sites, and the remaining ones are those close to the active sites. Studies suggest that a change of amino acid within the binding pockets of the active site of the OPH enzyme results in a significant change in the stereo selectivity and reactivity of the enzyme. Within the active sites, there are three different binding pockets: (a) the small subsite, G60, I106, L303, S308, C59, and S61; (b) the large subsite, H254,

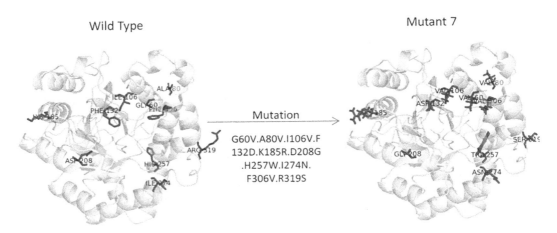

FIGURE 13.4 Depiction of single-point mutations in organophosphorus hydrolase (which resulted in a 77-fold increase in the specific activity) for the improved biodegradation of the organophosphate pesticide malathion.

H257, L271, and M317; and (c) the leaving group subsite, W131, F132, F306, and Y309. On this basis, Schofield and DiNovo (2010) performed saturation mutagenesis at each active site residue, that is, H257, H254, W131, F132, C59, I106, L271, G60, L303, Y309, M317, S61, F306, and S308. Based on the results of saturation mutagenesis, an increase in specific activity (µmoles hydrolyzed/min/mg protein) of mutant cell lysates against demeton-S methyl and malathion was observed in the laboratory experimentation (amino acid changes at H257W, F132C, F132D, I106V, G60V, F306V, and S308L) as well.

To better understand these mutation studies, in the present section, we provide a structural depiction and comparison of the mutant with the wild type in Figure 13.4. The figure highlights the previously discussed single-point mutations, which resulted in a 77-fold increase in the specific activity of the enzyme for malathion degradation.

13.5 CONCLUSION AND FUTURE PROSPECTS

The present chapter offers its readers a comprehensive perspective on the health consequences and bioremediation of EDCs, with a specific focus on phthalic acid esters and the organophosphate pesticide malathion. Further, the role of microbial enzymes, especially organo phosphorus hydrolases, as potent candidates for *in silico* and *in vivo* protein engineering for enhanced malathion biodegradation has also been discussed. Future research should also be focused on resolving bottlenecks that hinder the success of efficient EDC-degrading microorganisms under *ex-situ* conditions.

13.6 ACKNOWLEDGMENTS

The authors gratefully acknowledge Dr. Shalini Verma and Dr. Madhulika Kushwaha for providing relevant information, which helped in writing this chapter. Dr. Rishi Mahajan gratefully acknowledges the funding of research in his laboratory by the Indian Council of Agricultural Research (ICAR) through the National Agricultural Higher Education Project (NAHEP), Centre for Advanced Agricultural Sciences and Technology (CAAST) on Protected Agriculture and Natural Farming—a World Bank–funded project.

REFERENCES

Abo El-Atta, H. M. H., and A. K. El-Hawary. 2021. "Is Childhood Obesity a Result of Toxic Exposure to Cadmium or Malathion? An Observational Pilot Egyptian Study." *Toxicology Communications* 5, no. 1: 11–4. https://doi.org/10.1080/24734306.2020.1869898.

Aja, P. M., J. N. Awoke, P. C. Agu, A. E. Adegboyega, E. M. Ezeh, I. O. Igwenyi *et al.* 2022. "Hesperidin Abrogates Bisphenol A Endocrine Disruption Through Binding with Fibroblast Growth Factor 21 (FGF-21), α-Amylase and α-Glucosidase: An *In Silico* Molecular Study." *Journal of Genetic Engineering & Biotechnology* 20, no. 1: 1–14.

Alfaro-Lira, Susana, María Pizarro-Ortiz, and Gloria M. Calaf. 2012. "Malignant Transformation of Rat Kidney Induced by Environmental Substances and Estrogen." *International Journal of Environmental Research & Public Health* 9, no. 5: 1630–48. https://doi.org/10.3390/ijerph9051630.

Arowolo, Olatunbosun, J. Richard Pilsner, Oleg Sergeyev, and Alexander Suvorov. 2022. "Mechanisms of Male Reproductive Toxicity of Polybrominated Diphenyl Ethers." *International Journal of Molecular Sciences* 23, no. 22: 14229. https://doi.org/10.3390/ijms232214229.

Chatterjee, S., and T. K. Dutta. 2008a. "Metabolic Cooperation of *Gordonia* sp. strain MTCC 4818 and *Arthrobacter* sp. strain WY in the Utilization of Butyl Benzyl Phthalate: Effect of a Novel Co-culture in the Degradation of a Mixture of Phthalates." *Microbiology* 154, no. 11: 3338–46. https://doi.org/10.1099/mic.0.2008/021881-0.

Chatterjee, S., and T. K. Dutta. 2008b. "Complete Degradation of Butyl Benzyl Phthalate by a Defined Bacterial Consortium: Role of Individual Isolates in the Assimilation Pathway." *Chemosphere* 70, no. 5: 933–41. https://doi.org/10.1016/j.chemosphere.2007.06.058.

Cosci, Ilaria, Andrea Garolla, Anna Cabrelle, Stefania Sut, Stefano Dall'Acqua, Alberto Ferlin, Carlo Foresta, and Luca De Toni. 2022. "Lipophilic Phthalic Acid Esters Impair Human Sperm Acrosomal Reaction Through the Likely Inhibition of Phospholipase A2-Signaling Pathway." *Biochemical Pharmacology* 205: 115249. https://doi.org/10.1016/j.bcp.2022.115249.

Eales, J., A. Bethel, T. Galloway, P. Hopkinson, K. Morrissey, R. E. Short, and R. Garside. 2022. "Human Health Impacts of Exposure to Phthalate Plasticizers: An Overview of Reviews." *Environment International* 158: 106903. https://doi.org/10.1016/j.envint.2021.106903.

Hamid, Naima, Muhammad Junaid, and De-Sheng Pei. 2021. "Combined Toxicity of Endocrine-Disrupting Chemicals: A Review." *Ecotoxicology & Environmental Safety* 215: 112136. https://doi.org/10.1016/j.ecoenv.2021.112136.

Hu, Chenyan, Yachen Bai, Jing Li, Baili Sun, and Lianguo Chen. 2023. "Endocrine Disruption and Reproductive Impairment of Methylparaben in Adult Zebrafish." *Food & Chemical Toxicology* 171: 113545. https://doi.org/10.1016/j.fct.2022.113545.

Jo, Y. S., H. S. Ko, A. Y. Kim, H. G. Jo, W. J. Kim, and S. K. Choi. 2022. "Effects of Polycyclic Aromatic Hydrocarbons on the Proliferation and Differentiation of Placental Cells." *Reproductive Biology & Endocrinology* 20, no. 1: 1–7.

Jung, Da-Woon, Da-Hyun Jeong, and Hee-Seok Lee. 2022. "Endocrine Disrupting Potential of Selected Azole and Organophosphorus Pesticide Products Through Suppressing the Dimerization of Human Androgen Receptor in Genomic Pathway." *Ecotoxicology & Environmental Safety* 247: 114246. https://doi.org/10.1016/j.ecoenv.2022.114246.

Kokai, Dunja, Bojana Stanic, Biljana Tesic, Dragana Samardzija Nenadov, Kristina Pogrmic-Majkic, Svetlana Fa Nedeljkovic, and Nebojsa Andric. 2022. "Dibutyl Phthalate Promotes Angiogenesis in EA. hy926 Cells Through Estrogen Receptor-Dependent Activation of ERK1/2, PI3K-Akt, and NO Signaling Pathways." *Chemico-Biological Interactions* 366: 110174. https://doi.org/10.1016/j.cbi.2022.110174.

Konieczna, A., A. Rutkowska, and D. Rachoń. 2015. "Health Risk of Exposure to Bisphenol A (BPA)." *Roczniki Panstwowego Zakladu Higieny* 66, no. 1: 5–11.

Kumar, Vinay, Neha Sharma, and S. S. Maitra. 2017. "Comparative Study on the Degradation of Dibutyl Phthalate by Two Newly Isolated *Pseudomonas* sp. V21b and *Comamonas* sp. 51F." *Biotechnology Reports* 15: 1–10. https://doi.org/10.1016/j.btre.2017.04.002.

Lauretta, Rosa, Andrea Sansone, Massimiliano Sansone, Francesco Romanelli, and Marialuisa Appetecchia. 2019. "Endocrine Disrupting Chemicals: Effects on Endocrine Glands." *Frontiers in Endocrinology* 10: 178. https://doi.org/10.3389/fendo.2019.00178.

Mahajan, R., S. Verma, and S. Chatterjee. 2021. "Biodegradation of Organophosphorus Pesticide Profenofos by the Bacterium *Bacillus* sp. PF1 and Elucidation of Initial Degradation Pathway." *Environment & Technique* 9: 1–9.

Mahajan, R., S. Verma, S. Chandel, and S. Chatterjee. 2022. "Organophosphate Pesticide: Usage, Environmental Exposure, Health Effects, and Microbial Bioremediation." In *Microbial Biodegradation and Bioremediation*. 2nd ed., edited by S. Das and H. R. Dash: 473–90. Amsterdam: Elsevier.

Mahajan, Rishi, Shalini Verma, Madhulika Kushwaha, Dharam Singh, Yusuf Akhter, and Subhankar Chatterjee. 2019. "Biodegradation of di-n-Butyl Phthalate by Psychrotolerant *Sphingobium yanoikuyae* strain P4 and Protein Structural Analysis of Carboxylesterase Involved in the Pathway." *International Journal of Biological Macromolecules* 122: 806–16. https://doi.org/10.1016/j.ijbiomac.2018.10.225.

Metcalfe, C. D., S. Bayen, M. Desrosiers, G. Muñoz, S. Sauvé, and V. Yargeau. 2022. "An Introduction to the Sources, Fate, Occurrence and Effects of Endocrine Disrupting Chemicals Released into the Environment." *Environmental Research* 207: 112658. https://doi.org/10.1016/j.envres.2021.112658.

Miranda, R. A., B. S. Silva, E. G. de Moura, and P. C. Lisboa. 2022. "Pesticides as Endocrine Disruptors: Programming for Obesity and Diabetes." *Endocrine*: 1–11.

Moreira, Sílvia, Ricardo Silva, David F. Carrageta, Marco G. Alves, Vicente Seco-Rovira, Pedro F. Oliveira, and Maria de Lourdes Pereira. 2022. "Carbamate Pesticides: Shedding Light on Their Impact on the Male Reproductive System." *International Journal of Molecular Sciences* 23, no. 15: 8206. https://doi.org/10.3390/ijms23158206.

Morsi, N., M. Wassem, M. Badr, and A. K. Abdel Abdel Latif. 2022. "Biochemical, Molecular and Histopathological Studies on Malathion Toxicity on Some Vital Organs of Male Rats." *Egyptian Academic Journal of Biological Sciences, B. Zoology* 14, no. 1: 193–207. https://doi.org/10.21608/eajbsz.2022.231218.

Mukherjee, Urmi, Anwesha Samanta, Subhasri Biswas, Soumyajyoti Ghosh, Sriparna Das, Sambuddha Banerjee, and Sudipta Maitra. 2022. "Chronic Exposure to Nonylphenol Induces Oxidative Stress and Liver Damage in Male Zebrafish (*Danio rerio*): Mechanistic Insight into Cellular Energy Sensors, Lipid Accumulation and Immune Modulation." *Chemico-Biological Interactions* 351: 109762. https://doi.org/10.1016/j.cbi.2021.109762.

Otênio, Joice Karina, Karine Delgado Souza, Odair Alberton, Luiz Rômulo Alberton, Karyne Garcia Tafarelo Moreno, Arquimedes Gasparotto Gasparotto Junior, Rhanany Alan Calloi Palozi, Emerson Luiz Botelho Lourenço, and Ezilda Jacomassi. 2022. "Thyroid-Disrupting Effects of Chlorpyrifos in Female Wistar Rats." *Drug & Chemical Toxicology* 45, no. 1: 387–92. https://doi.org/10.1080/01480545.2019.1701487.

Ramhøj, Louise, Karen Mandrup, Ulla Hass, Terje Svingen, and Marta Axelstad. 2022. "Developmental Exposure to the DE-71 Mixture of Polybrominated Diphenyl Ether (PBDE) Flame Retardants Induce a Complex Pattern of Endocrine Disrupting Effects in Rats." *PeerJ* 10: E12738. https://doi.org/10.7717/peerj.12738.

Schofield, D. A., and A. A. DiNovo. 2010. "Generation of a Mutagenized Organophosphorus Hydrolase for the Biodegradation of the Organophosphate Pesticides Malathion and Demeton-S." *Journal of Applied Microbiology* 109, no. 2: 548–57. https://doi.org/10.1111/j.1365-2672.2010.04672.x.

Street, Maria Elisabeth, Sabrina Angelini, Sergio Bernasconi, Ernesto Burgio, Alessandra Cassio, Cecilia Catellani, Francesca Cirillo, Annalisa Deodati, Enrica Fabbrizi, Vassilios Fanos, Giancarlo Gargano, Enzo Grossi, Lorenzo Iughetti, Pietro Lazzeroni, Alberto Mantovani, Lucia Migliore, Paola Palanza, Giancarlo Panzica, Anna Maria Papini, Stefano Parmigiani, Barbara Predieri, Chiara Sartori, Gabriele Tridenti, and Sergio Amarri. 2018. "Current Knowledge on Endocrine Disrupting Chemicals (EDCs) from Animal Biology to Humans, from Pregnancy to Adulthood: Highlights from a National Italian Meeting." *International Journal of Molecular Sciences* 19, no. 6: 1–44. https://doi.org/10.3390/ijms19061647.

Tiwari, N., A. Kumar, A. Pandey, and A. Mishra. 2022. "Computational Investigation of Dioxin-Like Compounds as Human Sex Hormone-Binding Globulin Inhibitors: DFT Calculations, Docking Study and Molecular Dynamics Simulations." *Computational Toxicology* 21: 100198. https://doi.org/10.1016/j.comtox.2021.100198.

Vasseghian, Yasser, Yasser, Fares Almomani, Masoud Moradi Le Van Thuan, and Elena-Niculina Dragoi. 2022. "Decontamination of Toxic Malathion Pesticide in Aqueous Solutions by Fenton-Based Processes: Degradation Pathway, Toxicity Assessment and Health Risk Assessment." *Journal of Hazardous Materials* 423, no. A, no. A: 127016: 127016. https://doi.org/10.1016/j.jhazmat.2021.127016.

Witczak, Agata, Anna Pohoryło, and Hassan Abdel-Gawad. 2021. "Endocrine-Disrupting Organochlorine Pesticides in Human Breast Milk: Changes During Lactation." *Nutrients* 13, no. 1: 229. https://doi.org/10.3390/nu13010229.

Xu, Q., Jian Li, Shang Cao, Guangcai Ma, Xianglong Zhao, Qiuyi Wang, Xiaoxuan Wei, Haiying Yu, and Zhiguo Wang. 2023. "Thyroid Hormone Activities of Neutral and Anionic Hydroxylated Polybrominated Diphenyl Ethers to Thyroid Receptor β: A Molecular Dynamics Study." *Chemosphere* 311, no. 1: 136920. https://doi.org/10.1016/j.chemosphere.2022.136920.

Zeng, Jia-Yue, Y. Miao, Chong Liu, Yan-Ling Deng, Pan-Pan Chen, Min Zhang, Fei-Peng Cui, Tian Shi, Ting-Ting Lu, Chang-Jiang Liu, and Qiang Zeng. 2022. "Serum Multiple Organochlorine Pesticides in Relation to Testosterone Concentrations Among Chinese Men from an Infertility Clinic." *Chemosphere* 299: 134469. https://doi.org/10.1016/j.chemosphere.2022.134469.

Zhang, Nian-Jie, Yuanwei Zhang Zhang, Shuo Yin, Du-Ji Ruan, Nian He, X. Chen, and Xue-Feng Yang. 2022. "Nonylphenol Promoted Epithelial–Mesenchymal Transition in Colorectal Cancer Cells by Upregulating the Expression of Regulator of Cell Cycle." *Chemical Research in Toxicology* 35, no. 9: 1533–40. https://doi.org/10.1021/acs.chemrestox.2c00180.

Zhou, Qinghua, Jinyuan Chen, Junfan Zhang, Feifei Zhou, Jingjing Zhao, Xiuzhen Wei, Kaiyun Zheng, Jian Wu, Bingjie Li, and Bingjun Pan. 2022. "Toxicity and Endocrine-Disrupting Potential of PM2. 5: Association with Particulate Polycyclic Aromatic Hydrocarbons, Phthalate Esters, and Heavy Metals." *Environmental Pollution* 292, no. A, no. A: 118349: 118349. https://doi.org/10.1016/j.envpol.2021.118349.

14 Bacterial Ammonia Oxidation
A Way towards Environment Remediation

Vijaylakshmi, Arti Chamoli, Anne Bhambri, Neetu Pandey, and Santosh Kumar Karn

14.1 INTRODUCTION

Ammonia, which contributes significantly to air pollution, is released into the atmosphere by industrial and agricultural processes. Ammonia is extensively used in various industries, such as chemical and manufacturing, and as a leading preservative in plastics manufacturing, explosives, and fertilizer, this chemical is essential to manufacturing industries, but it can also pose a threat to the environment when large amounts of ammonia are combined with wastewater and released into the sewage system.

Particularly, widespread use of ammonia and its derivatives, such as nitrogen fertilizers for agriculture, has greatly increased ammonia emissions into the atmosphere, leading to a variety of environmental issues (Finlay, 2002; Bhambri *et al.*, 2021). These include the creation of fine particulate matter in the atmosphere, the acidification of soils, the eutrophication of semi-natural ecosystems, and modifications to the global greenhouse gas balance. Ammonium salts, which are ultra-fine particles that can irritate the respiratory system, damage the lungs, and cause breathing difficulties, are created when the nitrogen/hydrogen complex combines with other airborne contaminants.

The ecosystem is being degraded as a result of increased industrial and agricultural sector advancement. Conventionally, to support plant growth and development, a variety of chemical-based fertilizers are used that have various detrimental consequences for the ecosystem. The current agricultural system mostly relies on synthetic nitrogen fertilizers (N fertilizers) to maintain crop yield and fulfill the world's growing food demand. The release of fertilizers causes groundwater to have fluctuating concentrations of nitrogen and ammonia, and is harmful to the marine environment. Natural environmental processes produce ammonia, which is contained in soil and serves as a vital supply of nitrogen for plants. Because soil bacteria metabolize ammonium quickly, it does not usually concentrate in soils. The primary sources of nitrogen for soil-dwelling bacteria are ammonia and ammonium, which are essential energy substrates (Daebeler *et al.*, 2014; Bhambri and Karn, 2021). Nitrification is the process by which microorganisms oxidize ammonia or ammonium. The primary stage of the nitrogen cycle in soil is nitrification. In agricultural systems, where ammonia fertilizer is often administered, nitrification is crucial. Nitrifying microorganisms live in the phyllosphere, in the soil surrounding the roots, inside the plants as endophytes, and on the surfaces of the plants as epiphytes. They are also associated with the rhizosphere of plants.

The biological conversion of ammonia to nitrite and then nitrite to nitrate, which is absorbed by plants, is known as nitrification. The rate-limiting phase in nitrification is often the conversion of ammonia to nitrite. Since nitrate is more water soluble than ammonia, converting this ammonia to nitrate increases nitrogen leaching. The main source of nutrients for plants is nitrates. Certain species and habitats are particularly vulnerable to ammonia pollution, and its discharge has a severe impact on biodiversity. Nitrate is an excellent component for plant growth and development that is

Bacterial Ammonia Oxidation

essential to a plant's healthy operation. Nitrates can be forced below the plant root zone in the soil by the effects of heavy rainfall.

The characteristics of the underlying soil and/or bedrock, as well as the depth of the groundwater, determine whether or not nitrates continue to migrate downhill and into groundwater. This is a typical nitrogenous waste, particularly for aquatic life. Despite being toxic, ammonia helps plants to develop (Givan, 1979; Bhambri *et al.*, 2023b). Nitrates found in ammonia-containing wastewater are harmful for the environment and unhealthy for people. Consequently, it's imperative to discover strategies for reducing ammonia's harmful impacts by understanding how ammonia poisoning affects aquatic and soil species.

14.1.1 AMMONIA IN SOIL AND ITS EFFECTS

Ammonia is produced through both natural and anthropogenic sources. Natural sources of ammonia include the breakdown of organic waste or decomposition, gas exchange with the environment, forest fires, human and animal waste, and nitrogen fixation processes. The primary cause of anthropogenic ammonia pollution is agriculture, which includes heavy fertilizer use, inappropriate handling of plant and animal waste, sewer disposal facilities, municipal and industrial wastewater outflow, and landfills.

Ammonia is toxic to the plants as it can seriously harm plants when it is fed to their roots, and it can also be harmful when applied topically or by leaf fumigation or misting. The damage manifests as yellowing of the leaves, necrosis, decreased growth (especially of the roots), or in extreme situations, tissue death (Givan, 1979). If the roots of most plants are submerged in a solution that contains more than 0–1 mM of ammonium, it will negatively impact their growth (Mehrer and Mohr, 1989). Toxicity caused due to ammonium leads to inhibition of root and shoot growth that is confederated with leaf chlorosis, ionic imbalances, and disturbance of pH gradients across plant membranes, or oxidative stress (Esteban *et al.*, 2016, Bhambri *et al.*, 2023a).

Ammonia affects grassland herbs, heathlands, trees, epiphytic mosses, and lichens. The eutrophication-induced alterations in plant susceptibility to biotic and abiotic stimuli may cause additional disturbances to these ecosystems. These include increased vulnerability to drought, insects, and fungi, as well as frosts (Führer, 1990).

14.1.2 AMMONIA IN AQUATIC WATER AND ITS EFFECTS

Ammonia is one of the predominant environmental pollutants in aquatic systems (Xia *et al.*, 2018). It enters the aquatic environment from several sources such as sewage effluent, industrial waste, agricultural run-off, and the decomposition of biological waste, causing pernicious toxic effects on aquatic organisms and soils (Attah *et al.*, 2021). In natural surface waters, ammonia is present in two forms: ionized ammonia, NH_4^+, and unionized ammonia, NH_3^+ (Francis-Floyd *et al.*, 2009). Excessive ammonia can cause a reduction in growth performance, tissue erosion and degeneration, immune suppression, and high mortality in aquatic animals, and exceeding ammonia levels in blood and tissues leads to toxicity (Li *et al.*, 2014). The ammonia and nitrates are the major groundwater contaminants existing in the rural areas. Unionized ammonia (NH_3) is the most hazardous form of ammonia because it is a neutral molecule and can pass through the epithelium membranes of aquatic species far more readily than the charged ammonium ion. Ammonia in both gaseous and liquid forms can be harmful to the eyes, respiratory tract, and skin due to its alkaline nature. The biological effects of ammonia and nitrates in humans after severe exposures are dose-dependent. If the levels of ammonia in the water increase by a certain amount, it will lead to ammonia poisoning which may cause slow growth, tissue erosion, oxidative stress in the body, decreased immunity, and even the death of aquatic animals (Cheng *et al.*, 2015). Ammonia poisoning is considered to be the major contributing factor of fish disease and mortality in aquaculture.

14.2 METHODS USED IN AMMONIA REMOVAL FROM WASTEWATER

The most typical contaminant introduced into water streams is ammonia. Ammonia is often released from home, commercial, and agricultural wastewater. Most often, excessive concentrations of ammonia are released into water by sectors including food processing, rubber processing, textile and leather production, fertilizer plants, etc. (Capodaglio *et al.*, 2015). Prevalent methods of separation of ammonia from wastewater are biological denitrification, stripping, ion exchange, break-point chlorination, and chemical precipitation. Some of the other techniques are photocatalytic and electrochemical oxidation (Kim *et al.*, 2006).

14.2.1 AMMONIA STRIPPING AND DISTILLATION

Air or steam stripping can achieve the separation of ammonia from a highly concentrated environment. It has substantial applications in concentrated wastewater treatment such as landfill leachate, supernatant of anaerobic digestion processes, specific flows of the petrochemical industry, and very few in sewage treatment. In this process, lime or some other caustic substances is eventually added to the wastewater until pH reaches 10.8–11.5. Magnesium chloride is the frequently used reagent. Air stripping is more prevalent and frequently used than steam stripping since it needs working temperatures greater than 95°C (Tchobanoglus *et al.*, 2003).

14.2.2 BREAKPOINT CHLORINATION FOR REMOVAL OF AMMONIA

Treatment of drinking water and wastewater with aqueous chlorine separates both ammonia and other oxidizable substances (Stumm and Morgan, 2012). Aqueous chlorine is a combination of Cl_2, HOCl, and OCl^-, depending on the pH, and is known accordingly as "free available chlorine" (FAC). In the oxidation of ammonia, concentration of total nitrogen and positive chlorine remains high up to ratios of $FAC:NH_3$ ¼:1 due to the intermediate formation of chloramine NH_2Cl, called "combined chlorine" (Jafvertand Valentine, 1992). Beyond $FAC:NH_3$ ¼:1 $NHCl_2$ is formed and denitrification is attained through the reaction between NH_2Cl and $NHCl_2$. Breakpoint oxidation of ammonia is pH dependent both ammonia and hypochlorous acid exist as acid-base conjugates, with the fastest reaction between undissociated HOCl and unprotonated ammonia. This occurs near pH 8.5, midway between the pKa values for HOCl and NH_4^+ (Edzwald, 2011).

14.2.3 ION EXCHANGE METHOD

Ion exchange methods are used to remove dissolved ions from solutions and replace them with ions that have comparable or similar charges. Anion exchangers interchange negatively charged ions (anions), whereas cation exchangers exchange positively charged ions (cations). The use of an ion exchange method to treat wastewater has the important benefit of being able to handle wide variations in temperature and ammonia content. (Imchuen *et al.*, 2016). Researchers have looked at the use of ion exchange technology to remove ammonia from aqueous solutions for wastewater treatment.

14.3 BIOLOGICAL OXIDATION OF AMMONIA

An intricate web of microorganisms that interact with one another mediates the biological N cycle. One example of this is the relationship between anammox processes and aerobic ammonia oxidation, which is demonstrated by the subsequent generation and use of nitrite and the elimination of fixed nitrogen from the system. Ammonia is the most common inorganic nitrogen source in influent material, and nitrification and denitrification are the processes that inevitably cause nitrogen loss.

Bacterial Ammonia Oxidation

Nitrogen removal from waste treatment is crucial for the environment since releasing untreated waste can cause catastrophic eutrophication, especially in the vicinity of densely inhabited areas. Additionally, nitrification helps prevent environmental contamination with potentially harmful ammonia salts in states where treatment is unsuccessful in achieving denitrification (Painter, 1986). The oxidation of ammonia to nitrite and the absorption of inorganic carbon provide bacterial ammonia oxidizers with energy. They are quickly enriched from the majority of soils, and the cultured strains' abundance and cell activity emerge in sufficient numbers to support the measured rates of soil nitrification.

Ammonia can be oxidized by a variety of microorganisms, including bacteria, archaea, and even fungi. The soil is heavily fertilized with ammonia oxidizing bacteria (AOB). Wastewater from cities and industries is treated using biological technologies that are low-cost and simple to manage. One cannot overstate the importance of nitrogen in plants. It is the single most important nutrient required by plants since it is an essential part of nucleic acids, which comprise all living things' DNA. It also contains one of the components of photosynthesis, chlorophyll. It has an important effect on the amount of land acquired, the growth of plants, and their chemical behavior (Wu *et al.*, 2016). Certain plants can use atmospheric nitrogen in the same way as ammonium and nitrate; most plants can only use inorganic forms of nitrogen.

Ammonia-oxidizing bacteria (AOB) and ammonia-oxidizing archaea (AOA) are the main operators that are liable for the conversion of N into usable forms. Nitrification plays an essential role in the removal of nitrogen from municipal wastewater. It is a microbial process that aerobically converts ammonia to nitrate, and joins nitrogen fixation and denitrification as leading functions of the global nitrogen cycle (Stein *et al.*, 2016). During the conversion of nitrogen cyanobacteria will first convert nitrogen into ammonia and ammonium, during the nitrogen fixation process. Nitrogen fixation is imposed according to the following reaction:

$$N_2 + 3H_2 \rightarrow 2NH_3$$

Ammonia-oxidizing bacteria (AOB), ammonia oxidizing archaea (AOA) and comammox bacteria are the three specific groups of aerobic autotrophic microorganisms that oxidize ammonia (complete oxidation of ammonia to nitrate). Ammonia oxidizing bacteria use ammonia compounds as the key source of energy and carbon dioxide as a carbon source. Decomposition process by AOB play an essential role in nature to preserve environment quality and these bacteria can survive in environments with low oxygen levels (Arp *et al.*, 2003). In nitrogen cycle nitrification is one of the most essential steps, including the microbial oxidation of ammonia to nitrite and later to nitrate. As the rate-limiting step in nitrification, ammonia oxidation is catalyzed by a series of phylogenetic- and physiological-related microorganisms (Treusch *et al.*, 2005).

The oxidation of ammonia to nitrite is eventually brought off by the chemolithotrophic AOB (Rotthauwe *et al.*, 1997). During ammonia oxidation, ammonia is oxidised to hydroxylamine by ammonia monooxygenase (AMO), a membrane bound enzyme that belongs to a superfamily of ammonia, methane and alkane monooxygenases. AMO is an essentially key enzyme in ammonia oxidation and the only enzyme of the ammonia oxidation pathway that is shared by all three major groups of ammonia oxidizing microorganisms. Moreover, amoA (the gene encoding of the A subunit of the AMO) is enormously the most abundantly used functional marker gene for ammonia oxidation and has been essential in understanding the ecology of ammonia oxidizers (Leininger *et al.*, 2006). The beta proteobacterial ammonia oxidizers are further divided according to their 16S rRNA gene sequence (Pjevac *et al.*, 2017). The abundance and activity of autotrophic AOB positively correlate to the degree of nitrogen saturation in the soils. Widely distributed AOB in the environment are *Nitrosomonas* and *Nitrosospira* belonging to beta-proteobacteria (Stephen *et al.*, 1996). Until recently, it was believed that just a few different species of bacteria of the genera *Nitrosomonas*, *Nitrosospira*, and *Nitrosococcus* oxidize ammonia, However, an archaea that could also oxidize ammonia was found in 2005 (Könneke *et al.*, 2005). Since this discovery, it has regularly been observed that ammonia-oxidizing archaea surpass ammonia-oxidizing bacteria in numerous ecosystems. The abundance of ammonia-oxidizing archaea in soils, salt marshes, and seas during the past several years suggests that these recently discovered organisms play a significant part in the nitrogen cycle. *Nitrosococcus*, which is associated with the Gamma-proteobacteria, is found especially in marine environments. *Nitrosospira* spp. generally influence the AOB community in soil environments while *Nitrosomonas* spp. are frequent in environments of high nitrogen loading such as sewage treatment plants and eutrophic lakes (Norton *et al.*, 2008).

Ammonia from wastewater and industries can be remediated by heterotrophic nitrifying bacteria using carbon as an external source. Although heterotrophic nitrifiers have great potential in future bioremediation systems (Yang *et al.*, 2011). Therefore it is required to find efficient primitive organisms for ammonia oxidation and reduce their toxicity. Nitrite (NO_2^-) oxidation is the second step in nitrification where it is transformed into nitrate (NO_3^-). A distinct group of prokaryotes namely nitrite-oxidizing bacteria performs this action. The genera that participate in nitrite oxidation include *Nitrospira*, *Nitrobacter*, *Nitrococcus*, and *Nitrospina*. In marine environments, AOA are the dominant microbes and play a main role, probably accounting for over 70% of total ammonia oxidation (Lam *et al.*, 2007). Also, in terrestrial ecosystems, some studies indicated that AOA was more important than AOB (Verhamme *et al.*, 2011). Several studies have used molecular biology techniques to assess the effect of various factors on AOB community structure in WWTPs (Pholchan *et al.*, 2010). To date, however, the relative influence of specific deterministic environmental factors to AOB community dynamics in WWTP (with associated concurrent changes in a multitude of environmental parameters) is uncertain (Wells *et al.*, 2009). Different populations of the AOB community normally coincide in WWTPs, but temperature changes can change the composition of the AOB community (Siripong and Rittmann, 2007). Furthermore, temperature has been reported as the most significant factor influencing the AOB community structure when compared with other environmental variables (Park *et al.*, 2009). In addition to the influence of temperature, the growth rate and activity of AOB can be negatively affected by other environmental factors and process parameters such as decreased pH, low dissolved oxygen (DO) concentration, toxic compounds, sludge retention time (SRT), and high organic load (Hallin *et al.*, 2005).

Due to the division of the nitrification metabolism, two functional microbial populations with overlapping ecological niches have become specialized. Recently, it was discovered that complete ammonia oxidizers (comammox), which are capable of oxidizing ammonia to nitrate, occur

TABLE 14.1
List of ammonia-oxidizing bacteria (Boncristiani *et al.*, 2009)

Genus	Phylogenetic Group	Habitat
Nitrosococcus	Gamma	Freshwater, marine
Nitrosomonas	Beta	Soil, sewage, freshwater, marine
Nitrospira	Beta	Soil

TABLE 14.2
List of nitrite-oxidizing bacteria (Boncristiani *et al.*, 2009)

Genus	Phylogenetic Group	Habitat
Nitrobacter	Alpha	Soil, Freshwater, Marine
Nitrococcus	Gamma	Marine
Nitrospira	Delta	Marine, Soil

in enrichment cultures containing the species *Nitrospira* (Palomo *et al.*, 2016). According to the findings of recent studies, comammox bacteria are extensively dispersed in the environment (Shi *et al.*, 2018) and in man-made areas (Wang *et al.*, 2017). All comammox organisms identified to date belong to *"Candidatus Nitrospiranitrosa"*, *"Candidatus Nitrospiranitrificans"*, *"Candidatus Nitrospirainopinata"*, and *Nitrospira* sp. strain Ga0074138 (Camejo *et al.*, 2017).

14.3.1 THE CONTRIBUTION OF ANAMMOX BACTERIA TO N CYCLING

The dominant source of fixed nitrogen in most municipal wastewater treatment plants (WWTPs) is NH_4^+ which can reach concentrations of up to a few mmol L^{-1} and would lead to strong eutrophication if released untreated into aquatic environments. Conventionally, NH_4^+ is removed in municipal WWTPs via biological nitrogen removal (BNR) through alternating nitrification and denitrification in nitrification/denitrification tanks of the mainstream (Van Hulle *et al.*, 2010). Biological techniques based on a sequence of nitrification and denitrification leading to the release of molecular nitrogen are currently the most widely used methods for the industrial-scale removal of ammonium compounds, which are among the main pollutants of the water bodies intended for partially treated wastewater.

But in recent years, sidestream and mainstream wastewater treatment plants have been utilizing alternative "shortcut" nitrogen removal pathways because they provide the possibility of cost reduction for both external carbon for denitrification and aeration energy for nitrification (Hu *et al.*, 2010; Lackner *et al.*, 2014; Laureni *et al.*, 2016; Van Dongen *et al.*, 2001; Winkler *et al.*, 2012). The coupling of partial nitritation (fractional aerobic oxidation of influent ammonia to nitrite by AOB) with anaerobic ammonia oxidation (anammox) represents one such shortcut process.

The bacteria carrying out the anammox process were subsequently identified as planctomycetes. The first discovered anammox microorganism was named *Candidatus "Brocadia anammoxidans"* (Kuenen, 2001). All known anammox bacteria are members of the same monophyletic order Brocadiales, belonging to *Planctomycetes*. To date, five genera of anammox bacteria have been described in the status of *Candidatus*, with 16S rRNA identities of the species ranging between 87% and 99% (Jetten *et al.*, 2009). Four genera have been enriched from activated sludge: *Candidatus* "Kuenenia" (Schmid *et al.*, 2000; Strous *et al.*, 2006), *Ca.* "Brocadia" (Kuenen, 2001; Strous *et al.*,

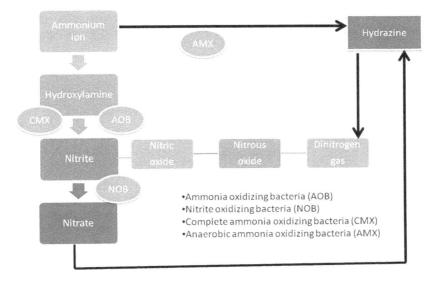

FIGURE 14.1 Nitrogen cycling in engineered biological nitrogen removal (BNR) systems.

1999; Kartal *et al.*, 2008), *Ca.* "Anammoxoglobus" (Kartal *et al.*, 2007; Liu *et al.*, 2008), and *Ca.* "Jettenia" (Quan *et al.*, 2008; Nikolaev *et al.*, 2015). The fifth anammox genus, *Ca.* "Scalindua" (Kuypers *et al.*, 2003; Schmid *et al.*, 2003; Van De Vossenberg *et al.*, 2008), has often been detected in natural habitats, especially in marine sediments and low-oxygen zones (Dalsgaard *et al.*, 2005; Penton *et al.*, 2006; Schmid *et al.*, 2007; Woebken *et al.*, 2008). In 2013, the discovery of a novel genus of anammox bacteria, *Ca.* "Anammoxomicrobium", was reported (Khramenkov *et al.*, 2013).

Autotrophic ammonia-oxidizing bacteria (AOB) are the drivers in the nitrogen cycle that convert ammonium (NH_4^+) to nitrite (NO_2^-) through hydroxylamine (NH_2OH), and nitrite-oxidizing bacteria (NOB) oxidize nitrite to nitrate (NO_3^-). Ordinary heterotrophic denitrifiers (OHO) consists of sequential reductive reactions from NO_3 to NO_2, nitric oxide (NO), N_2O and finally to nitrogen gas (N_2), carried out by heterotrophs (Pan *et al.*, 2013; Desloover *et al.*, 2012; Wunderlin *et al.*, 2012; Law *et al.*, 2012).

On the other hand, anaerobic ammonia-oxidizing bacteria (AMX) oxidize ammonium and reduce nitrite to dinitrogen gas through hydrazine (N_2H_4). Complete ammonia-oxidizing bacteria (CMX) have only recently been described and studied, and are capable of converting ammonium to nitrate through hydroxylamine and nitrite by the single organism (Annavajhala *et al.*, 2018; Jhala *et al.*, 2018). Nitrogen removal from the anaerobic sludge digestion liquor in a side-stream process has become a common practice in wastewater treatment plants (WWTPs) (Kampschreur *et al.*, 2008; Mulder *et al.*, 2001).

14.4   CONCLUSION

Ammonia-oxidizing bacteria and archaea are the key players that participate in the conversion of ammonia to nitrates. Nitrate is the only form of nitrogen that can be utilized by most plants. Nitrification, or the removal of ammonia, is crucial in the environment and in wastewater treatment plants also. Nitrogen cycle in a wastewater treatment system needs further assessment. The investigation of microbial communities is essential to determine their collaborative, competitive, and inhibitive nature in actual wastewater systems. Effective nitrification-denitrification in wastewater treatment systems depends on partial ammonia removal followed by the conversion of residual ammonia and nitrite to nitrogen gas by anamox bacteria. The interaction of comamox *Nitrospira* with anamox bacteria requires further examination and the evaluation of the effect of comammox-anammox interaction on nitrification performance, stability, and efficiency must be considered. Modern tools of molecular microbiology (metagenomics) enable detection, identification, and estimation of the ammonia-oxidizing microorganisms bacteria content in biomass samples.

14.5   ACKNOWLEDGMENTS

The authors are thankful to Dr. Gaurav Deep Singh Chancellor, SBS University, Dehradun, Uttarakhand for providing, space, facility, and resources to complete this work.

REFERENCES

Annavajhala, Medini K., Vikram Kapoor, Jorge Santo-Domingo, and Kartik Chandran. 2018. "Comammox Functionality Identified in Diverse Engineered Biological Wastewater Treatment Systems." *Environmental Science & Technology Letters* 5, no. 2: 110–16. https://doi.org/10.1021/acs.estlett.7b00577.

Arp, Daniel J., and Lisa Y. Stein. 2003. "Metabolism of Inorganic N Compounds by Ammonia-Oxidizing Bacteria." *Critical Reviews in Biochemistry & Molecular Biology* 38, no. 6: 471–95. https://doi.org/10.1080/10409230390267446.

Attah, U. E., O. C. Chinwendu, C. P. Chieze, O. H. Obiahu, and Z. Yan. 2021. "Evaluating the Spatial Distribution of Soil Physicochemical Characteristics and Heavy Metal Toxicity Potential in Sediments of Nworie River Micro-watershed Imo State, Southeastern Nigeria." *Environmental Chemistry & Ecotoxicology* 3: 261–8. https://doi.org/10.1016/j.enceco.2021.08.001.

Bacterial Ammonia Oxidation

Bhamri, Anne, and Santosh Kumar Karn. 2021. "Nitrate Problems and Its Remediation." In *An Innovative Approach of Advanced Oxidation Process in Wastewater Treatment*, edited by Maulin P. Shah. New York: Nova Science Publishers, Inc. https://doi.org/10.52305/DBMP3480; ISBN: 978-1-68507-235-3.

Bhambri, Anne, Santosh Kumar Karn, and Arun Kumar. 2023a. "Application or Utilization of Algae and Bacteria in Aquaculture." In *Applications in Agriculture, Food and Environment*, edited by Charles Oluwaseun Adetunji, Julius Kola Oloke, Naveen Dwivedi, Sabeela Beevi Ummalyma, Shubha Dwivedi, Juliana Bunmi Adetunji, and Daniel Ingo Hefft. Beverly, MA: Scrivener Publishing. https://doi.org/10.1002/9781119857839.ch6. ISBN10: 1119857813.

Bhambri, Anne, Santosh Kumar Karn, and Arun Kumar. 2023b. "Regulation and Measurement of Nitrification in Terrestrial Systems." In *Anaerobic Ammonium Oxidation*, edited by Maulin P. Shah. Berlin: De Gruyter. https://doi.org/10.1515/9783110780093-004.

Bhambri, Anne, Santosh Kumar Karn, and R. K. Singh. 2021. "In-situ Remediation of Nitrogen and Phosphorus of Beverage Industry by Potential Strains BK1 and BK2." *Scientific Report* 11, no. 1: 1–11. https://doi.org/10.1038/s41598-021-91539-y.

Boncristiani, H. F., M. F. Criado, E. Arruda, and M. Schaechter. 2009. *Encyclopedia of Microbiology*: 500–518. Amsterdam: Elsevier. doi: 10.1016/B978-012373944-5.00314-X.

Camejo, Pamela Y., K. D. McMahon, and D. R. Noguera. 2017. "Genome-Enabled Insights into the Ecophysiology of the Comammox Bacterium 'Candidatus Nitrospira Nitrosa'." *mSystems* 2, no. 5: e00059–17. https://doi.org/10.1128/mSystems.00059-17.

Capodaglio, A. G., P. Hlavínek, and M. Raboni. 2015. "Physico-chemical Technologies for Nitrogen Removal from Wastewaters: A Review." *Revista Ambiente e Água* 10: 481–98.

Cheng, Chang-Hong, Fang-Fang Yang, Ren-Zhi Ling, Shao-An Liao, Yu-Tao Miao, Chao-Xia Ye, and An-Li Wang. 2015. "Effects of Ammonia Exposure on Apoptosis, Oxidative Stress and Immune Response in Pufferfish (Takifugu obscurus)." *Aquatic Toxicology* 164: 61–71. https://doi.org/10.1016/j.aquatox.2015.04.004.

Daebeler, Anne, Paul L. E. Bodelier, Zheng Yan, Mariet M. Hefting, Zhongjun Jia, and Hendrikus J. Laanbroek. 2014. "Interactions Between Thaumarchaea, Nitrospira and Methanotrophs Modulate Autotrophic Nitrification in Volcanic Grassland Soil." *ISME Journal* 8, no. 12: 2397–410. https://doi.org/10.1038/ismej.2014.81.

Dalsgaard, Tage, B. Thamdrup, and Donald E. Canfield. 2005. "Anaerobic Ammonium Oxidation (Anammox) in the Marine Environment." *Research in Microbiology* 156, no. 4: 457–64. https://doi.org/10.1016/j.resmic.2005.01.011.

Desloover, Joachim, Siegfried E. Vlaeminck, Peter Clauwaert, Willy Verstraete, and Nico Boon. 2012. "Strategies to Mitigate N2O Emissions from Biological Nitrogen Removal Systems." *Current Opinion in Biotechnology* 23, no. 3: 474–82. https://doi.org/10.1016/j.copbio.2011.12.030.

Edzwald, J. 2011. *Water Quality and Treatment: A Handbook on Drinking Water*. New York: McGraw-Hill Education.

Esteban, R., I. Ariz, C. Cruz, and J. F. Moran. 2016. "Review: Mechanisms of Ammonium Toxicity and the Quest for Tolerance." *Plant Science* 248: 92–101. https://doi.org/10.1016/j.plantsci.2016.04.008.

Finlay, M. R. 2002. "Vaclav SMIL, Enriching the Earth: Fritz Haber, Carl Bosch, and the Transformation of World Food Production." *British Journal for the History of Science* 35, no. 1: 97–123. Cambridge, MA and London: MIT Press. ISBN: 0-262-19449-X.£-23-95 (hardback)2001: xvii + 338.

Francis-Floyd, R., C. Watson, D. Petty, and D. B. Pouder. 2009. "Ammonia in Aquatic Systems." *EDIS* 2009, no. 6: FA16, FA031, rev. 2.

Führer, E. 1990. "Forest Decline in Central Europe: Additional Aspects of Its Cause." *Forest Ecology & Management* 37, no. 4: 249–57. https://doi.org/10.1016/0378-1127(90)90094-R.

Givan, C. V. 1979. "Metabolic Detoxification of Ammonia in Tissues of Higher Plants." *Phytochemistry* 18, no. 3: 375–82. https://doi.org/10.1016/S0031-9422(00)81870-1.

Hallin, S., P. Lydmark, S. Kokalj, M. Hermansson, F. Sörensson, A. Jarvis, and P. E. Lindgren. 2005. "Community Survey of Ammonia-Oxidizing Bacteria in Full-Scale Activated Sludge Processes with Different Solids Retention Time." *Journal of Applied Microbiology* 99, no. 3: 629–40. https://doi.org/10.1111/j.1365-2672.2005.02608.x.

Hassan, W., M. Hussain, S. A. S. J. Bashir, A. N. Shah, R. Bano, and J. David. 2015. "ACC-Deaminase and/or Nitrogen Fixing Rhizobacteria and Growth of Wheat (Triticum aestivum L.)." *Journal of Soil Science & Plant Nutrition* 15, no. ahead: 232–48. https://doi.org/10.4067/S0718-95162015005000019.

Hu, B. L., Ping Zheng, C. J. Tang, J. W. Chen, Erwinvan der Biezen, Lei Zhang, B. J. Ni, Mike S. M. Jetten, Jia Yan, Han-Qing Yu, and Boran Kartal. 2010. "Identification and Quantification of Anammox Bacteria in Eight Nitrogen Removal Reactors." *Water Research* 44, no. 17: 5014–20. https://doi.org/10.1016/j.watres.2010.07.021.

Imchuen, N., Y. Lubphoo, J. M. Chyan, S. Padungthon, and C. H. Liao. 2016. "Using Cation Exchange Resin for Ammonium Removal as Part of Sequential Process for Nitrate Reduction by Nanoiron." *Sustainable Environment Research* 26, no. 4: 156–60. https://doi.org/10.1016/j.serj.2016.01.002.

Jafvert, C. T., and R. L. Valentine. 1992. "Reaction Scheme for the Chlorination of Ammoniacal Water." *Environmental Science & Technology* 26, no. 3: 577–86. https://doi.org/10.1021/es00027a022.

Jetten, Mike S. M., L. V. Niftrik, Marc Strous, Boran Kartal, Jan T. Keltjens, and Huub J. M. op den Camp. 2009. "Biochemistry and Molecular Biology of Anammox Bacteria." *Critical Reviews in Biochemistry & Molecular Biology* 44, no. 2–3: 65–84. https://doi.org/10.1080/10409230902722783.

Kampschreur, Marlies J., Wouter R. L. van der Star, Hubert A. Wielders, Jan Willem Mulder, Mike S. M. Jetten, and Mark C. M. van Loosdrecht. 2008. "Dynamics of Nitric Oxide and Nitrous Oxide Emission During Full-Scale Reject Water Treatment." *Water Research* 42, no. 3: 812–26. https://doi.org/10.1016/j.watres.2007.08.022.

Kartal, Boran, Jayne Rattray, Laura A. van Niftrik, Jackvan de Vossenberg, Markus C. Schmid, Richard I. Webb, Stefan Schouten, John A. Fuerst, Jaap Sinninghe Damsté, Mike S. M. Jetten, and Marc Strous. 2007. "Candidatus '*Anammoxoglobus propionicus*' a New Propionate Oxidizing Species of Anaerobic Ammonium Oxidizing Bacteria." *Systematic & Applied Microbiology* 30, no. 1: 39–49. https://doi.org/10.1016/j.syapm.2006.03.004.

Kartal, Boran, Laura Van Niftrik, Jayne Rattray, Jack L. C. M. Van De Vossenberg, Markus C. Schmid, Jaap Sinninghe Damsté, Mike S. M. Jetten, and Marc Strous. 2008. "Candidatus 'Brocadiafulgida': An Autofluorescent Anaerobic Ammonium Oxidizing Bacterium." *FEMS Microbiology Ecology* 63, no. 1: 46–55. https://doi.org/10.1111/j.1574-6941.2007.00408.x.

Khramenkov, S. V., M. N. Kozlov, M. V. Kevbrina, A. G. Dorofeev, E. A. Kazakova, V. A. Grachev, B. B. Kuznetsov, D. Y. Polyakov, and Y. A. Nikolaev. 2013. "A Novel Bacterium Carrying Out Anaerobic Ammonium Oxidation in a Reactor for Biological Treatment of the Filtrate of Wastewater Fermented Sludge." *Microbiology* 82, no. 5: 628–36. https://doi.org/10.1134/S002626171305007X.

Kim, Kwang-Wook, Young-Jun Kim, In-Tae Kim, Gun-Ii Park, and Eil-Hee Lee. 2006. "Electrochemical Conversion Characteristics of Ammonia to Nitrogen." *Water Research* 40, no. 7: 1431–41. https://doi.org/10.1016/j.watres.2006.01.042.

Könneke, Martin, Anne E. Bernhard, José R.de La Torre, Christopher B. Walker, John B. Waterbury, and David A. Stahl. 2005. "Isolation of an Autotrophic Ammonia-Oxidizing Marine Archaeon." *Nature* 437, no. 7058: 543–6. https://doi.org/10.1038/nature03911.

Kuenen, J. G. 2001. "Extraordinary Anaerobic Ammonium-Oxidizing Bacteria." *ASM News* 67: 456–63.

Kuypers, Marcel M. M., A. Olav Sliekers, Gaute Lavik, Markus Schmid, Bo Barker Jørgensen, J. Gijs Kuenen, Jaap S. Sinninghe Damsté, Marc Strous, and Mike S. M. Jetten. 2003. "Anaerobic Ammonium Oxidation by Anammox Bacteria in the Black Sea." *Nature* 422, no. 6932: 608–11. https://doi.org/10.1038/nature01472.

Lackner, Susanne, Eva M. Gilbert, Siegfried E. Vlaeminck, Adriano Joss, Harald Horn, and Mark C. M. van Loosdrecht. 2014. "Full-Scale Partial Nitritation/Anammox Experiences–An Application Survey." *Water Research* 55: 292–303. https://doi.org/10.1016/j.watres.2014.02.032.

Lam, Phyllis, Marlene M. Jensen, Gaute Lavik, Daniel F. McGinnis, Beat Müller, Carsten J. Schubert, Rudolf Amann, B. Thamdrup, and Marcel M. M. Kuypers. 2007. "Linking Crenarchaeal and Bacterial Nitrification to Anammox in the Black Sea." *Proceedings of the National Academy of Sciences of the United States of America* 104, no. 17: 7104–9. https://doi.org/10.1073/pnas.0611081104.

Laureni, Michele, Per Falås, Orlane Robin, Arne Wick, David G. Weissbrodt, Jeppe Lund Nielsen, Thomas A. Ternes, Eberhard Morgenroth, and Adriano Joss. 2016. "Mainstream Partial Nitritation and Anammox: Long-Term Process Stability and Effluent Quality at Low Temperatures." *Water Research* 101: 628–39. https://doi.org/10.1016/j.watres.2016.05.005.

Law, Yingyu, Liu Ye, Yuting Pan, and Zhiguo Yuan. 2012. "Nitrous Oxide Emissions from Wastewater Treatment Processes." *Philosophical Transactions of the Royal Society of London. Series B, Biological Sciences* 367, no. 1593: 1265–77. https://doi.org/10.1098/rstb.2011.0317.

Leininger, S., T. Urich, M. Schloter, L. Schwark, J. Qi, G. W. Nicol, J. I. Prosser, S. C. Schuster, and C. Schleper. 2006. "Archaea Predominate Among Ammonia-Oxidizing Prokaryotes in Soils." *Nature* 442, no. 7104: 806–9. https://doi.org/10.1038/nature04983.

Li, Ming, N. Yu, Jian G. Qin, Erchao Li, Zhenyu Du, and Liqiao Chen. 2014. "Effects of Ammonia Stress, Dietary Linseed Oil and *Edwardsiella ictaluri* Challenge on Juvenile Darkbarbel Catfish *Pelteobagrus vachelli*." *Fish & Shellfish Immunology* 38, no. 1: 158–65. https://doi.org/10.1016/j.fsi.2014.03.015.

Liu, Sitong, Fenglin Yang, Zheng Gong, Fangang Meng, Huihui Chen, Yuan Xue, and Kenji Furukawa. 2008. "Application of Anaerobic Ammonium-Oxidizing Consortium to Achieve Completely Autotrophic

Bacterial Ammonia Oxidation

Ammonium and Sulfate Removal." *Bioresource Technology* 99, no. 15: 6817–25. https://doi.org/10.1016/j.biortech.2008.01.054.

Mehrer, I., and H. Mohr. 1989. "Ammonium Toxicity: Description of the Syndrome in Sinapisalba and the Search for Its Causation." *Physiologia Plantarum* 77, no. 4: 545–54. https://doi.org/10.1111/j.1399-3054.1989.tb05390.x.

Mulder, J. W., M. C. M. Van Loosdrecht, C. Hellinga, and R. Van Kempen. 2001. "Full-Scale Application of the SHARON Process for Treatment of Rejection Water of Digested Sludge Dewatering." *Water Science & Technology* 43, no. 11: 127–34. https://doi.org/10.2166/wst.2001.0675.

Nikolaev, Y. A., M. N. Kozlov, M. V. Kevbrina, A. G. Dorofeev, N. V. Pimenov, A. Y. Kallistova, V. A. Grachev, E. A. Kazakova, A. V. Zharkov, B. B. Kuznetsov, and E. O. Patutina. 2015. "Candidatus 'Jettenia Moscovienalis' sp. nov., a New Species of Bacteria Carrying Out Anaerobic Ammonium Oxidation." *Microbiology* 84: 256–62.

Norton, Jeanette M., Martin G. Klotz, Lisa Y. Stein, Daniel J. Arp, Peter J. Bottomley, Patrick S. G. Chain, Loren J. Hauser, Miriam L. Land, Frank W. Larimer, Maria W. Shin, and Shawn R. Starkenburg. 2008. "Complete Genome Sequence of Nitrosospira multiformis, an Ammonia-Oxidizing Bacterium from the Soil Environment." *Applied & Environmental Microbiology* 74, no. 11: 3559–72. https://doi.org/10.1128/AEM.02722-07.

Painter, H. A. 1986. "Nitrification in the Treatment of Sewage and Wastewaters." *Nitrification* 185–211.

Palomo, Alejandro, S. Jane Fowler, Arda Gülay, Simon Rasmussen, Thomas Sicheritz-Ponten, and Barth F. Smets. 2016. "Metagenomic Analysis of Rapid Gravity Sand Filter Microbial Communities Suggests Novel Physiology of Nitrospira spp." *ISME Journal* 10, no. 11: 2569–81. https://doi.org/10.1038/ismej.2016.63.

Pan, Yuting, Bing-Jie Ni, Philip L. Bond, Liu Ye, and Zhiguo Yuan. 2013. "Electron Competition Among Nitrogen Oxides Reduction During Methanol-Utilizing Denitrification in Wastewater Treatment." *Water Research* 47, no. 10: 3273–81. https://doi.org/10.1016/j.watres.2013.02.054.

Park, Hee-Deung, Seung-Yong Lee, and Seokhwan H. Hwang. 2009. "Redundancy Analysis Demonstration of the Relevance of Temperature to Ammonia-Oxidizing Bacterial Community Compositions in a Full-Scale Nitrifying Bioreactor Treating Saline Wastewater." *Journal of Microbiology & Biotechnology* 19, no. 4: 346–50. https://doi.org/10.4014/jmb.0806.399.

Penton, C. Ryan, Allan H. Devol, and James M. Tiedje. 2006. "Molecular Evidence for the Broad Distribution of Anaerobic Ammonium-Oxidizing Bacteria in Freshwater and Marine Sediments." *Applied & Environmental Microbiology* 72, no. 10: 6829–32. https://doi.org/10.1128/AEM.01254-06.

Pholchan, Mujalin K., J. D. C. Baptista, Russell J. Davenport, and Thomas P. Curtis. 2010. "Systematic Study of the Effect of Operating Variables on Reactor Performance and Microbial Diversity in Laboratory-Scale Activated Sludge Reactors." *Water Research* 44, no. 5: 1341–52. https://doi.org/10.1016/j.watres.2009.11.005.

Pjevac, Petra, Clemens Schauberger, Lianna Poghosyan, Craig W. Herbold, Maartje A. H. J.van Kessel, Anne Daebeler, Michaela Steinberger, Mike S. M. Jetten, Sebastian Lücker, Michael Wagner, and Holger Daims. 2017. "AmoA-Targeted Polymerase Chain Reaction Primers for the Specific Detection and Quantification of Comammox Nitrospira in the Environment." *Frontiers in Microbiology* 8: 1508. https://doi.org/10.3389/fmicb.2017.01508.

Quan, Zhe-Xue, Sung-Keun Rhee, Jian-E. Zuo, Yang Yang, Jin-Woo Bae, Ja Ryeong Park, Sung-Taik Lee, and Yong-Ha Park. 2008. "Diversity of Ammonium-Oxidizing Bacteria in a Granular Sludge Anaerobic Ammonium-Oxidizing (Anammox) Reactor. "*Environmental Microbiology* 10, no. 11: 3130–9. https://doi.org/10.1111/j.1462-2920.2008.01642.x.

Rotthauwe, J. H., K. P. Witzel, and W. Liesack. 1997. "The Ammonia Monooxygenase Structural Gene amoA as a Functional Marker: Molecular Fine-Scale Analysis of Natural Ammonia-Oxidizing Populations." *Applied & Environmental Microbiology* 63, no. 12: 4704–12. https://doi.org/10.1128/aem.63.12.4704-4712.1997.

Schmid, M., U. Twachtmann, M. Klein, M. Strous, S. Juretschko, M. Jetten, J. W. Metzger, K. H. Schleifer, and M. Wagner. 2000. "Molecular Evidence for Genus Level Diversity of Bacteria Capable of Catalyzing Anaerobic Ammonium Oxidation." *Systematic & Applied Microbiology* 23, no. 1: 93–106. https://doi.org/10.1016/S0723-2020(00)80050-8.

Schmid, Markus C., Nils Risgaard-Petersen, Jack Van De Vossenberg, Marcel M. M. Kuypers, Gaute Lavik, Jan Petersen, Stefan Hulth, B. Thamdrup, Don Canfield, Tage Dalsgaard, Søren Rysgaard, Mikael K. Sejr, Marc Strous, Huub J. M. Opden Camp, and Mike S. M. Jetten. 2007. "Anaerobic Ammonium-Oxidizing Bacteria in Marine Environments: Widespread Occurrence but Low Diversity." *Environmental Microbiology* 9, no. 6: 1476–84. https://doi.org/10.1111/j.1462-2920.2007.01266.x.

Schmid, Markus, Kerry Walsh, Rick Webb, W. Irene C. Rijpstra, Katinkavan de Pas-Schoonen, Mark Jan Verbruggen, Thomas Hill, Bruce Moffett, John Fuerst, Stefan Schouten, Jaap S. Sinninghe, S. S. Damsté, James Harris, Phil Shaw, Mike Jetten, and Marc Strous. 2003. "Candidatus 'Scalindua Brodae', sp. nov., Candidatus 'Scalindua wagneri', sp. nov., Two New Species of Anaerobic Ammonium Oxidizing Bacteria." *Systematic & Applied Microbiology* 26, no. 4: 529–38. https://doi.org/10.1078/072320203770865837.

Shi, X., H. W. Hu, J. Wang, J. Z. He, C. Zheng, X. Wan, and Z. Huang. 2018. "Niche Separation of Comammox Nitrospira and Canonical Ammonia Oxidizers in an Acidic Subtropical Forest Soil Under Long-Term Nitrogen Deposition." *Soil Biology & Biochemistry* 126: 114–22. https://doi.org/10.1016/j.soilbio.2018.09.004.

Siripong, Slil, and Bruce E. Rittmann. 2007. "Diversity Study of Nitrifying Bacteria in Full-Scale Municipal Wastewater Treatment Plants." *Water Research* 41, no. 5: 1110–20. https://doi.org/10.1016/j.watres.2006.11.050.

Stein, Lisa Y., and Martin G. Klotz. 2016. "The Nitrogen Cycle." *Current Biology* 26, no. 3: R94–8. https://doi.org/10.1016/j.cub.2015.12.021.

Stephen, J. R., A. E. McCaig, Z. Smith, J. I. Prosser, and T. M. Embley. 1996. "Molecular Diversity of Soil and Marine 16S RRNA Gene Sequences Related to Beta-Subgroup Ammonia-Oxidizing Bacteria." *Applied & Environmental Microbiology* 62, no. 11: 4147–54. https://doi.org/10.1128/aem.62.11.4147-4154.1996.

Strous, M., J. A. Fuerst, E. H. Kramer, S. Logemann, G. Muyzer, K. T. van de Pas-Schoonen, R. Webb, J. G. Kuenen, and M. S. Jetten. 1999. "Missing Lithotroph Identified as New Planctomycete." *Nature* 400, no. 6743: 446–9. https://doi.org/10.1038/22749.

Strous, Marc, Eric Pelletier, Sophie Mangenot, Thomas Rattei, Angelika Lehner, Michael W. Taylor, Matthias Horn, Holger Daims, Delphine Bartol-Mavel, Patrick Wincker, Valérie Barbe, Nuria Fonknechten, David Vallenet, Béatrice Segurens, Chantal Schenowitz-Truong, Claudine Médigue, Astrid Collingro, Berend Snel, Bas E. Dutilh, Huub J. M. Op den Camp, Chrisvan der Drift, Irina Cirpus, Katinka T. van de Pas-Schoonen, Harry R. Harhangi, Lauravan Niftrik, Markus Schmid, Jan Keltjens, Jackvan de Vossenberg, Boran Kartal, Harald Meier, Dmitrij Frishman, Martijn A. Huynen, Hans-Werner Mewes, Jean Weissenbach, Mike S. M. Jetten, Michael Wagner, and Denis Le Paslier. 2006. "Deciphering the Evolution and Metabolism of an Anammox Bacterium from a Community Genome." *Nature* 440, no. 7085: 790–4. https://doi.org/10.1038/nature04647.

Stumm, W., and J. J. Morgan. 2012. *Aquatic Chemistry: Chemical Equilibria and Rates in Natural Waters*. New York: John Wiley & Sons.

Tchobanoglus, G., F. Burton, and H. D. Stensel. 2003. "Wastewater Engineering: Treatment and Reuse." *Journal of the American Water Works Association* 95, no. 5: 201.

Treusch, Alexander H., Sven Leininger, Arnulf Kletzin, Stephan C. Schuster, Hans-Peter Klenk, and Christa Schleper. 2005. "Novel Genes for Nitrite Reductase and Amo-Related Proteins Indicate a Role of Uncultivated Mesophilic Crenarchaeota in Nitrogen Cycling." *Environmental Microbiology* 7, no. 12: 1985–95. https://doi.org/10.1111/j.1462-2920.2005.00906.x.

Van De Vossenberg, Jack, Jayne E. Rattray, Wim Geerts, Boran Kartal, Laura Van Niftrik, Elly G. Van Donselaar, Jaap S. Sinninghe Damsté, Marc Strous, and Mike S. M. Jetten. 2008. "Enrichment and Characterization of Marine Anammox Bacteria Associated with Global Nitrogen Gas Production." *Environmental Microbiology* 10, no. 11: 3120–9. https://doi.org/10.1111/j.1462-2920.2008.01643.x.

Van Dongen, L. G. J. M., M. S. M. Jetten, and M. C. van Loosdrecht. 2001. *The Combined SHARON/Anammox Process*. London: IWA Publishing.

Van Hulle, S. W. H., H. J. P. Vandeweyer, B. D. Meesschaert, P. A. Vanrolleghem, P. Dejans, and A. Dumoulin. 2010. "Engineering Aspects and Practical Application of Autotrophic Nitrogen Removal from Nitrogen Rich Streams." *Chemical Engineering Journal* 162, no. 1: 1–20. https://doi.org/10.1016/j.cej.2010.05.037.

Verhamme, Daniel T., James I. Prosser, and Graeme W. Nicol. 2011. "Ammonia Concentration Determines Differential Growth of Ammonia-Oxidising archaea and Bacteria in Soil Microcosms." *ISME Journal* 5, no. 6: 1067–71. https://doi.org/10.1038/ismej.2010.191.

Vitousek, P. M., J. D. Aber, R. W. Howarth, G. E. Likens, P. A. Matson, D. W. Schindler, W. H. Schlesinger, and D. G. Tilman. 1997. "Human Alteration of the Global Nitrogen Cycle: Sources and Consequences." *Ecological Applications* 7, no. 3: 737–50. https://doi.org/10.1890/1051-0761(1997)007[0737:HAOTGN]2.0.CO;2.

Wang, Yulin, Liping Ma, Yanping Mao, Xiaotao Jiang, Y. Xia, K. Yu, Bing Li, and Tong Zhang. 2017. "Comammox in Drinking Water Systems." *Water Research* 116: 332–41. https://doi.org/10.1016/j.watres.2017.03.042.

Wells, George F., Hee-Deung Park, Chok-Hang Yeung, Brad Eggleston, Christopher A. Francis, and Craig S. Criddle. 2009. "Ammonia-Oxidizing Communities in a Highly Aerated Full-Scale Activated Sludge

Bioreactor: Betaproteobacterial Dynamics and Low Relative Abundance of Crenarchaea." *Environmental Microbiology* 11, no. 9: 2310–28. https://doi.org/10.1111/j.1462-2920.2009.01958.x.

Winkler, M. K., R. Kleerebezem, and M. C. M. Van Loosdrecht. 2012. "Integration of Anammox into the Aerobic Granular Sludge Process for Main Stream Wastewater Treatment at Ambient Temperatures." *Water Research* 46, no. 1: 136–44. https://doi.org/10.1016/j.watres.2011.10.034.

Woebken, Dagmar, Phyllis Lam, Marcel M. M. Kuypers, S. Wajih, A. W. A. Naqvi, Boran Kartal, Marc Strous, Mike S. M. Jetten, Bernhard M. Fuchs, and Rudolf Amann. 2008. "A Microdiversity Study of Anammox Bacteria Reveals a Novel Candidatus Scalindua Phylotype in Marine Oxygen Minimum Zones." *Environmental Microbiology* 10, no. 11: 3106–19. https://doi.org/10.1111/j.1462-2920.2008.01640.x.

Wu, Q., G. Xia, T. Chen, X. Wang, D. Chi, and D. Sun. 2016. "Nitrogen Use and Rice Yield Formation Response to Zeolite and Nitrogen Coupling Effects: Enhancement in Nitrogen Use Efficiency." *Journal of Soil Science & Plant Nutrition* 16: 999–1009. https://doi.org/10.4067/S0718-95162016005000073.

Wunderlin, Pascal, Joachim Mohn, Adriano Joss, Lukas Emmenegger, and Hansruedi Siegrist. 2012. "Mechanisms of N_2O Production in Biological Wastewater Treatment Under Nitrifying and Denitrifying Conditions." *Water Research* 46, no. 4: 1027–37. https://doi.org/10.1016/j.watres.2011.11.080.

Xia, H., Ting Song, L. Wang, Liangsen Jiang, Qiting Zhou, Weimin Wang, Liangguo Liu, Pinhong Yang, and Xuezhen Zhang. 2018. "Effects of Dietary Toxic Cyanobacteria and Ammonia Exposure on Immune Function of Blunt Snout Bream (Megalabrama amblycephala)." *Fish & Shellfish Immunology* 78: 383–91. https://doi.org/10.1016/j.fsi.2018.04.023.

Yang, Xin-Ping, Shi-Mei Wang, De-Wei Zhang, and Li-Xiang Zhou. 2011. "Isolation and Nitrogen Removal Characteristics of an Aerobic Heterotrophic Nitrifying–Denitrifying Bacterium, *Bacillus subtilis* A1." *Bioresource Technology* 102, no. 2: 854–62. https://doi.org/10.1016/j.biortech.2010.09.007.